ELSEVIER

CMP

Acknowledgement is made to Z.H. Zhong for the use of Figure 3c on page 297, which appears on the front cover of this book.

STRUCTURES UNDER SHOCK AND IMPACT

Proceedings of the first International Conference, Cambridge, Massachusetts, USA, July 1989

Edited by

P.S. Bulson

ELSEVIER
Amsterdam - Oxford - New York - Tokyo 1989

Co-published with

COMPUTATIONAL MECHANICS PUBLICATIONS
Southampton - Boston 1989

P.S. Bulson
Computational Mechanics Institute
Wessex Institute of Technology
Ashurst Lodge
Ashurst
Southampton
SO4 2AA
UK

Also:
Mott MacDonald Group
Capital House
48 Andover Road
Winchester
SO23 7BH
UK

Distribution of this book is being handled by:

ELSEVIER SCIENCE PUBLISHERS B.V.
Sara Burgerhartstraat 25, P.O. Box 211
1000 AE Amsterdam, The Netherlands

Distributors for the United States and Canada:

ELSEVIER SCIENCE PUBLISHING COMPANY INC.
655 Avenue of the Americas
New York, N.Y. 10010, U.S.A.

British Library Cataloguing in Publication Data

Structures under shock and impact.
1. Structures. Effects of impact
I. Bulson, P.S. (Philip Stanley), *1925-*
624.1'76
ISBN 1-85312-036-7

Library of Congress Catalog Card Number 89-61081

ISBN 0-444-88024-0 Elsevier Science Publishers B.V.
ISBN 1-85312-036-7 Computational Mechanics Publications Southampton
ISBN 0-945824-21-1 Computational Mechanics Publications Boston

Published by:

COMPUTATIONAL MECHANICS PUBLICATIONS
Ashurst Lodge, Ashurst
Southampton, SO4 2AA, U.K.

Printed in Great Britain by The Bath Press Ltd, Avon

CONTENTS

PREFACE

The response of structures and structural elements to explosive shock or high speed impact is of growing interest to scientists, research engineers and designers throughout the world. Attacks on civil and military buildings, bridges and transport vehicles by terrorists mean that more attention has to be given at the design stage to the effects of explosions and impact, and to the high speed penetration of attack weapons. New design codes of practice need to be based on scientific research, both analytical and experimental, and there is a need for the cross-fertilisation of ideas between military and civil experts. This book brings together the experience of specialists in the behaviour of concrete and metal structures, both above and below ground, to the actions of blast, penetration and high speed collisions.

The book contains the edited versions of most of the papers presented at the first International Conference on Structures under Shock and Impact, held in Cambridge, Massachusetts, USA, in July 1989. This multidisciplinary meeting was held to bring together workers in a wide range of engineering activities, who employ common analytical and experimental methods in their estimation of structural response. The results of the meeting are of world-wide interest, and will help to stimulate future research and analysis.

P.S. Bulson
July 1989

SECTION 1 - CONCRETE STRUCTURES

Formation of Full Depth Cracks in Concrete Slabs subjected to Hard Impact

S.H. Perry*, G. Dinic** and P.H. Bischoff***
Trinity College Dublin, Republic of Ireland
***Davy McKee Ltd., UK*
****Institut fur Massivbau und Baustoff/technologie, Karlsruhe University, West Germany*

ABSTRACT

This paper describes the likely causes of hard impact, then discusses briefly the various modes of impact damage found in concrete slabs. Experimental results from laboratory tests, undertaken by the authors on a large number of reinforced concrete slabs, are presented and the damage phenomena discussed in detail. The influence of the maximum contact force and the imput energy, measured during impact, on the extent of damage are discussed, and primary and secondary phases are identified. Further work by the authors on the provision of protection against hard impact by the use of low strength fender layers is summarized.

INTRODUCTION

Preliminary design of civil engineering structures assumes, customarily, that loading is uniformly distributed. However, attention must sometimes be given to statically concentrated loads and, occasionally, to impact loading [1]. This is particularly the case when the consequences of local damage or possible penetration of a structural member are deemed to be sufficiently serious [2] and, also, when the risk of such loading is significant. Reinforced concrete slabs could be susceptible to intense hard impact loading in the low velocity range when, for example, rocks fall onto roof protection shelters in mountainous areas, the solidchassis of a forklift truck collides against a vertical column, or when objects are accidentally dropped onto offshore sub-surface storage tanks.

Several incidents, some of them major, have already been reported on oil platforms in the North Sea where objects such as drill collars, riser pipes, drill casings, drain caissons, mud pumps, reinforcing bars, scaffolding poles and even cranes have been dropped onto the roofs of base caissons [3,4,5]. Near normal, or vertical, impact of slender objects travelling at velocities up to 26m/s and weighing as much as 50 tonnes have been identified as being the most critical. Naturally, the velocity of the dropped object will depend on whether the terminal velocity has been reached during its free fall in water, and end-on impact velocities between 5 and 24m/s for objects weighing less than 20 tonnes are more likely to occur [5]. Impact loads at higher velocities (say 200m/s to 1000m/s - corresponding to aircraft and ballistic impact, respectively) can also affect protection requirements for aircraft shelters and nuclear reactor installations; but only the lower velocity range will be considered in this paper. Nonetheless, even a low-velocity impact can lead to environmental damage and complex repairs costing far more in monetary terms than the initial cost incurred for an adequate design. However, in order to be able to design structures that can readily withstand impact loading, a better understanding of load/-structure interaction, and the process of damage development within the structural material (in this case, concrete), must be obtained. Additionally, methods of providing protection against impact damage should be given consideration.

Hard impact of concrete is characterised by a sharply peaked force/time relationship (1 to 3×10^{-3} s), regardless of the approach velocity. This can lead to spalling, concrete plug ('shear' plug) formation, back face scabbing and, in the limit, complete perforation of both reinforced and prestressed concrete slabs. An extensive series of impact tests on concrete slabs and fender materials (in the 0-10m/s velocity range) was undertaken at the Imperial College of Science, Technology and Medicine under the general direction of Professor S.H. Perry [6] in order to investigate the different possible modes of damage, examine the parameters which govern internal concrete plug formation and study ways of furnishing adequate protection. Results, reported in detail elsewhere, are summarized here for tests on circular flat slabs [7,8] (1.5m diameter, 60-80mm thick and with two-way reinforcement in both faces) and lightweight concrete fender materials [9, 10].

CLASSIFICATION OF IMPACT DAMAGE

A standardized description of the different modes of damage caused by impact loading is necessary to allow accurate comparison between various experiments. Brown and Perry [6] define the following modes (Fig 1), spalling, cratering damage on the struck face; penetration, of the missile into the slab; scabbing, fracturing and expulsion of concrete, often at high velocities, from the face opposite the struck face; perforation, occurs when the missile passes completely through the slab; a concrete plug (sometimes refered to as a shear plug, and often in the form of a conical frustum) is formed by inclined cracking through the thickness of the slab. Scabbing may occur before plug formation or as a result of plug movement.

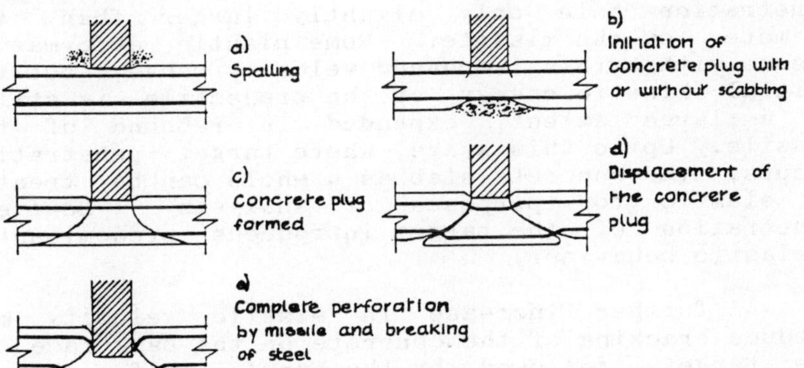

Figure 1. Progressive development of full-depth cracks, and subsequent damage.

The form and degree of impact damage is heavily influenced by the ratio of slab thickness to missile diameter (t/d). Hence, using a small missile diameter in comparison with the slab thickness (high t/d values, for example, of the order of 6) will most likely cause penetration and spalling. Smaller t/d values, of about unity, are more likely to cause scabbing and concrete plug formation; all localized forms of failure. Impact will occur over a relatively large area of the slab for even lower values of t/d, and behaviour will be dominated by the global flexural response of the slab, although scabbing is still possible. Impactor shape, also, can affect the type of failure, and results for a pipe-shaped projectile have been discussed elsewhere [6]. Other factors influencing impact damage include the deformability of the projectile, the velocity of impact, the relative characteristics of the target and missile and also the severity of impact [11].

PROGRESSIVE DEVELOPMENT OF DAMAGE

Relatively low contact velocities with small (compared to that of the target) missile mass will lead to small contact forces and exercise the structure in the elastic range which, depending on the flexibility of the structure, velocity and missile mass, can be followed by a rebound of the impactor. This type of impact will generally not cause any perceptable damage to the target.

As the impact velocity increases, spalling, in the form of a crater with an area larger than the cross-sectional area of the missile, may occur (Fig 1). Further increases of missile velocity will result in penetration; with a depth greater than the depth of the spall crater, and a cylindrical penetration hole only slightly larger than the diameter of the missile. Some plastic deformation occurs within this increased velocity range, but the initial kinetic energy of the projectile is still, to a large extent, expended in rebound of the missile. Up to this stage, where target penetration occurs, the concrete slab as a whole can be treated as elastic for purposes of analysis. However, penetration of the target introduces predominantly inelastic behaviour.

A further increase in missile velocity may produce cracking of the concrete on the back face of the target, followed by the expulsion of concrete from the back surface (scabbing), initiated by the reflected tensile wave effect. Initially, impact loading creates a compressive wave which travels through the target; on reaching the free surface, the wave is reflected as a tensile wave of similar magnitude [12]. Generally, the area of scabbing will be much larger than the spall area, but not so deep; often only of the same depth as the concrete cover to the bottom steel [13]. Once scabbing begins, the depth of penetration will increase rapidly, usually causing a concrete plug to be formed by inclined cracking through the remaining thickness of the slab. However, scabbing may occur either before, or as a result of, the concrete plug movement.

Depending on the initial velocity, once the concrete plug is formed and displaced, the missile may break through the reinforcement, or may only deform it. This will depend, also, on the amount of, and size of, the reinforcement contained within the concrete plug. Finally, with a further increase in velocity, complete perforation of the target will occur.

Apart from the major damage caused at the point of contact during hard impact, radial cracks spreading from the penetration crater are usually formed on the back surface of a slab [13]. Some circumferential cracks can be formed, also, when the flexural tensile stresses produced by the vibrational response of the overall structural element (in this case, the slab) exceed the tensile strength of the concrete.

IDENTIFICATION OF THRESHOLDS FOR CRACK INITIATION

An extensive series of carefully controlled hard impact tests on reinforced concrete slabs (summarized in [8], detailed in [15]) has allowed identification of the actual contact force required for formation of an incipient concrete plug. This was achieved by cutting slabs through the point of impact revealing, in some cases, the presence of full-depth cracks (Fig. 2, scale in cm). Interpretation of the masses and velocities involved suggests that shear plug formation is not only dependent on the peak contact force but, also, on the input kinetic energy. This was demonstrated by comparing the peak contact force with the input kinetic energy at impact, $(mv^2)/2$. The results and observations of these tests imply that both requirements, namely, the peak contact force and the input impact energy, have to reach a certain, critical, value for damage to occur. Clearly, if a critical peak contact force is reached, yet the input impact energy is less than the critical value required, (which, for these tests, was approximately 500 J), the plug will not form. If, however, both requirements are satisfied, damage of the slab will occur by formation of a full-depth crack which will determine the concrete plug.

Figure 2. Section through a concrete slab after hard impact, showing an incipient concrete plug.

Examination of the results also showed that the contact force/time relationship for normal aggregate concrete (dome 6, Fig. 3) can be broken down into two distinct phases. The primary phase is characterised by a steeply ascending curve to the maximum load which, typically, occurs 0.2 to 0.3ms after contact. The descent is almost as rapid, and this phase is over in less than 1.0ms. The actual duration is largely "hardness" dependent; there is some evidence that it is influenced separately by contact velocity, impactor mass and target thickness, and that when penetration occurs, the duration of the primary phase decreases. Additionally, other parameters not examined in detail in this study, such as impactor nose shape, concrete strength and slab surface roughness, will certainly affect the primary phase behaviour. Nevertheless, it can be postulated that

$$t = f\ [v,\ m,\ d,\ damage]$$

where t is the duration of impact up to maximum contact force, v is the velocity, m is the mass of impactor, and d is the thickness of slab.

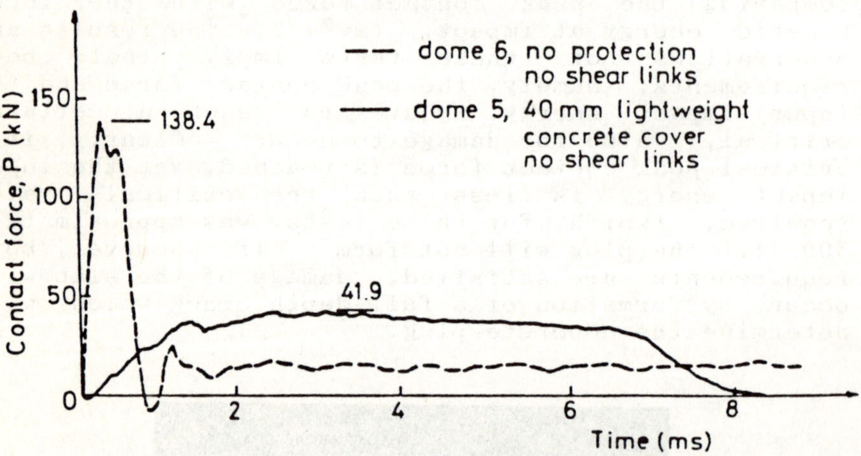

Figure 3. Contact force/time relationships for pipe impactor (mass of 37.85kg, velocity of 8m/s, dome thickness of 40mm).

The secondary phase describes the remaining duration of the impact event. Clearly, this also depends on the extent of damage caused by the impactor. When only cracking of the back face occurs the overall impact duration is approximately 1.75ms to 2.0ms. In this case, the slab reacts to the impact loading with a stiff response, representative of hard impact; therefore, the impact duration is not prolonged. However, when the concrete plug has been formed, the overall impact duration is increased. This, to a certain extent, resembles more a soft impact, where either the impactor or the structure, by means of deformation, absorbs the impact energy and prolongs the secondary response.

PROTECTION AGAINST IMPACT DAMAGE

Some method for improving the impact resistance of concrete is necessary when serious damage, such as a local punching failure, could result from a possible accident. One obvious way to resist full depth cracking would be to improve the shear resistance within the structure. An alternative would be to alter the structure's response, by some means, into overall flexural behaviour. It has been shown [16] that the use of shear reinforcement can change a local impact failure into an overall bending mechanism. Increasing the flexibility of the slab might reduce the magnitude of the contact force [11] to a certain extent. However, even if practicable, this alone would not normally be sufficient to prevent local failure. Another method is to provide an energy absorbing "fender" layer on any vulnerable face of a structure. This would be effective both for the design of new structures and for strengthening existing structures. Indeed, the placing of a 50mm thick sand cement screed to protect cables and pipes during the construction phase onto a caisson roof appeared to provide substantial protection and lead to minimal damage when an 18.9 tonne drain caisson was dropped onto the roof 110m below sea level [5]. The use of a lightweight concrete fender layer during hard impact tests on model reinforced concrete domes and concrete blocks has had the visible effect of preventing structural damage that would have occurred without its presence [6,9,17], and led to considerable reduction in the maximum steel strains.

Although an extra thickness of normal aggregate concrete has been recommended as a practical method of reducing impact loads on offshore structures [5], energy absorbing materials (of which lightweight aggregate concrete is one) have been shown to be much more efficient, and this type of layer has been used in construction for some recent offshore platforms [6, 18]. One other requirement when choosing a material for marine applications is its ability to withstand hydrostatic pressure, which increases by 1MPa per 100 metres depth. For offshore structures, perhaps 200 to 300 metres high, this demands a material crushing strength of 4 or 5 MPa. Typical lightweight aggregate concretes are stronger than this. Water saturation does not appear to affect test results significantly [6].

An impact situation comparable to those anticipated on North Sea concrete gravity platforms was modelled by dropping a solid shaped 31.6kg cylindrical mass at velocities up to 8m/s onto different thicknesses (20, 40 and 80mm) of concrete made with various lightweight aggregates such as expanded shale [9] or polystyrene beads [10]. These layers were rigidly supported on a concrete test block. A few tests, also, are reported for a pipe shaped impactor [6].

Of all the materials tested, polystyrene concrete seems to exhibit the most benefical properties required for an ideal energy absorbing material. That is, low strength combined with a large deformation capacity to ensure that impact loads transmitted to the structural concrete are minimal. In experiments at 10m/s, peak contact loads were reduced from about 300 to 400 kN for an unfendered structural test block to between 100 and 200 kN for lightweight concrete made from expanded clay [9] and, finally, as low as 60 kN for the polystyrene aggregate concrete [10]. Requirements for replica scale modelling are given in Table 1. The impactor used in this test programme represented a prototype 20 to 30 tonne solid shaped steel mass, scaled down by a factor of about 10, being dropped at velocities up to 10m/s onto fender layers as thick as 800mm. A material such as polystyrene concrete, having a crushing strength of about 7.5 MPa (or load capacity of 60kN for the scaled down model) and compressibility of 0.45, needed a model thickness or about 40mm (corresponding to a full scale prototype thickness of 400mm) to reduce damage to a minimum. This is a similar order of thickness (500-1200mm) to that normally used for the structural slab concrete of offshore caisson roofs.

Variable	Geometry	Material Properties	Mass	Velocity	Time	Contact Load	Contact Stress
Scale	S_L	1	S_L^3	1	S_L	S_L^2	1

Table 1 Replica scale modelling requirements

Ideally, actual design of a fender layer should take into account velocity, mass and shape of the assumed impactor. Shape can influence impact load, contact duration and penetration depth [16]. Pipe-shaped impactors have been identified as one of the most likely shapes to be dropped from offshore platforms [5], but tests have shown that the contact load from a solid impactor is greater than that generated by one that is pipe-shaped [6], although the latter will result in a greater penetration depth.

CONCLUSIONS

Experimental evidence suggests that the initial impact duration, the primary phase, depends on the velocity and mass of the impactor, thickness of impacted structure and the extent of local damage incurred by the structure. The secondary phase is governed by the overall response of the structure and, for the case when perforation occurs, the amount and distribution of the reinforcement. Furthermore, the tests undertaken show that concrete plug (or 'shear' plug) formation is governed by both the magnitude of the peak contact force and the input impact energy, and that each of these has to reach a certain critical value in order for the plug to form. The use of an energy absorbing fender layer of lightweight concrete reduced considerably the contact loads transmitted to the structure. This had the effect of minimizing, and sometimes entirely eliminating, impact damage.

ACKNOWLEDGEMENTS

The experimental work described in this paper was undertaken in the Concrete Laboratories of the Imperial College of Science, Technology and Medicine, London, and would not have been possible without the invaluable assistance given by Ian Brown and the late Gyebi Kufour. The assistance of Anne Martin (Trinity College Dublin) with the preparation of the manuscript is gratefully acknowledged.

REFERENCES

1. Perry, S.H., Dinic, G. and Brown, I.C. Design of structural concrete slabs against extreme hard impact at intermediate velocities (up to 25m/s). Proc. 1st Int Conf. on Concrete for Hazard Protection, pp 215-226. The Concrete Society, Edinburgh, 1987.

2. Barr, P. Guidelines for the design and assessment of concrete structures subjected to impact - 1987 edition. Safety and Reliability Directorate, SRD R 439, UKAEA, 1987.

3. Fjeld, S. and Roland, B. In-service Experience with Eleven Offshore Concrete Structures. Proc. Offshore Technology Conf, Houston, 1982, 3 No. OTC4358.

4. Faulds, E.C., Phillip, M.G. and Neffgen, J.M. Operating experience and repair case histories of five offshore concrete structures. Proc. Conf. on Development in the Design and Construction of Offshore Structures, Institution of Civil Engineers, 1983, No. 12.

5. Wimpey Laboratories. Review of object types and velocities, and of available empirical formulae. Part A, The Assessment of impact damage caused by dropped objects on concrete offshore platforms. pp 7-78, HMSO, London, 1989.

6. Brown, I.C. and Perry, S.H. Development of a new assessment method. Part B, Assessment of impact damage caused by dropped objects on concrete offshore structures. pp 79-151, HMSO, London, 1989.

7. Perry, S.H., Brown, I.C. and Dinic, G. Factors influencing the response of concrete slabs to impact. Proc. Int. Conf. and Exposition on Structural Impact and Crashworthiness. London, 2, 617-627, 1984.

8. Dinic, G. and Perry, S.H. Shear plug formation in concrete slabs subjected to hard impact. Engineering Fracture Mechanics, 1989.

9. Perry, S.H. and Dinic, G. Fender layer protection of sub-surface concrete caissons. Proc. 4th Int. Conf. on Behaviour of Offshore Structures (BOSS), Delft, pp 889-898, 1985.

10. Bischoff, P.H., Yamura, K. and Perry, S.H. Polystyrene aggregate concretes subjected to hard impact. To be published, 1989.

*11. Kennedy, R.P. A review of procedures for the analysis and design of concrete structures to resist missile impact effects. Nuclear Engineering and Design, 37, pp 183-203, 1976.

12. Johnson, W. Impact strength of materials. Edward Arnold, London, 1972.

13. Sliter, G.E. Assessment of empirical concrete formulae. Proc. American Society of Civil Engineers - Journal, Structural Division, 106, No. ST5, pp 1024-1045, 1980.

14. Perry, S.H. and Brown, I.C. Resistance of prestressed concrete slabs to extreme loads. Proc. 3rd. Int.Conf. on Behaviour of Offshore Structures (BOSS), Boston, 2, pp 874-883, 1982.

15. Dinic, G. Hard impact of reinforced concrete domes, circular slabs and fender materials. PhD thesis, Imperial College, London, p 263, 1986.

16. Eibl, J and Kreuser, K. Durchstanzfestighkeit von Stahlbetonplatten unter dynamisher Beanspruchung. Abschlussbericht zum Forschungsvorhaben, Lehrstuhl fur Beton und Stahlbeton, Universitat Dortmund, Dortmund, 1982.

17. Jenson, J.J. and Hoiseth, K. "Impact of dropped objects on lightweight concrete", Nordic Concrete Research, Publication No. 2, The Nordic Concrete Federation, pp 102-113, 1983.

18. Hoff, G.,C. The service record of concrete offshore platforms in the North Sea, Marine Concrete, Int. Conf. on Concrete in the Marine Environment, Concrete Society, London, pp. 131-142, 1986.

Scaling Explosive Damage to Reinforced Concrete Beams and Slabs

A.J. Watson, B. Hobbs and S.J. Wright

Department of Civil and Structural Engineering, University of Sheffield, Mappin Street, Sheffield S1 3JD, UK

ABSTRACT

This paper identifies and measures the damage caused to reinforced concrete T-beams and slabs, by contact and close proximity explosive charges using different areas of contact and angles of inclination for the explosive charge. Comparisons are made between prototype specimens with a span of 2.5m and 1/2.5 scale models of these specimens.

INTRODUCTION

Some of the possible benefits in being able to use small scale models for experimental investigations of the behaviour of structures are obvious. Within any time and cost constraints, many more experiments can be carried out at small scale than at large scale, so more variables can be studied and confidence limits established. The benefits are even greater when the loading is produced by explosives. This form of loading can be a hazard in the laboratory and is easier to contain at small scale.

Because the cost and time advantages of small scale experiments are so desirable it seems inevitable that work will continue to try and establish confidence in using scale models to predict prototype behaviour. This paper attempts to contribute to that end and is a sequel to an earlier paper by the same authors[1].

Few establishments have adequate facilities for carrying out explosive experiments and so relatively few results have been published. However the work has important applications for industrial and public hazard reduction and for the fundamental understanding of structural and material behaviour under load.

This paper reports the experimental results that were obtained when an explosive impulse load was applied to the mid span region of reinforced concrete T-beams and slabs. A study of scaling effects is made by comparing the damage caused to a prototype series of single T-beams and one way spanning slabs, with similar model specimens where all linear dimensions were scaled at 1/2.5 of the corresponding prototype dimensions.

TEST ARRANGEMENTS

Supports and test bay

The supports for the prototype and model specimens were steel bars which gave vertical support without unduly restricting the rotation of the bar. One support also allowed horizontal displacement. The general arrangement for prototype and model specimens is shown in fig. 1. All specimens were strapped to the supports during the test to prevent uplift.

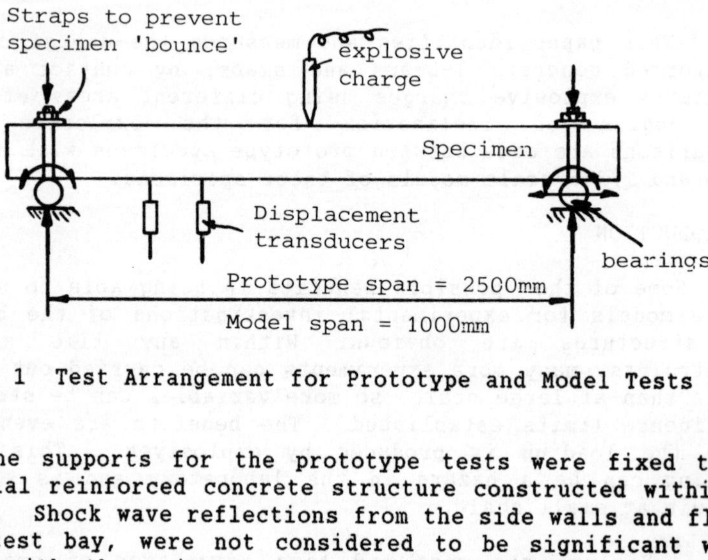

Fig. 1 Test Arrangement for Prototype and Model Tests

The supports for the prototype tests were fixed to a substantial reinforced concrete structure constructed within a test bay. Shock wave reflections from the side walls and floor of the test bay, were not considered to be significant when compared with the primary shock loading on the prototype or model specimens.

Instrumentation

The axial strain in the main longitudinal reinforcement at the bottom of the prototype and model specimens was monitored using electrical resistance strain gauges, Type PL2, gauge length 6mm or 2mm respectively, bonded to one of the bars at midspan. The deflection of the beam and slab specimens was measured at 1/5 and about 1/3 span positions, fig. 1, using linear variable differential transformers with a 75mm stroke, recording both upward and downward deflections.

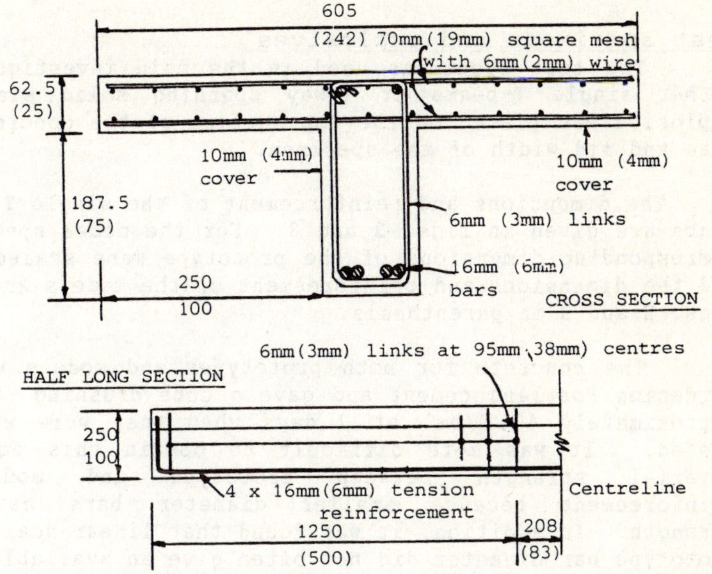

Figure 2 Details of Prototype (and Model) Single Web T-Beams

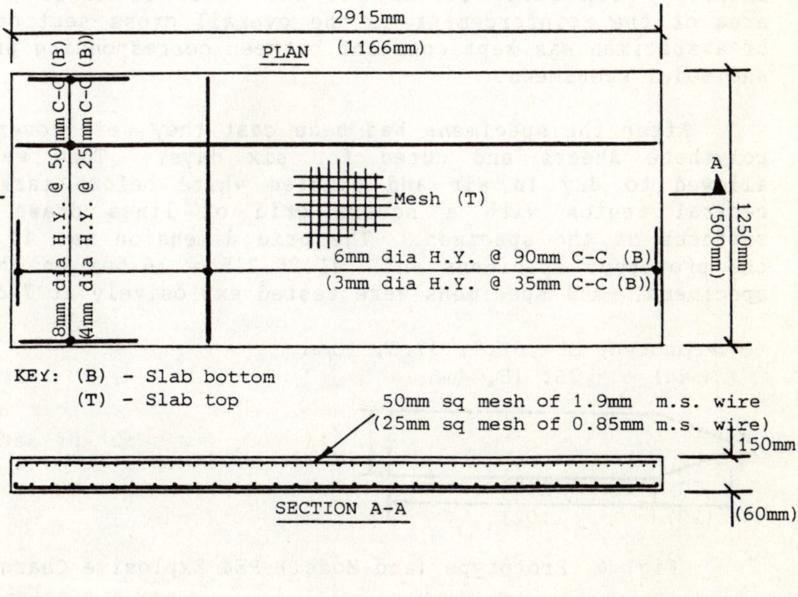

Figure 3 Details of Prototype (and Model) Solid Slab

Some of the model specimens were photographed at an effective framing rate of 12000pps using a half frame high speed rotating prism cine camera.

Test specimens and explosives

The test specimens used in the main investigation were either single T-beams or 1-way spanning solid slabs. The explosive was placed on the top surface of the concrete at mid span and mid width of the specimen.

The dimensions and reinforcement of the single T-beams and slabs are given in figs. 2 and 3. For the model specimens all corresponding dimensions of the prototype were scaled at 1/2.5 and the dimensions and reinforcement of the models are given in figs. 2 and 3 in parenthesis.

The concrete for both prototypes and models used rapid hardening Portland cement and gave a cube crushing strength of approximately 40 N/mm^2 at 7 days when they were explosively tested. It was more difficult to obtain this equality of material strength between prototype and model steel reinforcement because smaller diameter bars have higher strength. In addition, it was found that linear scaling of the prototype bar diameter did not often give an available diameter for the model bar. For these reasons it was necessary to allow the bar diameters and bar spacing to deviate from the linear scale factor of 1/2.5 in order to keep the reinforcement force resistance to the area scale factor of $1/(2.5)^2$. The cylindrical surface area of the bars in the model, which determined the bond resistance, could not be kept to the $1/(2.5)^2$ area scale factor but the ratio of cross sectional area of the reinforcement to the overall cross sectional area of a specimen was kept constant between corresponding prototype and model specimens.

After the specimens had been cast they were covered with polythene sheets and cured for six days. They were then allowed to dry in air and painted white before marking the central region with a square grid of lines drawn on the surfaces of the specimen. The grid dimension was 41.25mm on the prototype specimens and 41.25/2.5 = 16.5mm on the model specimens. All specimens were tested explosively at 7 days.

Prototype d = 62.5, 37.5, 10mm
Model d = 25, 15, 4mm

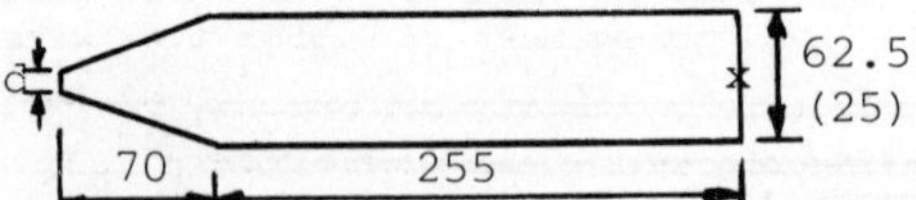

x point of detonation

Fig. 4 Prototype (and Model) PE4 Explosive Charge

The shape of the uncased PE4 explosive charge used on the prototype and model specimens was cylindrical with one end a truncated cone. Dimensions in millimetres and the detonation point are given in fig. 4 and the prototype charge mass was 1359g. The linear dimensions of the prototype charge are scaled by 1/2.5 and so the volume scale factor of $1/(2.5)^3$ gives a model charge mass of $1359/(2.5)^3 = 87g$. The charges were formed from PE4 explosive in machined perspex moulds which could be separated into two halves to release the moulded charge.

The properties of the explosive may be size dependent but both the prototype and model charges are assumed to have a detonation velocity of 8000m/s; a detonation pressure ($\bar{p}$) = 30kN/mm²; and a mass specific energy of 5000 KJ/kg. For an explosion in air, Baker[2] shows that pressure waves propagate faster in directions perpendicular to the axis of a cylindrical charge than they do in the axial direction from the ends of the charge. In the near field region this produces aspherical pressure fronts and according to Henrych[3], the shock front propagates at a maximum velocity in air of about 7500m/s.

Assuming plane wave detonation, then at any instant, the pressure at all contact points between the reacting explosive charge and the concrete surface will be the same, with a peak value equal to the detonation pressure. The loading then occurs as a time varying pressure p(t), of short duration τ, producing an instantaneous specific impulse;

$$i = \int_0^\tau p(t)dt$$

According to Henrych[3] the peak pressure exerted on a surface by a charge with a standoff distance a ≤ 5 charge diameters is $p = \bar{p}(R_w/R)\cos^2 \alpha$. Here R_w, is the charge radius, R is the distance vector from the centre of the charge to a point X on the surface, and α is the angle of incidence for the distance vector R at position X.

From Henrych[3] the specific impulse produced on the surface by a cylindrical charge is: $i = (2A_o W_c/a)\cos^3\alpha$. Here W_c is the mass/length of the cylindrical charge and $A_o \approx 1000$m/s.

<u>Test variables</u>

(1) The conical end of the explosive charge was truncated at different diameters, d, and for all contact tests the tip of the explosive was placed on the concrete. When the charge axis was perpendicular to the concrete, ($\Theta = 90^o$), the area of contact varied depending on the tip diameter d. When truncating the cone, the explosive removed was replaced on the back of the charge so that the mass remained constant at 1359g for the prototype and 87g for the model specimens.

(2) The inclination θ of the charge was either 21°, 45°, 60° or 90° to the horizontal when the charges were in contact with the concrete. Those with $\theta \neq 90°$ were inclined parallel to the main axis, or span, of the specimen and had a constant cone tip diameter d. The contact between concrete and explosive charge was then at mid span but only at one point on the edge of the circular end of the truncated cone. For the model single T-beam, one test was carried out with no contact at all between the explosive charge and the concrete. The charge axis was placed parallel to the beam axis, ($\theta = 0°$), and the standoff distance was a = 2.1 x charge diameter = 52.5mm, between the charge axis and the surface of the concrete. In this position the centroid of the charge was at mid span and at the same distance from the surface of the beam as it was when using the same charge vertically and in contact with the concrete.

The span : overall depth ratio for the prototypes and models was kept constant at 10 for the single T-beams and at 16.7 for the slabs. The span : overall width ratio of the slabs was kept constant at 2.

TEST RESULTS AND DISCUSSION

Following the explosive test, the damage to the specimens was carefully measured. The most apparent damage was cratering and cracking of the concrete, deformation and sometimes fracture of the top steel reinforcement, fig. 5. Photographs of all the specimens were taken to record the crack distribution and the cratering. Profiles were also measured on the cross section of each specimen which had lost the greatest area of concrete and consequently had become the weakest section of the beam.

Qualitative similarity

For a constant mass of explosive, variables d and θ had a great influence on the amount of damage. Qualitatively, the results were the same in both the prototype and model specimens and are as follows:

(1) Contact area between explosive and concrete
- The area of concrete lost on the most-damaged cross section of the prototype or model beams when $\theta = 90°$, increased with the diameter d, i.e. area of contact between explosive and concrete, fig. 6.

(2) Inclination θ of charge axis to concrete surface
- Inclining the explosive charge at $\theta = 45°$ on the single T-beams T2, PT2, produced more damage than any test with the explosive charge at $\theta = 90°$, even though the contact between the charge and the concrete was then only at one point, fig. 7.

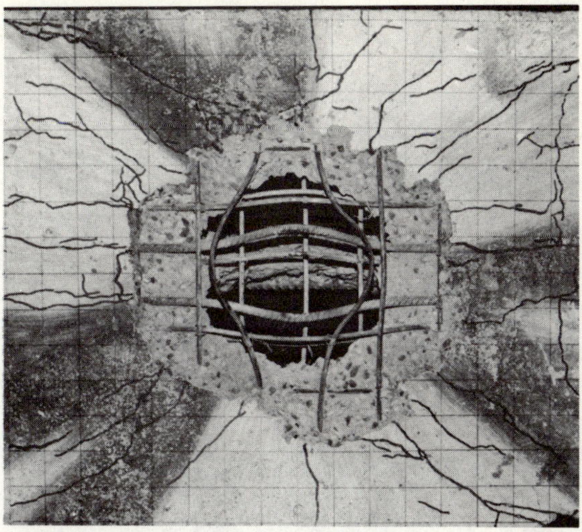

ABOVE

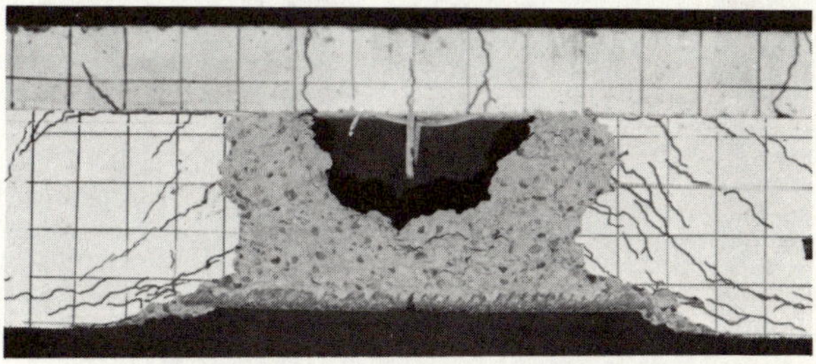

SIDE

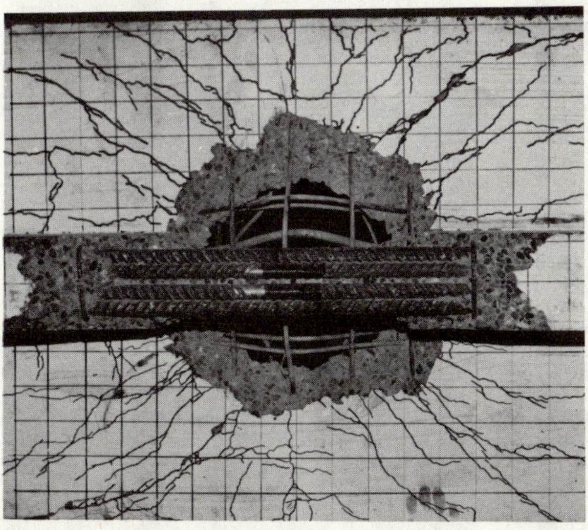

BELOW

Figure 5 Typical Prototype T-Beam (PT1, $\theta = 90^{\circ}$, d = 62.5mm)

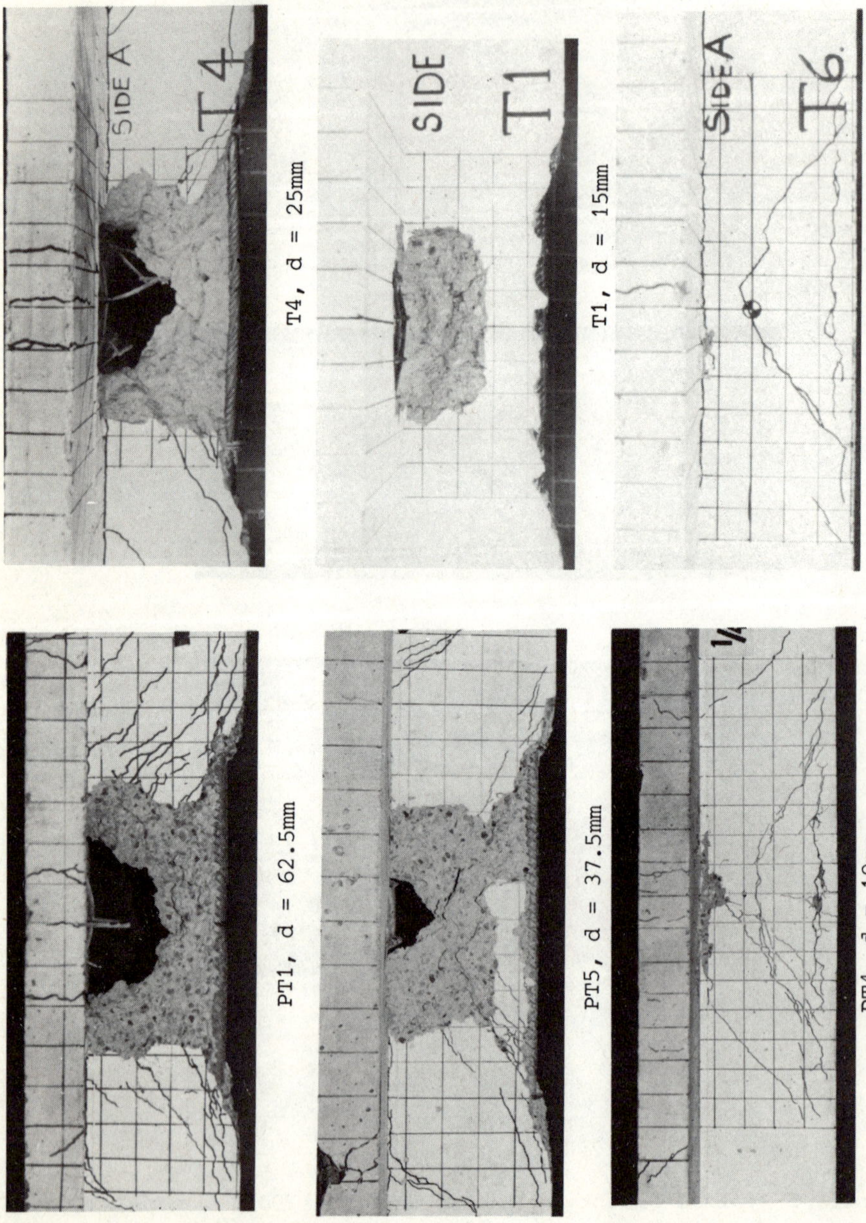

PT1, d = 62.5mm

T4, d = 25mm

PT5, d = 37.5mm

T1, d = 15mm

PT4, d = 10mm

T6, d = 4mm

Figure 6 Influence of Contact Area with θ = 90° on Prototype and Model Single T-Beams

- Maintaining the point contact between explosive and concrete but decreasing the inclination angle θ from 45° to 21° on the prototype and model slabs, increased the area of concrete lost, figs. 8 and 9, respectively.
- When the explosive charge had no contact with the concrete, and was placed with its axis horizontal (θ = 0°) and at 2.1 charge diameters above the model T-beam (T8), there was more damage, fig. 10, than on any of the other model T-beams where the explosive charges were in contact and vertical, (θ = 90°), fig. 6.

These results demonstrate a qualitative similarity for a scale factor of 1/2.5.

Quantitative similarity

To obtain one measure of the quantitative similarity, the max % cross sectional area of concrete removed is determined on the most-damaged cross section, Table 1. This criteria was selected because loss of cross sectional area has a most critical influence on the bending and shear strength of that section. The shear or bending moment capacity of the most-damaged section, then limits the load carrying capacity of the damaged beam.

Other quantitative measures of similarity are determined by measuring those characteristic dimensions of damage on the surfaces of the specimen, that are critical to the load capacity.

On the top (charged) surface of the prototype and model single web T-beams and slabs, a crater is formed with many cracks radiating from the centre of the crater which is at the point where the explosive and concrete made contact, figs. 5, 8, 9. In most of the tests, whatever the inclination of the explosive charge, the crater rim was roughly circular and the diameter of the crater on the surface of the beam is given in Table 1.

On the bottom surface of the prototype and model specimens, the concrete had in every case either spalled off the reinforcement or cracked at the level of the reinforcement, fig. 6. This is not regarded as important for the short term load carrying capacity of the damaged specimen. Of greater significance are the inclined cracks below the crater which form the boundaries of a more or less intact cone in the slabs, and wedge in the T-beam, fig. 6. The widest part of this concrete plug was at the bottom surface of the specimen. The largest dimensions of the plug were measured on the sides of the T-beam and the bottom surface of the slabs, Table 1. In the slab tests where the charge was inclined at θ = 21°, the base of the concrete plug was elliptical on the bottom surface of the slab and so both the major and minor diameters are

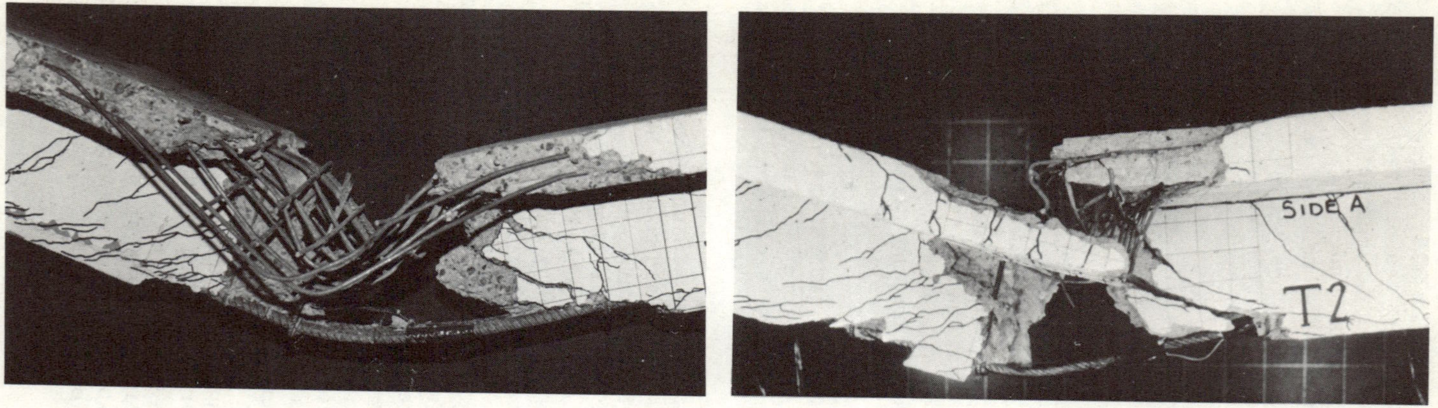

PT2, d = 10mm

T2, d = 4mm

Figure 7 Influence of Inclination θ=45° (//$^{\ell}$to span) of Explosive Charge on Prototype and Model Single T-Beam

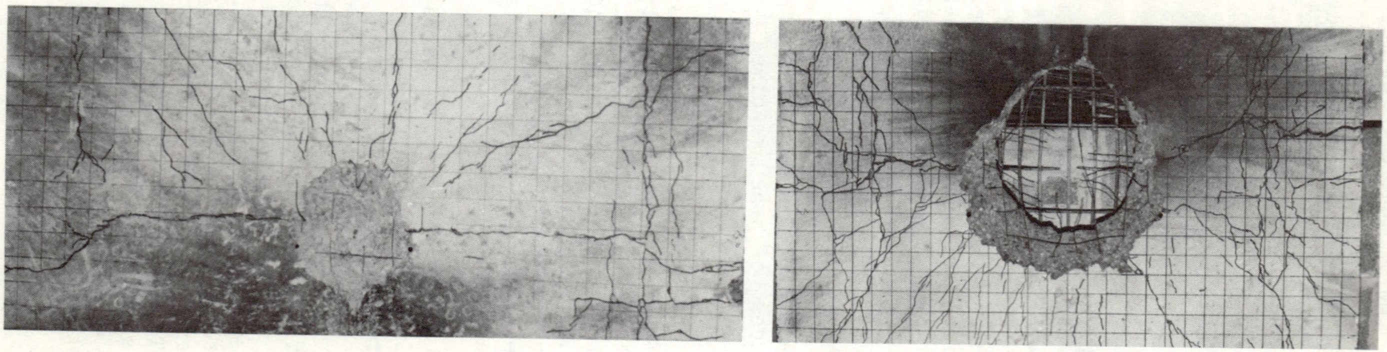

PS1, θ = 45°

PS2, θ = 21°

Figure 8 Influence of Inclination θ (//$^{\ell}$ to span) of Explosive Charge on Prototype Solid Slabs (d = 10mm)

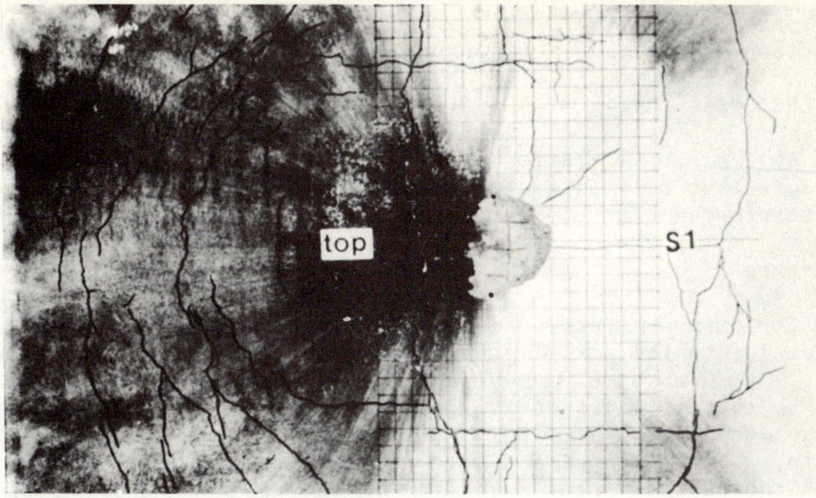

S1, $\theta = 45^{o}$

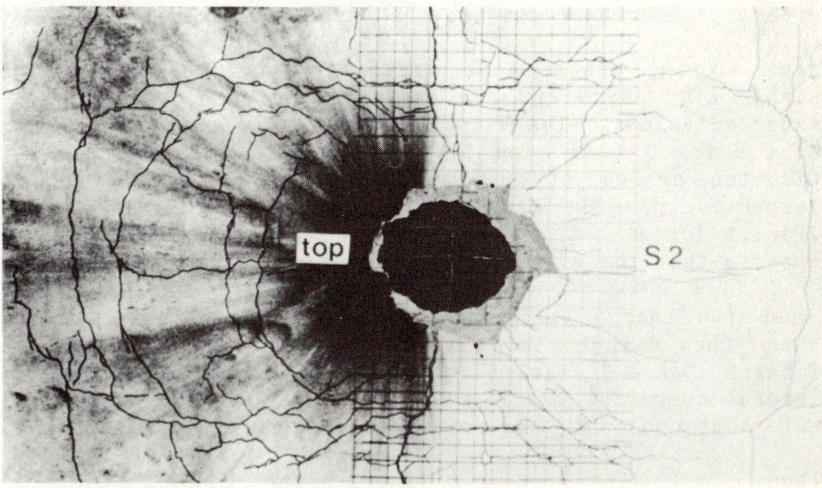

S2, $\theta = 21^{o}$

Figure 9 Influence of Inclination ($//^{\ell}$ to span) of Explosive Charge on Model Solid Slab (d = 4mm)

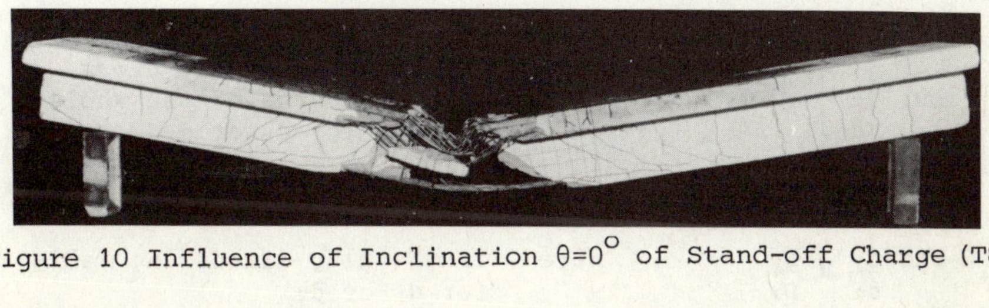

Figure 10 Influence of Inclination $\theta=0^{o}$ of Stand-off Charge (T8)

recorded in Table 1. The major diameter was in the same direction as the inclined charge.

CONCLUSIONS

1. At a scale ratio of 1/2.5 there is good qualitative agreement between the damage caused to model and prototype reinforced concrete single T-beams and slabs by a contact or close proximity cylindrical/cone-ended, explosive charge.

2. For charges perpendicular to the concrete surface, the damage to the scaled pairs of single T-beams, Fig. 6, Table 1, depends upon the area of explosive-concrete contact. The perfect ratio of 1.0 for the % loss in c/s area of the scaled pairs, varied in the experiments between 1.1 and 2.0 for the web and 1.1 and 4.0 for the whole section. Although the linear scale factor was 2.5 between model and prototype, the scale of damage varied between 2.2 and 3.1 for the crater top diameter and between 1.9 and 2.4 for the plug base dimensions. In general it appears that the greater the area of contact the better the scaling.

3. For sloping charges the damage to the scaled pairs of slabs, Figs. 8, 9 and Table 1 has good agreement with the expected values. The % loss in cross sectional area scales at 0.9 for $\theta = 45°$ and 1.0 for $\theta = 21°$. The crater top diameter scales at 2.0 and 2.6 for $\theta = 45°$ and $21°$ respectively. The plug base dimension scales at 2.5 and 2.4/3.1 for $\theta = 45°$ and $21°$ respectively. In general it appears that the greater the slope the better the scaling.

4. When the charge is perpendicular to the concrete surface then the damage to either model or prototype single T-beams, Table 1, Fig. 6, can be considered separately. In general comparing results where charge and beam dimensions were kept constant, and contact diameters are d_1 and d_2:

then $\dfrac{d_1}{d_2} \approx \dfrac{D_1}{D_2}$ — Here D is the crater top diameter for d_1 or d_2

and $\left[\dfrac{d_1}{d_2}\right]^2 \approx \dfrac{A_1}{A_2}$ — Here A is the %age loss in cross sectional area of web for d_1 or d_2

but $\left[\dfrac{d_1}{d_2}\right]^2 \neq \dfrac{A_1}{A_2}$ — Here A is the %age loss in cross sectional area of the whole section for d_1 or d_2

and $\dfrac{d_1}{d_2} \neq \dfrac{D_1}{D_2}$ — Here D is the plug dimension, Table 1 for d_1 or d_2

It appears that D is mainly controlled by the gap in the web reinforcement, Fig. 2, since $\frac{D_1}{D_2} \approx 1.0$.

REFERENCES

1. Hobbs, B., Watson, A.J., Wright, S.J. Explosive Tests on Model Concrete Bridge Elements. Fourth International Symposium on the Interaction of Conventional Munitions with Protective Structures, Florida, April 1989. (in press)
2. Baker, W.,E., Cox, P.A., Westine, P.S., Kulesz, J.J. and Strehlow, R.A. Explosion Hazards and Evaluation, (Chapter 2), Elsevier, Amsterdam - Oxford - New York, 1983.
3. Henrych, J. The Dynamics of Explosion and its Use, (Chapter 2), Elsevier, Amsterdam - Oxford - New York, 1979.

ACKNOWLEDGEMENTS

The work described in this paper is part of a research project in collaboration with the Denver Research Institute, University of Denver, USA, and the authors gratefully acknowledge the support of the staff of that establishment.

SPECIMEN CODE (SCALED PAIRS)		% LOSS IN C/S AREA				CRATER TOP DIA		PLUG BASE DIMENSION	
PROTOTYPE	MODEL	WEB	WHOLE	RATIO WEB	RATIO WHOLE	mm	SCALE FACTOR	mm	SCALE FACTOR
PT1	T4	81 77	57 50	1.1	1.1	358 118	3.0	495 267	1.9
PT3	T1	69 38	48 25	1.8	1.9	268 94	2.8	471 214	2.2
PT4	T6	3.8 1.9	9.1 2.3	2	4	56 25	2.2	453 214	2.1
PT5	T1	58 38	45 25	1.5	1.8	290 94	3.1	507 214	2.4
PS1	S1	N/A	2.0 2.3	N/A	0.9	190 94	2.0	701 280	2.5
PS2	S2	N/A	33.2 32.0	N/A	1.0	410 156	2.6	620x1337 263x429	2.4/3.1

TABLE 1 SCALED RESULTS

Concrete in Impact Tensile Tests

J. Weerheijm* and H.W. Reinhardt**

**Prins Maurits Laboratory TNO, P.O. Box 45, 2280 AA, Rijswijk, The Netherlands*

***Darmstadt University of Technology, Alexanderstrasse 5, 6100, Darmstadt, West Germany*

ABSTRACT.

In a test programme on concrete under axial impact tensile loading and lateral compression, uniaxial impact tests were also performed. The scope of the biaxial programme constrained to a prismatic geometry of the specimen and a local cross section reduction by two saw cuts.

In the paper the results of the uniaxial impact tests and the consequences of the structural response are discussed.

The influence of the structural response of the specimen on the observed strength, deformation capacity and the fracture energy appeared to be significant. A method to estimate the influence of a section reduction on the observed impact strength as a function of the loading rate is suggested.

INTRODUCTION.

At the Prins Maurits Laboratory (PML) a research programme is being performed on the prediction of the response of concrete structures to impact and explosive loading. The behaviour of concrete as a function of the loading rate and the loading condition appeared to be insufficiently known. In cooperation with the Delft University of Technology (DUT) an experimental programme has been carried out using the Split Hopkinson Bar (SHB) technique. The loading conditions were dynamic axial tension with or without lateral compression. The biaxial test conditions constrained to a prismatic specimen with a local section reduction. This geometry significantly differs from the usual cylindrical geometry in impact tests. The differences offered the opportunity to examine the influence of this geometry on the uniaxial test results by a comparison with former results obtained with the same Split Hopkinson bar apparatus at the DUT.

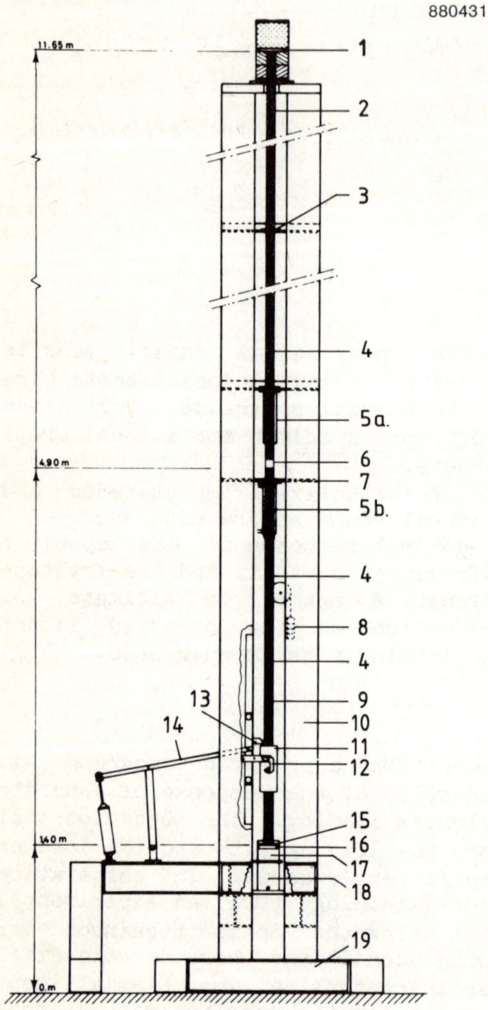

1 buffer
2 upper bar
3 guide
4 strain gauge
5a upper cooling jacket
5b lower cooling jacket
6 concrete test specimen
7 working platform
8 counterweight
9 lower bar
10 frame
11 drop weight
12 coupling
13 uncoupling
14 lifting device
15 damping material
16 anvil
17 guide tube
18 pneumatic jack
19 frame base.

Figure 1. Split Hopkinson bar equipment, schematically.

UNIAXIAL EXPERIMENTS.

The objectives of the experimental programme are the determination of the strength, the energy dissipated during the fracture process and a study of the process itself.
The experiments were performed with the Split-Hopkinson bar apparatus (SHB) at the DUT. In former years many uniaxial impact tests were successfully carried out, Reinhardt [1].
In the current programme the geometry of the specimen is prismatic and notches are applied to fix the failure plane. The strains outside the failure zone, the axial deformation of the whole specimen and the strains in the incident and transmitter bar of the SHB have been registered as a function of time. With these signals the strength can be determined and the failure process can be reconstructed. In former programmes cylindrical specimens were used and only the deformation of the whole specimen was registered besides the strain at the transmitter bar.

In Figure 1 the SHB is depicted and the specimen with the location of the strain gauges is given in Figure 2.

In the programme a high (A) and a low (B) quality concrete was tested to examine the influence of the internal damage on the strength and the fracture process. The loading rate of the incident pulse was not varied in the programme and chosen as high as possible. The applicable loading rate is limited by the notches because the stress concentrations in the adjacent zones will cause brittle failure beyond a certain loading rate, as will be discussed later. Unfortunately the initially applied notches of 5 mm * 5 mm proved to be insufficient to fix the failure plane in the biaxial tests, so the depth had to be increased to 7 mm.
The dimensions of all specimens are as follows: length 100mm, width 80mm, depth 50mm. The notch depth, the concrete quality and the loading rate in the critical cross section of the tested specimen are listed in Table 1. The mechanical properties of the two concrete qualities are given in Table 2.

Table 1. Different types of specimen.

type	1	2	3	4	5
notch depth (mm)	5	7	0	5	7
concrete quality	A	A	B	B	B
loading rate (GPa/s)	17.7	11.5	10.0	13.8	12.6

Table 2. Mechanical properties and mix proportions

		Quality A	Quality B
cube compressive strength	(MPa)	40.	27.
splitting strength	(MPa)	2.3	1.7
portland cement	(kg/m3)	375.	200.
water-cement ratio	(-)	0.5	0.6
air content	(%)	2.1	2.5

The strain and displacement signals were recorded on tape with a SELABS SE 3000 (14 channels) recorder at a speed of 60ips and a bandwidth of 40kHz. Afterwards all the signals were digitized simultaneously with a 14-channel AD converter (DIFA Benelux TS-9000; 250kHz) and stored on the VAX 11/750 of the Prins Maurits Laboratory. The signals were analysed by means of an extensive software programme for data processing.

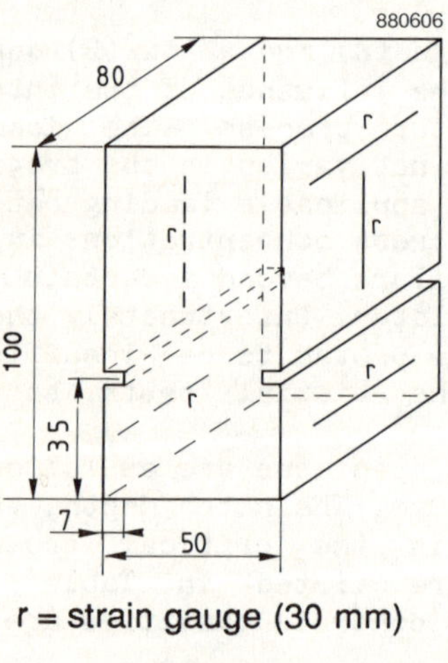

Figure 2. Geometry of specimen and location of strain gauges.

TEST RESULTS.

In this section the main results will be summarized. These results concern:

- the ultimate strength (f_d)
- the total deformation (d_{td}) at ultimate strength
- the deformation of the fracture zone (d_{fd})
- the fracture energy in the whole specimen (G_{td})
- the fracture energy in the fracture zone (G_{fd})
- the total energy absorbed during the failure process (G_f).

Figure 3 shows the shape of the stress-deformation curve in- and outside the failure zone. The data are listed in Table 3.

Table 3. Test results.

	type	1	2	3	4	5
f_d	(MPa)	4.55	4.39	2.72	2.61	2.60
d_{td}	(0.01mm)	1.17	1.66	2.03	1.00	1.42
d_{fd}	(0.01mm)	0.61	1.08	-	0.42	0.94
G_{td}	(N/m)	26.5	30.6	30.0	15.6	23.3
G_{fd}	(N/m)	15.7	22.4	-	10.5	17.3
G_f	(N/m)	98.8	93.2	110.0	77.7	90.0

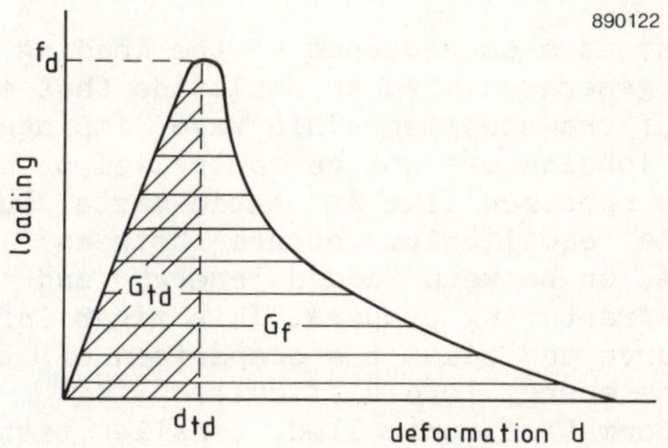

Figure 3. Shape of the load-deformation curve.

For the high quality concrete the plane of fracture consisted of bond, matrix and aggregate fracture. In case of the low quality concrete the plane was formed by bond and mortar fracture while in several cases multiple fracture occurred. In spite of the uniaxial loading pulse, the specimen did not fail in a symmetrical way. Rotation occurred during the failure process for all types of experiments.
In the next sections the test results and the influence of the structural response on these results will be discussed.

STRUCTURAL RESPONSE.

The response of specimen and loading device will always obscure the true material response in a greater or less degree. Due to the structural response, the stress distribution in the specimen changes. This might change the failure process in such a way that the observed response of the specimen does not reflect the true material behaviour.

For the current tests the following effects must be mentioned:

- due to the notches the failure zone is localized
- stress concentrations occur adjacent to the notches
- the tests are not deformation controlled
- rotation of the material adjacent to the failure zone.

The first point must be mentioned because in former tests the height of the cylindrical specimens was the measuring length. Because of multiple cracking several fracture zones could be present resulting in a relatively large deformation capacity of the material. The current tests should give information about the deformation capacity of one failure zone.

The effect of the notches will be a more brittle response when the redistribution of stresses, due to microcracking, is insufficient to eliminate the stress concentrations. The effect on the observed strength is discussed in the section about the dynamic strength.

The third point is a consequence of the loading device. A loading wave is generated with an amplitude that exceeds the ultimate strength of the specimen. This wave impinges on the specimen and the loading can not be controlled or adjusted to the response of the specimen like in static tests. Due to this effect an unstable equilibrium occurs between loading and (material) response, or between added energy and dissipated energy during the fracturing process. This might influence the load-deformation curve and makes the comparison of the static and dynamic fracture energy more difficult.

In static, deformation controlled, tensile tests it was observed that the boundaries of the softening zone do not remain parallel but rotate when the maximum strength is reached somewhere in the failure zone, Hordijk et al.[5]. Due to this rotation a lower state of energy content is reached, so unless rotation is prevented by e.g. supports or inertia, it will occur and the deformation during the failure process increases.

In the following sections the test results and the influence of the mentioned effects will be discussed shortly. An extensive description and discussion of the results is given in Weerheijm [7].

DISCUSSION OF TEST RESULTS.

The deformation capacity.

Concerning the experiments of type 3, without notches, the results correspond with former tests (d_{td} was about 0.015mm, Kormeling [3]; the failure zone is not fixed and multiple cracking can occur, resulting in a large axial deformation of the specimen. Comparison of the data from the other tests with those of type 3, show that the deformation capacity and the fracture energy G_{fd} appear to be over-estimated in the tests of type 3 and former tests, see Table 3. In the tests, with one fracture zone, the deformation of this zone appears to be about 60% of the deformation of the whole specimen. From this result it emerges that the deformation capacity observed in former tests, with possible multiple cracking, should at least be reduced to 60%.

These results illustrate the carefulness that is necessary in calculations when the deformation capacity or the fracture energy for an element has to be defined in which one or more failure zones can occur, Weerheijm et al.[4].

Rotation during fracture process.

As mentioned before, rotation of the material adjacent to the failure zone and the increase of deformation due to the rotation have been observed in uniaxial static tests.

Non-uniform failure was also observed in the impact tests, as is illustrated in Figure 4. Apparently rotation was not prevented by the stiffness and the inertia of the surrounding material. The specimen could respond in a mode of lower energy content resulting in a larger deformation at ultimate loading. Non-uniform cracking and rotation was observed from the measurements for all types of tests. The rotation was the strongest for the "7mm" tests, which is reflected in the larger average deformation d_{fd}.

In the dynamic tests, the non-uniform response starts before the ultimate load is reached and the failure process becomes more ductile. This explains the larger deformation for the "7mm" tests but also the lower loading rate in the tests with notches (see Table 1) as registered in the transmitter bar of the SHB.

The fracture energy.

The fracture energy is given by the integral for the area under the load-deformation curve for the failure zone. This energy is mainly determined by the descending branch beyond ultimate loading (see Table 3).

The energy supply in dynamic impact tests is not coupled to the deformation or the fracture process and the observed curve corresponds with an unstable equilibrium. Consequently, the curve does not represent a material property and will differ from the curves determined in deformation controlled tests. The conclusion is that the load-deformation curve depends on the loading conditions which also have to be considered in dynamic and static calculations.

The curves obtained from the impact tests reflect the separation of both specimen parts including the fracture process and the structural response with inertia effects. Complete separation occurred at a crackwidth of about 0.1mm and corresponds with static results as it should be. The influence of the structural response on the shape of the descending branch cannot be determined with the test results. To study the load-deformation curve, the impact experiment will be simulated in a finite-element calculation, as was successfully done for the mentioned static tests, Hordijk et al.[6].

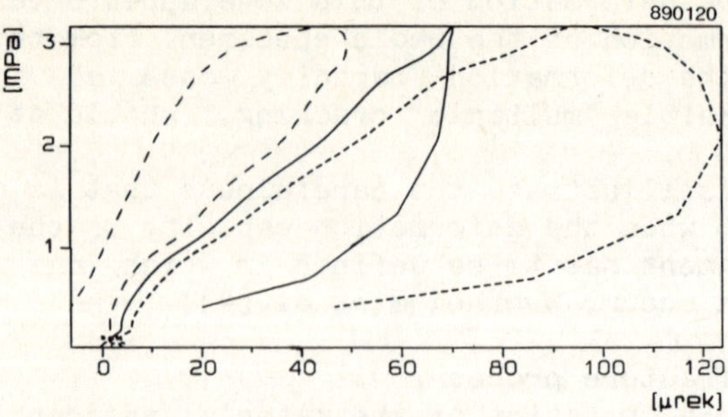

(a)

----- strain front side
- - - strain rear side
----- mean longitudinal strain

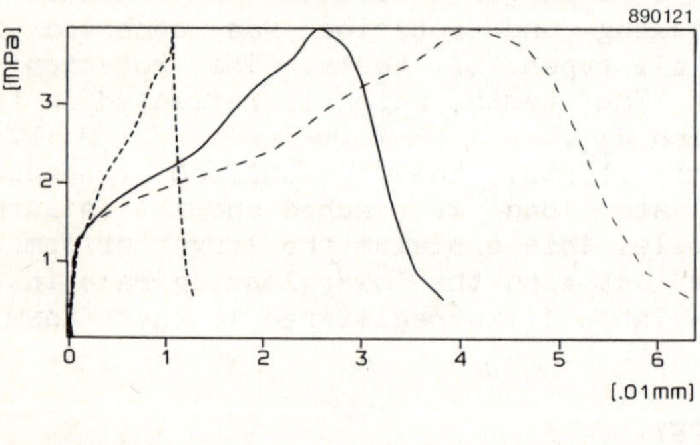

(b)

----- deformation front side
- - - deformation rear side
----- mean deformation (crack width)

Figure 4. Stress-deformation curves outside and inside the fracture zone ("a" and "b" resp.)

The dynamic tensile strength.
Whether the notches influenced the strength or not, can be determined by comparison with results of former programmes. These programmes (Zielinski,[2]) led to an empirical relation for the dynamic tensile strength. For the mixes A and B this relation predicts a strength $f_d(A) = 4.92$(MPa) and $f_d(B) = 2.88$(MPa). Comparison with the experimental results (Table 3) and the corresponding confidence levels of 90%, leads to the conclusion that the strength was affected by a structural response. In spite of the increased loading rate the strength decreases for the tests with notches, so the explanation should be found in the stress concentrations. The theoretically

expected, more brittle failure due to these concentrations was not observed in the "7mm notch" tests, probably because of the mentioned rotation.

In the next section the influence of notches is discussed theoretically and a method is suggested to estimate the influence on the strength of the specimen.

SECTION REDUCTION AND DYNAMIC STRENGTH.

The stress concentrations adjacent to the notches will not affect the ultimate strength of the specimen when the stresses are redistributed due to microcracking. Microcracking occurs beyond a stress level of about 60% of the static tensile strength fs. To estimate the size r of the zone in which the stress exceeds the threshold for microcracking the approach used by Dugdale, to determine the extent of the plastic zone in metals, is applied.

The notch is replaced by a crack of same depth a. With the Dugdale approach the size of the "plastic" zone of microcracking can be determined and is denoted by R. When sufficient microcracking occurs the stress inside the zone R equals a "yield stress" f_{mc} and r equals R. However, the extension of the zone of sufficient microcracking is limited to a certain velocity C_{mc} because the velocity of crack extension is limited. Consequently, when the extension velocity of r exceeds the velocity of microcracking C_{mc} the redistribution will lag behind and the fracture process will become more brittle and the strength of the specimen will decrease. The criterion for the application of section reduction follows from the equations:

$$r = a\,(1 - \cos y)/\cos y$$

$$\dot{r} = \frac{n.\dot{p}.a.\sin y}{2.f_{mc}.\cos^2 y}$$

with $y = n.\,p / 2.f_{mc}$

f_{mc} = stress level micro cracking

p = loading

$\dot{p}$ = loading rate

These equations show the influence of the various parameters. Those situations have to be considered in which the extension rate $\dot{r}$ exceeds the velocity C_{mc} for values of r smaller than about two times the crack size 2a. For r-values larger than 2a the stress is smaller than p and the validity of the equations is doubtful. The condition r equals 2a is equivalent with y equals 1.231(rad.). The criterion for the application of notches of depth a follows from:

$$a < 2.C_{mc}.f_{mc}.\cos^2 y \,/\, n.\dot{p}.\sin y$$

$$\text{for } y < 1.231$$

In this equation the velocity C_{mc} and the stress f_{mc} are unknown. A conservative approximation for f_{mc} is the level of bond cracking under static loading (0.6fs).

During the time that the microcracking lags behind, the stresses in front of the microcracking zone will not be redistributed and damage occurs. Without notches this damage would not occur, so the effect on the strength can be estimated with the load increase during the time interval between the moment C_{mc} is reached and the moment the size of the microcracking zone equals two times the notch depth. When for this time interval the boundaries for R are given by R1 and R2, the influence on the strength Df can be estimated by:

$$Df = \frac{(R_2 - R_1)^2 . \dot{p}}{C_{mc} . (D - 2.a)}$$

with all parameters in SI-units and D the diameter of the specimen.

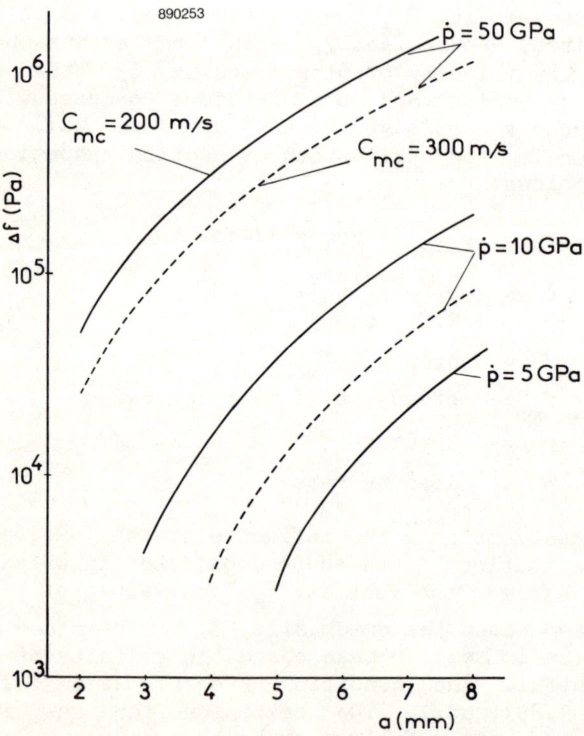

Figure 5. The strength decrease of a specimen as a function of the notch depth at various loading rates. Specimen diameter is 50mm.

For two values of C_{mc} and the loading rates, applied in the experimental programme for the low quality concrete, the effect on the observed strength is given in Figure 5 for various notch depths and two values of C_{mc}.

To examine the applicability of the approach, the data for the B-quality and 5mm notch were used to determine C_{mc} and the 7mm tests to control the predicted influence of the notch depth. The observed decrease in tensile strength due to the 5mm notch is 0.11 Mpa at a loading rate, outside the failure zone, of 1.1 GPa/s. This would be predicted with C_{mc} = 130 m/s. In the "7mm" tests the loading rate is smaller due to the extra deformation during the fracture process. The loading rate in these tests was 0.9 GPa/s. With C_{mc} = 130 m/s the predicted decrease in strength is only 0.08 MPa. For a loading rate of 1.0 GPa/s the predicted decrease in strength is the same as previously calculated for the 5mm notch and observed in the tests.

These calculations and the reasonable feature of the results depicted in Figure 5, show that with the described approach the influence of notches on the dynamic strength can be estimated.

CONCLUDING REMARKS.

The determination of material properties appears to be very difficult, and is probably impossible, due to the structural response of the specimen in the loading device. The results of the discussed impact tests emphasize this statement. The conclusions are:

- The deformation capacity at ultimate loading, as determined in former tests, is too large due to the inclusion of more than one cracking zone in the measuring length of 100mm. It should probably be reduced to at least 60% when a single failure zone is taken into consideration.
- The load-deformation curve is influenced by rotation of the material adjacent to the failure zone. Due to the rotation the average deformation increases.
- The load-deformation curve is influenced by the loading conditions. The obtained curve in impact tests and the derived fracture energy cannot be regarded as a material property and directly compared with results obtained in static tests.
- Application of notches leads to a more brittle failure and a lower ultimate strength of the specimen for loading rates beyond a certain level. The threshold is about 5GPa/s for the applied geometry.
- A method has been given to estimate the effect of notches on the ultimate strength as a function of the notch depth and the loading rate.

To study the true material response under impact loading, a combination of experiments and numerical simulation of the experiments will be very fruitful. The structural and material response can be separated in this way.

REFERENCES.

1. Reinhardt, H.W. et al. The Split Hopkinson Bar, a versatile tool for the impact testing of concrete. Materiaux et Constructions, Vol.19, No 109, 1986, pp 55-63.
2. Zielinski A.J. Fracture of concrete and mortar under uniaxial impact loading. Delft University Press, Delft,1982.
3. Kormeling, H.A. Impact tensile strength of steel fibre concrete. Stevin Report 5-84-8, Delft, 1984.
4. Weerheijm, J.,Van Zantvoort, P.J.H. and Opschoor, G. The Application of the FE-Technique to Dynamic Failure Analysis of Concrete Structures. Proc. 23rd DoD Explosive Safety Seminar, Atlanta, Georgia, 9-11 August 1988.
5. Hordijk, D.A., Reinhardt, H.W.,Cornelissen, H.A.W. Fracture Mechanics Parameters of concrete from uniaxial tensile tests as influenced by specimen length. In: "Fracture of concrete and rock", eds. S.P. Shah, S.E. Swartz, Houston, 1987, pp 138-149.
6. Hordijk, D.A., Rots, J.G., Reinhardt, H.W. Finite element supported fracture testing of concrete. Proc. IABSE Colloquium, Delft 1987, pp 189-196.
7. Weerheijm, J. Dynamic properties of concrete 4: Uniaxial Impact tensile Tests. PML 1989- , to be published in 1989.

Nonlinear Finite Element Analysis of Impulse Loaded RC Beams

David J. Stevens* and Theodor Krauthammer**

**Department of Civil and Environmental Engineering, Clarkson University, Potsdam, New York 13676, USA*

***Department of Civil and Mineral Engineering, University of Minnesota, Minneapolis, Minnesota 55455, USA*

INTRODUCTION

The analysis of the response of impulse and blast loaded structures has received a continuous but varying level of attention over the past 40 years. Due to the infinite number of permutations of structural types and anticipated threats, the amount of available experimental data, while broad, is also scant relative to any particular combination of structure and impulse load. Thus, the development of analytical/computational tools are required in order to: 1) interpret the complex nonlinear response observed experimentally, 2) perform parametric studies, and 3) develop design guidelines.

One such analytical tool is the Finite Element Method, in which geometric nonlinearities can be modeled with the well-known kinematic formulations of large displacement analyses (Lagrangian, Eulerian, Updated Lagrangian) and material nonlinearities are accommodated through the constitutive models. In the case of Reinforced Concrete (RC) beams, these materials are steel, for which adequate models are readily available, and plain concrete, for which the development of an accurate, consistent, and computationally efficient model is still an active area of research. Since the quality of the numerical solution strongly depends upon the material models, a rate-independent, strain softening, nonlocal Continuum Damage/Plasticity (CDP) model for plain concrete is presented in this paper. Also, since rotatory inertia and shear effects significantly alter the response of structural elements that are abruptly loaded, a layered, large strain, Timoshenko beam element is developed. The proposed concrete model is then combined with the Timoshenko beam element and a plasticity model for the reinforcement steel to analyze two simply supported, impulse loaded, RC beams tested by Feldman and Siess [1].

A CONSTITUTIVE MODEL FOR PLAIN CONCRETE

Given the complex behavior of concrete, it is not surprising that a large number of distinctly different constitutive models for concrete have been proposed over the years. It is beyond the scope of this paper to even lightly touch upon all of these and, thus, only Continuum Damage Mechanics/Plasticity models and models for strain softening are discussed.

Physically, the nonlinear response of concrete is created through the combination of microcrack growth and frictional slip. The theory of Continuum Damage Mechanics accounts for the phenomena of strain softening and stiffness degradation due to microcrack growth; the plastic flow and prepeak nonlinearity of concrete, due to frictional slip, can be modeled with the theory of Plasticity. Recently, the two theories have been combined to form a number of successful approaches for modeling plain concrete (Bazant and Kim [2], Han and Chen [3], and Yazdani and Schreyer [4]). Of interest are the models of Simo et al. [5], and Frantziskonis and Desai [6] in which a scalar damage parameter approach was combined with plasticity theory to yield effective models that match experimental results for a number of load paths.

Unfortunately, the application of such softening, rate-independent constitutive models in the treatment of initial boundary value problems leads to mesh dependency, localization of strains, and erroneous predictions of energy dissipation in numerical calculations; thus, the potential for such negative side effects exists for each of the models mentioned in the previous paragraph. Since strain softening in concrete is the result of the formation of internal structure, concrete can no longer be represented as a continuum, and some form of representation of the internal structure must be considered. To avoid the impracticality of modeling the mortar and individual aggregate particles, various methods for enforcing finite energy dissipation and removing the mesh dependency have been proposed (Willam et al. [7], Lasry and Belytschko [8], Schreyer and Chen [9]). One method of particular interest involves the use of "nonlocal" constitutive laws, in which the dependent variable at a point is a function of independent variables in the neighborhood of that point. Recently, Pijaudier-Cabot and Bazant [10] successfully applied such an averaging technique to a nonlocal definition of the damage variable in a Continuum Damage Mechanics model.

The concrete model developed here is a combination of Plasticity and Continuum Damage Mechanics theories, and it incorporates a nonlocal definition of a scalar damage variable. The Continuum Damage Mechanics portion of the model is based loosely on the approach proposed by Frantziskonis and Desai [6], in which the strength and stiffness degradation are governed by a scalar damage parameter. It can rightfully be argued that a complete three dimensional treatment of microcracking effects on the response of concrete requires a higher order definition of the damage variable, and, first, second, fourth and even eighth order tensors have been proposed. However, in cases where damage directionality is not a dominant feature, such as in beams, frames, and slabs, a scalar parameter is sufficient.

The key assumption in this approach is that the material behavior can be decomposed into two constituents, an undamaged portion (or topical portion, to use the nomenclature of Frantziskonis and Desai [6]) and a damaged portion. During loading, the topical material is transformed into damaged material. The effective stress tensor $\underline{\underline{\sigma}}$ is the damage weighted sum of the damaged stress $\underline{\underline{\sigma}}^o$ and topical stress $\underline{\underline{\sigma}}^t$ or

$$\underline{\underline{\sigma}} = (1 - d)\underline{\underline{\sigma}}^t + d\,\underline{\underline{\sigma}}^o \tag{1}$$

where $d \equiv A_o/A$ is the damage parameter and A_o and A are the damaged and total areas, respectively. It is assumed that the topical and damaged stress tensors are symmetric, and, that the strains in the damaged and topical components are equal to the average strains ($\underline{\underline{\epsilon}} = \underline{\underline{\epsilon}}^t = \underline{\underline{\epsilon}}^o$) (Ortiz [11]).

It is further assumed that microcracking reduces the area of contact over which both shear and normal stresses act; this concept of area reduction is the principle upon which many Continuum Damage Mechanics theories are based. Therefore, the damaged material cannot support shear stresses or confining pressures, i.e., $\underline{\underline{\sigma}}^o = \underline{\underline{0}}$, and, the average stress is now defined as

$$\underline{\underline{\sigma}} = (1 - d)\underline{\underline{\sigma}}^t \tag{2}$$

Equation 2 determines both the reduction in stiffness of the damaged concrete as well as the development of strain softening. To remove the mesh dependency and erroneous energy dissipation associated with strain softening models, a nonlocal definition of the damage variable is used and it is calculated as a weighted average, using the error density function as the weight function as suggested by Pijaudier-Cabot and Bazant [10]. The value of the characteristic length, over which the averaging is performed, is based on considerations of the proposed application of the constitutive model. As Pijaudier-Cabot and Bazant [10] show, the use of the Euler hypothesis in a beam analysis requires that the averaging length be greater or equal to the beam depth. Therefore, since this constitutive model will be implemented into a Timoshenko beam element, which is based on the Euler principle, the characteristic length is taken as the beam depth. The use of the beam depth for averaging the damage can be validated physically by examining the hinge formations in beams loaded to failure (Corley [12]); in most cases, the hinge length is approximately 85 to 95 percent of the depth.

In the proposed approach, the evolution equations for the local damage parameter, d, are parameterized using the concept of equivalent tensile strain (Simo et al. [5]), which is defined as follows. The positive projection of the strain tensor, $\underline{\underline{\epsilon}}^+$, may be written as

$$\underline{\underline{\epsilon}}^+ = \sum_{a=1}^{3} \langle\epsilon^{(a)}\rangle^+ \, \underline{e}^{(a)} \otimes \underline{e}^{(a)} \tag{3}$$

where $\epsilon^{(a)}$ are the eigenvalues of $\underline{\underline{\epsilon}}$, $\underline{e}^{(a)}$ are the normalized eigenvectors of $\underline{\underline{\epsilon}}$, and $\langle x\rangle^+ \equiv (x + |x|)/2$. The equivalent tensile strain is then defined as the norm of the positive strain tensor projection or $\tau = [\mathrm{tr}(\underline{\underline{\epsilon}}^+\underline{\underline{\epsilon}}^+)]^{1/2}$, where tr is the trace operator.

The state of damage is characterized by a strain based damage criterion G with the following form (Simo et al. [5]):

$$G = \tau - r \le 0 \tag{4}$$

where r is the damage threshold. Thus, damage is created when the tensile strain, τ, exceeds the current damage threshold r. The rate equations, loading/unloading conditions, and damage consistency conditions associated with Eq. 4 are given in Stevens and Krauthammer [13].

The response of the topical material is controlled by the plasticity portion of the CDP model, which employs a continuous, smooth yield surface that combines a strain hardening modified Drucker-Prager failure surface with a curved cap; this model was developed by Schreyer and Bean [14], who refer to it as a "Prager-Drucker" cap model to differentiate it from the Drucker-Prager cap model commonly used for soils. The modified Prager-Drucker yield surface is plotted in the $J_2^{½}$, P plane in Fig. 1 (J_2 is the second invariant of the topical deviatoric stress tensor and P is the hydrostatic pressure). As Fig. 1 shows, the flow surface is continuous and has a continuous derivative everywhere so corner algorithms are not required.

Over the majority of the yield surface, the plastic strains develop through use of an associated flow rule; however, when the model is specialized from three dimensional to a plane strain state, a nonassociated flow rule provides more reasonable results in regions of tensile hydrostatic pressures and low compressive hydrostatic pressures. Under plane strain conditions, the associated flow rule generates dilation, which creates, in turn, an unrealistic confinement stress in the out-of-plane direction; since concrete is pressure sensitive, this out-of-plane compressive stress can lead to erroneous predictions of shear and tensile strengths. To avoid this, the CDP model uses a nonassociated flow rule in regions of tensile or low compressive (less than 0.1 f_c') hydrostatic pressure to reduce the dilational strain predictions while allowing adequate predictions of strength. The nonassociated flow rule is implemented with a plastic potential g that is chosen as a nearly horizontal Drucker-Prager surface, as shown in Figure 1.

The main criticism leveled against the use of nonassociated flow rules is the potential for spontaneous energy generation. This unpleasant feature was first discussed by Il'iushin [15], who showed that energy generation is possible whenever the predicted elastic stress increment falls into the "wedge" that is above the yield surface but below the plastic potential surface. To remove the potential for spontaneous energy generation, the flow rules are modified as follows: during loading from the yield surface and for any predicted elastic state that falls into the "wedge", the associated flow rule will be used, while for other loading conditions in which the yield surface is violated, the nonassociated flow surface is implemented.

The proposed concrete model (using the local definition of the damage variable) was evaluated successfully, on a material level, through comparisons with three sets of concrete test data; for a full presentation of the results, the reader is directed to Stevens and Krauthammer [13].

TIMOSHENKO BEAM ELEMENT

The kinematic relations for the layered Timoshenko beam element are based on the usual assumptions: 1) plane sections originally normal to the neutral axis remain plane after deformation but not necessarily normal and 2) the shear strain is constant over the beam's cross section and may be defined as a rotation of the cross section. Assuming small to moderate rotations and small axial strains and, using the definition of the Green-Lagrange strain tensor, the nonzero, nonlinear, strain displacement relations may be written as

$$\epsilon_{11} = U' - \xi_1(\theta'-\gamma') + \frac{1}{2} V'^2; \qquad \epsilon_{12} = \frac{1}{2} [\gamma + U'(\gamma-\theta)] \tag{5}$$

where $U \equiv U(\xi_1)$ is the axial displacement of the neutral axis along the beam, $\gamma \equiv \gamma(\xi_1)$ is the shear deformation, ß is the total rotation of the cross section, $V \equiv V(\xi_1)$ is the transverse displacement of the neutral axis along the beam, ξ_i are the local beam coordinates (i = 1,2), $\theta \equiv V'$, and $(\ )' \equiv$ derivative w.r.t. the independent variable. In this formulation, the axial displacement and the shear rotation are interpolated with linear shape functions acting on the nodal variables, and the transverse displacement is interpolated using the standard cubic Hermetian polynomials acting on the total rotations. The shear rotations are retained as independent degrees of freedom and continuity of shear across elements is preserved. Also, a convected coordinate formulation is used to remove rigid body rotation effects from the strain calculations (Belytschko, Welch, and Bruce [16]).

In the implementation of the Finite Element method using central difference time integration, the internal force vector of each element must be calculated. This force vector is equivalent, in a virtual work sense, to the element stresses and is defined as

$$\{F_{INT}\} \equiv \int_{Vol} [B]^T [\sigma] \ dVol \tag{6}$$

where $\{F_{INT}\}$ is the internal force vector equivalent to the element's stress state, $[B]^T$ is the three dimensional strain displacement matrix, and $[\sigma]$ is the Cauchy stress tensor (matrix). For integration through the depth, the element is subdivided into 10 layers and the integration is performed by a summation of the stresses at each layer; the stresses in the steel and concrete are calculated with the models discussed above and a perfect bond is assumed between the reinforcement and concrete. The integration over the length of the element is performed using three point Gaussian integration. Also, the movement of the neutral axis due to the material nonlinearities is tracked with an iterative technique developed by Puglisi and Krauthammer [17].

BEAM ANALYSIS AND DISCUSSION

The Timoshenko beam element is combined with the CDP concrete model and a uniaxial elastic/strain hardening plastic model for the reinforcement steel (see Park and Paulay [18]); assuming that the shear contribution of the steel reinforcement is small compared to the shear force resisted by the concrete, the steel response may be taken as uniaxial. The resulting approach is applied to the analysis of simply supported RC beams, which were subjected to large amplitude, short duration loads at midspan (Feldman and Siess [1]). Figure 2 presents the dimensions, cross section and instrumentation of beams H1 and C1. Figure 3 presents linear approximations of the applied loads of beams H1 and C1. During testing, both beams developed considerable damage and large permanent deflection. The residual midspan displacements were approximately 8 inches and 2.5 inches in beams H1 and C1, respectively.

Due to the symmetry of the beams, loads, and boundary conditions, only half of the beam is modeled, using coarse and refined meshes, as shown in Figure 4. At the right node, the horizontal displacement, shear rotation, and bending rotation are restrained. The equations of motion are integrated temporally with the conditionally stable central difference technique and the integration time step was chosen as half the critical time step.

As is typical with most test reports, only the compressive strength of the concrete and the yield strain and stress of the steel were reported. Based on the concrete compressive strengths of beams H1 and C1, "typical" values for medium strength concrete are used (Stevens and Krauthammer [13]). Also, typical values for the reinforcing steel are assumed.

Comparisons between the experimental and calculated displacement histories of Gages 3, 4, and 5 of beam H1 are shown in Figure 5 for the coarse and refined meshes. As this figure shows, there is good agreement between peak displacement, residual displacement and the time when peak displacements occur.

The results shown in Figure 5 also appear to be independent of the mesh refinement used, which, considering that a significant amount of strain softening occurs at midspan, shows that the computational difficulties normally present with strain softening models have been alleviated with the nonlocal CDP model. In order to verify the benefits of the nonlocal approach, beam H1 is reanalyzed using the coarse and refined meshes <u>and</u> the CDP model <u>without</u> damage averaging; the computed histories of the midspan displacement for the local and nonlocal, coarse and refined mesh analyses are plotted in Figure 6(a). As shown, the <u>local</u> strain softening model overestimates the displacement with the coarse mesh, and, with the refined mesh, the displacement appears to continue to infinity. An examination of the output for these local analyses shows that the damage localizes into the element directly next to the load; for the refined mesh, this element becomes a kinematic hinge and hence the increasing displacement.

In order to further investigate the effects of the nonlocal approach, the internal inelastic work history of beam H1 is calculated using the local and nonlocal, coarse and refined meshes, and these histories are plotted in Figure 6(b). The nonlocal model curves are the same shape and reach approximately the same value of internal work. The local model curves have slightly different form but the peak values are much different. Interestingly, the local refined mesh curve predicts almost the same internal work as the nonlocal approaches, yet, as shown in Figure 6(a), the displacement at midspan continues to grow with time.

Comparisons between the experimental and calculated displacement histories of Gages 3, 4, and 5 of beam C1 are shown in Figure 7 for the coarse mesh (essentially identical results were found with the refined mesh and, hence, only the coarse mesh results are shown). As with beam H1, there is good agreement between peak displacement, residual displacement and the times at peak displacements.

Comparisons between the measured reaction at the left end of the beam and the calculated reaction using the coarse mesh for beams H1 and C1 are shown in Figure 8. Again, there is very good agreement with the peak reaction and the time of peak reaction, as well as the time when the reaction force returns to zero. However, after approximately 60 msec, the calculated reaction oscillates with a higher frequency than the measured reaction; this difference is due to the fact that the unloading stiffness predicted with the CDP model is too large (as discussed in Stevens and Krauthammer [13]), resulting in a stiffer beam with a higher natural frequency. A correction to the constitutive model would improve the results for the response after 60 msec; however, the key point is that the peak value is correctly predicted as are the times at which the peak occurs and the times when the reaction returns to zero.

Lastly, the experimental and predicted compressive steel strain histories using the coarse mesh for beams H1 and C1 are presented in Figure 9. In these figures, compression is taken as negative. As with the reaction history, there is very good agreement until 60 msec; after 60 msec, the large unloading stiffness of the concrete model "catches" the beam and keeps the strain in the reinforcement lower than that measured. In the real beam, the more flexible concrete forces the reinforcement to carry the load and strain. After 60 msec, the experimental value of strain in beam H1 reaches a high tensile value; the reason for this is unclear but the authors view it as an anomaly in the test data.

CONCLUSIONS

A nonlocal Continuum Damage/Plasticity model is developed and implemented for the analyses of RC beams. The analyses of the two impulse loaded beams are found to agree well with the experimental data. The analyses predict the displacement, reaction, and strain histories very closely. The observed mesh independence is particularly pleasing considering the shortcomings usually associated with strain softening constitutive models.

In conclusion, it appears that the combination of the Timoshenko beam element with the nonlocal CDP model and the uniaxial steel model is capable of representing the dynamic behavior of RC beams quite accurately.

ACKNOWLEDGEMENTS

This work was sponsored by the Defense Nuclear Agency under Contract No. DNA001-85C-0410 with the University of Minnesota. The authors wish to thank Dr. H. Schreyer and Mr. J. Bean of the New Mexico Engineering Research Institute for supplying the Prager-Drucker cap model and for their helpful discussions.

REFERENCES

1. Feldman, A. and Siess, C.P. Investigation of Resistance and Behavior of Reinforced Concrete Members Subjected to Dynamic Loading, Part II, AFSWP No. 1088, U. of Illinois, Urbana, IL, 1958.
2. Bazant, Z.P., and Kim, S.S. Plastic-Fracturing Theory for Concrete, J. Eng. Mech., ASCE, 105(3), 407-428, 1979.
3. Han, D.J., and Chen, W.F. Constitutive Modeling in Analysis of Concrete Structures, J. Eng. Mech., ASCE, 113(4), 577-593, 1987.
4. Yazdani, S., and Schreyer, H.L. (1987). "A Unified Plasticity and Damage Mechanics Model for Plain Concrete." U. of New Mexico Report, Unpublished.
5. Simo, J.C., Ju, J.W., Taylor, R.L., and Pister, K.S. On Strain-Based Continuum Damage Models: Formulation and Computational Aspects, in Constitutive Laws for Engineering Materials: Theory and Applications, Ed. by C.S. Desai et al., Elsevier Science, 233-245, 1987.
6. Frantziskonis G., and Desai, C.S. Elastoplastic Model with Damage for Strain Softening Geomaterials, Acta Mechanica 68, 151-170, 1987.

7. Willam, K.J., Bicanic, N., and Sture, S. Constitutive and Computational Aspects of Strain-Softening and Localization in Solids, in Constitutive Equations: Macro and Computational Aspects, ed. by K.J. Willam, ASME, New Orleans, LA, 233-252, 1984.
8. Lasry, D., and Belytschko, T.B. Localization Limiters in Transient Problems, sub. to Int. J. Solids Struct., Northwestern University, Evanston, IL, 1987.
9. Schreyer, H.L., and Chen, Z. One-Dimensional Softening with Localization, J. App. Mech., ASME, 53, 791-797, 1986.
10. Pijaudier-Cabot, G., and Bazant, Z.P. Nonlocal Damage Theory, Report No. 86-8/428n, revised Jan. 1987, Center for Concrete and Geomaterials, Northwestern University, IL, 1987.
11. Ortiz, M. A Constitutive Theory for the Inelastic Behavior of Concrete, Mech. of Mat., 4, 67-93, 1985.
12. Corley, W.G. Rotational Capacity of Reinforced Concrete Beams, J. Str. Eng., ASCE, 92(5), 121-146, 1966.
13. Stevens, D.J., and Krauthammer, T. Development of an Advanced Computational Approach for the Analysis of Buried Reinforced Concrete Structures Subjected to Severe Stress Transients, Civil and Min. Eng. Rep. ST-88-05, University of Minnesota, Minneapolis, MN, 1988.
14. Schreyer, H.L., and Bean, J.E. Plasticity Models for Soils, Rock and Concrete, Rep. to the New Mexico Engrg. Res. Inst., Albuquerque, NM, 1987.
15. Il'iushin, A.A. On the Postulate of Plasticity, Prikl. Mat. Mekh., 25(3), 503-507, 1961.
16. Belytschko, T., Welch, E., and Bruce, R. Large Displacement Nonlinear Transient Analysis by Finite Elements, Proc. Int. Conf. Vehicle Struct. Mech., SAE, 188-197, 1974.
17. Puglisi, R.D. and Krauthammer, T. Dynamic Response Analysis of Shallow-Buried Reinforced Concrete Arches." Civil and Min. Eng. Rep. ST-87-06, University of Minnesota, Minneapolis, MN, 1987.
18. Park R., and Paulay, T. Reinforced Concrete Structures, Wiley-Interscience, John Wiley and Sons, NY, 1975.

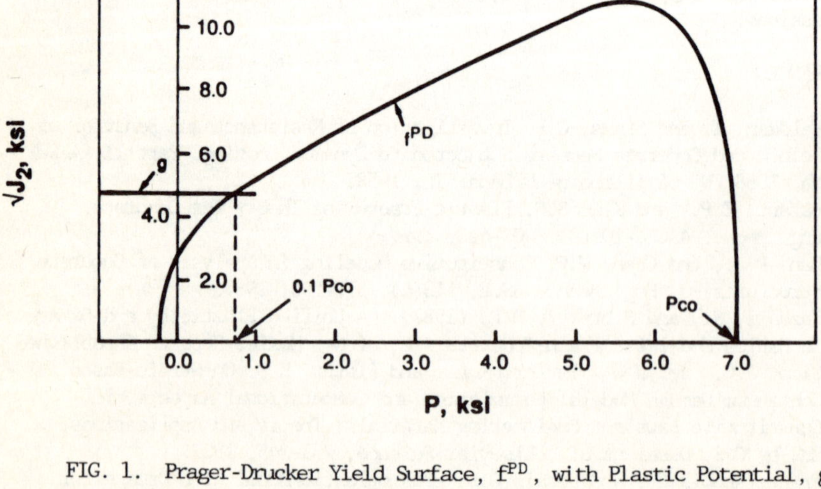

FIG. 1. Prager-Drucker Yield Surface, f^{PD}, with Plastic Potential, g.

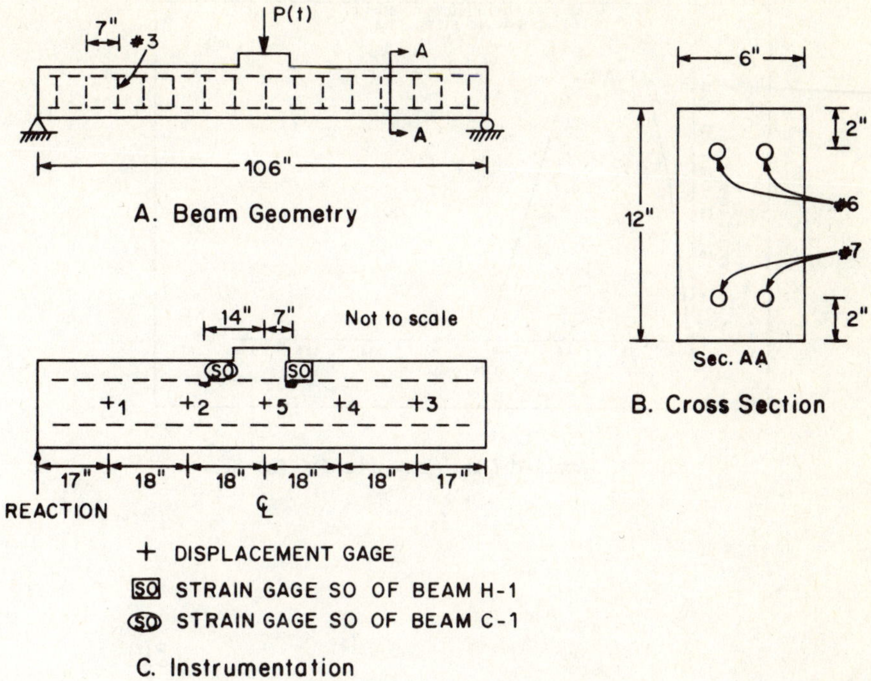

FIG. 2. Geometry and Instrumentation of the Impact Loaded Beams (Feldman and Siess [1]).

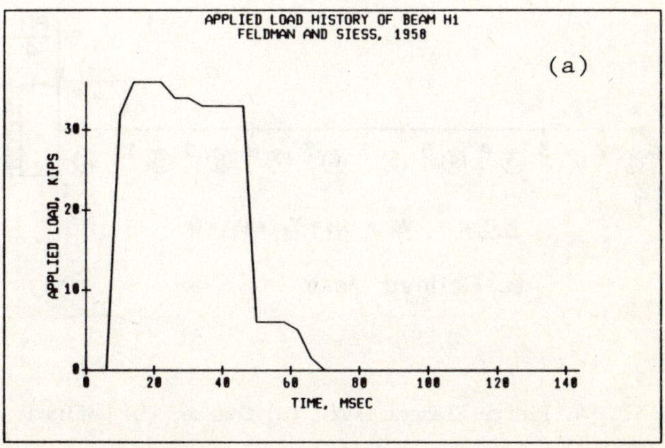

FIG. 3. Applied Load (Linearized): (a) H1.

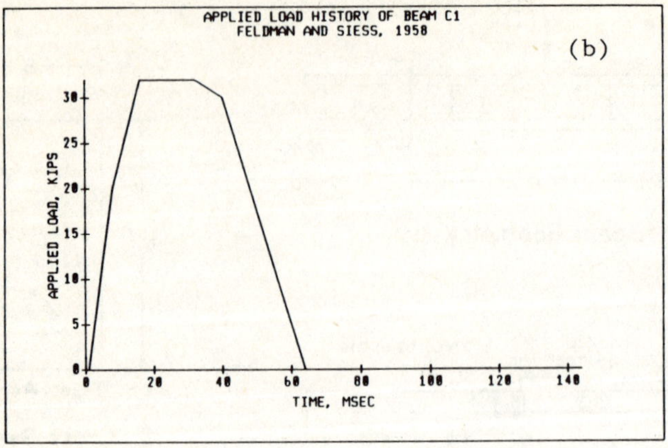

FIG. 3. Applied Load (Linearized): (b) Cl.

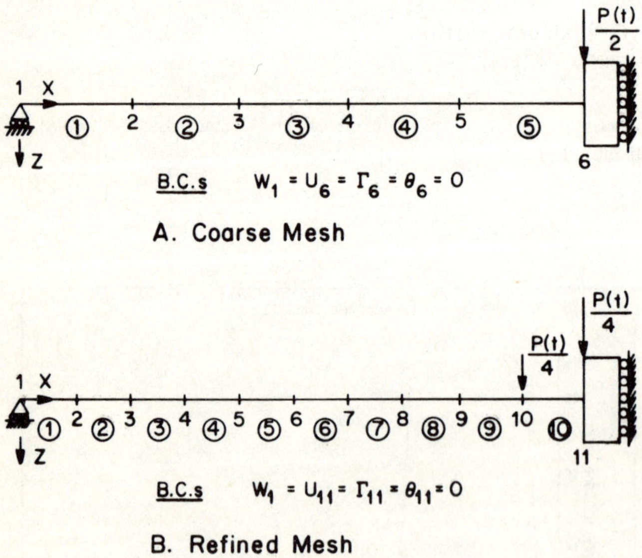

FIG. 4. Finite Element Mesh, (a) Coarse, (b) Refined.

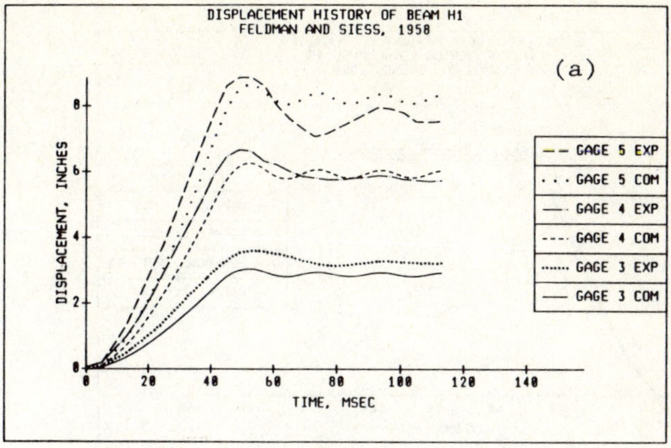

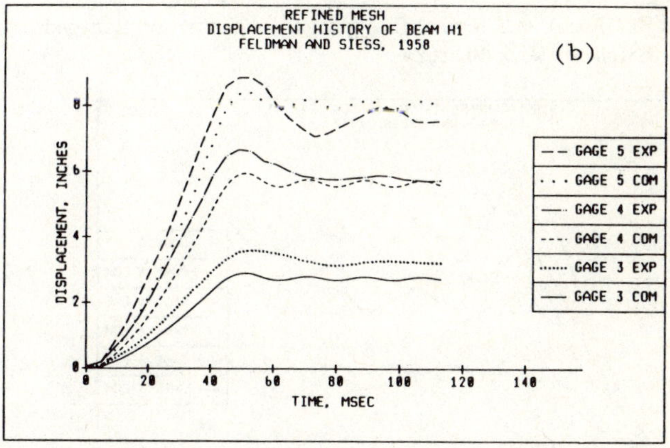

FIG. 5. Displacement History of Beam H1: (a) Coarse Mesh; (b) Refined Mesh.

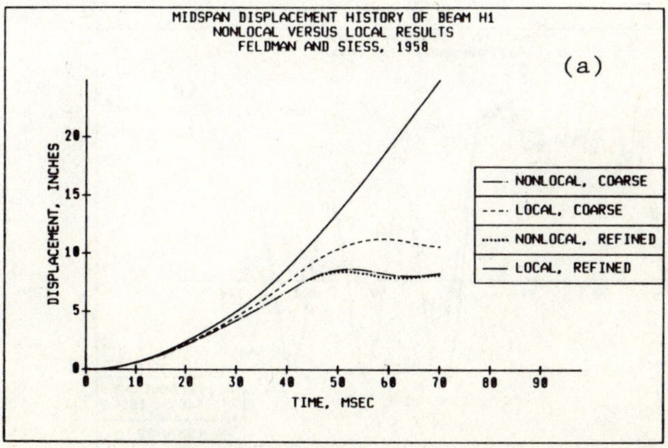

FIG. 6. Beam H1, Local vs. Nonlocal Approach, Coarse and Refined: (a) Displacement History.

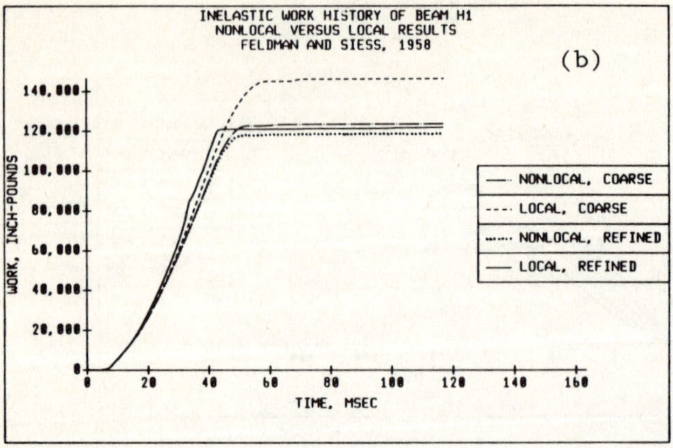

FIG. 6. Beam H1, Local vs. Nonlocal Approach, Coarse and Refined: (b) Inelastic Work History.

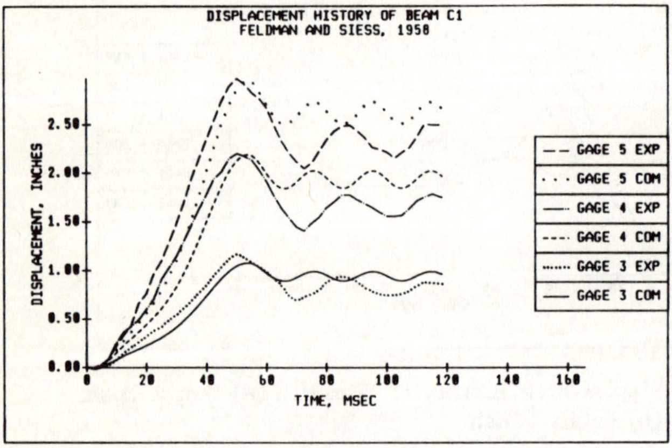

FIG. 7. Displacement History of Beam C1, Coarse Mesh.

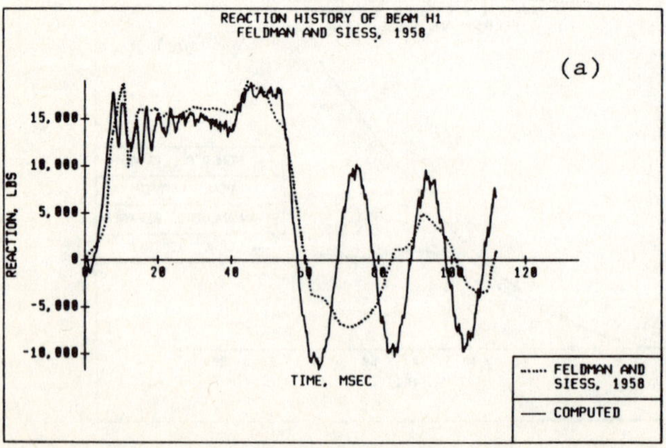

FIG. 8. Reaction Histories, Coarse Mesh: (a) H1.

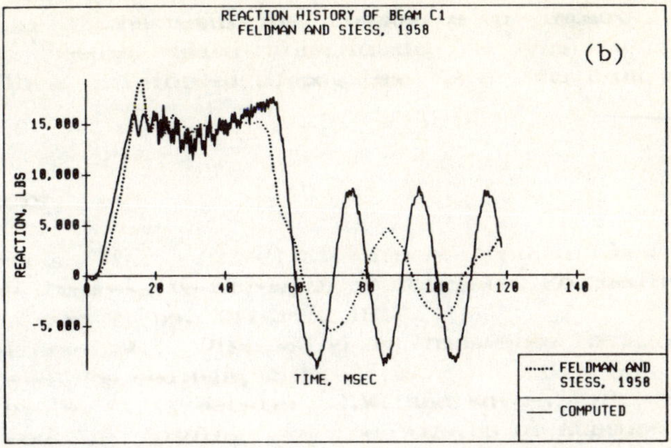

FIG. 8. Reaction Histories, Coarse Mesh: (b) C1.

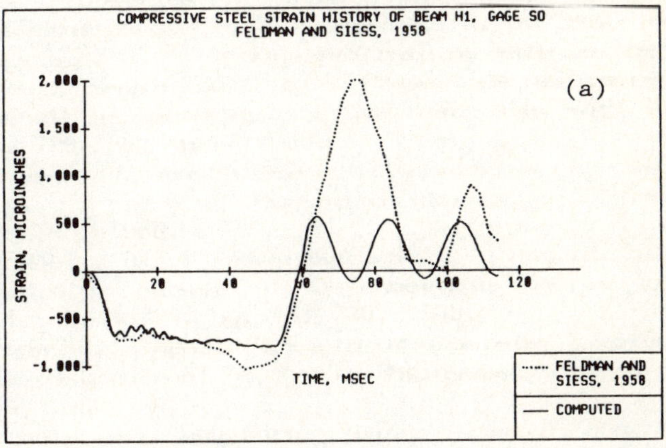

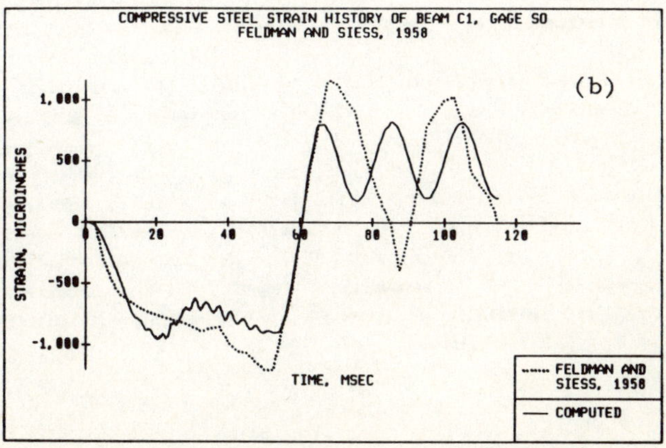

FIG. 9. Reinforcement Steel Strain Histories, Coarse Mesh: (a) H1; (b) C1.

Concrete Damage Assessment under Blast Loading

C.C. Fleischer, N.S. Mistry and A.K. Welch
Taylor Woodrow Construction Limited, 345 Ruislip Road, Southall, Middlesex UB1 2QX, UK

ABSTRACT

The controlled use of explosives, for demolition, is a technique being developed to cut and remove selected parts of concrete structures without impairing the overall structural integrity. This technique is particularly applicable for the safe controlled removal of activated and/or contaminated parts of nuclear facilities during plant decommissioning. This controlled removal will precede the general demolition of the non-active parts of the containment structure which will be relied on to provide containment during the above operation. To substantiate and add credibility to the development work being carried out, analytical studies have been performed in which the effects of firing an explosive buried in a concrete mass have been investigated.

This paper concentrates on demonstrating the application of finite element methods to this class of problem with emphasis on numerical modelling of the concrete materials. In order to compare with field trials an axisymmetric formulation was considered, use being made of the computer program Adina. The general concrete material model was based on a hypoelastic formulation with cracking and crushing criteria. However, in the regions close to the explosive charge, a plastic flow material model with compaction limitations was used instead of the general concrete model.

The paper presents the basis for the theoretical formulation of the problem together with the criteria for modelling the behaviour of the concrete. The analytical results are presented and compared with experimental results from field trials. Conclusions that can be drawn from the analyses are given together with recommendations to optimise on the results.

INTRODUCTION

The use of explosives for the controlled cutting and partial removal of concrete structures is a technique being developed for the decommissioning of nuclear facilities. Here the technique is being developed for the controlled removal of activated and/or contaminated materials from the insides of biological shielding structures whilst relying on the outer layers of non-activated or contaminated materials to contain the activity. The development programme which has formed part of the Commission of European Communities' (CEC) programme of research on decommissioning (CEC[1]) has concentrated on the use of experimental tests to study the cratering aspects of explosive charges buried in concrete. A detailed description of the preliminary programme and results obtained are fully presented by Fleischer[2].

In order to increase applicability, substantiate and add credibility to the test results, investigations have been started into demonstrating the application of finite element methods to this class of problem with emphasis on numerical modelling of the concrete materials. In order to compare with field trials an axisymmetric formulation was considered, use being made of the computer program ADINA.

Earlier evaluation of the concrete material model supplied in ADINA-81 code had identified some shortcomings. Modifications were therefore carried out on this code to derive an updated version referred to as ADINA-TW (Gedling et al[3]). This updated version has been used to investigate the response of a concrete block to a theoretical pressure pulse loading. The latest version of ADINA (5.0) incorporates most of the modified concrete model (Bathe et al[4,8]), but its application to this work has not as yet been extensively performed.

Earlier work had identified that the principal limitation to analyses of this type arises from the inability of the concrete model to correctly represent the very localised pulverised layer adjacent to the charge. Thus, several layers of elements surrounding the charge were represented by a "compaction" material model, essentially an isotropic crushable material (see Figure 1). For the first problem analysed, the mathematical model was represented by the concrete mass above the charge with a symmetrical boundary being introduced at the level of the charge. This analysis was introduced to obtain some initial understanding of the problem. Subsequently a second analysis was carried out in which the effect of the concrete mass ignored in the first analysis was included.

THEORETICAL MODEL DESCRIPTION

An axisymmetric formulation was used for the analyses. The first problem, defined in Figure 1, investigated local effects following the firing of a charge. The assumed boundary of symmetry at the level of the charge introduces unrealistic stress fields due to reflected waves arriving from the fictitious boundary at the wrong times. Therefore, the analysis after the time when such reflected waves have arrived at the areas under consideration is not representative of physical test results. In view of this, the analysis was carried out only up to 500 micro-seconds, when the reflected waves from the fictitious boundary were known to have already arrived within the area of damage inducing unrepresentative behaviour.

Using previous experience gained from the analyses of wave propagation problems, the model was discretized as shown in Figure 2. The firing of a buried explosive charge in the concrete mass would set up incident waves propagating out radially. The instantaneous stress waves induced by this charge would dissipate through the concrete mass causing damage, hence a suitable finite element model which can be adapted to capture the energy and simulate its effects on the material is essential.

When non linearities are considered, extra precautions are necessary to avoid unrepresentative stress disruptions which may lead to premature failure, hence producing inaccurate structural responses. To cover the various areas of interest the use of a spherical mesh around the charge and a cylindrical mesh near the surfaces is required, as shown in Figure 2. The transition from one type of mesh to the other and changes in element sizes can give rise to stress disruptions.

MATERIAL MODELLING

Initial studies established that the concrete material model in all versions of Adina (Bathe et al [4,7]) cannot be used satisfactorily for the region close to the charge. This region experiences crushing and compaction under high impulsive load and the characteristics of the concrete material model in Adina-TW cannot be used for capturing the appropriate responses. Therefore, two material models have been used in the analyses. A Drucker-Prager model for the compaction region and the Adina concrete model for the rest of the structure. In order to derive material constants for the analyses, information on material behaviour in the form of a pressure against volumetric strain relationship in the two regions was prepared.

a) For the region close to the charges (compaction region).

The values given in the Hull Hydrocode model were used. They consist of a pressure against volumetric strain relationship as shown in Table 1. A single deviatoric yield surface was evaluated by combining deviatoric failure curves for compression and tension meridians.

P (MP_a)	ln ($^v/_{vo}$)
0	0
13.3	.000631
150.0	.49
340.0	.095
2700.0	.182
4200.0	.223
7100.0	.262
11200.0	.336
15200.0	.405
19000.0	.470

Table 1

The deviatoric surface is represented by

$$\frac{J_2}{\sigma_2{}^2} = 0.491\,(^P/_{\sigma c})^2 + 0.745\,(^P/_{\sigma c}) + 0.02 \qquad (1)$$

where P (mean pressure) $= \dfrac{\sigma_1 + \sigma_2 + \sigma_3}{3}$

$J_2 = \dfrac{1}{6}\,\{(\sigma_1-\sigma_2)^2+(\sigma_2-\sigma_3)^2+(\sigma_3-\sigma_1)^2$, where

J_2 (second invariant of stress deviation), σ_1, σ_2 and σ_3 are three principal stresses and σ_c - crushing limit of the material.

b) For the outer region, the volume compaction curve is:

P (MP_a)	ln ($^v/_{vo}$)
0	0
13.3	.000631
64.0	.01
100.0	.03
152.0	.05
224.0	.07
340.0	.095
2700.0	.182
4200.0	.223

Table 2

An Ottosen failure surface was recommended for use by Chen [5].

$$A\frac{J_2}{\sigma_c^2}+\lambda\frac{\sqrt{J_2}}{\sigma_c}+B\frac{I_1}{\sigma_c}-1=0 \qquad (2)$$

$$\text{where } \lambda=K_1\cos\left\{\frac{1}{3}\arccos(K_2\cos 3\theta)\right\};\ \cos 3\theta>0$$

$$\lambda=K_1\cos\left\{\frac{\pi}{3}-\frac{1}{3}\arccos(-K_2\cos 3\theta)\right\};\ \cos 3\theta<0$$

I_1(first invariant of stresses) = $\sigma_1+\sigma_2+\sigma_3$

and A, B, K_1, K_2 and 0 are constants
with A = 0.7214 These constants are suitable
B = 11.1309 in the situation where:
K_1= 11.8988 $\frac{\sigma_t}{\sigma_c}$ = 0.1, in which σ_t
K_2= 0.9677 is the tensile limit.

a. DRUCKER-PRAGER MODEL IN ADINA

Adina provides a Drucker-Prager model for the analysis of granular material, like crushed concrete. It is a three-dimensional generalization of the Mohr-Coulomb model. Moreover, Adina provides a simple plane cap model which contains a plane cap and a tension cutoff limit. This cap model can control the dilatancy which is overestimated when using only the Drucker-Prager model. Adina[6] provides general description of the material as available in Adina code. In the analyses, the cap model was used. The yield function given in eq (2) was used for deriving necessary material constants required for this model in Adina.

b. CONCRETE MODEL IN ADINA-TW

Features of the concrete material model enhanced in Adina-TW are given by Gedling etal[3]. The material model uses the failure envelopes to establish the uniaxial stress-strain law accounting for multiaxial stress conditions and to identify whether tensile or crushing failure of the material has occurred. In this paper, only compressive failure curves are described briefly to appreciate their significance in this application.

An Ottosen failure surface has been suggested to represent the concrete as given in equation 2. The equation was used to find sets of values of σ_1, σ_2 and σ_3 that would satisfy the yield criterion. The values (σ_1/σ_c) on the tensile meridian (θ = 0°, $\sigma_1 > \sigma_2 = \sigma_3$) and the corresponding values on the compressive meridian ($\theta = 60°$, $\sigma_1 = \sigma_2 > \sigma_3$) were evaluated using equation 2. The values of σ_1, σ_2 and σ_3 evaluated from eq. 2 were similar to that produced by Ottosen in Fig 3. Because of the nature of the problem, the compressive stresses will be high near to the blasting load, hence a set of failure curves were required to take account of larger values of compressive stresses. Fig 4 shows the triaxial compressive failure envelope used in this study.

Material Properties:

$E = 38.0E9\ N/m^2$
$\nu = 0.2$
$\sigma_c = 40.0E6\ N/m^2$
$\sigma_t = 4.0E6\ N/m^2$

SOLUTION PROCEDURE

The following dynamic characteristics were used in the analysis.

1) Implicit integration using the Newmark method was used to obtain solution of the incremented finite element equilibrium equation.

2) Equilibrium iterations were not performed but stiffness matrix was reformed for each time increment.

3) Mass matrix was formed using the lumped mass option.

4) 2 x 2 integration rule.

5) No damping factors used.

Even though equilibrium iteration is recommended with implicit time integration, one cannot choose values for tolerances with confidence. Hence the reason for avoiding equilibrium iteration for each time step. Moreover use of equilibrium iteration in concrete model using cracking criterion may not be suitable.

EXAMPLE 1

In order to establish the dominant crack path, principal tensile strain vector diagrams were used at various time steps for vector lengths above 100 micro-strains and 1000 micro-strains. Plots with the higher strains (1000 micro-strains) can provide evidence of major crack path while the plots at 100 micro-strain show the extent of damage due to compressive and tensile wave fronts.

Although predictions were made at various times only the results at 300 micro-second are shown here. Fig 5 shows onset of major cracks with a change in direction of cracking from 45° to 60° as the cracks approached the free surface.

EXAMPLE 2

As indicated earlier the second example considered the full concrete structure. Using the lessons learnt from the first example from reflected waves, the decision was taken to analyse the structure up to 1100 micro-seconds. In presenting the results 1000 micro-strain was chosen as the cut-off limit. Figure 6 shows the positions for the major strain vectors at 1100 micro-seconds.

CONCLUSIONS AND RECOMMENDATIONS

1. The analyses carried out have demonstrated the capabilities of Adina to follow the effect of blast loading on concrete and the analyses show the effects of reflecting waves.

2. Without economic constraints, a more accurate response prediction should be possible by adopting a regular mesh of sufficiently small element size.

3. Positive conclusions can only be made through investigations on other test cases. Therefore further studies together with test results are necessary to increase knowledge and understanding of such problems.

4. Adina has demonstrated the feasibility of analysing this type of problem. Further improvements in the material modelling aspects and the consideration of other test cases should prove beneficial in increasing the confidence in this type of analytical work.

5. Having demonstrated that this class of problem is capable of solution with non-linear finite element, it is recommended that topics such as mesh densities, damping and material modelling should be further investigated by analysing further examples of field trials. Also some consideration needs to be given to the interpretation of damage and prediction of crack widths.

REFERENCES

1. Commission of European Communities. Proceedings of European Conference on Decommissioning Nuclear Power Plants. Luxembourg May 1984.

2. Fleischer, C. C. A study of explosive demolition techniques for heavy reinforced and prestressed concrete structures. CEC Report No EUR 9862 EN.

3. Gedling, J.S., Mistry, N. S. and Welch, A.K. Evaluation of Material Models for Reinforced Concrete Structures - Computers and Structures - Computers and Structures Vol 24 No 2 pp 225-232 1986.

4. Bathe, K. J and Sundburg, J. A concrete material model - Chapter 4. Computational Modelling of Reinforced Concrete Structures. Hinton, E and Owen, R. Linderidge Press 1986.

5. Chen, W.F. Plasticity in Reinforced Concrete. McGraw-Hill. 1982.

6. ADINA - Theory and modelling Guide ARD 87-8.

7. Bathe, K. J. and Ramaswany S. On three-dimensional non-linear analysis of concrete structures. Nuclear Eng. and Design 52 pp 385-409 1979.

8. Bathe, K. J., Walczak, K., Welch, A.K. and Mistry, N. Non-linear Analysis of Concrete Structures. 7th Adina Conference, 1989.

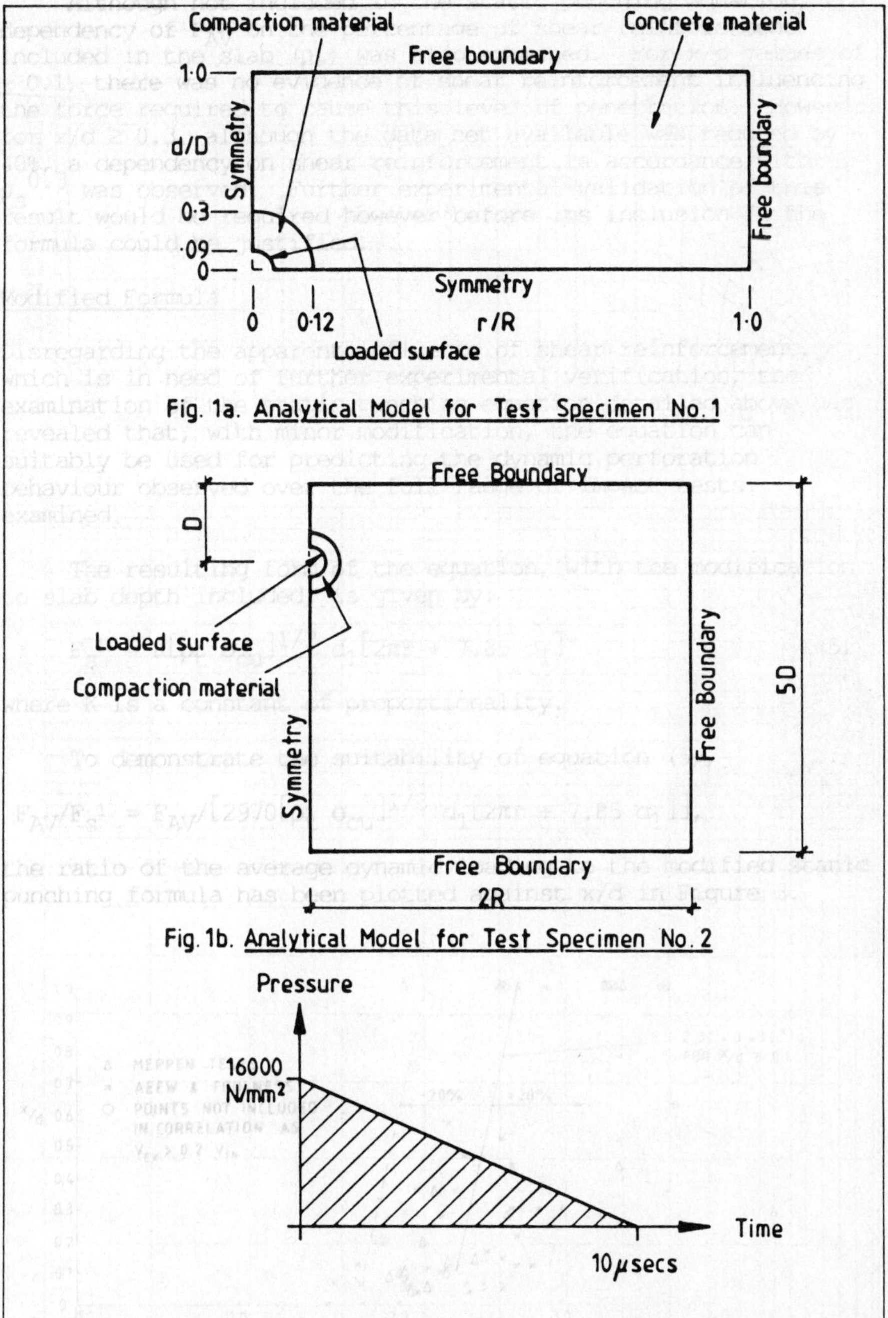

Fig. 1a. Analytical Model for Test Specimen No. 1

Fig. 1b. Analytical Model for Test Specimen No. 2

Fig. 1c. Pressure Transient Applied to Loaded Surface

Fig. 1. Theoretical Modelling

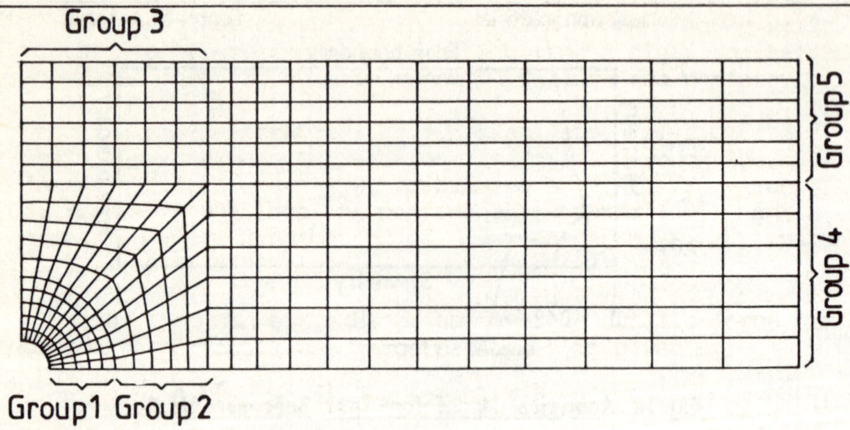

Fig. 2a. Case 1

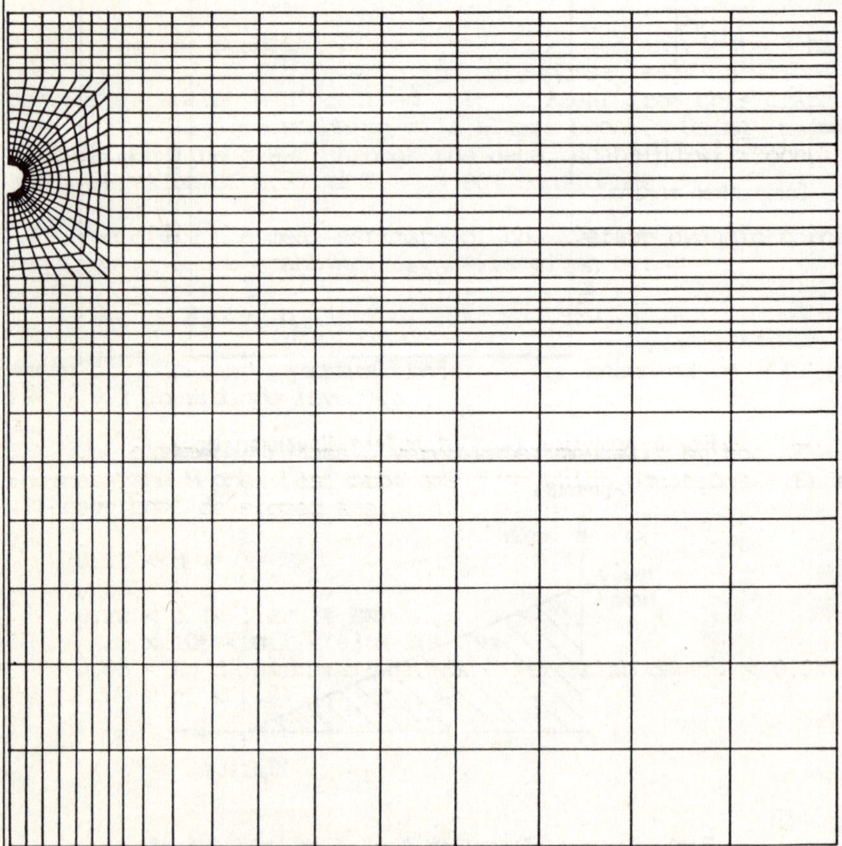

Fig. 2b. Case 2

Fig. 2. Finite Element Discretization

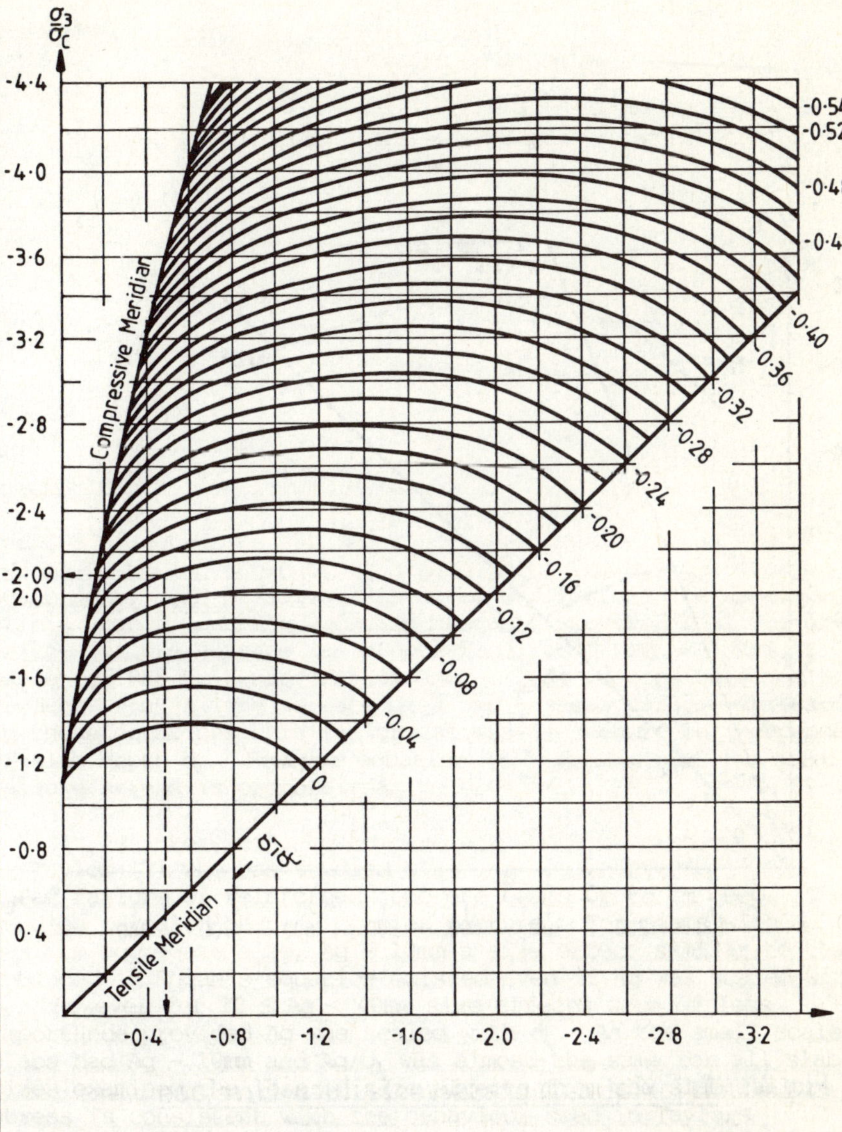

Fig. 3. Design Chart for Ottosen Criterion for Compressive Stresses; Failure Stresses, $\sigma_1 \geq \sigma_2 \geq \sigma_3$ Tensile Positive, all Stresses Compressive.

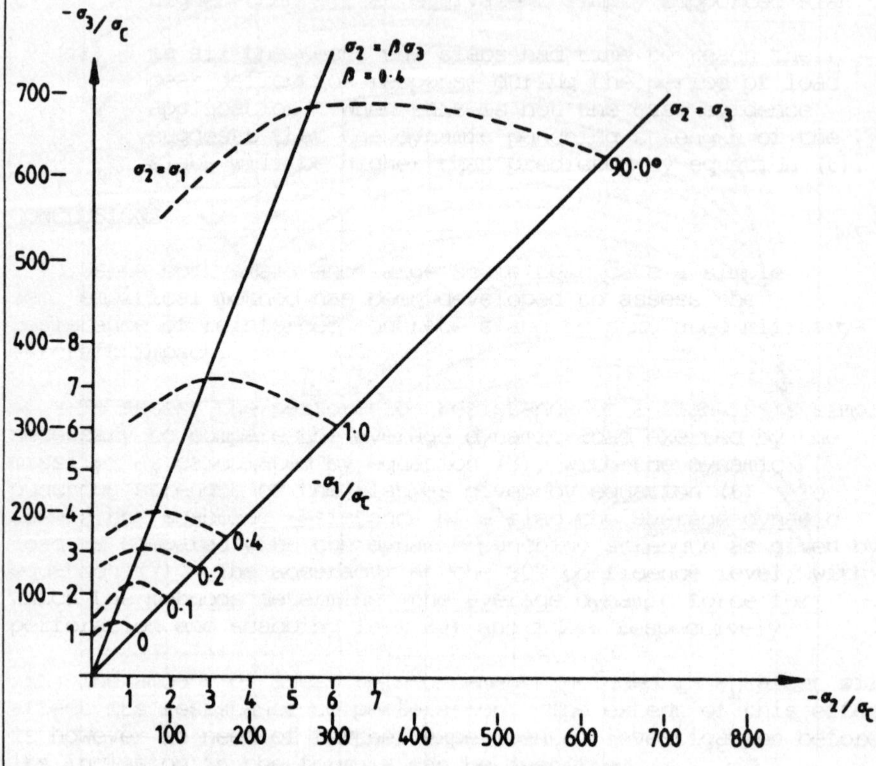

Fig. 4. Triaxial Compressive Failure Envelope

Fig. 5. Strain Vectors at T = 300 micro-seconds

Fig. 6. Strain Vectors at T = 1100 micro-seconds

The Response of Reinforced Concrete Slabs to Non-Nuclear Blast Loading

R.S. Crouch and T.M. Chrisp

Special Services Division, Mott MacDonald Group,
20 - 26, Wellesley Road, Croydon, England, CR9 2UL

ABSTRACT

This paper outlines two techniques recently developed to simulate the deformation response of reinforced concrete slab elements to non-nuclear remote blast loading. The frameworks behind both the single degree of freedom (SDoF) and multi degree of freedom (MDoF) schemes are discussed. The simplified SDoF model employs decoupled shear and flexural analyses and addresses shear plug size by consideration of either an empirical scaling based on Mach Stem formation, or the position of the maximum momentum gradient. The MDoF is an advanced finite element code (based on work by others), modified to incorporate spatial and temporal descriptions of the blast load acting on planar or curved surfaces.

The model predictions have been compared with impact data and are currently undergoing validation for blast trials.

1. INTRODUCTION

The blast wave from an explosion damages a reinforced concrete structure by causing it to deform. The deformation may range from the trivial to that corresponding to total destruction.

Near-field blast effects on high inertia structures may be dominated by rapid stress wave transit through the material of the structure. These effects can make a significant contribution to the damage produced in the loaded area. Nevertheless, near-field effects have been neglected in the present studies and effort has been concentrated on more remote blast (scaled distances greater than 0.8 m.kg$^{-1/3}$) where the loading period increases and intensity decreases.

A structure and an impinging blast wave forms a dynamic system with the blast acting as a forcing function to cause deformation. This deformation is resisted by the inherent stiffness of the structure, its damping capacity and its inertial resistance. The computation of a blast-imposed displacement, with resulting irrecoverable deformation and damage, becomes highly complex for a real structure. However, considerable simplifications for this problem are available.

There are essentially two methods by which a suitable model for dynamic analysis may be set up; the empirical generation of lumped mass, spring and dashpot models, or the finite element idealization. The former tends to lend to fewer degrees of freedom but relies heavily on the experience and expertise of the analyst. The finite element approach on the other hand replaces some of the engineering experience by a rigorous calculation of stiffness and inertial properties of a structure. The implementation of the finite element idealization of complex three dimensional structures usually results in a large number of degrees of freedom, the handling of which by a direct integration scheme can become prohibitively expensive. To use this method therefore, the reduction of a number of degrees of freedom is important and desirable. However, it is important that the dynamic characteristics of the three–dimensional model are not significantly affected by any reduction procedure.

2. SDoF MODEL

The SDoF model is derived by assuming a kinematic equivalence for displacement between the disturbed element and the SDoF system. That is, this simplification involves the representation of the system by an equivalent one–mass, spring, dashpot system with just one motional direction. This representation considers that the effective mass acts at a single point in the centre, and that the structural resistance is a unique function of the displacement of this centre. The equation of motion for such a system may be written as:

$$M(U_{flex}, t_{shear}).\ [\ddot{U}] + C.\ [\dot{U}] + R\ [U_{shear}, U_{flex}] = P(t) \qquad \text{... (1)}$$

where M is the active mass which is a function of time or displacement and $\ddot{U}$ is the acceleration. The term $M.\ddot{U}$ represents the inertia effect. C is the damping coefficient and $\dot{U}$ the velocity. The term R represents the inherent resistance of the structure and P(t) is the forcing function as determined from the blast loading.

For the shear analysis, it is assumed that the active mass is given by the total mass of the responding plug. This infers that the plug is treated as a rigid body sliding between the two adjacent rigid support bodies.

For the flexural analysis, the active mass for the unyielded structure may be determined from modal considerations. For example, assuming a simply supported beam deforms approximately sinusoidally under a uniformly distributed load, simple integration gives an effective mass factor of 0.64. Once the beam has yielded, it may be thought to act like two rigid bodies linked together by a plastic hinge. The effective mass factor for such a system is taken as 0.5.

Several procedures exist for studying the response of SDoF systems under general excitation. The common procedures for solving this system are:

- o the Duhamel integral method (time–domain solution),
- o the Laplace transform method
- o the Fourier transformation method (frequency–domain solution).

For simple forms of excitation these expressions can be evaluated in closed form.

However, the equation of motion for a generalised non–linear SDoF system is non–homogeneous and has no simplified closed form solution. For these more complex problems a numerical solution is required and the non–linear differential equation is solved by step by step integration. Either of two general numerical approaches may by used:

- o interpolation of the excitation or interpolation of the integrand of the Duhamel integral,
- o approximation of the derivatives in the differential equation.

Both approaches may be used for linear systems, while only the latter is applicable to non–linear systems. The model described here employs the approximate average acceleration method (also referred to as the Newmark $\beta = 1/4$ method). This method is particularly suited to non–linear systems. The Newmark β method uses equilibrium conditions at time $t + \Delta t$ and is referred to as an implicit integration method. The time–stepping scheme adopted follows the procedure given in Craig (1981)[1] and may be summarised as follows:

(i) Input $K, R, M, U_o, \dot{U}_o, P(t_i), \; i = 0, 1, ..., n$

where R, M and P(t) have the same meaning as in Eqn. (1), K is the inherent stiffness of the element, U_o is the initial displacement and $\dot{U}o$ the initial velocity (always zero). Note: Uo may be non–zero at the start of a rerun if the structure was damaged in a previous run.

(ii) Solve for $\ddot{U}$ using

$$\ddot{U}_i = (1/M).\,(P_i - R) \qquad ...\ (2)$$

(iii) Input Δt_i

(iv) Form K^*_i using

$$K^*_i = K_i + (4.M/(\Delta t_i)^2) \qquad ...\ (3)$$

(v) Form ΔP^*_i using

$$\Delta P^*_i = \Delta P_i + (4.M/\Delta t_i).\,\dot{U}_i + 2.M.\ddot{U}_i \qquad ...\ (4)$$

(vi) Solve for ΔU_i using

$$\Delta U_i = \Delta P^*_i / K^*_i \qquad ...\ (5)$$

(vii) Solve for $\Delta \dot{U}_i$ using

$$\Delta \dot{U}_i = (2/\Delta t_i).\,\Delta U_i - 2.\dot{U}_i \qquad ...\ (6)$$

(viii) Solve for U_{i+1} using

$$U_{i+1} = U_i + U_i \qquad ...\ (7)$$

(ix) Solve for $\dot{U}_{i+1}$ using

$$\dot{U}_{i+1} = \dot{U}_i + \Delta \dot{U}_i \qquad ...\ (8)$$

(x) Go to step (iii)

The shear and flexural analyses share the same solution routine, yet a constant time step for flexure would prove far too large for the shear analysis. The program overcomes this difficulty by starting with a small time step and then scaling the time interval by a linear factor after a predefined number of steps, provided the displacement increments are seen to increase uniformly (within a specified tolerance).

Analysis of RC elements depends not only on the numerical procedures but also on assumption with regard to structural behaviour (mode shape). Severe blast loading of structures invariably causes excursions of stresses well into the inelastic range. The influence of changes in geometry on the response can also be significant. In such cases it is necessary to assume the shape of the collapse mechanism (Figure 1). The displacement function (deformation or shape pattern) which must satisfy the geometric boundary conditions of the real structure is assumed to be the same as that which would result from the static application of the load to the structure.

Resistance Functions

The main concern with resistance functions is the derivation of structural functions that are accurate, yet simple enough for incorporation into the analyses. The inherent resistance is provided by a combination of bending and shear which are invariably functions of the dynamic displacement.

The problem of shear strength for reinforced concrete sections under very rapid rates of loading is not well understood. Here, use is made of a semi-empirical method used by Krauthammer (1986)[2] for predicting the direct shearing resistance of slabs near their supports. The resistance function is in the form of a five-stage linear model which addresses the elastic and inelastic shear contributions from the concrete and the reinforcement. The flexural yield deformation and ultimate plastic resistance moment for a doubly reinforced section are determined using conventional RC beam and slab theory as given by Warner et al. (1982)[3]. However, to date no satisfactory expression exists for the ultimate flexural deformation in beam or slabs. In the developed model a wholly empirical ductility expression is used, based on the tensile steel percentage.

Load Definition

The load is assumed to be acting symmetrically about the point of detonation. At the beginning of each run, the maximum reflected pressure (P_o) experienced directly below the charge is calculated from the Bode expression given by Kinney and Graham (1985)[4]. The peak pressure at all other points on the blast surface is given by:

$$P_j = P_o \, (Z/S_j)^3 \qquad \text{...(9)}$$

where Z is the stand-off distance and S_j is the slant distance from the centre of the charge to the point on the structure. At each of these points the shock-wave velocity and positive pulse duration (t_d) are also determined from expressions given by Kinney and Graham (Figure 2). The shock arrival time (t_a) is simply given from the front velocity and the slant distance. All that remains to be done in order to define the pressure at any point, at any time, is to employ a decay function of the form:

$$P\,(i,\,j) = P_j.\,(1 - ((t - t_a)/t_d)).e^{-A.(t-t_a)/t_d} \qquad \text{... (10)}$$

where A is a decay parameter, currently fixed at 0.1 (Figure 3).

Shear damage is known to be more probable for charges placed close–in. At these stand–offs, a Mach Stem formation, at an incidence angle of approximately 40°, essentially reduces the reflected pressure experienced by a rigid target. This discontinuity observation suggests a maximum effective boundary to the shear plug. That is, only forces acting within the Mach Stem perimeter are considered. The situation is reversed for more remote detonations where there may be a pressure rise across the discontinuity. The actual plug size is assumed to be given by the point of maximum impulse gradient. This will vary at each time step but is fixed for a series of intervals until the new position lies outside a pre–defined bandzone (typically equal to the structural element thickness, Figure 3).

3. MDoF MODEL

The MDoF model has been built upon an existing non-linear dynamic finite element algorithm for reinforced concrete slabs and shells supplied by Rockfield Software Ltd, The University of Swansea, Wales. This explicit code was based on the original scheme of Lui and Owen (1986) [5]. The explicit technique is attractive since it involves relatively little computational effort; requiring no formal matrix factorization. However, it is conditionally stable and very small time steps are often required. The critical time step is effectively controlled by the material deformation moduli and the smallest mesh dimension.

The program gives consideration to both material (plastic strain–hardening) and geometric (total Lagrangian) non–linearity. Eight–noded Serendipity or nine–noded Lagrangian degenerate shell elements are used with a reduced integration scheme and a layered formulation to represent the steel reinforcement and simulate progressive cracking through the concrete. Although material behaviour under dynamic loading is complex and experimental information scarce, some general statements may be made. For example, it is well established that the instantaneous peak concrete stress is significantly influenced by the rate of straining, particularly under tension fields. For a material like concrete, with limited ductility, the strain sensitivity can give rise to a more brittle behaviour under rapid rates of deformation. The dynamic biaxial concrete, and uniaxial steel, yield functions are sensitive to the current strain rate in addition to being total plastic strain or work dependent. To complete the concrete material mode, both tensile cracking and compressive crushing of the concrete are included. The complete loss of compressive strength (crushing) is determined from limiting strain invariants in the concrete layer and a smeared crack approach adopted to simulate post–fracture tensile behaviour. Interfacial shear transfer across cracks is assumed to be a function of the strain over the crack face and bond deterioration effects are simulated by a tension stiffening mechanism.

The code currently runs on a VAX 11/750 but it is intended to switch to a transputer based system in order to speed up run time.

A pre–processor gives a choice of slab and arched structures, with typical default values for the concrete and steel material properties. The spatially and temporally variable loading is generated for each node point from the defined position and charge mass. Shadowed areas of the arched structures are catered for. The load application function is given by Eqn. (10). Work is currently underway on extending the pre-processor to include more complex structures such as multi-layered boxes and pinned segmental arches.

A post-processor generates graphical output producing deformed structure and stress plots. Damage parameters developed from cracking of the concrete layers allows the structures proximity to failure to be monitored and graphically produced.

4. SUMMARY

The simplified decoupled SDoF formulation provides a useful method for determining the response of beam and slab structures subject to explosions for optimisation studies, concept evaluation and preliminary designs. The major difficulties lie in the proper descriptions of the blast load, the interaction of local damage with flexural response and the interpretation of active ductility mobilised by the reinforced element. A more rigorous non-linear finite element approach can address the latter two difficulties and can provide a satisfactory predictor of flexural response but pays the time and cost penalty on data preparation and result output.

5. ACKNOWLEDGEMENTS

The authors would like to thank:

- o the Procurement Executive of the Ministry of Defence, for supporting the work described in this paper;
- o Dr. Alan Watson of The University of Sheffield, England, for his valuable comments upon the comparison of the computer model predictions with actual model test results;
- o Rockfield Software Ltd, The University of Swansea, Wales, for providing the original analysis routines (DYNSHL) on which the MDoF model is based.

6. REFERENCES

1. Craig, R.R., Structural Dynamics: An introduction to computer methods, John Wiley, 1981.
2. Krauthammer, T., Modified SDoF analysis of R.C. box-type structures, ASCE, J Strut E, 112, 4, 1986.
3. Warner, R.F., Rangan, B.V., and Hall, A.S., Reinforced Concrete, Pitman Australia, 1982.
4. Kinney, G.F., Graham K.J., Explosive shocks in air, 2nd Edition, Springer-Verlag, 1985.
5. Liu, G.Q., Owen, D.R.J., Ultimate load behaviour of reinforced concrete plates and shells under dynamic transient loading, Int. J. num. Meth. Eng., 22, 1986, 189-208.

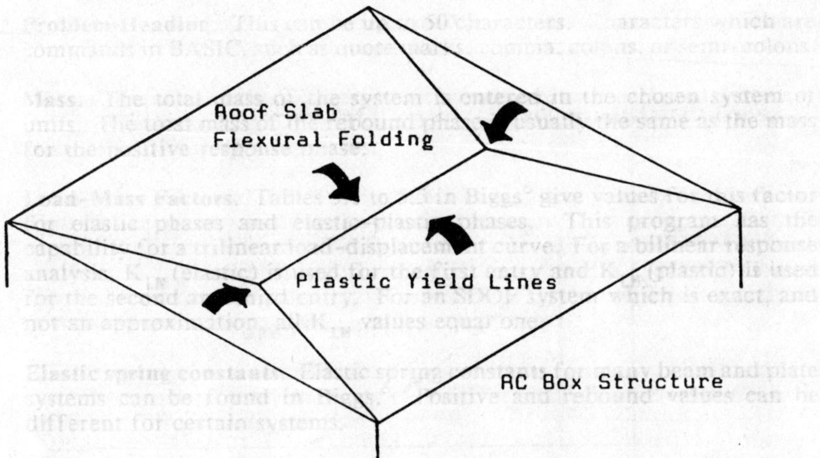

Figure 1 – Flexural Collapse Mode

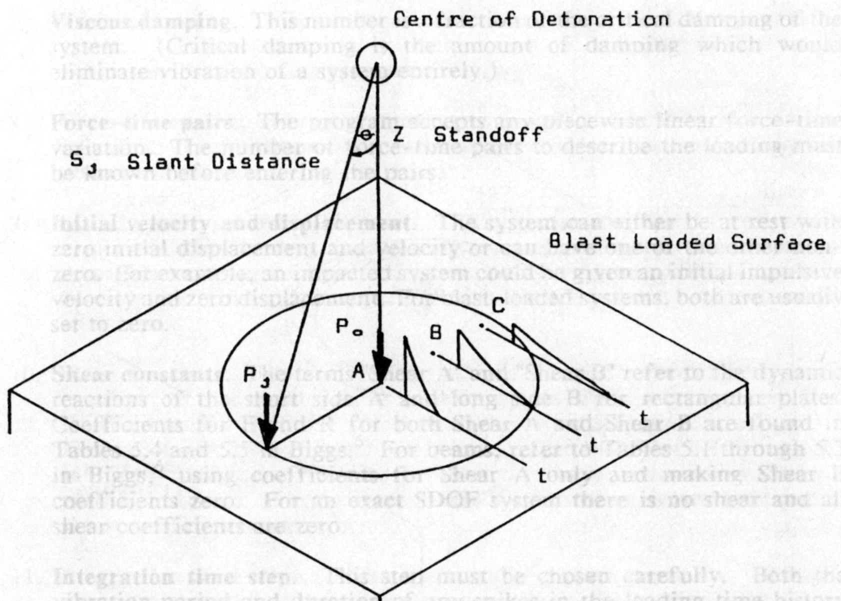

Figure 2 – Free Field Stresses

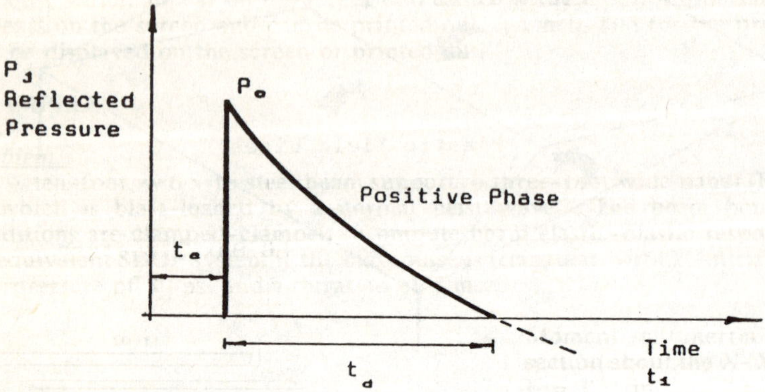

Figure 3 – Blast Pressure Decay Function

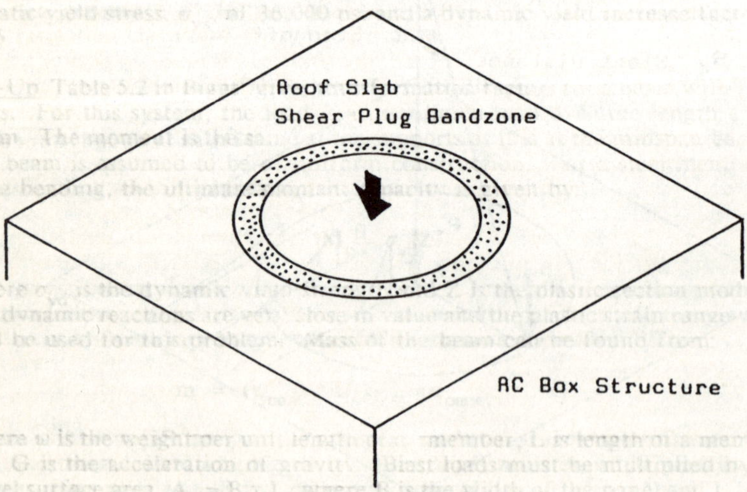

Figure 4 – Shear Collapse Mode

Nonlinear Analysis of Concrete Barrier Structures under Explosive Charges

W.M. Kuntze, P. Gruner and H. Alex
Gesellschaft für Reaktorsicherheit (GRS) mbH,
D5000 Köln 1, West Germany

ABSTRACT

Numerical analyses of concrete barrier structures loaded by an explosive charge are reported. Possibilities of assessment of protective capabilities of barrier structures are discussed on grounds of results from model calculations.

1. INTRODUCTION

Numerical investigations were performed to study protective capacities of concrete barrier structures under explosive charge attack /1,2/. The barrier structures are intended to protect safety related regions of industrial plants.

The barrier structures investigated consist of monolithic wall segments of reinforced concrete with different wall thicknesses and of a two-folded barrier structure as well, formed by two reinforced concrete slabs with an air gap in between. The explosive charge which was specified by authorities is attached to the surface of the respective walls.

The numerical analyses are verified to some extent by experimental results. Within reasonable limits of uncertainty they seem to predict rather well the degree of damage of concrete barrier structures loaded by blasting charges. Hence, numerical methods may be suited to help in avoiding or at least reducing large expenditures for experimental investigations.

2. NUMERICAL PROCEDURES

The analyses were performed using the HEMP Code which is a general Finite Difference program for nonlinear analysis of fluids and structures subjected to strong dynamic and transient phenomena; eg. shocks, explosions and impacts. The code originated with Dr. Mark L. Wilkins, Lawrence Livermore Laboratory /3/ and was used in the extended version HEMP/ESI /4/.

2.1 Material models

Material models for explosive, concrete, steel reinforcement, and the one for air with respect to the two-folded barrier structure were employed. A short description of the modelling of the nonlinear material behavior of reinforced concrete is given here.

Table 1: Material Data for Concrete Model POSTRU

Data	Value
Density	2.5 g/cm^3
Shear Modulus	0.15358 Mbar
σ_a	$1.7 \cdot 10^{-4}$ Mbar
σ_e	0.0 Mbar
σ_b	$2.15 \cdot 10^{-4}$ Mbar
σ_f	$4.55 \cdot 10^{-5}$ Mbar
ε'_a	$4.41 \cdot 10^{-4}$
ε'_f	$4.473 \cdot 10^{-3}$
σ_3^1	0.0 Mbar
σ_3^2	$7.0 \cdot 10^{-5}$ Mbar
θ_1	0.0149
P_1	$3.8 \cdot 10^{-3}$ Mbar
θ_2	0.1329
P_2	0.0138 Mbar
σ_t	$2.1 \cdot 10^{-5}$ Mbar

(The meaning of the abbreviations used is shown in Fig. 1)

The material model POSTRU /5/ is a constitutive model for the dynamic analysis of concrete subject to short term mechanical loads (e.g. high pressure peaks) which takes into account the following damage modes:

- tensile cracking
- shear damage
- hydrostatic compaction damage

The theory of plasticity is used to simulate the inelastic behavior of each of the damage modes.

The material data of the concrete model POSTRU used in the calculations are compiled in Table 1. The material parameters were adopted to concrete qualities corresponding approximately to the German Standard DIN 1045 class Bn350 taking into consideration experimental results with high impact loadings. The

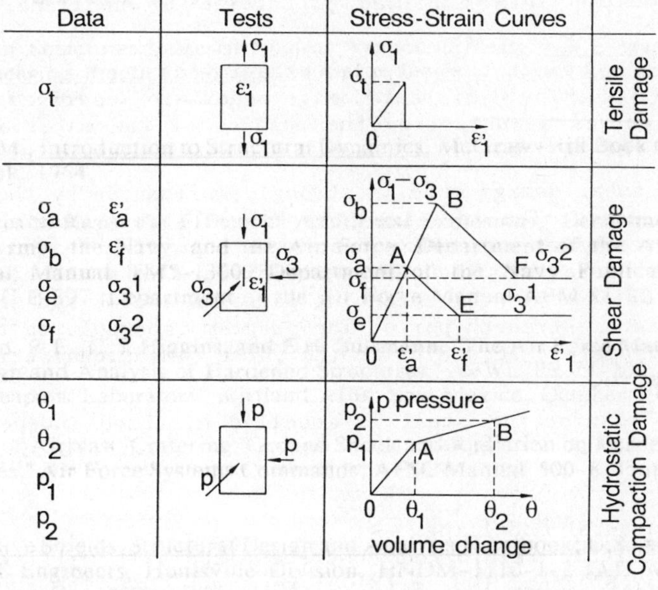

Fig. 1: Data for Concrete Model POSTRU

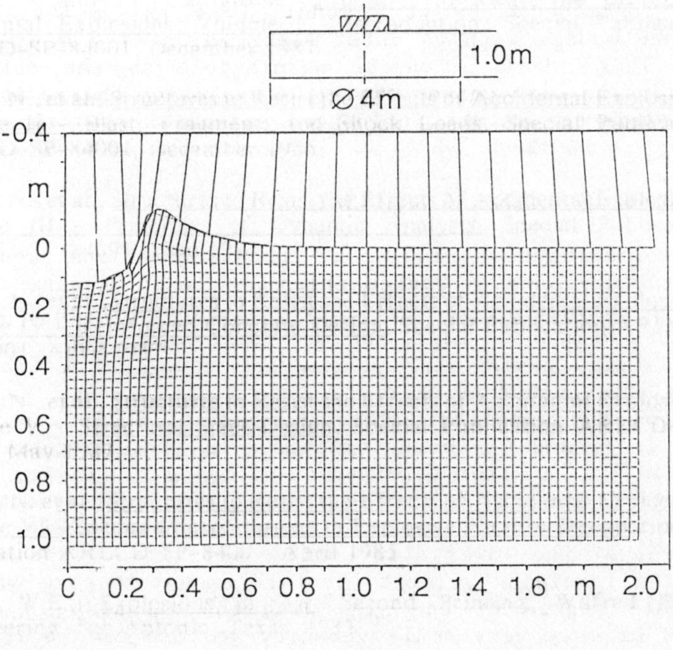

Fig. 2: Deformations of Concrete Barrier D100
(Wall Thickness 1m) at Time t = 1000 μs (After Ignition)

data is temporary and is to be updated, if improved and extended measurements yield modified properties of nonlinear concrete behavior.

Fig. 1 depicts the material behavior as used in the concrete model POSTRU. The different damage modes are described by stress-strain curves illustrating the meaning of the material parameters listed in Table 1.

The shear damage mode is strongly influenced by the confinement stresses. The idealized respective stress-strain curves in Fig. 1 represent two shear tests corresponding to two confining stresses σ_3^1 and σ_3^2. The curves are divided into the elastic parts OA and OB, the strain softening parts AE and BE and the remaining residual strength parts. The program interpolates the actual shear parameters according to the three dimensional stress state in the structure.

The reinforcing steel is modelled by three independent orthogonal sets of steel bars using an elastoplastic material model.

2.2 Structural modelling

Two monolithic reinforced concrete slabs with wall thicknesses of 70 cm (B70) and 100 cm (B100) were modelled. The square slabs have a side length of 4 m. The steel reinforcement consists mainly of the upper and lower bending reinforcement (steel BSt 420/450, Ø 14 mm, e = 15 cm in both orthogonal directions) and a strong confining reinforcement at the edge of the slabs.

The concrete slabs were modelled as axisymmetric Finite-Difference models with an outer slab radius of 2 m as shown in Fig. 2. This idealization is permissible as the interesting damages in the slab center where the explosive charge is applied on the surface, are not influenced essentially by this simplification.

3. RESULTS OF NUMERICAL ANALYSES

In the following some results of the numerical analyses of barrier structures are presented which led to an assessment of the extent of damages in the protective slabs resulting from a certain kind and mass of explosive.

3.1 Monolithic concrete barriers

Results of the monolithic concrete barrier B 100 are shown in Figures 2, 3, and 4. The deformations of the concrete slab are plotted in Fig. 2 after termination of major blast wave loadings illustrating the shape of the crater. The development of the velocity field over the wall thickness is shown in Fig. 3 at the same time. At the lower slab surface opposite to the explosive charge the remaining velocity vectors amount to some 40 m/s. In the middle region of the slab with an extension in depth of about 40 cm only very small velocity vectors can be recognized which may be taken as a first hint that the slab is not completely destroyed but contains a rest of load carrying capacity.

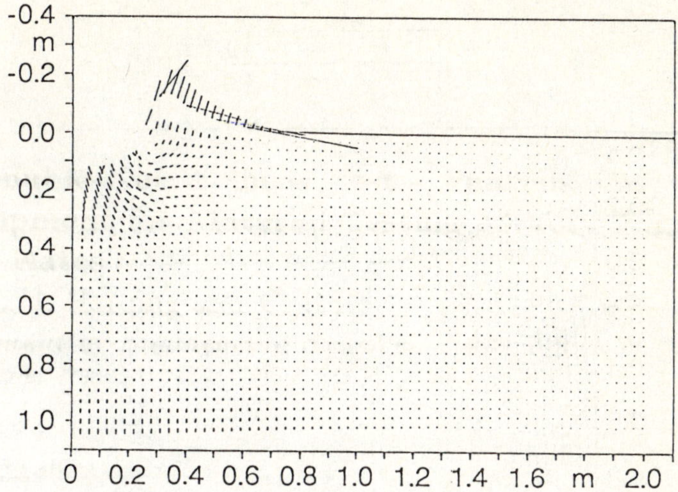

Fig. 3: Field of Velocity Vectors in Concrete Barrier B100 (Wall Thickness 1m) at Time t = 1000 μs (After Ignition)

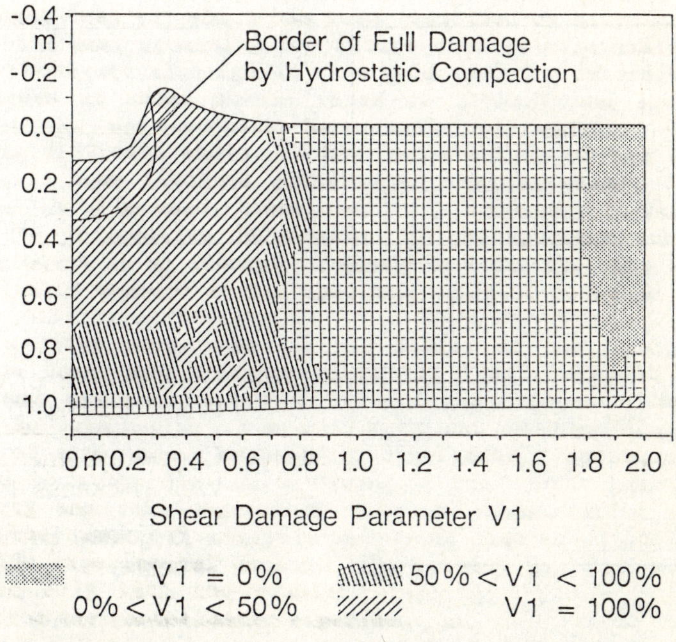

Fig. 4: Material Damage of Concrete Barrier B100 (Wall Thickness 1m) at Time t = 1000 μs (After Ignition)

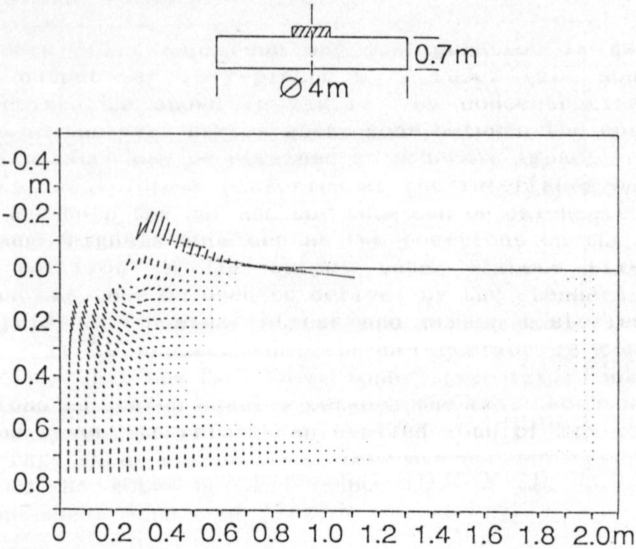

Fig. 5: Field of Velocity Vectors in Concrete Barrier B70 (Wall Thickness 0,7 m) at Time t = 1000 μs (After Ignition)

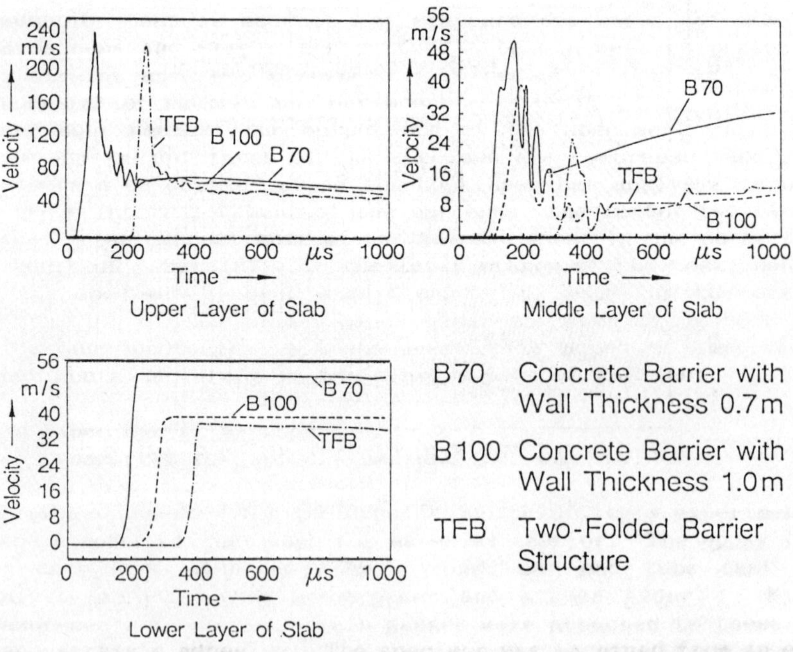

Fig. 6: Velocity Time Histories at Various Positions in Concrete Barriers (Below Blast Center)

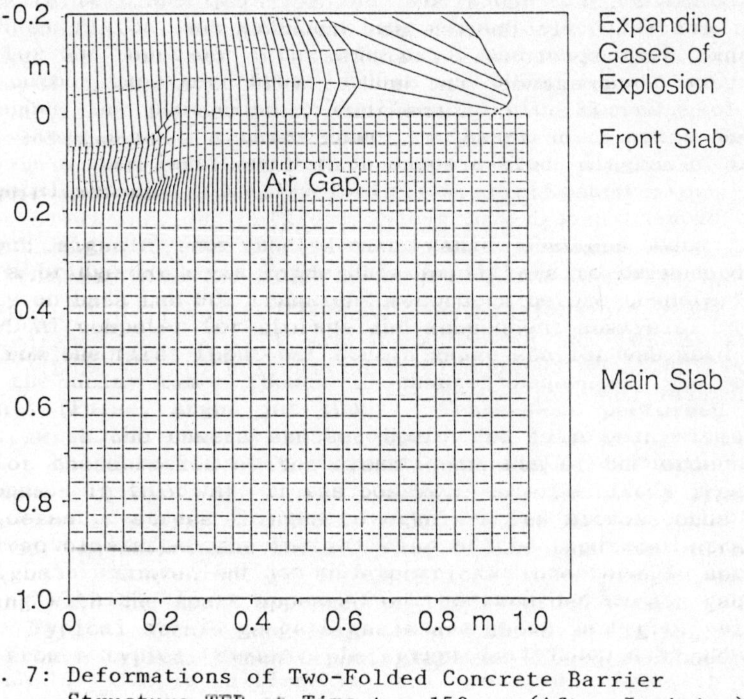

Fig. 7: Deformations of Two-Folded Concrete Barrier Structure TFB at Time t = 150 μs (After Ignition)

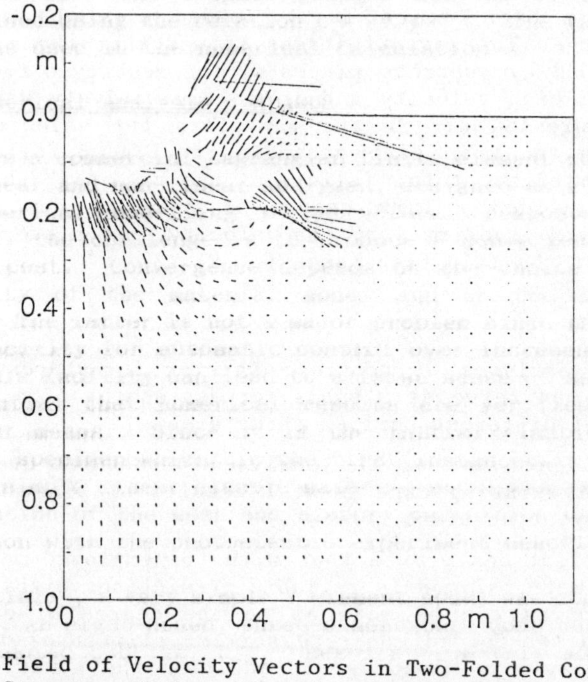

Fig. 8: Field of Velocity Vectors in Two-Folded Concrete Barrier Structure TFB at Time t = 600 μs (After Ignition)

This statement is confirmed by the resulting material damage in the concrete slab which is shown in Fig. 4. Though tension cracks have spread all over the slab the area of hydrostatic compaction damage is confined to a depth of about 30 cm as can be seen from Fig. 4. On the other hand the load carrying capacity in shear is obviously not exhausted in the lower third of the slab. From the interpretation of the velocities in Fig. 3 and the damage modes in Fig. 4, it follows that the explosive blast has not broken through the protective slab but that there remains a restricted region in the lower half of the slab which has preserved some parts of its load carrying capacity.

The velocity field of a concrete slab with 0,7 m wall thickness (B 70) can be compared from Fig. 5. The continuous mass flow indicates the complete destruction of the middle slab region over the whole depth.

Time histories of velocities at various depths in the analysed concrete slabs are shown in Fig. 6 from which some knowledges about the material behavior can be derived. The high velocity values at the upper layer of the slabs mean complete material damage. In the middle of the slabs the lower velocity range of concrete barrier B 100 can be distinguished from the higher values of barrier B 70 (factor ~ 3). This difference is to be assessed as a global criterion of destruction or resistance of a concrete barrier. At the lower surfaces of all slabs the diagrams show the effect of spalling of material particles.

3.2 Two-folded concrete barrier structures

The design of the two-folded concrete barrier structure can be recognized from Fig. 7 when it is loaded with the specified explosive charge. The investigations showed that the main slab is mainly loaded by concrete pieces of the destroyed front slab impacting with high velocity and to a comparatively very small amount by compressed air in the gap. From the velocity field developping in the main slab according to Fig. 8 follows that there is no immediate total failure of the main slab though the calculations yielded a high degree of material damage. This statement is also confirmed by the velocity time history in the middle layer of the main slab (see Fig. 6).

4. CONCLUSION

The numerical analyses of concrete barrier structures yielded results which can serve as a basis for assessment of protective capabilities of different barrier structures.

The global behavior of the analysed structures is in good agreement with existing experimental results of blast tests of similar barrier structures.

The investigations proved that with the employed mathematical tools and material models reliable predictions within tolerable uncertainties can be made as to the degree of damage of monolithic and two-folded concrete barrier structures which are subjected to explosive loading. In addition the numerical analyses give valuable information about the extent and formation of various failure modes.

5. ACKNOWLEDGEMENT

The work was sponsored by the German Federal Ministry of Environmental, Natural Conservation and Reactor Safety (BMU) to whom the authors express their gratitude.

6. REFERENCES

/1/ Alex, H., Gruner, P. and Kuntze, W.M. Untersuchungen zur Barrierewirkung von Stahlbetonstrukturen bei Sprengstoffeinwirkungen, Berechnung monolithischer Stahlbetonplatten und Vergleich mit Versuchsergebnissen, GRS-A-1206 (März 1986), Gesellschaft für Reaktorsicherheit (GRS)mbH, Köln, FRG

/2/ Alex, H., Gruner, P., and Kuntze, W.M. Untersuchungen zur Barrierewirkung von Stahlbetonstrukturen bei Sprengstoffeinwirkungen, Berechnung einer zweischichtigen Vorsatzplattenkonstruktion, GRS-A-1286 (Juli 1986), Gesellschaft für Reaktorsicherheit (GRS)mbH, Köln, FRG

/3/ Wilkins, M.L. Calculation of Elastic-Plastic Flow, UCRL-7322, Rev. 1, 1969, Lawrence Radiation Laboratory, Livermore, CA, USA

/4/ HEMP/ESI, User's Manual, Vol. 1: Theory, Vol. 2,3: Input Data Cards, 1982, Engineering System International GmbH, Eschborn, FRG

/5/ POSTRU, A Material Model for Concrete Behaving under High Pressure which Includes Post Rupture, 1982, Engineering System International GmbH, Eschborn, FRG

The Behaviour of Concrete under Explosive Shock in Relation to its Compaction under Uniaxial Strain Loading

A.J. Sheridan*, W.H. Stephens* and S.H. Perry**

**Royal Aerospace Establishment, Farnborough, Hampshire, UK*

***Trinity College, Dublin, Ireland*

ABSTRACT

This paper presents a means whereby direct comparisons can be made between the static and dynamic properties of concrete. The upper limit, on the strain rate that can be applied under dynamic conditions, is fixed by the thickness and velocity of the front of a hydrodynamic shock wave. This would be a unique function of the peak wave-normal stress component for any specified concrete. The material immediately behind a shock would have been subjected to dynamic uniaxial strain loading - ie. there would be no lateral strain, through the shock, due to inertial confinement. Concrete shock data can be provided by measuring transients produced by sensors cast into concrete targets, which are subjected to contact explosive loading. The static equivalent of the load condition, behind a hydrodynamic shock, would require the application of a uniaxial stress to a test specimen under conditions whereby any lateral strain was prevented from occurring. Ideally the specimen should be confined by a rigid boundary. However, it has been demonstrated that the essential data can be produced by casting a cylindrical concrete specimen within a steel tube. Furthermore, the elastic properties of the confining tube can be used to accurately evaluate the stress state of the specimen, up to the point where the concrete is observed to behave hydrostatically. A relationship between static and dynamic material properties is sought because of the relative ease and low cost of static testing in the laboratory.

INTRODUCTION

This paper describes part of the research being carried out at the Royal Aerospace Establishment to investigate the fundamental behaviour of concrete in response to conventional weapons effects (Sheridan [1]). The research is directed at investigating and developing experimental methods for measuring high speed transients within concrete and generating theoretical methods for analysing concrete material behaviour. One of the principal objectives is to produce material equations of state and constitutive properties for application to computer hydrocode methods.

THEORY

The governing equations for the propagation of hydrodynamic shocks are developed from the equations of continuum mechanics for conservation of mass, momentum and energy. It is shown, in the section describing hydrodynamic shocks, that these equations rely only upon the wave-normal parameters across the shock. This being the case, the manner in which the concrete is deformed is directly comparable with uniaxial strain loading.

The relationship between wave-normal strain ε and particle velocity **u**, behind a shock of velocity **U**, can be simply demonstrated by considering the dynamic stress σ applied to the free end of a bar, at time t_o=0, shown in fig.1. After an infinitesimal time dt the wave front would travel a distance **U**dt. Simultaneously, the mass behind it would flow at velocity **u**, with which would be associated a deformation of length **u**dt. The strain, induced within the wave, relates this deformation to the original length of that part of the bar over which the stress had been transmitted:

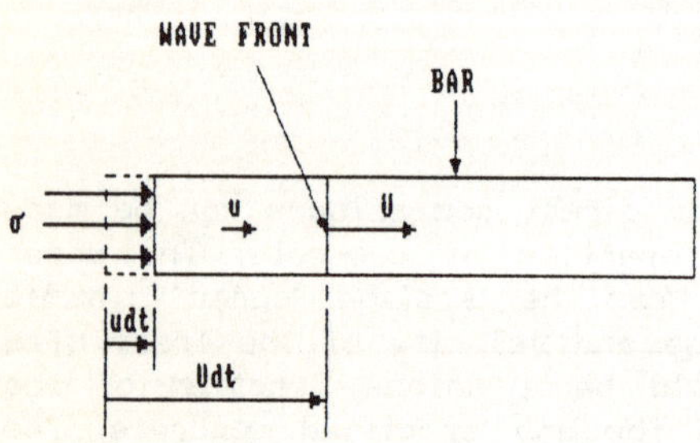

Fig. 1 Wave Induced Strain

Wave induced strain, $$\varepsilon = \frac{\mathbf{u}dt}{\mathbf{U}dt} = \frac{\mathbf{u}}{\mathbf{U}} \quad (1)$$

If one considers such a wave to be propagated within a large solid mass, as shown in fig.2, it can be seen that any material element behind the wave front would be laterally supported by similar adjacent elements. The result is that during shock loading, such elements are unable to fail, though it is recognized that any curvature of the wave front will contribute to some divergent mass flow. However, in the following sections it will be shown that this does not effect the Rankine-Hugoniot equations for shocks and that hydrodynamic stress conditions prevail, independently of the strain state.

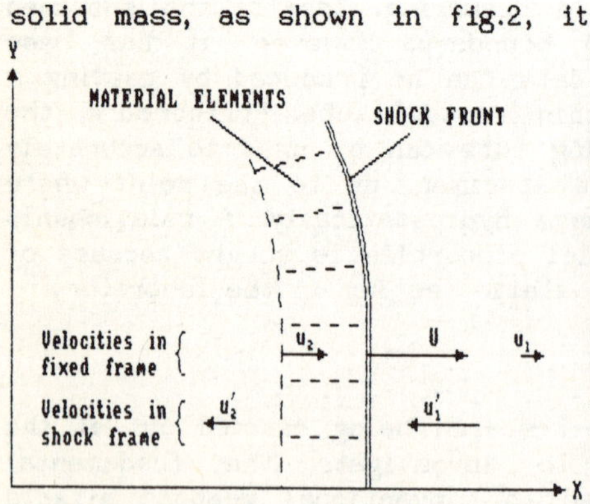

Fig. 2 Shock Wave

Hydrodynamic Shocks

It can be shown that the Rankine Hugoniot relations, for conditions either side of a shock, are not dependent on the curvature of the shock front - from their derivation:

The continuum equations for a fluid or solid are often referred to as the conservation equations for mass, momentum, and energy. However, with the exception of the mass equation, this is not strictly true. They are more precisely described as 'source' equations. When considering the supply of some physical property X (eg. mass, momentum, energy), in the shock frame of reference, that supply of X is expressed by the following source term:

$$\mathcal{S}(X) = \frac{\partial X}{\partial t} + \nabla.(X\mathbf{u}) \tag{2}$$

where **u** is the velocity of the material, with respect to the shock, and the time derivative is that relating to the material frame of reference. The change in some physical property X across a steady shock discontinuity is due only to the divergent flow of X. Whilst X would change through the shock, the upstream and downstream values (ie. X_1 and X_2 respectively) would not change as a function of time in the material frame of reference - ie.

$$\frac{\partial}{\partial t} = 0 \tag{3}$$

The basic assumption for shocks is that only the wave-normal component - defined by unit vector $\hat{\mathbf{n}}$ - of the divergence term need be considered:

$$\mathcal{S}(X) = \nabla_{\hat{n}}.(X\mathbf{u}) \tag{4}$$

Then, since the shock represents a discontinuous change in X this may be expressed as:

$$\mathcal{S}(X) = \hat{\mathbf{n}}.[X\mathbf{u}] \tag{5}$$

$$\text{where} \qquad [X\mathbf{u}] = (X_2\mathbf{u}_2 - X_1\mathbf{u}_1) \tag{6}$$

For the mass conservation equation - often referred to as the continuity equation - the physical property that is represented by X, in the source equation, is the mass density ρ. If mass is neither created nor destroyed within the system then the source equation is equal to zero. For mass conservation across a shock:

$$\hat{\mathbf{n}}.[\rho\mathbf{u}] = 0 \tag{7}$$

If the scalar u represents the wave-normal component of velocity **u**, then this can be simplified to:

$$[\rho u] = 0 \tag{8}$$

and this is used to define mass continuity:

$$[m] = [\rho u] = 0 \qquad (9)$$

Since m has a unique value, this can be rewritten, in terms of the specific volume τ, as:

$$m[\tau] = [u] \qquad (10)$$

Referring to the shock shown in fig.2 - which has a velocity U, with respect to the fixed frame, and an upstream material velocity of u_1, with respect to the fixed frame - the following relationships are defined:

$$u_0' = U \qquad (11)$$

$$u_1' = U - u_1 \qquad (12)$$

$$u_2' = U - u_2 \qquad (13)$$

Then equation (10) becomes:

$$\rho_0 U(\tau_2 - \tau_1) = u_1 - u_2 \qquad (14)$$

which is the Rankine-Hugoniot relation for the conservation of mass across a shock - demonstrating that the only assumptions associated with its development are those of equations (3) and (4) which are independent of the curvature of the wave front.

Similarly, the equation for conservation of momentum:

$$\underline{\hat{\mathbf{n}}}.[\rho \mathbf{u}\mathbf{u}] = -\underline{\hat{\mathbf{n}}}.[\underline{\sigma}] \qquad (15)$$

yields the Rankine-Hugoniot relation:

$$U^2 = \tau_0{}^2 \frac{\sigma_1 - \sigma_2}{\tau_2 - \tau_1} \qquad (16)$$

and the equation for conservation of energy:

$$\underline{\hat{\mathbf{n}}}.[\rho(i+\tfrac{1}{2}u^2)\mathbf{u} = -\underline{\hat{\mathbf{n}}}.[\underline{\sigma}.\mathbf{u}] \qquad (17)$$

yields the Rankine-Hugoniot relation:

$$(i_2 - i_1) = \tfrac{1}{2}(\sigma_2 + \sigma_1)(\tau_1 - \tau_2) \qquad (18)$$

A further constraint on shock development is that the entropy s must increase through the shock:

$$m[s] > 0 \qquad (19)$$

A consequence of the increase in entropy in the downstream region is the internal generation of heat energy q per unit mass. By the first law of thermodynamics:

$$[i] = [q] + [w] \qquad (20)$$

This can be readily developed to the form:

$$[q] + [w] + [½u^2] + [\sigma\tau] = 0 \qquad (21)$$

where the conservation of energy is separated into its energy components of heat production [q], work done [w], increase in kinetic energy [½u²], and the total energy to maintain the shock [στ]. Equation (21) is presented diagrammatically in Fig.3 and shows that the difference between static and dynamic material properties is associated with the production of 'heat' energy in the latter. Such energy is probably largely associated with random thermal type motion of the coarse aggregate, in the first instance - but, irrespective of its mechanism, it is this 'heat' energy that the RAE research programme aims to quantify.

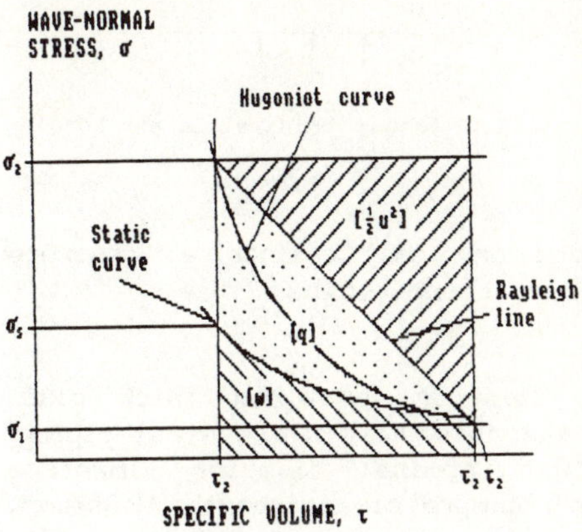

Fig. 3 Pressure-Specific Volume Curves

Uniaxial Strain

During normal uniaxial stress loading, as shown in fig.4, a concrete specimen undergoes axial compressive deformation and lateral tensile deformation. In general, the load path is determined by the rate at which the concrete is loaded in relation to the rate at which it can unload itself during progressive failure. It has been shown, above, that, within a shock wave, lateral failure cannot occur. So, if static material properties are to be related to those which occur during shock loading, similar lateral constraints must be applied to the static test. Under pure uniaxial strain loading, no lateral components of strain would be allowed to develop and, consequently, the specimen would not be able to fail.

This condition could not be reproduced exactly in the laboratory test described below, where lateral deformation was minimized as shown in fig.5. However, the deviation from pure uniaxial strain was not considered to be significant and additionally provided the means for determining triaxial data.

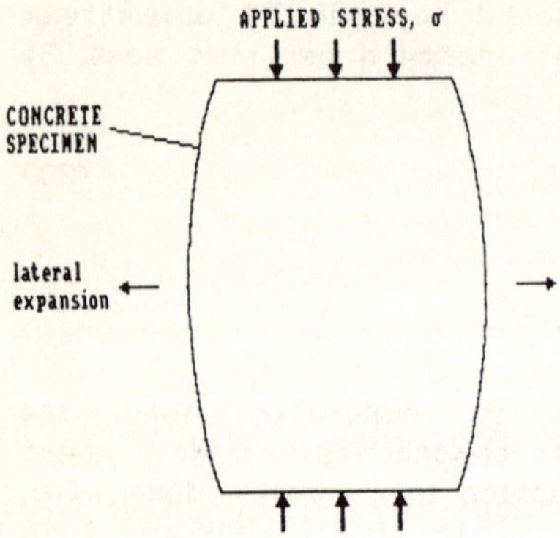

Fig. 4 Static Uniaxial Stress Loading

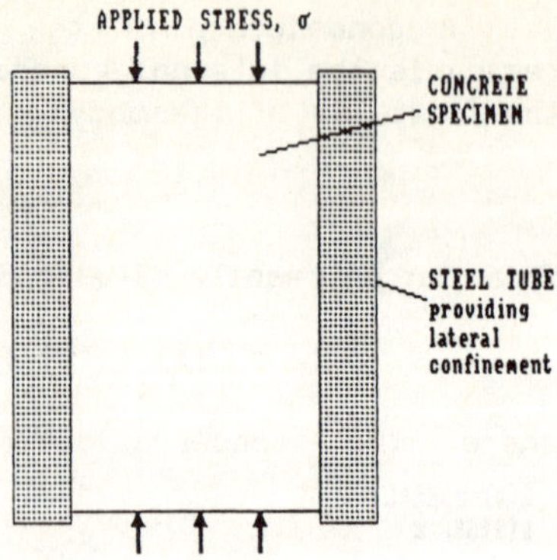

Fig. 5 Static Uniaxial Strain Loading

EXPLOSIVE TRIALS ON CONCRETE

The explosive trials were performed on small cylindrical concrete targets on the RAE's Ball Hill Range at Farnborough.

Trials Description

The concrete targets were 0.5m diameter by 0.25m thick plain concrete, produced by casting the mix into 13mm thick steel piping. The concrete was made using 10mm nominal diameter limestone aggregate and had density 2360kg/m³, compressive strength 71.3N/mm², and dynamic modulus 40.5kN/mm². Each target incorporated a quantity of 'cast-in' sensors for measuring strain, particle velocity or stress transients. Targets were loaded by detonating 70g of Composition B explosive in contact with the surface. The data was captured and stored using a 0.5MHz, 12-bit Datalab transient recorder, operated from a Hewlett Packard 9816S microcomputer.

Shock Parameters

Due to the simplification that shock theory provides to the analysis of material states, during appropriate loading, it is possible to determine features of the material Hugoniot curve directly from the transients (see fig.3). Shock velocities can be determined based on transit times between gauge locations, within the concrete, and only the peak value of either stress, strain or particle velocity need be measured from the transient to describe the load path on the dynamic stress strain curve. For example, the velocity transient in fig.6 shows a peak velocity of 45m/s and had a time of arrival of 14μs at a mutual inductance velocity gauge 50mm beneath the concrete surface. Equation (14)

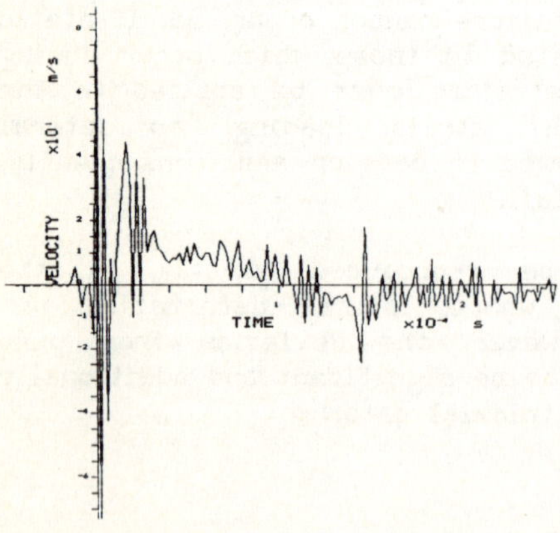

Fig. 6 Velocity Transient

can be used to evaluate a peak strain of 1.3% and eliminating $(\tau_2-\tau_1)$ from equations (14) and (16) gives a peak stress of 380N/mm².

LABORATORY TESTS ON CONCRETE SPECIMENS

The laboratory tests described in this paper were developed at Imperial College, London by Perry and Kotsovos [2] under the direction of RAE. The analysis has subsequently been extended by RAE as part of the development of this research.

Description of Experiment

The concrete specimens were produced by casting the same concrete mix as used in the explosive trials into steel tubes, which were greased and lined with polythene sheet. The steel tubes were 75mm inside diameter, 13mm thick and 150mm long (refer to fig.5).

The load was provided by a 3000 kN Amsler testing machine, through a spherical platen seating unit at a rate of 14 N/mm² per minute, to the ends of the concrete cylinder only. The uniaxial strain was determined from the platen to platen displacement measured by 4 LVDTs. Lateral and axial deformation of the steel tube was measured by using pairs of electrical resistance strain gauges placed at mid-height and diametrically opposite each other.

Uniaxial Stress-Strain Curves

The uniaxial stress-strain curves, produced by end loading concrete specimens under conditions of passive lateral confinement, possess the following features. After an initial period of elastic deformation the concrete yields. The majority of the strain is then taken up by the collapse of the mortar, particularly within the 'contact' regions between adjacent pieces of coarse aggregate. Sectioned test specimens have indicated that to a much lesser extent, coarse aggregate crushing also contributes to this process. As the inter-aggregate 'contact' regions stabilize, the concrete is observed to stiffen. Fig.7 shows a sectioned specimen after testing and was photographed using a flourescent dye to reveal the cracks under ultraviolet light. Fig.8 shows the same type of failure, but to a greater extent, for the concrete below an explosively formed crater. The uniaxial stress-strain curve for the concrete used in the explosive trials is shown in fig.9.

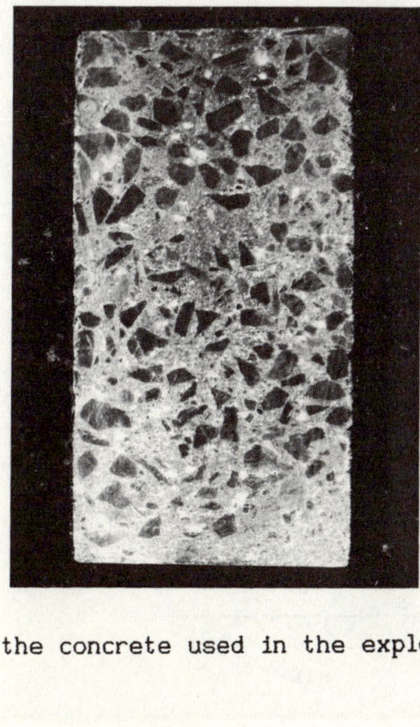

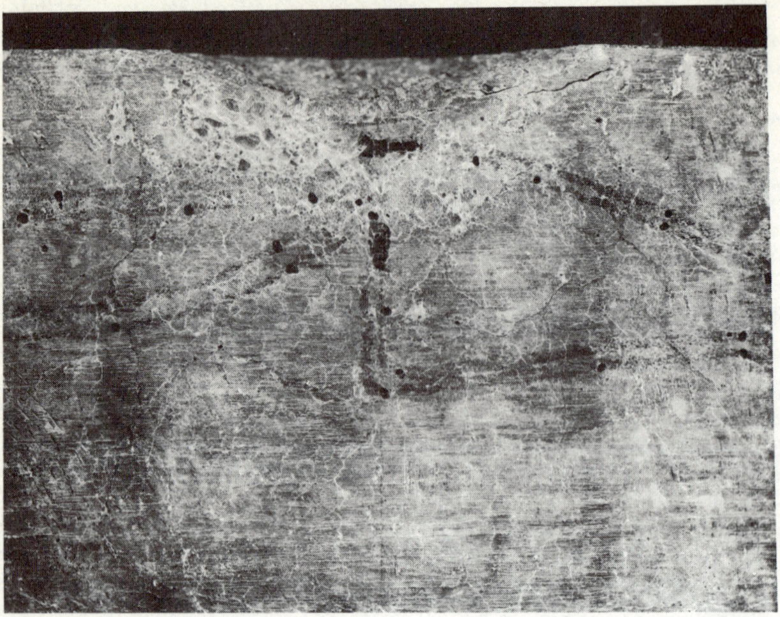

Fig. 8 Section Through Cratered Concrete

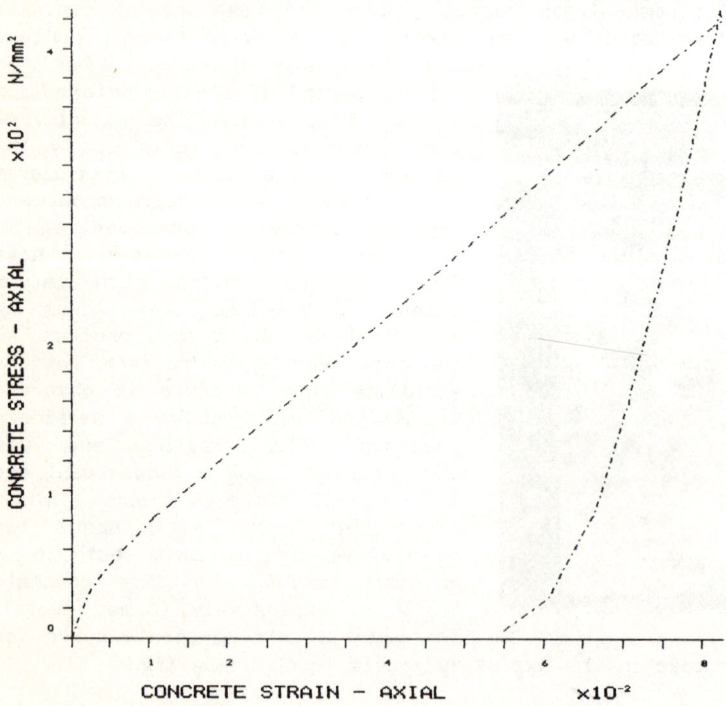

Fig. 9 Uniaxial Stress-Strain Curve

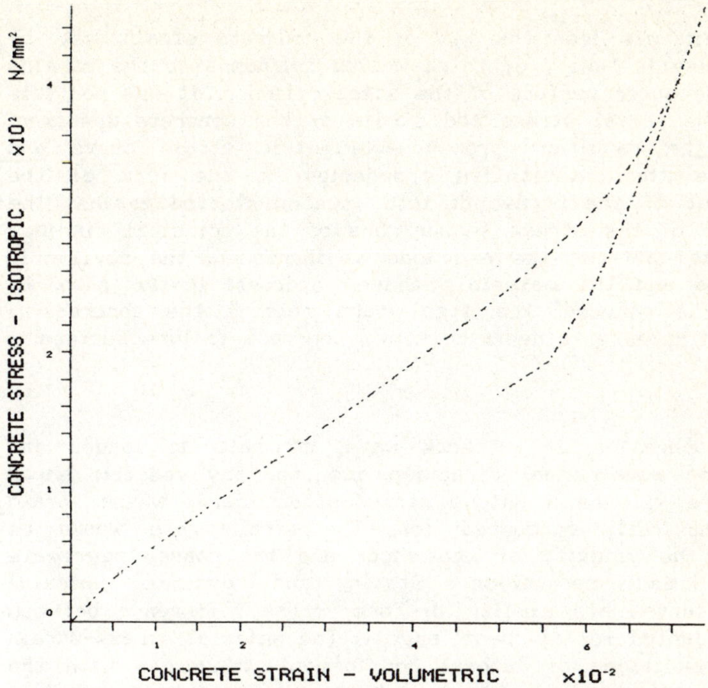

Fig. 10 Pressure-Volumetric Strain Curve

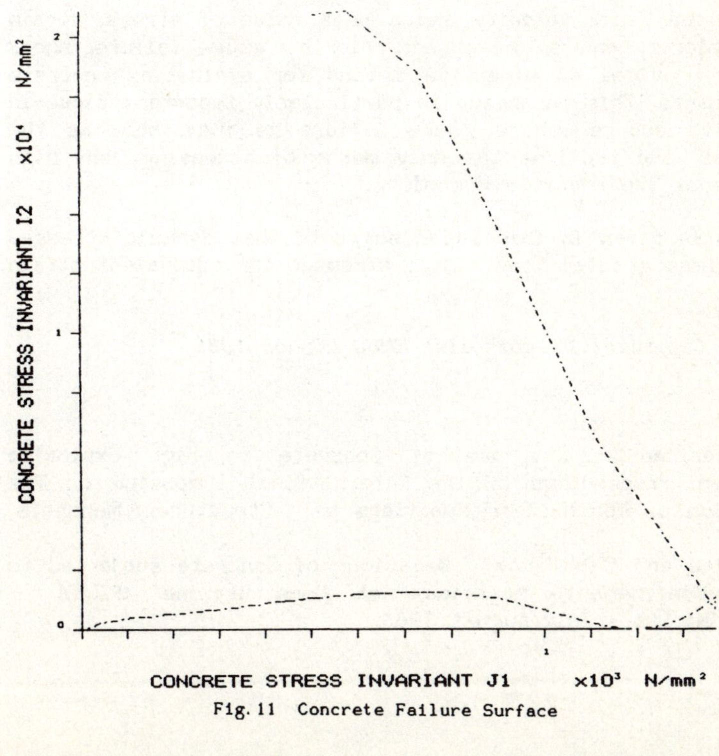

Fig. 11 Concrete Failure Surface

Triaxial Data

The steel strain was less than 1% of the concrete strain, yet, by applying the elastic theory of thick walled cylinders to the strains measured on the outer surface of the steel cylinder, it was possible to calculate the radial stress and strain of the concrete specimen. Fig.10 shows the resultant pressure/volumetric strain curve and fig.11 presents the constitutive properties in the form of the second invariant of the stress deviator tensor plotted against the first invariant of the stress tensor. One of the principal findings of this test is that, as the end load is increased, the deviatoric stress vanishes and the material behaves hydrostatically. Also, as the end load is reduced, the steel tube reloads the concrete - laterally - and appears to describe a new concrete failure surface.

CONCLUSIONS

During the propagation of a shock wave, concrete is loaded and deformed in the wave-normal direction and, for any specific shock amplitude, there will be a unique stress-strain curve where strain rate is automatically accounted for. The strain rate would be determined by the velocity of the shock and the coarse aggregate average size and separation. Static and dynamic uniaxial stress-strain curves are similar in form - the difference between them being accounted for by 'heat' energy. The uniaxial stress-strain curve, under conditions of lateral confinement, converges with the pressure-volumetric strain curve at high stresses - independently of the strain state.

Provided the shock velocity and a peak value of stress, strain or particle velocity can be measured, prior to gauge failure, shock wave analysis provides an effective method for evaluating concrete equations of state. This technique is particularly important close-in to an explosive source, where gauge failure is inevitable as the crater develops, and provides the only means of accessing very high stress regions of the concrete Hugoniot.

The example given in this paper suggests that dynamic stresses can be many times greater than static stresses for equivalent strain states.

REFERENCES

1. A J Sheridan - Response of Concrete to High Explosive Detonation. Proceedings of the International Symposium on The Interaction of Non-Nuclear Munitions with Structures, Mannheim, 1987.
2. M Kotsovos and S H Perry - Behaviour of Concrete subjected to Passive Confinement. Materiaux et Constructions (RILEM) - Vol.19 - No.112 - July/August 1986.

Simulation of Concrete Beams under Explosive Loading with the Finite Element Method

R.M.M. van Wees and J. Weerheijm
TNO Prins Maurits Laboratory, PO Box 45,
2280 AA Rijswijk, The Netherlands

Abstract

For many years, the TNO Prins Maurits Laboratory has investigated the behaviour of structures under explosive or impact loading. In this paper, some methods used at PML to calculate the response of structures are reviewed.
A simple method which takes account of wave propagation effects is the Normal Mode Technique. Some applications of the NMT are given. The NMT is restricted to linear elastic problems. The Finite Element Method is more generally applicable. A FEM simulation of a concrete beam, loaded by a nearby explosion was made. A method to estimate the mesh refinement and the size of the time steps in the time integration method necessary to obtain an accurate solution is described.
For the simulation, a simple material model with tension softening was used. This was found to be sufficient for the later stages of the calculation, but not for the initial stage, where high strain rates were found. A material model that describes the behaviour of concrete at high strain rates is needed.

Introduction

For many years, the TNO Prins Maurits Laboratory has given attention to the behaviour of structures under explosive or impact loading. In this paper, some methods used at PML to calculate the response of structures are reviewed.

One of the simplest and most effective methods to calculate the response of dynamically loaded structures is to schematize the structure to a Single Degree Of Freedom system (Norris[6]). However, this method does not take into account that the force distribution in structures, subjected to dynamic loads of short duration, is quite different from that under static loading. This difference is visible as wave propagation effects and is most pronounced in the shear forces: very high local peaks can arise. Especially for structures of reinforced concrete this is important, because the shear force peaks might cause brittle collapse of the structure.

The Normal Mode Technique

A simple method which takes account of wave propagation effects is the normal mode technique. The technique is explained in detail by Karthaus and Leussink[3]. The place function of the loading is split up into a number of contributions which correspond with the normal modes of the structure.

Because of the orthogonality of the normal modes the value of each component is independent of the other ones, while each loading component will only activate the corresponding normal mode, not the others. In this way the response can be determined for each component. By adding up the individual response components the total response is found.

A set of programs was written at the Prins Maurits Laboratory that use the normal mode technique to evaluate the response (deflections, bending moments and shear forces) of beams and plates subjected to arbitrary dynamic loads.

Apart from calculating the response of beams and plates, we found other useful applications for these programs as well (Mercx[5]). For example, they were used to estimate the error made when the response is calculated with a SDOF schematization, and to see how the place where the maximum bending moments are present in a beam loaded by a blast wave shifts from the supports towards the middle of the beam.

But, the normal mode technique has one severe limitation; it can only handle linear elastic problems. For cases where a rigorous elasto-plastic analysis is required, other methods are necessary.

Linear elastic Finite Element simulation

For this purpose, commercially available finite element programs were tried (Diana and Abaqus). First, a linear elastic calculation was made, in order to identify and solve the problems associated with FE computations of this kind. As a matter of fact, it proved to be quite complicated to perform the computations in such a way that the accuracy was high while the computational cost remained acceptable. Establishing a method to keep the computational costs low is important in these cases, since this sort of problem has a tendency to require large amounts of computer time. This is caused by the fact that in the FEM, at a given point of time in the computation, the algorithm searches an equilibrium between internal stresses, external loads and momentum forces. This may require several iterations for non-linear problems. With the momentum forces, or in other words the accelerations, the deformed shape at the next point in time is computed, and the process repeats until the response over the entire period is found. Since the time steps are small, many computations are needed.

As was said above, the differences between the static and the dynamic response are the wave propagation effects. These effects become increasingly important as load duration decreases. To simulate the response of these structures numerically, care must be taken to represent these effects accurately. To do this, one would like to be able to predict the errors, but for wave propagation, this is difficult. However, as is demonstrated with the normal mode technique, describing the response accurately enough is equivalent to taking account of enough normal modes. The more normal modes are used, the better the wave propagation is described. The error made in describing a normal mode can be predicted.

The number of normal modes to be described depends on:

1 the desired accuracy of the solution

2 the type of response to be described. (Deflections can be described with only a few normal modes, bending moments need more and shear forces need the largest number of modes to be described.)

3 the duration of the load relative to the first normal frequency. (When the duration of the load decreases, more modes are excited.)

4 the load distribution. (The more concentrated the load is, the more normal modes are excited.)

The normal mode technique was used to demonstrate the influence of some of these parameters in Figure 1. With this knowledge, the number of normal modes necessary to describe the response accurately can be estimated.

Choice of mesh and time step

The highest normal mode to be described is the most susceptible to numerical errors. Therefore, mesh and time step of the numerical integration method must be chosen fine enough to keep the numerical errors of this mode below the desired level over the entire duration of the simulation. To choose the mesh, a convergence study of the model or some part of it could be used, to see how the accuracy of the frequencies is related to the element size.

To choose the time step, one must look at the numerical errors the integration scheme can produce, in particular the numerical dispersion. Time integration methods can produce quite different relative period errors. From graphs of these errors, the time step can be read (Hughes[2], Van Wees[7]).

This method was followed to choose the mesh and time step for the simulation of the beam subjected to a blast load. The results were then compared with the results produced by the normal mode technique. When beam elements were used in the simulation, the results were almost indistinguishable. This proves the FEM can produce the same results as an analytical method. But also when continuum elements were used, the results looked very much alike, thus confirming the applicability of the normal mode technique.
The computation effort needed for this problem was not very high, but already hundreds of time steps were needed, so a more complicated, nonlinear problem would take much computer time. Therefore, an effort made to choose the optimum mesh and time step is well spent.

Non linear simulation

At the PML the failure mode due to impulsive loading of concrete slabs which are simply supported at two or four edges has been investigated experimentally. The slabs were loaded by a shock wave of very short duration originating from an exploding charge at a short distance (Harmanny[1]). These tests were performed with small slabs (50x50x6 cm3) in a cubical set-up. The sides were formed by the slabs to be tested, while the top was open and the bottom consisted of a steel plate. The charge was detonated inside the cube. The stiffness of the supports was varied by the application of thin strips of wood. The slabs were made of high quality concrete and were reinforced in two directions on both sides of the slab.
From these tests, one series was selected to be simulated with the FEM (Weerheijm et al.[8]). In these series, the slabs were

supported at two edges, so the slabs could be regarded as beams. The slabs failed in a bending mode, not unlike the static failure mode.

The normal mode technique was applied to estimate the number of normal modes to be represented in the FE simulation. The best material model available in the program was used, a model which describes the loss in strength after cracking. In Figure 2 this model is depicted for an uniaxial stress state.

The material model ignores the rise in strength that concrete exhibits when it is subject to high strain rates. This proved to be fatal in the simulation.

The simulation predicted correctly that a pressure wave would run from the top of the beam to the bottom after the load was applied. It also predicted correctly that this pressure wave would reflect from the bottom of the beam, causing a tensile wave. It then predicted massive horizontal cracking, and collapse of the beam. In the experiments, this spalling was not observed. The cause for this incorrect prediction was the static material model. The strain rates observed in the simulation were high enough to cause a significant increase in strength.

Later, this problem was circumvented. Instead of applying the load to the top of the beam, the load was distributed over the height of the beam. This caused the strain rate to remain below the level where the material strength rises sharply.

During the simulation, three distinguishable crack patterns occurred after one another:

1 A dense pattern of arising and arresting cracks due to the expanding of blast load and initiatory deformations. (See Figure 3.)

2 A pronounced pattern of shear crack near the supports. The deformation exceeded the ultimate deformation e_u. (See Figure 4.)

3 A pronounced pattern of completely developed bending cracks near the mid-section of the slab. Again the ultimate deformation e_u was exceeded. (See Figure 5.)

Although the shear cracks showed excessive crack deformation the final failure mode proved to be the bending mode. This is well illustrated by the fact that the largest amount of deformation took place around the mid-section. The resulting total deformation of the mid-section was close to the experimental data. However, the applied deformation capacity e_u proved to be of great influence on the ultimate failure mode. If a smaller e_u was chosen, the beam failed in the shear mode (Weerheijm et al.[8]). (This mode was observed in other test series.)

Conclusion

A whole range of methods to calculate the response of structures to shock loads is available. All have their strengths and weaknesses, and all must be used with care. The Finite Element Method looks most promising for problems where an elasto-plastic analysis is required. When the model is carefully set up, computing costs can be kept to a minimum. Results obtained with the FEM compare well with experiments. However, much work remains to be done incorporating the high strain rate behaviour into the material models, especially of the strength and the deformation capacity.

References

1. Harmanny, A.
Reinforced concrete plates under static and dynamic loading.
Symposium: Military Applications of Blast Simulation.
Oxford, England, September 23-27, 1985.

2. Hughes, T.J.R.,
The Finite Element Method. Linear static and dynamic finite

element analysis.
Prentice-Hall, Englewood Cliffs, 1987.

3. Karthaus, W. and J.W. Leussink,
Dynamic Loading: more than just a dynamic load factor.
First Symposium on the interaction of non-nuclear munitions with structures, Colorado, 1983.

4. Mercx, W.P.M. and A. Harmanny,
The Normal Mode Technique: A useful tool to investigate the behaviour of structures subjected to local impulsive loading.
Symposium: Interaction of Conventional Munition with Protective Structures.
Mannheim, March 9-13, 1987.

5. Mercx, W.P.M.,
The Normal Mode Technique: applications for the analysis of transient loaded structures.
Proceedings of the Symposium on the interaction of non-nuclear munitions with structures, Palm Beach, Florida, 1989.

6. Norris, C.H., et al.
Structural design for dynamic loads,
McGraw-Hill, 1959.

7. Wees, R.M.M. van,
Simulation of the dynamic response of structures for very short load durations.
Second annual Abaqus user's conference, Stresa, Italy, 1989.

8. Weerheijm, J., P.J.H. van Zantvoort, G. Opschoor,
The applicability of the FE-technique to dynamic failure analysis of concrete structures.
23rd DoD Explosives Safety Seminar, Atlanta, Georgia, 1988

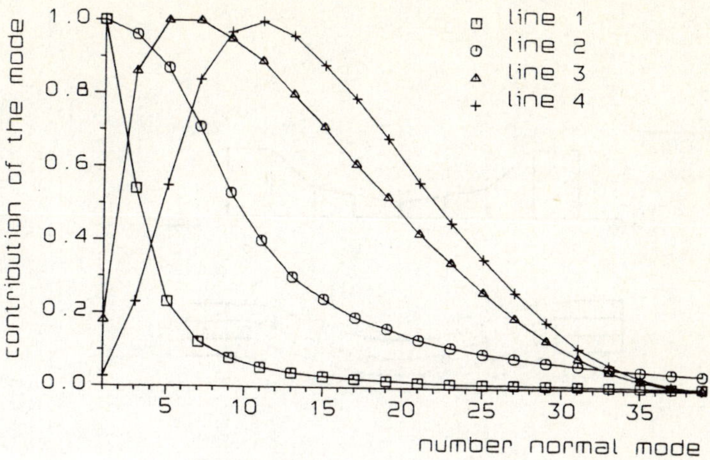

Figure 1. Contribution of the normal modes to the response of the beam.

line 1: distributed load, duration 0.1 x first normal frequency

line 2: distributed load, duration 0.01 x first normal frequency

line 3: concentrated load, duration 0.1 x first normal frequency

line 4: concentrated load, duration 0.01 x first normal frequency

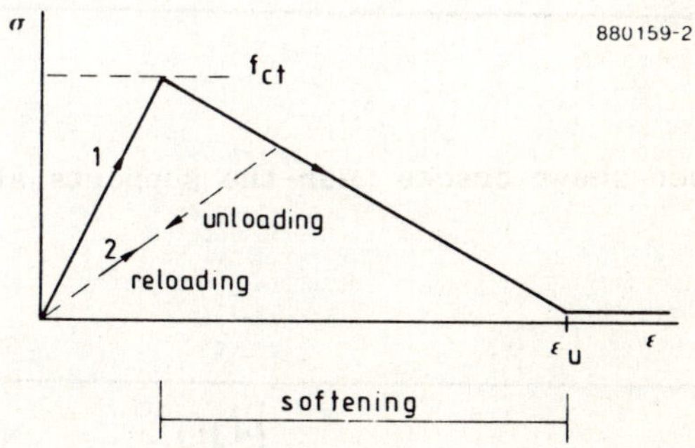

Figure 2. Static tension-softening representation of concrete.

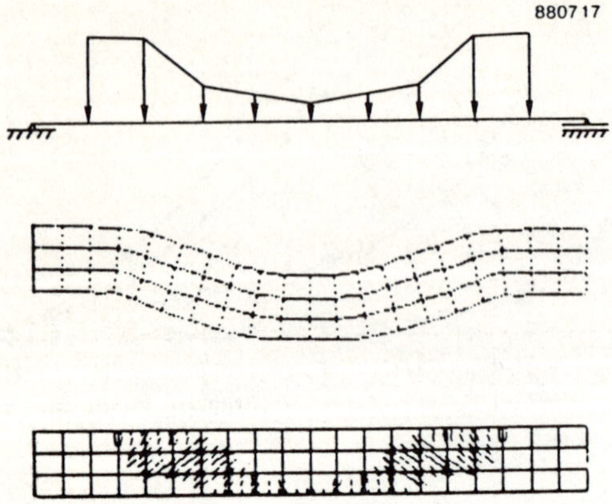

Figure 3. Load, deformation and initiatory dense crack pattern at time 36 μs.

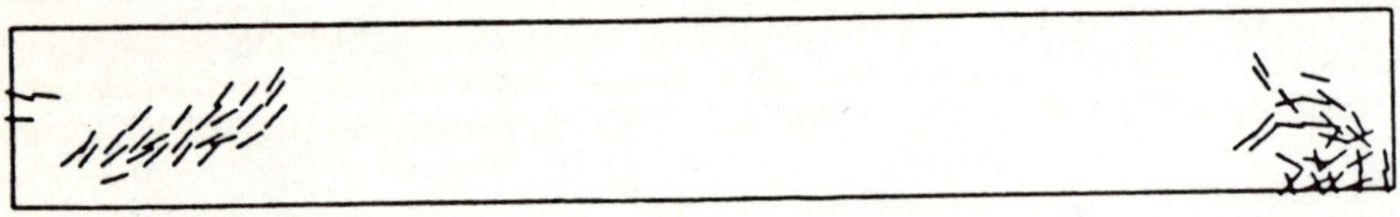

Figure 4. Open shear cracks near the supports at time 240 μs.

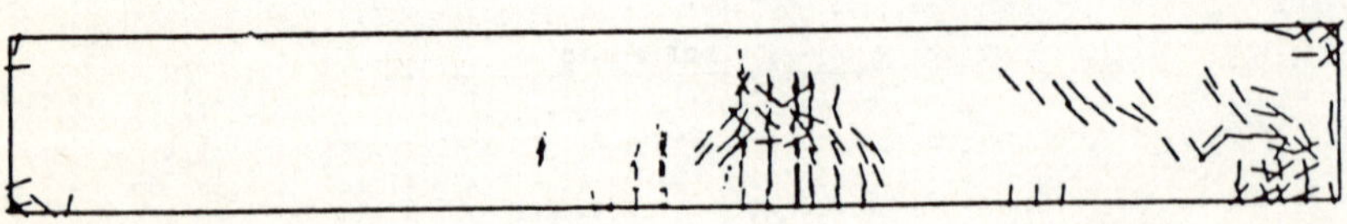

Figure 5. Open bending cracks near the mid-section at time 630 μs.

Dynamic Behaviour of Concrete and its Fibre Reinforced Composites: A Review and Future Research Needs

Nemkumar Banthia and Michel Pigeon

Department of Civil Engineering, Laval University, Ste-Foy, Quebec, G1K 7P4, Canada

ABSTRACT

A historical perspective of the research conducted towards assessing the behaviour of concrete and its fibre reinforced composites under high stress rates is presented. The state of our present knowledge is summarized and the areas where further research needs exist are discussed.

INTRODUCTION

During its life span, a concrete structure may be subjected to the various rates of load application ranging from static to dynamic. For a rational design approach, a knowledge of the properties of concrete at the various rates of load application is necessary. It is well known now that concrete is a stress rate sensitive material, i.e., its mechanical properties depend on the rate at which it is stressed. While the behaviour of concrete at static or quasi-static rates of load application is fairly well understood, there is insufficient experience and data regarding its behaviour at high stress rates, those associated with impact. It is the purpose of this paper to present a historical perspective of the research conducted so far in this area, the state of our present knowledge and the future research needs.

MECHANICAL PROPERTIES OF CONCRETE AT HIGH STRESS RATES: A CHRRONOLOGICAL SURVEY

The first attempt to study the properties of concrete at high stress rates was that of D.A. Abrams in 1917 [1] who tested concrete cylinders under compression with variable cross head speeds between 0.006 inch per minute and 0.15

inch per minute. There was not much interest generated in the topic for about a decade or so after the work of Abrams until Teller [2] picked up the topic again in 1927 followed by Klettle, Webster and White [3] in 1931. With some other studies in between, the first complete study was reported by Evans [4] in 1942, which was complimented by a study by Watstein [5] in 1953. With sufficient interest generated in this topic by now, the first major symposium in this area was held in 1955, where an excellent review on the the state-of-the-rate was presented by McHenry and Shideler [6] who summarized the findings as

"... higher rates of loading increase the strength and lower rates decrease the strength" and

"... strength is a direct function of the logarithm of the rate of loading"

Since the early sixties, this domain of concrete research has received considerable attention. In 1963, McNeely and Lash [7] subjected plain concrete beams to varying rates of loading and determined the moduli of rupture elastically at various rates of loading. In 1964, Green [8] used a ballistic pendulum to assess the impact behaviour of concrete using 100 mm cubes. They concluded that the impact strength, as determined by the number of the repeated blows, increased with the compressive strength and age. Atchely and Furr [9] also conducted impact tests in compression by using a drop hammer and concluded that the compressive strength and the energy absorbed increased with the rate of loading with the evidence of becoming a constant value at higher rates of loading. Goldsmith, Kenner and Ricketts [10] used the Hopkinson's pressure bar technique to propagate dynamic stress pulses through concrete like mixtures who, apart from noting the exceptional energy absorptive capacity of the igneous rock diorite, also concluded that the aggregates played a far more important role under impact than did the bonding material. Birkimer [11] and later Birkimer and Lindemann [12] measured the dynamic tensile strength of concrete and suggested that the critical fracture strain energy provided a meaningful fracture criterion. They also concluded that the fracture strain should vary as the one-third power of the straining rate. In 1972, Hughes and Gregory [13] dynamically loaded concrete up to a strain rate of 30 per sec in compression using "low friction pads". They reported the ratio of dynamic to static strength as 1.9, and concluded that of all the concrete variables, the type of the course aggregate plays the most important role in the resistance to impact. Later, in 1973, Sparks and Menzies [14] also noted the importance of the aggregate stiffness in the impact resistance of concrete and concluded that a reduction in the aggregate stiffness increases the sensitivity of concrete to the rate of loading in compression. Evans [15] in 1974 made an attempt to explain the stress rate sensitivity of concrete on the basis of the Linear Elastic Fracture Mechanics and the phenomenon of subcritical crack growth. He showed that the rate of loading may be related to the fracture stress through the crack growth parameter n (slope of log V. - log K_I plot). Later, in 1977, Mindess and Nadeau [16] conducted tests to verify Evan's theory [15] and concluded that while the theory did apply to mortar, results obtained for cement paste could not be predicted using Evan's theory.

In 1978, contrary to the findings of Atchely and Furr [9] and Birkimer and Lindemann [12], Hughes and Watson [17] reported a decrease in the ultimate strain with an increase in the rate of compressive loading. They attributed this to the lack of creep strains in high rates of loading. They also observed more aggregate failures in impact tests than in the static tests in which case the cracks took a tortuous path around the aggregates. Kaplan [18] in 1980, investigated the effect of free water, age and curing conditions on the strength-loading rate ratio, and reported that the moisture content at the time of the test was the most influential of all the variables. The same year, Zech and Wittman [19] generated strain rates of about 2 per sec in flexure on mortar specimens. The ratio of the dynamic to static modulus of rupture (σ_d/σ_s) was found to be related to the ratio of the dynamic to static stress rates ($\dot{\sigma}_d/\dot{\sigma}_s$) through a material parameter ß,

$$\frac{\sigma_d}{\sigma_s} = \left(\frac{\dot{\sigma}_d}{\dot{\sigma}_s}\right)^{1/1+\beta} \qquad (1)$$

In 1982, Zielinski [20] conducted an extensive investigation to study the fracture of concrete and mortar under uniaxial impact tensile loading using a Hopkinson's split bar. Based on the experimental results, they arrived at the following relation between the tensile strength σ_t and the rate of stressing, $\dot{\sigma}_t$

$$\ln \sigma_t = 1.51 + 0.042 \ln \dot{\sigma}_t \qquad (2)$$

This equation is closely related to the one obtained by Zech and Wittman [19] [Eqn. (1)]. Zielinsky [20] also conducted repeated impact tensile tests and concluded that the upper stress limit $\sigma_{t,max}$ applied in a fatigue testing and the number of impact loading cycles, N are related through an equation of the type:

$$\sigma_{t,max} = 4.30 - 0.318 \ln N \qquad (3)$$

The same year, Hughes and Beeby [21] conducted 3-point impact flexural tests on concrete beams and used a simple beam vibration model to model the tests. Apart from recognizing the importance of two parameters, mass ratio and pulse ratio, they also concluded that an equivalent static force approach is inadequate.

Suaris and Shah [22] conducted instrumented flexural impact tests on concrete in which the drop weight striker as well as the support anvil were instrumented. They concluded that the equation

$$\sigma_f \ \alpha \ (\dot{\epsilon})^{1/1+n} \qquad (4)$$

where $\dot{\epsilon}$ is the strain rate, is not satisfactory as the value of the constant n appears to decrease with strain rate. A similar conclusion was earlier reached by Birkimer [11] in 1970. Dilger, Koch and Kowalczyk [23] investigated the effect of strain rate on confined concrete and concluded that although high rates of loading increased the strength, it did not have a pronounced effect on the shape of the stress-strain curve of confined concrete under compression. Mlakar, Vitaya-Udom and Cole [24] tested concrete under biaxial state of dynamic stress application and reported a decrease in the impact tensile strength with an increase of the simultaneously acting compressive stress.

In the years 1985-87, an extensive experimental investigation was conducted at the University of British Columbia, Vancouver, Canada, to assess the impact resistance of various concrete systems [25-32]. A drop weight instrumented impact machine with a hammer weighing 3385N was used to conduct 3-point flexural impact tests on beams measuring 100 mm x 150 mm x 1400 mm [32]. A schematic view of the test setup is shown in Figure 1. The following conclusions were drawn:

(1) Both normal strength (crushing strength < 50 MPa) and high strength (crushing strength > 80 MPa) plain concretes have higher dynamic fracture strengths than the static ones. Further, high strength concrete is more brittle than the normal strength concrete in that it absorbs lesser fracture energy that does normal strength concrete.

(2) The stress rate sensitivity of concrete increases with the stress rate, i.e. the parameter "n" [Eqn. (4)] decreases with increasing stress rate.

(3) Concrete absorbs substantially high fracture energy under impact than under static loading. This increase in the fracture energy under impact could directly be related

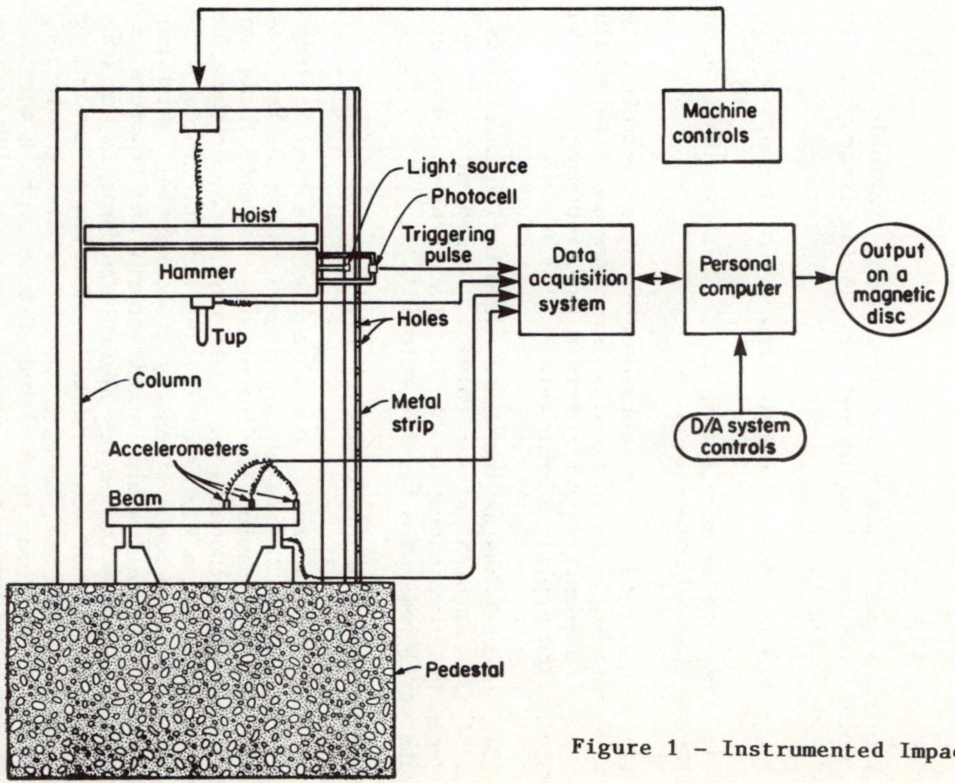

Figure 1 - Instrumented Impact Testing System (32)

to the increased microcracking in concrete under impact, and due to the propagation of cracks through zones other than the ones requiring lesser crack propagation energy.

(4) In the case of conventionally reinforced beams, use of a high strength concrete led to extremely low displacement ductilities under impact. In these cases, fractured reinforcing bars were a common occurrence. Normal strength concrete, on the contrary, enhanced the ductility of conventionally reinforced concrete beams under impact and reinforcing bar fractures were rare.

Summary

The behaviour of plain concrete under variable strain rates under tension, flexure and compression is graphically summarized in Figures 2.

The following general conclusions may be drawn:

(1) Concrete is more strain rate sensitive in tension and flexure than in compression.

(2) Concrete is a considerably stress/strain rate sensitive material. It is usually stronger at high strain rates and the sensitivity to strain rate increases with an increase in the applied strain rate.

(3) Concrete has an increased deformation capability, or ductility, at high strain rates. Consequently, it absorbs more fracture energy at higher strain rates. However, concrete still remains a brittle material.

FIBRE REINFORCED CONCRETE UNDER IMPACT

The brittle nature of concrete has resulted in scepticism with respect to its use in situations where the possibility of a dynamic loading exists. Dynamic loads may occur during an earthquake, during an explosion, or when the structure is hit by a flying or falling object. These loadings impart large amounts of energy to concrete, which may easily exceed the allowable strain energy and fracture may result. To increase the energy absorption capacity of concrete, the addition of short randomly distributed natural or artificial fibres has emerged as a promising solution. Fibres may range from low modulus fibres such as acrylic, polyethylene, polypropylene, nylon, etc. to high modulus fibres such as rockwool, steel and asbestos. Fibres are mixed with the rest of the concrete ingredients in the conventional manner in the mixer and a 3-dimensional random distribution of the fibres is achieved upon hardening.

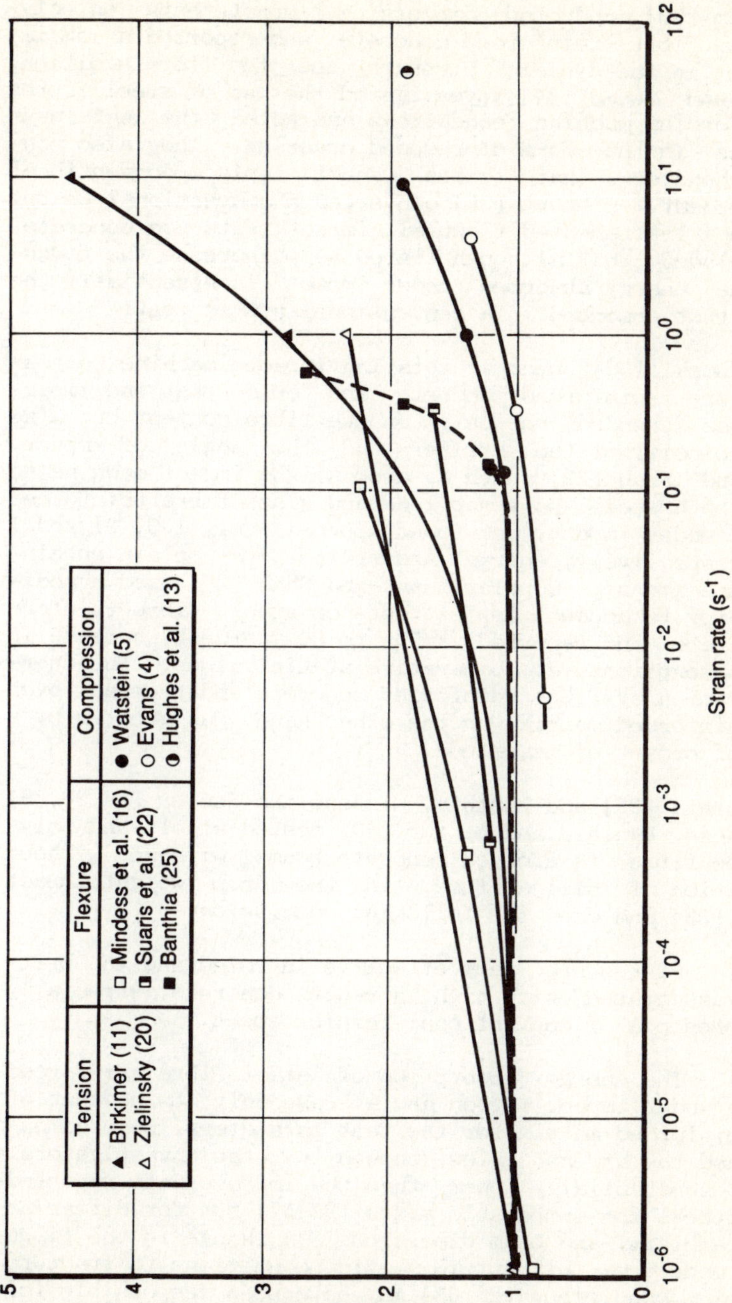

Figure 2 - Strain Rate Sensitivity of Concrete

Impact testing of fibre reinforced concrete is of a relatively recent origin in the literature. Bhargava and Rehnström [33] conducted compressive impact tests on polypropylene fibre reinforced concrete and reported a 15% improvement in the dynamic strength due to fibre addition. Jamrozy and Swamy [34] investigated the use of steel fibres in reinforcing machine foundations and noted the usefulness of fibres in the form of reduced cracking. They also concluded, however, that fibres cannot replace conventional reinforcement. Hibbert [35] converted a conventional Charpy to conduct instrumented flexural impact tests on concrete, and concluded that although fibres help increase the toughness, the energy absorbed under impact is essentially the same as that absorbed in a conventional static test.

Radomski [36] used a rotating impact machine to test steel fibre reinforced concrete and found that the impact resistance depended not only on the fibre content but also on the velocity of the striker and the angle of impact. Suaris and Shah [22] used a drop weight impact machine to subject both steel, polypropylene and glass fibre reinforced concrete under impact, and found approximately 100, 31 and 7 times, respectively, energy absorption over plain unreinforced specimens. Gopalaratnam and Shah [37] used a modified Charpy to conduct impact tests on steel fibre reinforced concrete and reported a 70 to 80 percent increase in energy absorption over comparative static values. The energy absorption of fibre reinforced concrete under impact over the unreinforced matrix, on the other hand, was greater by a couple of orders of magnitude.

Banthia [25] and Banthia, Mindess and Bentur [26,38,39] and Mindess, Banthia and Bentur [40] tested steel and polypropylene fibre reinforced concrete beams, with and without the addition of silica fume, in the drop weight impact machine [32] and drew the following conclusion:

(1) Fibres are very effective in remedying the brittleness associated with high strength concrete, especially in the presence of conventional reinforcement.

(2) The energy absorption of steel fibre reinforced concrete under impact is on an average only twice that of the unreinforced matrix for the test parameters, beam dimensions, and the stress rates chosen by the investigators. This is substantially lower than the improvements reported by the others for comparable mixes [22,37] but for different test parameters and beam dimensions. A change in the fibre failure mode from pull-out in static conditions to fracture in dynamic conditions [26,38] may have been responsible for this lowered resistance under impact.

One interesting fibre is the carbon fibre which is now cheaply available as "chop". With a proper dispersive agent (usually silica fume) and a high range water reducer, fibres upto 5% by volume may be properly mixed [41] (the corresponding upper limit for steel fibres is less than 2%). Carbon fibre reinforced cements have been tested under impact by Banthia and Ohama [42] and the results are encouraging.

CRACK PROPAGATION STUDIES UNDER IMPACT

There is some indication [43] that crack velocity is proportional to the stress rate. The tests conducted by Bhargawa and Rehnström [44], Alford [45] and Mindess, Banthia, Ritter, and Skalny [46] have indicated that the crack velocity under impact is about 100-200 m/s, which is only a fraction of the theoretical Rayleigh wave velocity [46].

John and Shah [47] monitored the pre-peak crack growth in notched concrete specimens under impact by using brittle Krak Gauges and concluded that the pre-peak crack growth decreases with an increase in the stress rate, and therefore, at high stress rates, Linear Elastic Fracture Mechanics (LEFM) may be valid. LEFM parameters like the critical stress intensity factor (K_{Ic}) under impact were measured by Mindess, Banthia and Yan [48] using notched beams of plain and fibre reinforced concretes. They reported higher K_{Ic} values at higher stress rates and found high strength concrete to be more notch-sensitive under impact than under static loading.

FUTURE RESEARCH NEEDS

(1) Standard Testing Technique: So far, a universally acceptable standard testing technique for conducting impact tests on concrete does not exist. Research is needed in this direction, so that the results from the various sources may be compared.

(2) High Strength Concrete: This attractive material is also known to be more brittle under impact. Attempts are needed to correctly assess its impact resistance and to devise techniques, like fibre reinforcement, etc., towards remedying its brittleness.

(3) Crack Propagation Studies: Nucleation and growth of cracks under impact loads, and their arrest by inclusions like the fibres and the aggregates needs further attention. Such a knowledge will be quite useful in devising analytical models to predict the response of structural elements under a prescribed dynamic load pulse.

27. Banthia, N., Mindess, S. and Bentur, A.; *High Stress Rate Testing of Concrete*, The Indian Concrete Journal, 60 (10), pp. 265-270, 1986.
28. Mindess, S., Banthia, N. and Bentur, A.; *The Impact Behaviour of Concrete in Bending*, Proc. Int. Conf. on Fract. Mech. of Concrete, Lausanne, Switzerland, 1985.
29. Bentur, A., Mindess, S. and Banthia, N.; *The Fracture of Reinforced Concrete under Impact*, Materials Research Society Symposium Proceedings, Vol. 64, 1986.
30. Bentur, A., Mindess, S. and Banthia, N.; *The Behaviour of Reinforced Concrete under Impact: The Effect of Concrete Strength*, Proc. SEM-RILEM Int. Conference on Fracture of Concrete and Rock, Houston, June 1987.
31. Banthia, N., Mindess, S. and Bentur, A.; *Energy Balance in Instrumented Impact Tests on Plain Concrete Beams*, Proc. SEM-RILEM Int. Conference on Fracture of Concrete and Rock, Houston, June 1987.
32. Banthia, N., Mindess, S., Bentur, A. and Pigeon, M.; *Impact Testing of Concrete Using a Drop Weight Impact Machine*, Experimental Mechanics (in press).
33. Bhargava, J. and Rehnström, A.; *Dynamic Strength of Polymer Modified and Fibre Reinforced Concrete*, Cement and Concrete Research, 7, pp. 199-207, 1977.
34. Jamrozy, Z. and Swamy, R.N.; *Use of Steel Fibre Reinforcement for Impact Resistance and Machinery Foundation*, Int. Journal of Cement Composites, 1 (2), pp. 65-76, 1979.
35. Hibbert, A.P.; *Impact Resistance of Fibre Concrete*, Ph.D. Thesis, University of Surrey, England, 1979.
36. Radomski, W.; *Application of the Rotating Impact Machine for Testing Fibre Reinforced Concrete*, Int. Journal of Cement Composites, Vol. 3, No. 1, 1981.
37. Gopalaratnam, V.S. and Shah, S.P.; *Properties of Fibre Reinforced Concrete under Impact Loading*, Journal of American Concrete Institute, 83 (1), pp. 117-126, 1986.
38. Banthia, N., Mindess, S. and Bentur, A.; *Steel Fibre Reinforced Concrete under Impact Loading*, Proceedings, Int. Conference on Fibre Reinforced Concrete, Madras, India, 1987.
39. Banthia, N., Mindess, S. and Bentur, A.; *Behaviour of Fibre Reinforced Concrete Beams under Impact Loading*, Proceedings, Sixth Int. Conf. on Composite Material, London, England, 1987.
40. Mindess, S., Banthia, N. and Bentur, A.; *The Response of Reinforced Concrete Beams with a Fibrous Concrete Matrix to Impact Loading*, Int. Journal of Cement Composites, Vol. 8, No. 3, August 1986.
41. Ohama, Y., Amano, M. and Endo, M.; *Properties of Carbon Fibre Reinforced Cement with Silica Fume*, Concrete International, March 1985.

(4) Steel Fibre Reinforced Concrete: **Research is needed towards optimizing the fibre geometry to maximize the resistance of the resulting composite to impact. Attempts** are also needed towards formulating the constitutive law of steel fibre reinforced composites as a function of stress rate.

REFERENCES

1. Abrams, D.A.; *Effect of Rate of Application of Load on the Compressive Strength of Concrete*, Proceedings, ASTM 17, Part 2, pp. 364-367, 1917.
2. Teller, L.W.; *Discussion on Transverse Testing of Concrete*, Proceedings, ASTM, 27(II), p. 418, 1927.
3. Klettle, A.J., Webster, O.W. and White, F.P.; *Effect of Speed of Loading on the Modulus of Rupture of Plain Concrete*, Thesis for Bach. of Science, Univ. of Wisconsin, 1933.
4. Evans, R.H.; *Effect of Rate of Loading on the Mechanical Properties of Some Materials*, Journal of the Institution of Civil Engineers, 18, p. 296, 1942.
5. Watstein, D.; *Effect of Straining Rate on the Compressive Strength and Elastic Properties of Concrete*, Journal of the American Concrete Institute, Vol. 49, No. 8, pp. 729-756, April 1953.
6. McHenry, D. and Shideler, J.J.; *Review of Data on Effect of Speed in Mechanical Testing on Concrete*, ASTM, STP-185, 1955.
7. MacNeely, D.J. and Lash, S.D.; *Tensile Strength of Concrete*, Journal of the American Concrete Institute, Vol. 60, No. 6, pp. 751-760, 1963.
8. Green, H.; *Impact Strength of Concrete*, Proceedings of the Institution of Civil Engineers, Vol. 28, pp. 383-396, 1964.
9. Atchely, B.L. and Furr, H.L.; *Strength and Energy Absorption Capacity of Plain Concrete under Dynamic and Static Loading*, Journal of American Concrete Institute, pp. 745-756, 1967.
10. Goldsmith, W., Kenner, V.H. and Ricketts, T.E.; *Dynamic Loading of Several Concrete Like Mixtures*, Proc. of American Society of Civil Engineers, Structural Division, pp. 1803-1827, 1968.
11. Birkimer, D.L.; *A Possible Fracture Criterion for the Dynamic Tensile Strength of Rock*, Proc. 12th Symp. on Rock Mechanics, pp. 573-589, 1971.
12. Birkimer, D.L. and Lindeman, R.; *Dynamic Tensile Strength of Concrete Materials*, Journal of the American Concrete Institute, 68, pp. 47-49, 1971.

13. Hughes, B.P. and Gregory, R.; *Concrete Subjected to High Rates of Loading in Compression*, Magazine of Concrete Research, 24 (78), pp. 25-36, 1972.
14. Sparks, P.R. and Menzies, J.B.; *The Effect of Rate of Loading Upon the Static and Fatigue Strength of Plain Concrete in Compression*, Magazine of Concrete Research, Vol. 25, No. 83, pp. 73-80, June 1973.
15. Evans, A.G.; *Slow Crack Growth in Brittle Materials under Dynamic Loading Conditions*, Int. J. Frac. Mech., 10, p. 251, 1974.
16. Mindess, S. and Nadeau, J.S.; *Effect of Loading Rate on the Flexural Strength of Cement and Mortar*, American Ceramic Soc. Bulletin, 56 (4), p. 429, 1977.
17. Hughes, B.P. and Watson, A.J.; *Compressive Strength and Ultimate Strain of Concrete under Impact Loading*, Magazine of Concrete Research, Vol. 30, No. 105, pp. 189-199, Dec. 1978.
18. Kaplan, S.A.; *Factors Affecting the Relationship between the Rate of Loading and the Measured Compressive Strength of Concrete*, Magazine of Concrete Research, 32 (111), 1980.
19. Zech, B. and Wittmann, F.H.; *Variability and Mean Value of Strength as a Function of Load*, Journal of the American Concrete Institute, Vol. 77, No. 5, pp. 358-362, Sept.-Oct. 1980.
20. Zielinsky, A.J.; *Fracture of Concrete and Mortar under Uni-axial Impact Tensile Loading*, Ph.D. Thesis, University of Delft, Netherlands, 1982.
21. Hughes, G. and Beeby, A.W.; *Investigation of the Effect of Impact Loading on Concrete Beams*, The Structural Engineer, 60B (3), 1982.
22. Suaris, W. and Shah, S.P.; *Properties of Concrete Subjected to Impact*, Journal of the American Society of Civil Engineers, Structural Division, Vol. 109, No. 7, pp. 1727-1741, July 1983.
23. Dilger, W.H., Koch, R. and Kowalczyk, R.; *Ductility of Plain and Confined Concrete under Different Strain Rates*, Journal of American Concrete Institute, p. 73, 1984.
24. Mlakar, P.F., Vitaya-Udom, K.P. and Cole, R.A.; *Dynamic Tensile-Compressive Behaviour of Concrete*, Journal of American Concrete Institute, pp. 484-491, July-August 1985.
25. Banthia, N.; *Impact Resistance of Concrete*, Ph.D. Thesis, University of British Columbia, Vancouver, Canada, 1987.
26. Banthia, N., Mindess, S. and Bentur, A.; *Impact Behaviour of Concrete Beams*, Materials and Stuctures, Vol. 20, pp. 293-302, 1987.

42. Banthia, N. and Ohama, Y.; *Dynamic Tensile Fracture of Carbon Fibre Reinforced Cements*, accepted by Int. Conf. on Recent Developments in Fibre Reinforced Cements and Concretes, Cardiff, Sept. 1989.
43. Shah, S.P. and John, R., Proc. of Int. Conf. on Frac. Mech. of Concrete, Lausanne, pp. 373-385, 1985.
44. Bhargava, J. and Rehnström, A.; Cement and Concrete Research 5, pp. 239-248, 1975.
45. Alford, N.McN.; *Dynamic Considerations for Fracture of Mortars*, Journal of Materials Science, 56 (3), pp. 279-287, 1982.
46. Mindess, S., Banthia, N., Ritter, A. and Skalny, J.P.; *Crack Development in Cementitious Materials under Impact Loading*, Material Research Society Symp. Proceedings, Vol. 64, 1986.
47. John, R. and Shah, S.P.; *Fracture of Concrete Subjected to Impact Loading*, Cement Concrete and Aggregates, Vol. 8, No. 1, pp. 24-32, Summer 1986.
48. Mindess, S., Banthia, N. and Cheng, Y.; *The Fracture Toughness of Concrete under Impact Loading*, Cement and Concrete Research, Vol. 17, pp. 231-241, 1987.

Soft Missile Perforation Analysis of Small and Large Scale Concrete Slabs

J. Jowett and K. Kinsella

Safety & Reliability Directorate, Wigshaw Lane, Culcheth, Warrington, WA3 4NE, UK

ABSTRACT

In the UK military combat aircraft pose the major aircraft hazard to nuclear facilities. To make a probabilistic assessment of the structural failure frequency arising from such impacts a methodology must account for the wide spread of impact characteristics which can be realised in an accident. For example, aircraft within the military combat category may have impact masses between 5 and 50 Te and impact velocities between 50 and 300 m/s. If the probabilistic methodology is to consider such a range of crash parameters a simple method of assessing the structural effects of such impacts is necessary.

This paper describes a semi-empirical method of assessing the resistance of reinforced concrete slabs to simulated military aircraft impact. The work is based on the results of a number of experimental programmes carried out in both the UK and the Federal Republic of Germany, in which a series of missiles and targets at scales from 1:15 to 1:2.7 of full size were used to simulate the dynamic loads and structural damage created by a military aircraft crash. By expressing the results of these tests in terms of a previously published static punching strength formula, it is shown that the dynamic failure response of the targets is well represented by the basic physical relationships observed for static failure. Furthermore, by making only a minor modification and including a suitable dynamic enhancement factor, the formula is shown to be capable of predicting the average dynamic force required to cause perforation to within ± 20%. By examining lesser damage states it is shown that the formula can also be used to define the loading threshold required to cause target scabbing.

INTRODUCTION

As part of external hazard assessment most regulators require nuclear operators to assess the frequency of aircraft crash. Often such assessments will not allow this hazard to be dismissed on low frequency grounds alone, so that crash consequences must be examined.

In the UK and other parts of Europe military combat aircraft provide the dominant contribution to the hazard from aircraft crash. For this reason, over recent years, many tests aimed at investigating the resistance of reinforced concrete slabs to simulated aircraft impact have been conducted in the UK and the Federal Republic of Germany (FRG). These soft missile impact tests used thin walled metal tubes to simulate the loading applied by the deformable airframe.

As military combat aircraft have a wide spread of impact characteristics (mass 5-50 Te, velocity 50-300 m/s), if a fully probabilistic evaluation of crash consequences is to be performed a simple method of predicting structural behaviour is needed. In the following, based on a wide range of test work, such a simplified method for predicting the soft missile perforation and scabbing resistance of reinforced concrete slabs is described.

EVALUATION OF IMPACT TEST RESULTS

Previous Studies

Jowett [1] has shown that the perforation resistance of the small scale (1:11, 1:15 full scale) reinforced concrete slabs tested in the UK [2,3] may be predicted by comparing the average dynamic load applied by the missile, F_{AV}, with the static punching shear strength of the slabs, F_S. Perforation was found to occur when $F_{AV} \geq 1.6\ F_S$, or equivalently, the average dynamic force for perforation F_{PAV} was given by:

$$F_{PAV} = 1.6\ F_S \tag{1}$$

where the static punching shear strength was calculated using

$$F_S = 2970[\rho_t\ \sigma_{cu}]^{1/3}\ d_1^{0.75}[2\pi r + 7.85\ d_1] \tag{2}$$

where

ρ_t = the average of the tensile reinforcement quantities in the two orthogonal directions expressed as a percentage of cross section area

σ_{cu} = compressive strength of concrete (cube) (Pa)

d_1 = depth of slab from load face to centre of rear face reinforcement (m)

r = radius of load area (m)

and $$F_{AV} = \frac{0.9I}{t_{0.9I}} \qquad (3)$$

where

I = the total impulse delivered by the missile to the slab (Ns)

and $t_{0.9I}$ = the time required to deliver the initial 90% of I (s)

The static punching strength equation noted above is that proposed by Regan [4] for normal weight aggregate concrete. This formula was derived by correlating the results of over a hundred static punching tests on reinforced concrete slabs up to 0.6m thickness. In slightly modified form the equation forms the basis for the assessment of punching strength in BS8110, the latest UK Code of Practice for the design of reinforced concrete structures.

The definition of F_{AV} given in equation (3) was adopted to prevent the relatively long tail end of the missile's load/time history from disproportionally affecting the estimation of the load level created during the most significant phase of the impact.

Within the parameter range covered by the small scale tests, which may be defined as

$0.11\ (m) \leq d_1 \leq 0.213\ (m)$
$0.35 \leq r/d_1 \leq 0.68$
$0.28 \leq \rho_t \leq 1.17$ (% EW)
$0.25 \leq \rho_s$ (shear reinforcement percentage) $\leq 0.6\ 3$ (%)
$26 \times 10^6 < \sigma_{cu} < 63 \times 10^6$ (Pa)

equation (1) was capable of predicting F_{PAV} to an accuracy of ±15%.

The ability of a static punching strength equation to predict the failures observed in the soft missile tests is accounted for as follows. In all the tests measurements of transient slab deflection showed that the slabs had sufficient time to reach their peak transient deflection during the period of load application. This fact implies that at failure the slabs will have adopted a deflection mode similar to that occurring under static loading.

The most important application of both scale model test work and analysis is its use for the prediction of full scale slab response. However in this regard the following cautionary note was given in reference 1. Soft missile studies carried out in collaboration between UK and FRG [3] indicated that between the 0.7m thick German slabs and quarter scale replicas tested in the UK linear scaling existed, a result which implies $F_{PAV} \; \alpha \; d_1^{\;2}$. However examination of equations (1) and (2) reveals that the perforation analysis method predicts $F_{PAV} \; \alpha \; d_1^{\;1.75}$. It was with a view to resolving this discrepancy and extending the validity of the methodology that the work described below was undertaken.

Data Correlation

The data base used in the study included all the small scale tests performed at Winfrith (AEEW) and Foulness in the UK, analysed previously [2,3] plus all the large scale Meppen tests for which results were readily available [5].

The method of data analysis used in reference [1] was as follows. A value of x/d, the relative penetration of the missile into the target, and F_{AV}/F_S was derived for each test. A plot of F_{AV}/F_S vs x/d was then made to obtain the mean value of F_{AV}/F_S at which x/d = 1.0. This plot for the UK data is shown in Fig 1 and indicates that for x/d ≥ 0.1 an approximately linear relationship exists between F_{AV}/F_S and x/d. It follows therefore that for a given x/d; $F_{AV} \; \alpha \; F_S$

$$\text{or} \quad F_{AV} \; \alpha \; [\rho_t \; \sigma_{cu}]^{1/3} \, d_1^{\,0.75} \, [2\pi r + 7.85 \, d_1] \tag{4}$$

To examine the suitability of this equation, for both the UK and FRG data, four groups of x/d values of 1.0, 0.5, 0.3 and 0.1 were chosen spanning the range of x/d levels between failure cone formation and complete perforation. Each group included data from tests of varying scale to ensure that any scaling dependencies were well represented. Details of these groupings, identifying the tests used and the corresponding levels of x/d are given in table 1.

To compute the average dynamic load, F_{AV}, load-time histories were obtained for all tests using the perfectly plastic soft missile formula developed by Riera [6]. From these the F_{AV} was calculated using equation (3).

The results of this calculation for each of the Meppen tests are given in table 2. Those for the AEEW and Foulness tests have been reported in the previous study [1].

Table 1

Groupings of Constant x/d ratio

	Test Reference	Measured x/d	
x/d = 1.0			
	M/5	1.0	)
	M/13	1.0	)
	A125	1.0	)
	A164	1.0	) $V_{Ex} > 0.2\ V_{in}$ tests not included
	A167	1.0	)
	A285	1.0	)
x/d ≈ 0.5			
	M/20	0.44	)
	A127	0.57	)
	A165	0.52	) 0.5 ± 0.07
	A289	0.47	)
x/d ≈ 0.3			
	M/16	0.29	)
	A166	0.28	)
	A286	0.29	) 0.29 ± 0.01
	F49	0.30	)
	F50	0.30	)
x/d ≈ 0.1			
	M/8	0.10	)
	M/11	0.13	)
	M/12	0.10	)
	M/21	0.09	)
	A126	0.11	)
	A129	0.10	) 0.11 ± 0.02
	A195	0.12	)
	A197	0.12	)
	A200	0.12	)
	A290	0.10	)
	F39	0.09	)

Table 2

Results of Average Loading Calculations
(Meppen tests 1-22)

Test No	Total Impulse (KN.s)	$t_{0.9I}$ (mS)	F_{AV} (MN)
1	278.6	26.2	9.57
2	-	-	-
3	-	-	-
4	278.6	26.2	9.57
5	260.7	25.5	9.20
6	261.0	22.0	10.68
7	247.2	25.0	8.90
8	261.2	25.5	9.22
9	261.2	25.5	9.22
10	278.7	26.4	9.50
11	244.4	25.0	8.80
12	275.0	26.5	9.34
13	277.5	26.4	9.46
14	278.6	26.2	9.57
15	247.2	24.1	9.23
16	278.2	26.3	9.52
17	-	-	-
18	259.1	27.7	8.42
19	250.9	23.4	9.65
20	229.9	23.3	8.88
21	264.9	25.5	9.35
22	-	-	-

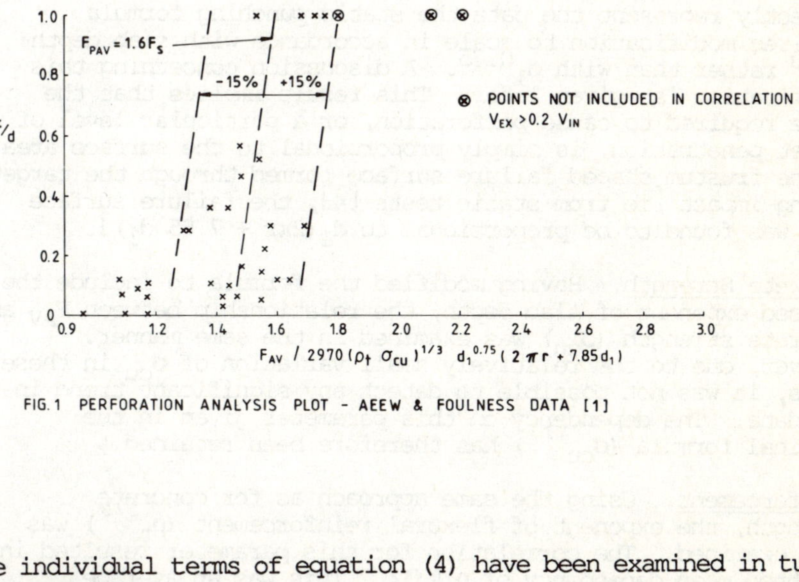

FIG. 1 PERFORATION ANALYSIS PLOT - AEEW & FOULNESS DATA [1]

The individual terms of equation (4) have been examined in turn as follows;

Slab Depth Effect. To investigate the relationship in the formula regarding slab depth, d_1, a graph of $F_{AV}/[(\rho_t\ \sigma_{cu})^{1/3}\ (2\pi r + 7.85\ d_1)]$ against d_1, has been plotted for all groupings of x/d (Figure 2). The exponents measured for each individual group lie within the range 0.97 - 1.22 with a mean value, weighted to the number of data points in each group, of 1.02. There is no evidence on this graph of any trend in the size of the exponent with varying x/d.

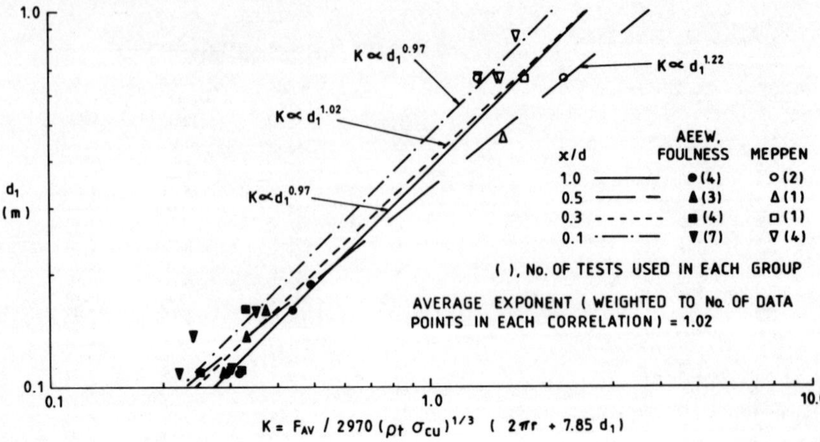

FIG. 2 SCALING RELATIONSHIP FOR SLAB DEPTH

The correlation shown in Figure 2 indicates that to correctly represent the data the static punching formula requires modification to scale in accordance with slab depth $d_1^{1.0}$ rather than with $d_1^{0.75}$. A discussion concerning this modification is given later. This result implies that the force required to cause perforation, or a particular level of target penetration, is simply proportional to the surface area of the frustum shaped failure surface formed through the targets during impact [ie from static tests [4], the failure surface area was found to be proportional to $d_1(2\pi r + 7.85\, d_1)$].

Concrete Strength. Having modified the formula to include the revised exponent of slab depth, the relationship between F_{AV} and concrete strength (σ_{cu}) was examined in the same manner. However, due to the relatively small variation of σ_{cu} in these tests, it was not possible to detect any significant trend in the data. The dependency on this parameter given in the original formula ($\sigma_{cu}^{1/3}$) has therefore been retained.

Reinforcement. Using the same approach as for concrete strength, the exponent of flexural reinforcement ($\rho_t^{1/3}$) was also examined. The correlation for this parameter resulted in a weighted mean dependency of $\rho_t^{0.42}$. This was an average exponent taken from the correlation of individual x/d groupings with no discernable evidence of any trend in the size of the exponent with increasing x/d.

However the scatter on the data about the $\rho_t^{0.42}$ dependence was sufficiently large that the ($\rho_t^{1/3}$) provided nearly as good a correlation. Hence the original form has again been retained.

Although not included in the static punching equation, the dependency of F_{AV} on the percentage of shear reinforcement included in the slab (ρ_s) was also examined. For x/d values of ~ 0.1, there was no evidence of shear reinforcement influencing the force required to cause this level of penetration. However, for x/d ≥ 0.3, although the data set available was reduced by ~ 40%, a dependency on shear reinforcement in accordance with $\rho_s^{0.3}$ was observed. Further experimental validation of this result would be required however before its inclusion in the formula could be justified.

Modified Formula

Disregarding the apparent influence of shear reinforcement, which is in need of further experimental verification, the examination of the static punching equation detailed above has revealed that, with minor modification, the equation can suitably be used for predicting the dynamic perforation behaviour observed over the full range of impact tests examined.

The resulting form of the equation, with the modification to slab depth included, is given by:

$$F_s^1 = K[\rho_t \; \sigma_{cu}]^{1/3} \; d_1[2\pi r + 7.85 \; d_1] \qquad (5)$$

where K is a constant of proportionality.

To demonstrate the suitability of equation (5),

$$F_{AV}/F_s^1 = F_{AV}/[2970[\rho_t \; \sigma_{cu}]^{1/3} \; d_1[2\pi r + 7.85 \; d_1]],$$

the ratio of the average dynamic loading to the modified static punching formula has been plotted against x/d in Figure 3.

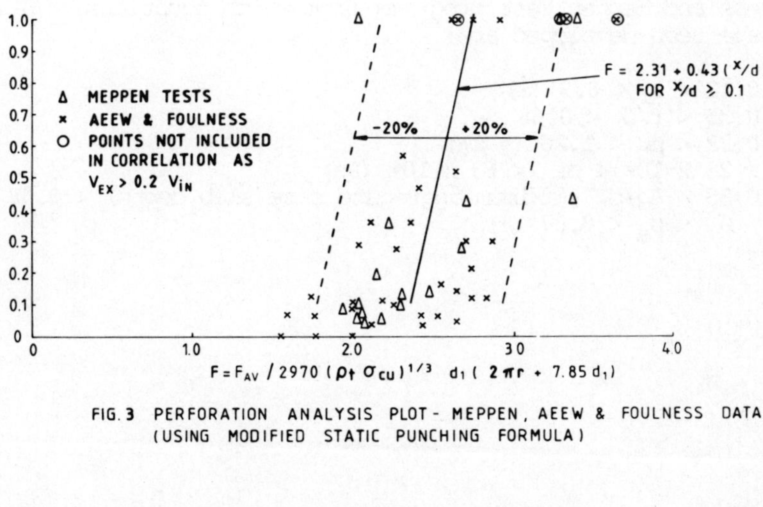

FIG. 3 PERFORATION ANALYSIS PLOT - MEPPEN, AEEW & FOULNESS DATA (USING MODIFIED STATIC PUNCHING FORMULA)

From this, it can be seen that the data from the three different test series (AEEW, Foulness and Meppen) agree well, with the steeply rising band of data giving a clear indication of both the onset of dynamic punching failure and perforation.

Predicting the Perforation Limit. Examination of Figure 3 reveals that the mean correlation line drawn through the data from x/d = 0.1 to x/d = 1.0, originates at an F_{AV}/F_S value of 2.36 (± 0.57 @ 90% level of confidence), at x/d = 0.1, and reaches x/d = 1.0 (indicating perforation) at an F_{AV}/F_S' value of 2.75 (± 0.57 @ 90% level of confidence).

Hence the mean estimate of the average dynamic force required to perforate the targets (F_{PAV}) is given by:

$$F_{PAV} = \beta p \, [\rho_t \, \sigma_{cu}]^{1/3} \, d_1 [2\pi r + 7.85 \, d_1] \quad (6)$$

where βp (constant of proportionality for perforation) = 8170 (± 20% @ 90% level of confidence).

Predicting the Scabbing Limit. The data used in Figure 3 has been replotted in Figure 4 to indicate those tests in which concrete scabbing had occurred. It is clear from this graph that for x/d ≥ 0.12 scabbing will always occur. Using the mean correlation line drawn through the data, scabbing will occur at an F_{AV}/F_Svalue of 2.37 (± 0.57 @ 90% confidence).

Therefore the mean estimate of the average dynamic force required to cause scabbing (F_{SAV}) is given by:

$$F_{SAV} = \beta_s \, [\rho_t \, \sigma_{cu}]^{1/3} \, d_1 \, [2\pi r + 7.85 \, d_1] \quad (7)$$

where β_s (constant of proportionality for scabbing) = 7040 (± 24% @ 90% confidence level).

The parameter ranges covered experimentally by the AEEW, Foulness and Meppen test programs from which equations (6) and (7) have been developed are:

$0.11 < d_1 < 0.9$ (m)
$0.35 < r/d_1 < 0.68$
$0.22 < \rho_t < 1.26$ (% EW)
$25 \times 10^6 < \sigma_{cu} < 63 \times 10^6$ (Pa)
$0.05 < Ag/d_1$ (maximum aggregate size/slab depth) < 0.07
$0 < \rho_s < 0.87$ (%)

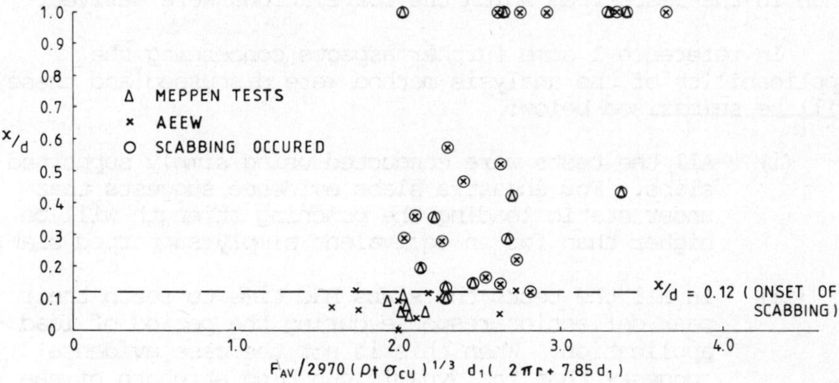

FIG. 4 ONSET OF SCABBING - MEPPEN & AEEW TESTS

DISCUSSION

The RHS of equations (6) and (7) may be considered as the dynamic punching strength of a slab which resists perforation or scabbing by soft missiles. As noted earlier, in the development of his static punching equation, Regan discovered that the area of the failure surface was proportional to $d_1(2\pi r + 7.85d_1)$. Hence the RHS of equations (6) and (7) may be considered as the product of a failure surface area and a shear failure stress. In these equations this failure stress is seen to be independent of slab depth d_1. However equation (2) implies that the shear failure stress is proportional to $d_1^{-0.25}$.

It is interesting to compare the above findings with those of Taylor [7] who has studied size and scale effects in the shear failure of reinforced concrete beams up to 1m deep. These studies revealed quite a complex picture. For beams with a maximum aggregate size, Ag $\leq$ 10mm a size effect similar to that included in Regan's equation existed even if Ag was scaled with d_1. However for 10 $\leq$ Ag < 40mm size effects were of less importance provided Ag was scaled with d_1. As the small scale slabs had Ag ~ 10mm and Ag/d_1 was almost the same for all slab sizes examined the loss of size effects from the slab failure stress is consistent with the behaviour seen in Taylors experiments.

The implication of these findings are that equations (6) and (7) are likely to depend upon Ag/d_1 so that care should be exercised in applying the equation particularly if Ag/d_1 $\leq$ 0.05.

In such cases the degree of aggregate interlock would be less than in the tests from which the correlations were derived.

In reference 1 some further aspects concerning the applicability of the analysis method were discussed and these will be summarised below:

(i) All the tests were conducted using simply supported slabs. For encastre slabs evidence suggests that under static loading the punching strength will be higher than for an equivalent simply supported slab.

(ii) In all the tests the slabs had time to reach their peak deflection response during the period of load application. When this is not the case evidence suggests that the dynamic punching strength of the slabs will be higher than predicted by equation (6).

CONCLUSIONS

i Using both small and large scale test data a simple semi-empirical method has been developed to assess the resistance of reinforced concrete slabs to simulated military aircraft impact.

ii To assess the perforation resistance of a slab it is simply necessary to compare the average dynamic load exerted by the missile, as calculated by equation (3), with the dynamic punching strength of the slab as given by equation (6). To assess the scabbing resistance of a slab the average dynamic load is compared with the dynamic punching strength as given by equation (7). The accuracy, at the 90% confidence level, with which the methods determine the average dynamic force for perforation and scabbing is ± 20% and ± 24% respectively.

iii The amount of shear reinforcement included in a target will affect its resistance to perforation. The extent of this effect is however in need of further experimental investigation before its inclusion in the formula can be justified.

REFERENCES

1. Jowett, J. Soft Missile Perforation Analysis of Model Scale Concrete Slabs. UKAEA Report No SRD R403, 1987.

2. Barr, P. et al. An Experimental Investigation of Scaling of Reinforced Concrete Structures under Impact Loading. Joint I. Struct. E/BRE Seminar on Dynamic Modelling of Structures, Watford, England, 1981.

3. Sage, F. and Pfeiffer, A. Response of Reinforced Concrete Targets to Impacting Soft Missiles, an FRGMRT-UKAEA

Co-operation in Tests to Validate Computer Codes and Scaling Laws. Paper J8/4, SMIRTS, 1979.

4. Regan, P.E. Behaviour of Reinforced Concrete Flat Slabs, CIRIA Report No 89, 1981.

5. Riech, H. and Rudiger, E. Versuchsergebrisse der Meppener Versuche II/11 bis II/22, II Technischer Bericht, Vorhaben 1500408 und 467. Hochtief AG, Kerntechmischer Ingenieurbau, Frankfurt, 1984.

6. Riera, J.D. A Critical Reappraisal of Nuclear Power Plant Safety against Accidental Aircraft Impact, Vol. J, paper J9/2, 1979.

7. Taylor, H.P.J. Shear Strength of Large Beams. ASCE, J.Struct.Div., Vol. 98, p.2473, 1972.

Nomenclature

A_g = Maximum aggregate size used in the concrete slab (m)

d_1 = Depth of slab from load face to centre of rear face reinforcement (m)

d = Full slab depth (m)

F_{AV} = Average dynamic load applied by missile as defined by equation (3) (N)

F_{PAV} = Value of F_{AV} needed to perforate slab (N)

F_{SAV} = Value of F_{AV} needed to cause scabbing damage in the slab (N)

F_s = Static punching shear strength of slab calculated using equation (2) (N)

I = Total impulse delivered by missile to slab (Ns)

r = Missile or load area radius (m)

$t_{0.9I}$ = Time needed to deliver initial 90% of total impulse to the slab (s)

x = Penetration depth of the missile into the slab (m)

ρ_t = Average of the tensile (rear face) reinforcement quantities in the two orthogonal directions expressed as a percentage of cross section area (% EW)

ρ_s = Shear reinforcement quantity expressed as a percentage of plan area (%)

σ_{cu} = Concrete compressive strength, measured on cubes (Pa)

V_{EX} = Exit velocity of missile from slab after perforation

V_{IN} = Incident velocity of missile.

SECTION 2 - ANALYSIS

BIGGS - Simplified Elastic-Plastic Dynamic Response

Wilfred E. Baker and Kathy H. Spivey

Wilfred Baker Engineering, San Antonio, Texas, USA

ABSTRACT

In encountering many analyses of structures under dynamic loads, a PC-compatible method of dealing with such problems is needed in order to facilitate the problem solution. BIGGS is a program written in BASICA which numerically solves dynamic response problems for structures loaded by transient pressures and/or with initial motions. The structures are represented as equivalent, elastic-plastic, single-degree-of-freedom (SDOF) systems. The methods upon which the program is based are found in Introduction to Structural Dynamics, by John M. Biggs.[6] The program yields both numeric and graphic outputs of SDOF dynamic responses. Topics included in the paper are: 1) discussion of methods for representing a structure as a simplified SDOF system; 2) explanation of inputs for the BIGGS program; and 3) example problems with problem set-up and computer results, including blast loading and non-penetrating impact analyses.

INTRODUCTION

Many textbooks present methods of analysis for dynamics of **elastic** simple systems and more complex structures. Reference 1 is one of the best presented and more thorough of these texts. It covers vibrations and response to dynamic loads of both simple, single-degree-of-freedom (SDOF) systems; more complex, lumped mass and stiffness systems; and still more complex distributed mass and stiffness systems.

But, in the late 1950s and the 1960s, a number of engineers realized that plastic deformations could and should be allowed in structures designed to withstand single, intense dynamic loads such as nuclear or conventional weapons blast loads, non-penetrating impact loads or seismic loads. Accordingly, analysis methods for dynamic elastic-plastic structural design were developed, and reported in several handbooks and textbooks (U.S. Army Corps of Engineers[2,3], Norris[4], ASCE[5], Biggs[6]). In most of these references, lumped or distributed parameter structures are treated as equivalent SDOF systems, because of the mathematical complexity of treating elastic-plastic deformations of the multi-degree-of-freedom systems. In essentially all later dynamic design manuals (U.S. Dept. of Army, Navy, Air Force[7], Crawford[8], U.S. Air Force Systems Command[9], U.S. Army Corps of Engineers[10], U.S. Army[11], Dobbs[12,13,14,15,16,17]), the equivalent SDOF methods developed and reported in References 2 through 6 are also employed.

Biggs[6] covers many more topics than dynamics of the SDOF approximation for elastic-plastic structures. But, his presentation of this topic is very clear, and it was used as the basis for the computer program reported in this paper. Other references are needed to help predict dynamic loads for the structures. For airblast and impact loads, References 18 through 21 (Baker[18,19,21], Kinney[20]) are useful sources.

This paper describes the approximate SDOF method for calculating dynamics of complex structures and our computer program which automates the method. Examples and conclusions complete the paper.

SDOF REPRESENTATION OF STRUCTURES

Real, distributed structures respond in a complex manner to dynamic loads (Timoshenko[1], Biggs[6]). Many normal modes are excited if the structure responds elastically, and mode shapes and periods must change if loads are intense enough to cause localized plastic deformations. But, for many types of real structures, the first or fundamental response mode dominates the response. This is particularly true if we wish to predict or observe dynamic **displacements** of the structure. Higher mode contributions to dynamic strains or stresses are often more significant so that single mode (SDOF) representation for these parameters is less accurate than for displacements.

The heart of the SDOF representation is the calculation of an elastic-plastic, SDOF system which is **dynamically** equivalent to the given multi-degrees-of-freedom structure. The actual structure has a total mass m_T, a fundamental elastic vibration period T, distributed stiffness parameters, and some maximum resisting force R_m which corresponds to fully plastic deformation. Dynamic loading F(t) is characterized by some (dimensionless) time function multiplied by a peak total force F_t. The fundamental vibration period of the real system can be used to define a spring constant k through the definition (Timoshenko[1]):

$$T = 2\pi\sqrt{m_T/k} \tag{1}$$

In the dynamically equivalent system, the actual vibration period is preserved, and **equivalent** values are chosen for peak force F_e, mass m_e, and maximum resistance R_{me} so that the deflection of some significant point on the real structure is the **same** as for the equivalent structure. The **ratios** of equivalent system parameters to corresponding real system parameters are termed **transformation factors**, e.g.

$$K_M = m_e/m_T \qquad K_L = F_e/F_t \qquad K_R = R_{me}/R_m \tag{2,3,4}$$

Methods for calculating these factors for beams and flat plates with some boundary conditions are described in detail in Reference 6, and are summarized in several tables in that reference. Additional transformation factors also appear in References 7, 10, and 16 for other boundary conditions.

In Biggs' method of development of equivalent SDOF systems, the K_L and K_R factors must be identical, and they serve to define an equivalent spring constant k_e, via

$$K_R = K_L = k_e/k \tag{5}$$

Finally, requiring the same vibration period for the two systems allows replacement of the load factor K_L and mass factor K_M with a single load-mass factor K_{LM} defined as

$$K_{LM} = K_M/K_L \tag{6}$$

As increasing loads are applied to structures such as beams or plates, they can proceed through several stages of deformation. At first, they deform elastically. Then, plastic hinges can form at one or more locations. For some boundary conditions, the structure becomes fully plastic at this stage (e.g. a simply-supported beam).

For others, the second stage of deformation is not fully plastic. In this elastic-plastic stage, plastic hinges form at maximum bending moment locations, but the remainder of the structure still deforms elastically (e.g. a clamped-clamped beam). The third and final stage of deformation for these boundary conditions occurs when one or more interior hinges form, and the structure becomes fully plastic.

A third and still more complex situation occurs for structures such as rectangular plates with asymmetric, clamped boundary conditions. Then, there can be four stages of deformation: 1) elastic, 2) first elastic-plastic, 3) second elastic-plastic, and 4) fully plastic.

Resistance-deflections for these three types of structures are shown graphically in Figure 1. In the SDOF approximations employed in References 6 through 12, all of these functions are converted to equivalent bilinear functions, by calculating an equivalent elastic deflection $(Y_{el})_E$ which preserves total strain energy up to the end of the final elastic-plastic deformation. From geometry in Figure 1, it can be easily shown that (Biggs[6],

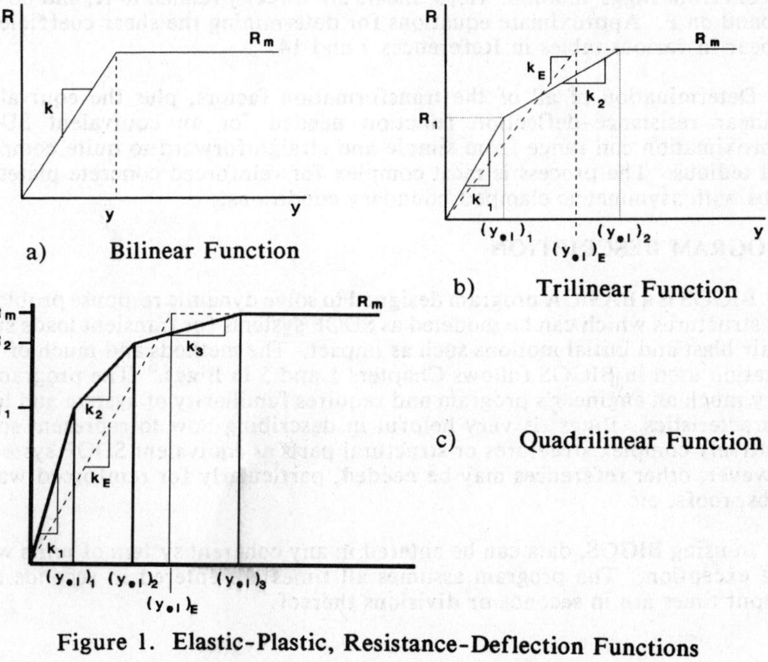

Figure 1. Elastic-Plastic, Resistance-Deflection Functions

U.S. Army, Navy, Air Force[7], Crawford[8], Air Force Systems Commands[9], U.S. Army Corps of Engineers[10], U.S. Army[11], Dobbs[12,13,14,15,16,17]), for the trilinear function,

$$(Y_{el})_E = (Y_{el})_1 + (Y_{el})_2\,(1 - R_1/R_m) \tag{7}$$

and for the quadrilinear function

$$(Y_{el})_E = (Y_{el})_1(R_2/R_m) + (Y_{el})_2\,(1 - R_1/R_m) + (Y_{el})_3\,(1 - R_2/R_m) \tag{8}$$

Once an equivalent bilinear function is determined, an equivalent elastic spring constant can easily be calculated from the equation

$$k_E = R_m/(Y_{el})_E \tag{9}$$

Transformation factors K_{LM}, formulas for maximum resistance R_m, and formulas for spring constants k for the various stages of deformation appear in tables in References 6 through 12 for beams and rectangular plates with a variety of boundary conditions, circular plates, and a few other structural elements.

Biggs[6] notes that dynamic reactions of real structural elements do not equal the spring force in the equivalent SDOF system. He develops an approximate method of determining dynamic reactions based on a distribution of inertia forces identical to the static deformed shape of the element for the specified load distribution. This method results in a dynamic reaction or shear force which is a function of time-varying resistance R and force F. The approximate coefficients appear in the tables in Reference 6 and others.

The method for approximating shear reactions in References 7 and 14 differs from Biggs' method. Here, shears are directly related to R, and do not depend on F. Approximate equations for determining the shear coefficients appear in various tables in References 7 and 14.

Determination of all of the transformation factors, plus the equivalent bilinear resistance-deflection function needed for an equivalent SDOF approximation can range from simple and straightforward to quite complex and tedious. The process is most complex for reinforced concrete plates or slabs with asymmetric clamped boundary conditions.

PROGRAM DESCRIPTION

BIGGS is a BASICA program designed to solve dynamic response problems for structures which can be modeled as SDOF systems for transient loads such as air blast and initial motions such as impact. The methods and much of the notation used in BIGGS follows Chapters 1 and 5 in Biggs.[6] The program is very much an engineer's program and requires familiarity of system and load characteristics. Biggs[6] is very helpful in describing how to represent some relatively complex structures or structural parts as equivalent SDOF systems; however, other references may be needed, particularly for reinforced walls, slabs, roofs, etc.

In using BIGGS, data can be entered in any coherent system of units with one exception. The program assumes all times are entered in seconds and output times are in seconds or divisions thereof.

The program will prompt the user for all of the data needed to define an equivalent elastic-plastic SDOF dynamic system and the dynamic loading. Entries for the program are as follows:

1. **Problem Heading.** This can be up to 50 characters. Characters which are commands in BASIC, such as quote marks, comma, colons, or semi-colons.

2. **Mass.** The total mass of the system is entered in the chosen system of units. The total mass of the rebound phase is usually the same as the mass for the positive response phase.

3. **Load-Mass Factors.** Tables 5.1 to 5.5 in Biggs[6] give values for this factor for elastic phases and elastic-plastic phases. This program has the capability for a trilinear load-displacement curve. For a bilinear response analysis, K_{LM} (elastic) is used for the first entry and K_{LM} (plastic) is used for the second and third entry. For an SDOF system which is exact, and not an approximation, all K_{LM} values equal one.

4. **Elastic spring constants.** Elastic spring constants for many beam and plate systems can be found in Biggs.[6] Positive and rebound values can be different for certain systems.

5. **Elastic-plastic spring constant.** The program allows for one change in stiffness with deflection. If an **equivalent** elastic spring constant has been entered in Step 4, then the elastic-plastic spring constant is zero.

6. **Yield loads.** A trilinear response function can be entered. R_1 is the load (force) to first yield, while R_m is the maximum load. Different rebound values are allowed. If an **equivalent** elastic limit has been calculated, it should be entered for both R_1 and R_m.

7. **Viscous damping.** This number is a fraction of the critical damping of the system. (Critical damping is the amount of damping which would eliminate vibration of a system entirely.)

8. **Force-time pairs.** The program accepts any piecewise linear force-time variation. The number of force-time pairs to describe the loading must be known before entering the pairs.

9. **Initial velocity and displacement.** The system can either be at rest with zero initial displacement and velocity or can have one or the other non-zero. For example, an impacted system could be given an initial impulsive velocity and zero displacement. For blast-loaded systems, both are usually set to zero.

10. **Shear constants.** The terms "Shear A" and "Shear B" refer to the dynamic reactions of the short side A and long side B for rectangular plates. Coefficients for F and R for both Shear A and Shear B are found in Tables 5.4 and 5.5 in Biggs.[6] For beams, refer to Tables 5.1 through 5.3 in Biggs,[6] using coefficients for Shear A only and making Shear B coefficients zero. For an exact SDOF system there is no shear and all shear coefficients are zero.

11. **Integration time step.** This step must be chosen carefully. Both the vibration period and duration of any spikes in the loading time history must be considered. A time step of 1/10 of the shorter of these two times should be adequate.

12. **Number of response cycles.** A choice of 1.5 to 2 cycles is usually reasonable. The minimum number of cycles should be chosen by the user taking into consideration whether the load "spikes" early (small number of cycles) or whether it is more of a quasi-static load with a long rise time (large number of cycles). More cycles may be needed if the system goes plastic. If this is apparent after a first run and the system has not reached maximum deflection, rerun the program with a greater number of cycles.

The program yields both numeric and graphic outputs of SDOF dynamic response which appear on the screen and can be printed out. Input data also appears on the screen and can be printed out. (A help file for the program can be displayed on the screen or printed out.)

EXAMPLES

Problem 1

A ten-foot, w 6 x 16 steel beam supports a three-foot wide panel (Figure 2) which is blast-loaded by a normal blast wave. The beam boundary conditions are clamped-clamped. Compute beam elastic-plastic response as an equivalent SDOF system if the blast pulse is triangular, with an initial peak overpressure of 30 psi and a duration of 5 msec.

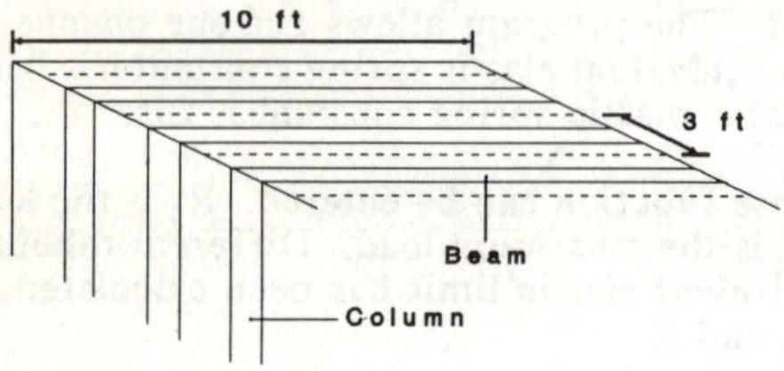

Figure 2.

- Moment of inertia of a section about the X-X axis, $I_x = 32.1$ in^4
- Plastic section modulus with respect to the major axis (X-X), $Z_x = 11.7$ in^3
- Nominal weight per foot, w = 16 lbs/ft.

The panel weights 5 lbs/ft. Assume the beam is made of A36 steel, with a static yield stress, σ_{ys}, of 36,000 psi and a dynamic yield increase factor of 1.25.

Set-Up Table 5.2 in Biggs[6] gives transformation factors for a beam with fixed ends. For this system, the load is a distributed load over the length of the beam. The moment is the same at the supports as it is at the midspan because the beam is assumed to be of uniform construction. For a steel member in pure bending, the ultimate moment capacity is given by:

$$M_p = \sigma_{yd} Z \tag{10}$$

where σ_{yd} is the dynamic yield strength and Z is the plastic section modulus. All dynamic reactions are very close in value and the plastic strain range value will be used for this problem. Mass of the beam can be found from:

$$m = (w_{beam} \cdot L/G) + (w_{panel} \cdot L/G) \tag{11}$$

where w is the weight per unit length of a member, L is length of a member, and G is the acceleration of gravity. Blast loads must be multiplied by the panel surface area, $A_p = B \times L$, where B is the width of the panel and L is the length. The elastic yield point can be found from

$$Y_E = R_m/k_E \tag{12}$$

where R_m is the maximum resistance and k_E is the equivalent effective spring constant. The ductility ratio, μ, is given by

$$\mu = Y_{max}/Y_E \tag{13}$$

where Y_{max} is the maximum deflection of the system under the applied load. A ductility ratio of less than one means the system remains elastic and a ratio of one or greater means the system goes plastic.

Program Variables In this problem, all rebound values are equal to the values for the positive response phase. It is assumed there is no damping.

Solution When BIGGS is run, it gives a maximum displacement of $Y_m = 1.341$ in. This is greater than the elastic yield of $Y_E = 0.411$ in. Thus, the system goes plastic with a ductility ratio of $\mu = 3.26$.

Problem 2

In pressure testing a reinforced concrete containment structure, there was concern that a sizeable section of the structure might reach the earth-covered test bunker if the containment failed. Structural response of the test bunker roof to the impact of a large fragment from the containment structure must be determined. The worst-case fragment is assumed as a disc-shaped piece of concrete with an aspect ratio of 10:1 (Figure 3). The bunker is a reinforced concrete structure. The rebar schedule is shown in Figure 4.

Properties:
Weight, $W_F = 6934$ lbs
Mass, $m_F = 17.96$ lb sec^2/in

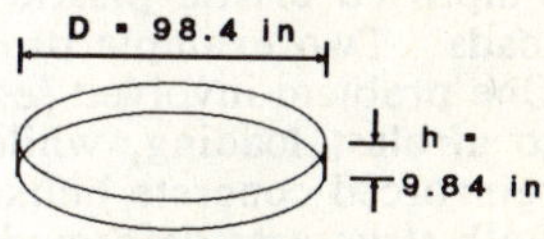

Figure 3.

Impact Conditions:
Angle, $\theta = 45°$
Velocity, $V_F = 2{,}990$ in/sec

Aspect Ratio of Bunker Roof:
Short Side A = 10 ft
Long Side B = 15 ft
A/B = 0.67
Earth Cover = 6 ft

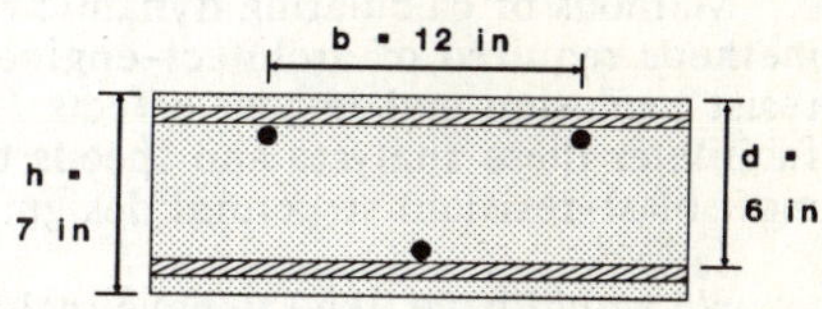

Figure 4.

Properties:
Dynamic yield strength of steel, $\sigma_{yd} = 44{,}000$ psi
Dynamic concrete compressive strength, $\sigma_{cd} = 7{,}500$ psi

Set-Up Table 5.5 of Biggs[6] gives the transformation factors for a two-way slab with four sides fixed. This table is for a uniform load over the slab. Due to the immense size of the fragment relative to the roof, impact will be considered a uniform load. The plastic moment for reinforced concrete and the moment of inertia of the roof slab can be calculated using methods in Reference 7. Care must be taken with units when using the equations from this reference since both plastic moment and moment of inertia for a slab are given per unit width. The mass of the system includes both the mass of the roof and the mass of the earth cover. However, only the strength properties of the concrete will be considered in system response since the earth cover contributes relatively little strength to the total.

For this problem, the impact of the fragment will be converted to an initial velocity of the roof. The initial velocity can be found from the conservation of momentum theory for this system:

$$m_{frag} V_{frag} = (m_{frag} + m_{concrete} + m_{earth\ cover}) V_{initial} \qquad (14)$$

where m is mass and V is velocity for the respective elements. A force-time history consisting of a very small force is needed only for program stability. The equivalent elastic yield can be found from:

$$(Y_{el})_E = (Y_{el})_1 + (Y_{el})_2(1 - R_1/R_m) \qquad (15)$$

The ductility ratios are found in the same way as in Problem 1.

Program Variables All rebound values are equal to the values for the positive response phase. Damping is zero. A nominal, very small, triangular force pulse is applied for stability. Initial displacement is zero, but initial velocity is large, as calculated from eq. (14).

Solution The BIGGS run gives a maximum displacement of Y_m = 4.932 in. Elastic yield displacement is Y_E = 0.1446 in, so ductility ratio is μ = 34.1, and there is large permanent deformation.

CONCLUSIONS

The paper describes the basis for and usage of a computer program for simplified elastic-plastic dynamic structural response to blast and impact loads. Two example problems were presented to illustrate program usage. One problem involved response of a section of structural steel roof structure to airblast loading, while the other illustrated response of the roof of a reinforced concrete bunker to impact from a large piece of concrete debris. Both structures deformed plastically, but neither collapsed.

Methods of calculating dynamic response in this paper are typical of the methods required of architect-engineers in dynamic design of structures to resist explosion and impact effects. Use of the BIGGS computer program facilitates these analyses and speeds the iterative design process required in most blast-resistant structural design.

We would have liked to have included printouts of typical program input and output listings and graphical output. But, limitations on paper length preclude this.

REFERENCES

1. Timoshenko, S.P., Vibration Problems in Engineering, Second Edition, D. Van Nostrand Co., New York, 1937.

2. "Design of Structures to Resist the Effects of Atomic Weapons. Principles of Dynamic Analysis and Design," U.S. Army Corps of Engineers, EM 1110-345-415, (AD 419 292) March 15, 1957.

3. "Design of Structures to Resist the Effects of Atomic Weapons. Structural Elements Subjected to Dynamic Loads," U.S. Army Corps of Engineers, EM 1110-345-416, U.S. Army Corps of Engineers, (AD 419 935) March 15, 1957.

4. Norris, C.H., et al., Structural Design for Dynamic Loads, McGraw-Hill Book Co., New York, 1959.

5. "Design of Structures to Resist Nuclear Weapons Effects, ASCE Manual of Engineering Practice, No. 42, American Society of Civil Engineers, New York, 1964.

6. Biggs, J.M., Introduction to Structural Dynamics, McGraw-Hill Book Co., New York, 1964.

7. "Structures to Resist the Effects of Accidental Explosions," Departments of the Army, the Navy, and the Air Force, Department of the Army Technical Manual TM5-1300, Department of the Navy Publication NAVFAC P-397, Department of the Air Force Manual AFM 88-22.

8. Crawford, R.E., C. J. Higgins, and E.H. Bultmann, "The Air Force Manual for Design and Analysis of Hardened Structures," AFWL TR 74-102, Air Force Weapons Laboratory, Kirtland AFB, New Mexico, October 1974.

9. "Effects of Airblast, Cratering, Ground Shock and Radiation on Hardened Structures," Air Force Systems Commands, AFSC Manual 500-8, January 1976.

10. Suppressive Shields, Structural Design and Analysis Handbook, U.S. Army Corps of Engineers, Huntsville Division, HNDM-1110-1-2 (AD A049 017), November 1977.

11. "Fundamentals of Protective Design for Conventional Weapons," U.S. Army, TM5-855-1, November 3, 1986.

12. Dobbs, N. and J.P. Caltigirone, Structures to Resist the Effects of Accidental Explosions. Volume I - Introduction, Special Publication ARLCD-SP-84001, December 1987.

13. Dobbs, N., et al., Structures to Resist the Effects of Accidental Explosions. Volume Ii - Blast, Fragment, and Shock Loads, Special Publication ARLCD-SP-84001, December 1986.

14. Dobbs, N., et al., Structures to Resist the Effects of Accidental Explosions. Volume III - Principles of Dynamic Analysis, Special Publication ARLCD-SP-84001, June 1984.

15. Dobbs, N., et al., Structures to Resist the Effects of Accidental Explosions. Volume IV - Reinforced Concrete Design, Special Publication ARLCD-SP-84001, April 1987.

16. Dobbs, N., et al., Structures to Resist the Effects of Accidental Explosions. Volume V - Structural Steel Design, Special Publication ARLCD-SP-84001, May 1987.

17. Dobbs, N. et al., Structures to Resist the Effects of Accidental Explosions. Volume VI - Special Considerations in Explosive Facility Design, Special Publication ARLCD-SP-84001, April 1985.

18. Baker, W.E., Explosions in Air, Second Printing, Wilfred Baker Engineering, San Antonio, Texas, 1983.

19. Baker, W.E., et al., Explosion Hazards and Evaluation, Elsevier Science Publishers, Amsterdam, 1983.

20. Kinney, G.F., Explosive Shocks in Air, Second Edition, Springer-Verlag, Berlin, 1985.

21. Baker, W.E., "A Manual for the Prediction of Blast and Fragment Loadings on Structures," DOE/TIC-11268, U.S. Department of Energy.

22. Manual of Steel Construction, 8th Edition, American Institute of Steel Construction Inc., Chicago, 1980.

The Double-Notch Shear Test - Analysis and Development for Material Testing at Very High Strain Rates

D. Ruiz, J. Harding and C. Ruiz
Department of Engineering Science, Parks Road, Oxford, OX1 3PJ, UK

1. Introduction

There is a growing interest in material behaviour at strain rates greater than $10^4 sec^{-1}$, for instance in the design of aero-engine turbine blades and of shaped charges. It is necessary, therefore, to develop material testing techniques that give well-defined information on mechanical behaviour in this very high strain-rate regime. A number of techniques are available, including the expanding ring test [1], a miniaturised compression Hopkinson bar technique using direct impact [2] and the double-notch shear test [3] which has been described by Nicholas [4] as "one of the most promising for future studies in dynamic plasticity". However, although it is a good test for determining the flow stress at shear strain rates of $10^4 sec^{-1}$ and above, the design of specimen used makes an accurate determination of strain extremely difficult while, in the later stages of the test the deformation mode involves rotation as well as shear. If this technique is to be used, therefore, it is necessary to examine in detail the progressive deformation of the specimen during the impact process in an attempt to assess how far the data obtained is a reliable measure of the specimen material response.

As a first step, the initial stages of the specimen deformation have been studied using the finite element package ABAQUS and assuming a perfectly elastic response. Numerical results so obtained are compared with and verified against the results of a dynamic photoelastic experiment. Subsequently it is hoped to study the later stages of the test, incorporating plasticity and rate dependency into the constitutive relation used in the numerical analysis and using high-speed photography to provide experimental backup. The present paper, however, is concerned only with the elastic stage of the test and gives results for the photoelastic experiment and the related numerical analysis.

2. The Double-Notch Shear Test

The standard split Hopkinson bar configuration is used except that the output bar is replaced by a tube. The double-notch shear specimen is shown in fig.1. By concentrating the deformation in the very narrow shear zone between the notches very high strain rates may be attained at moderate impact velocities. The stress and velocity histories at the interfaces between the specimen and both the bar and the tube can be calculated from the strain gauge signals obtained at two positions on the input bar and one position on the output tube. These stresses and velocities can then be used to derive, by the standard Hopkinson bar analysis, the nominal stress and strain histories for the specimen. It is assumed that the deformation is concentrated exclusively within the two "shear zones" (see fig.1) and that the applied load is shared equally between the two. For the specimen dimensions given, therefore, an applied load of 1kN corresponds to an average shear stress of 97.7MPa and a displacement between the ends of the input bar and output tube of 1mm corresponds to an average shear strain of 119%.

A typical stress-strain curve, calculated in this way, for a Ti6Al4V alloy specimen tested to failure at a mean strain rate of ~$2.6 \times 10^4 sec^{-1}$ is shown in fig.2. The apparent strain at yield is ~10%, corresponding to a relative displacement between the input bar and output tube of only 0.084mm and implying a dynamic elastic shear modulus of ~10GPa compared to the usual quasistatic value of ~45GPa. The principal source of inaccuracy is the assumption that is made in the standard Hopkinson bar analysis that stress and strain are uniform within a defined region and all strains are zero elsewhere. In reality, there is a stress concentration factor, the maximum stress being at the end of the shear zone nearest the output end of the specimen. Also, the input and output faces of the specimen are compressed and since the strain calculated using the Hopkinson bar analysis is based on the relative movement of the bar ends, the latter two factors result in an overestimate of the average shear in the shear zone. In addition, especially in the later stages of the test, bending of the specimen becomes significant. By studying the elastic deformation of the double-notch shear specimen using both numerical and photoelastic techniques it is hoped to improve the experimental accuracy in the elastic part of the test.

3. Experimental Methods

3.1 Dynamic Photoelastic Experiment

The impact assembly for the dynamic photoelasticity experiment is a scale model of that used for material testing. The epoxy loading bars were machined from a rough bar and tube cast from Araldite CT200, using a hardener and silica flour. Micro-Measurements EA-series strain gauges were attached to these with cyano-acrylate adhesive. The specimen was machined from an epoxy sheet 4.7mm thick, with all other dimensions four times those of the material test specimen (fig.1). The projectile, a swinging rod shown in fig.3a, was machined from a rough epoxy bar similar to that for the input bar. The lengths of the projectile, input bar and output tube were 600mm, 560mm and 500mm respectively. Fig.3a shows the photoelastic equipment, which consists of a circular polariscope, a delay timer, a spark unit and a 35mm camera with a 50mm lens, and fig.3b shows a block diagram of the experimental set up.

Output signals from the strain gauge stations were fed through a bridge box, via Fylde 351A amplifiers to be recorded digitally on Data Lab 902 transient recorders before transfer to an IBM PC AT computer for storage and subsequent analysis. The signal from the first input bar strain gauge station was used to trigger the delay timer (Bowen 10 channel sequencer) for the spark unit (Pulse, argon jet 100N). Tests were performed in darkness, with the camera shutter open, the film being illuminated for the duration of the flash which was of the order of nanoseconds. In this way, it was possible to effectively freeze the photoelastic stress fringes to within a few microseconds of any desired time after the initial rise of the input bar strain gauge signal. Conveniently, an electrical interference spike coinciding with the spark appeared on the recorded strain gauge signals. Typical strain gauge signals are shown in fig.4. Blue fringes from a typical dynamic photograph have been sketched in fig.5.

By measuring the wave speed between the two gauge stations on the input bar, the dynamic Young's modulus of the epoxy was ascertained using the relation $c = (E/\rho)^{0.5}$. The value obtained, 4GPa, was used in the numerical calculation.

3.2 Numerical Analysis : ABAQUS

ABAQUS is a commercial Lagrangian finite element code capable of both linear and non-linear analyses, designed as a flexible tool for numerical modelling of structural response. In most problems, the challenge is to produce a convergent solution at minimal cost. Convergence depends on the choice of mesh, the complexity of the material model and on the incrementation scheme. The latter is not a major problem since the ABAQUS code has a facility for automatic control over incrementation. However, this facility can lead to a false sense of security and it is essential that numerical results are verified by some independent means. Since it is the initial elastic response of the test specimen which, in the first instance, is being studied here, a simple linear elastic material model was implemented and verification of the mesh and history definition was obtained by comparison with the photoelastic experiment described above.

Initially, a very simple 7 element model was chosen, as shown in fig.6. An eight-noded, three-dimensional continuum element was used (denoted C3D8 in [5]). Results from this model suggested that there is a state of plane stress in the specimen. Subsequently, a 120 element mesh was chosen, see fig.7, using ABAQUS element type CPS8R - a continuum, plane-stress, eight-noded element with reduced integration. The Young's modulus and density of the specimen material were taken to be 4GPa and 1240 kg/m^3 respectively. For the boundary condition it was assumed that the output face was prevented from moving parallel to the direction of impact. Half the specimen was modelled, the centre-line parallel to the direction of impact being a line of symmetry. A uniformly distributed ramp load was applied to the input face. A cord was drawn between the origin and the specimen stress at the time of the photograph in photoelastic test PH-C-18: the gradient of this cord was used to define the ramp loading = $6.3GN/m^2$ per second.

Numerical data were obtained using ABAQUS for 1) the Tresca equivalent stress at all nodes along the notch centre-line, 2) the total displacement and velocity at all nodes bounding the shear zone, 3) the total displacement and velocity at all nodes at the specimen input face in the loading direction, the output face being fixed, and 4) the reaction force at all output face nodes.

4. Results

4.1 Dynamic photoelastic experiment

The raw data was analysed using the method of characteristics. It was ascertained that full equilibrium is reached in the specimen after about 14μs. Therefore, we were justified in taking the rise time of the force at the output bar strain gauge station and applying a similar ramp loading to the input face of the model in ABAQUS, ie. approximately $6.3GN/m^2\ sec^{-1}$. From the input and output face data we can also calculate strain and strain rate histories for the specimen. A typical calculated stress-strain curve is given in fig.8. It is clear that the Young's modulus is much lower than the true dynamic Young's modulus. Using the standard Hopkinson bar analysis, the average strain and strain rate for the specimen have been calculated, ie. 2.5% and $76sec^{-1}$.

The photograph from test PH-C-18 is studied here. This photograph was taken 516μs after the initial loading of the specimen. The fringes sketched from photograph PH-C-18 in fig.5 are proportional to the Tresca equivalent stress and show that the shear stress in the nominal "shear zone" is relatively constant. This figure also shows that the Tresca equivalent stress decreases away from the "shear zone" and that the notch corners are significant secondary sources which will have an effect on behaviour as the specimen approaches yield. At the upper right and lower left corners the fringe order is about twice that between the notches, which is itself more than twice that in the specimen away from the shear zone. By estimating the fringe order along the centre-line of the notches, A--B, an approximation to the magnitude of the maximum shear stress along this line may be found but it must be emphasised that there is no indication of the direction of the shear stress in general. The fringe order has accordingly been estimated and plotted in fig.9. Clearly, the key points on this figure are the integral values of fringe order, where a fringe clearly crosses the notch centre-line. A similar plot has been produced using a post-processing option in ABAQUS.

4.2 Numerical Analysis : ABAQUS

a) Tresca equivalent stress.
The specimen stress reached at the time of the photograph in photoelastic test PH-C-18 is 6.3MN/m^2. In the numerical analysis, this corresponds to the stress at increment 26 of the loading (20 us increments). In fig.10 the Tresca equivalent stress extrapolated to the notch centre-line nodes is plotted for the specific case of increment 26. The fringe values from fig.5, scaled up by a factor of 1.33, are also sketched in this figure. It is not practical to do a dynamic calibration of the photoelastic fringes but the scaling factor used can be verified by checking the agreement between the photoelastic test and the numerical analysis at other stages of loading.

b) Specimen strain and strain rate.
The mean displacement of the nodes at the upper boundary of the shear zone relative to those along the lower boundary is 0.012mm at increment 26;

=> mean shear strain across shear zone = 0.37%

The mean displacement of the input face is 0.065mm at increment 26;

=> average shear strain across specimen =
0.065mm/(0.84mmx4) x 100% = 1.96%

The ratio of these two values is 5.3. The corresponding strain rates calculated from relative mean velocities are 18.8sec^{-1} and 99.4sec^{-1} with the same ratio of 5.3.

c) Output face reaction force.
Approximately 340μs after initial loading of the specimen, tensile reaction forces are developed at the extreme end of the output face (top right in fig.5). However, the maximum tensile stress developed is less than 5% of the maximum compressive stress at the stage of the test under consideration.

5. Discussion

a) The matched curves of Tresca equivalent stress for the notch centre-line compare very well. At the ends of the line (0% and 100% in fig.10) the values of stress calculated using ABAQUS are about 1/3 of those observed in the photoelastic experiment. This is no cause for concern as the mesh in the ABAQUS shear zone is only 8 nodes wide and there will be edge effects in the photoelastic specimen. The trends elsewhere in this figure are very consistent.

b) The factor of 5.3 between the mean total displacements and the mean velocities of nodes at the specimen faces and at the shear zone boundaries show that the most significant source of calculation error in the elastic regime is due to the assumption that deformation is concentrated within the shear zone. The shear strain and strain rate calculated using the standard Hopkinson bar analysis for the photoelastic test specimen confirm this although numerical values differ slightly due, it is thought, to the coarseness of the ABAQUS mesh and, principally, to the approximation to linear ramp loading in the numerical model.

c) The observed development of a tensile reaction force at the extreme end of the specimen output face is consistent with bending of the specimen but is not critical in the elastic stage of the test. As far as the numerical model is concerned, this problem can be alleviated by redefining the boundary condition using, for instance, the *SURFACE CONTACT condition in ABAQUS.

6. Conclusions

The simple finite element model used here appears to be quite accurate as it stands. Good qualitative agreement has been obtained between the photoelastic experiment and the numerical model and the principal source of error in the elastic region of the double-notch shear test has been identified as the assumption that all deformation of the specimen is concentrated in the two shear zones. For the photoelastic specimen a calibration factor of 5.3 has been determined.

7. Future work

By observing the photoelastic fringes at the input face of the specimen the assumption of a uniform loading can be improved upon and the linear ramp loading replaced by the experimentally observed curve. Along with refinement of the boundary conditions and the mesh this should give improved agreement in the elastic region.

In the elastic region, as the stress concentration factors decrease, these amended boundary conditions should be adequate. However, it will be necessary to introduce more complex constitutive equations, catering for plasticity, strain rate and temperature effects. When completed this should allow the range of validity of the double-notch shear test data to be determined.

8. Acknowledgements

The authors wish to express their gratitude to P.Hardy for his

assistance in the laboratory and to M.Hogwood of RARDE Fort Halstead and MOD(PE) for financial support.

References

[1] A.E.Carden, P.E.Williams & R.R.Karpp, Int.Conf.metallurgical effects on high strain rate deformation and fabrication, Albuquerque, New Mexico (1980).
[2] D.A.Gorham, Measurement of stress-strain properties of strong metals at very high rates of strain, Inst.Phys.Conf.Ser. no.47 on the mechanical behaviour of materials at high rates of strain, ed.Harding,J., I.O.P. London & Bristol (1979), pp.16-24.
[3] Campbell,J.D. & Ferguson,W.G., The temperature and strain rate dependence of the shear strength of mild steel, Phil.Mag., Vol.24, pp.63-82, 1970
[4] Zukas,J.A., Swift,H.F., Nicholas,T. & Curran,D.R., Impact Dynamics, Chapter 8, Material behavior at high rates of strain, (Ed.Nicholas,T.), p.313, Wiley-Interscience, New York, 1971.
[5] Hibbitt,D., Karlsson,B.I. and Sorensen,P., ABAQUS User's manual, version 4.7, Hibbitt, Karlsson and Sorensen,inc., Rhode Island, 1988.

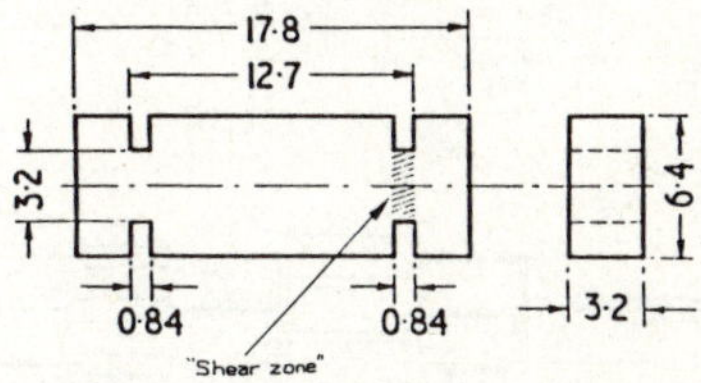

Figure 1 The double-notch shear specimen.

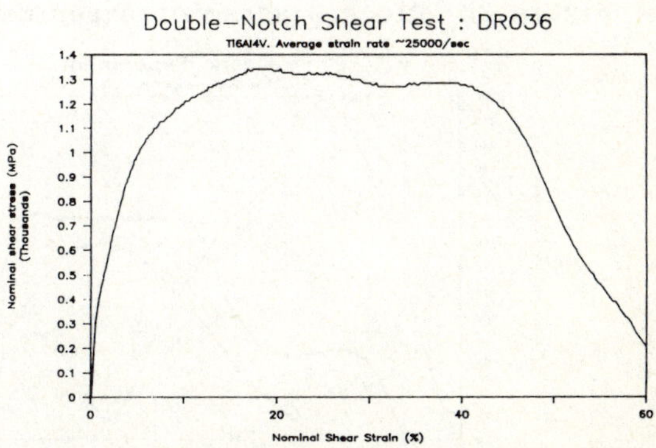

Figure 2 Typical stress-strain curve calculated for double-notch shear test on Ti6Al4V.

Figure 3a Photoelastic equipment: circular polariscope, spark unit, camera and epoxy Hopkinson bars.

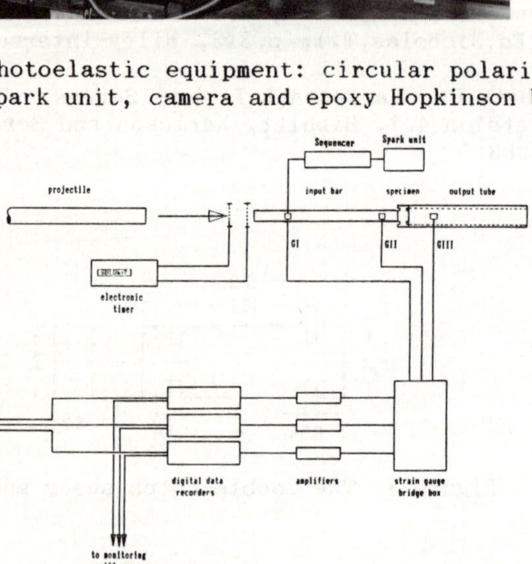

Figure 3b Block diagram of experimental set-up

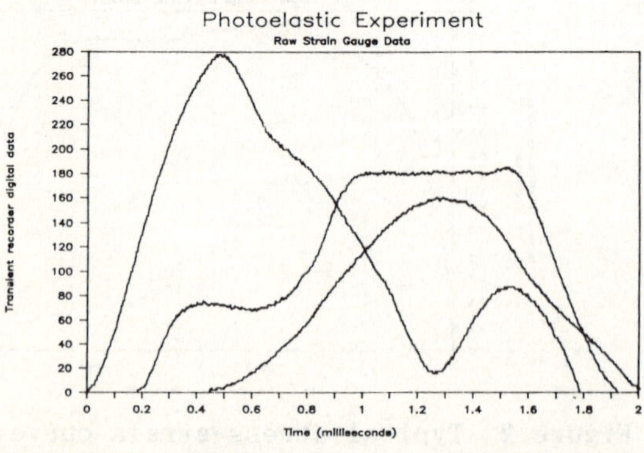

Figure 4 Typical strain gauge signals from photoelastic experiment.

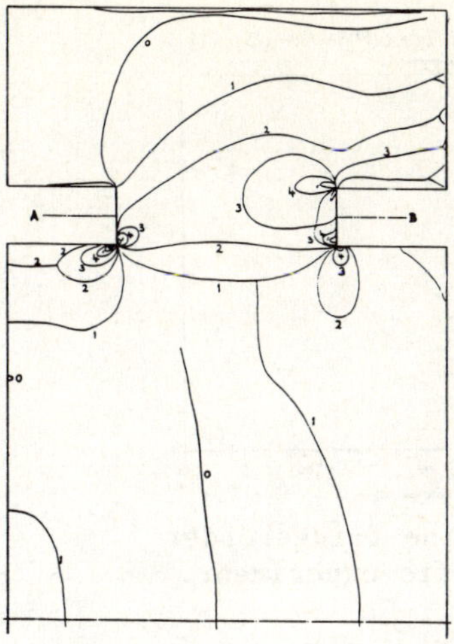

Figure 5 Blue fringes, sketched from photograph of dynamic photoelastic fringes.

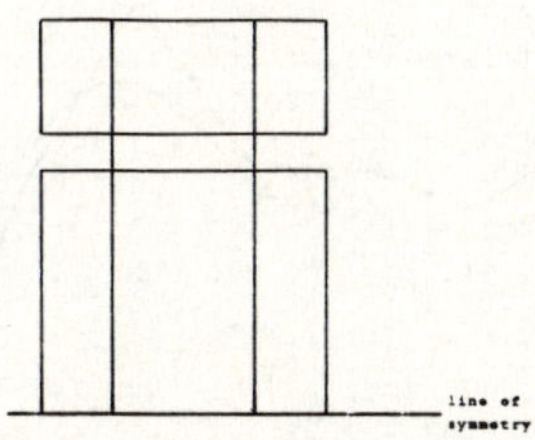

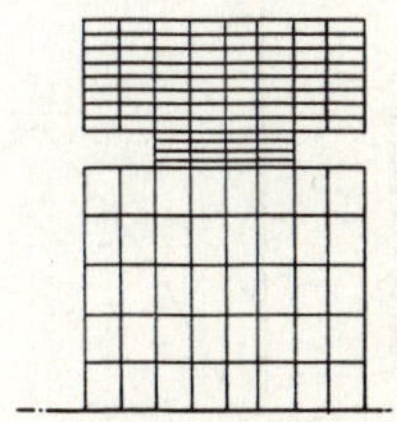

Figure 6 ABAQUS: Simple 7-element mesh.

Figure 7 ABAQUS: 120-element mesh.

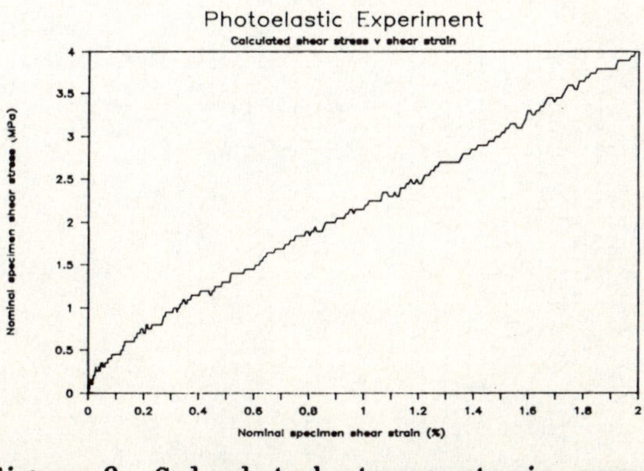

Figure 8 Calculated stress-strain curve for test PH-C-18 (epoxy bars).

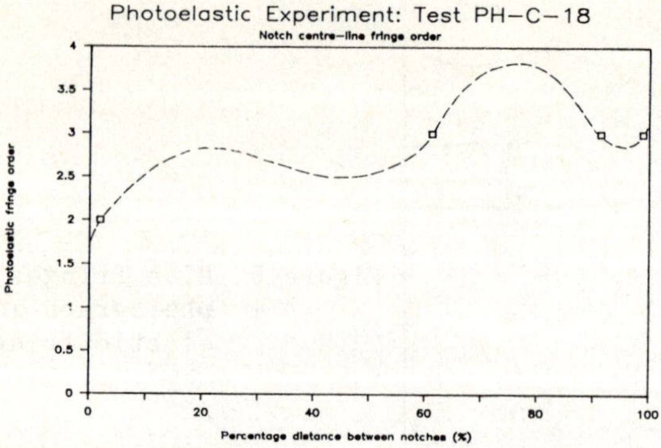

Figure 9 Notch centre-line fringe order from photoelastic experiment.

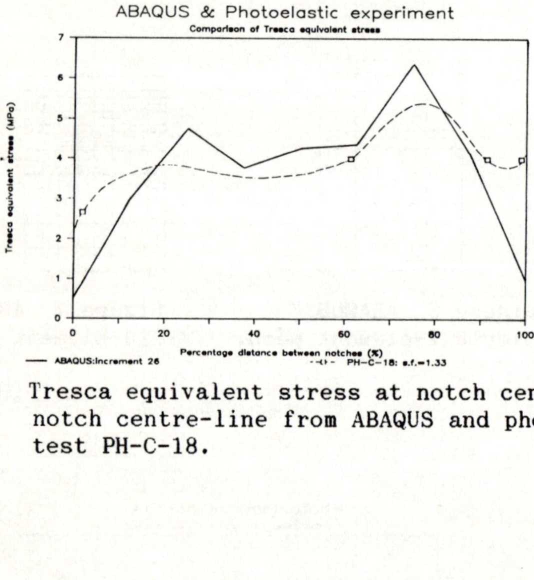

Figure 10 Tresca equivalent stress at notch centre-line notch centre-line from ABAQUS and photoelasti test PH-C-18.

Critical Time Step Estimation for Three-Dimensional Explicit Impact Analysis

R.F. Kulak

Reactor Analysis and Safety Division, Argonne National Laboratory, Argonne, Illinois 60439-4842, USA

ABSTRACT

A method to obtain a bound on the critical time step of impact problems that employ explicit integration and penalty type contact elements is presented. The method uses a Gerschgorin bound obtained from a partially assembled stiffness matrix for the nodes that belong to elements involved in the contact region. A numerical example is presented to illustrate the effectiveness of the method.

INTRODUCTION

The modeling of impact between reactor components must be taken into account in many reactor safety assessments that involve transient loadings. Finite element analysis provides a tractable approach to solving general nonlinear mechanics problems, and the use of adaptive penalty interface elements provide a means of taking into account the specific non-linearities that occur during impact between two bodies. For nonlinear transient problems, an explicit temporal integrator (e.g., central difference) is often used to obtain the temporal evolution of the impact problem. A disadvantage of explicit methods is that they are conditionally stable and, thus, require that the computational time step be smaller than the critical value.

For structural and continuum elements a bound on the critical time step can normally be obtained from the highest eigenvalue of the unconstrained elements. However, the penalty elements have zero mass and their resulting element frequencies are infinity. Thus a bound for the critical time step cannot be obtained from the unconstrained frequency for a penalty element.

Because of this ad hoc trial and error methods are often employed which can lead to overly conservative estimates that increase the computational cost. An approach by Plesha [1] uses the power method (e.g. Isaacson and Keller [2]) to determine the maximum eigenvalue for a finite element mesh.

Here, a method to obtain a bound on the critical time step is presented that uses a partially assembled global stiffness matrix assembled from the element stiffness matrices of the penalty elements and their contiguous neighbor elements. The method has been incorporated into the general-purpose, nonlinear, finite element code NEPTUNE [3]. One example is presented to illustrate the effectiveness of the method. The example is the axial impact of two bars.

GOVERNING EQUATIONS

The semidiscretized equations of motion for contacting bodies can be obtained by applying the principal of virtual power over the bodies (e.g., Kulak [4]) to give

$$m_{iIjJ}\,\ddot{u}_{iI} + f^{int}_{jJ} = f^{ext}_{jJ} \tag{1}$$

where m_{iIjJ} is a lumped (i.e. diagonal) nodal mass matrix, u_{iI} is the displacement, f^{int}_{jJ} is the internal nodal force, which includes contributions from the deformable bodies and the penalty contact elements, and f^{ext}_{jJ} is the external nodal force. The lower case subscripts refer to the global coordinate directions, and the upper case subscripts refer to node numbers.

Equation (1) is explicitly integrated in time using the central difference operator. The acceleration at time t is computed from equation (1) and the velocities and displacements are time advance according to

$$\dot{u}_{iI}\,(t + 0.5\,\Delta t) = \dot{u}_{iI}\,(t - 0.5\,\Delta t) + \Delta t\,\ddot{u}_{iI}\,(t) \tag{2}$$

$$u_{iI}\,(t + \Delta t) = u_{iI}\,(t) + \Delta t\,\dot{u}_{iI}\,(t + 0.5\,\Delta t)$$

The above algorithm is relatively simple. However, it is only conditionally stable, and for undamped linear systems the time step is limited to

$$\Delta t \le \Delta t_{crit} = \frac{2}{\omega_{max}} \tag{3}$$

where ω_{max} is the highest frequency of the discretized system.

Since we are interested in estimating a starting value for the time step, the materials are assumed to be linear, and geometric nonlinearities do not have to be considered. For a linear material, the stiffness form of equation (1) is given by

$$m_{iIjJ}\,\ddot{u}_{iI} + k_{iIjJ}\,u_{iI} = f^{ext}_{jJ} \tag{4}$$

where k_{iIjJ} is the global stiffness matrix.

The associated eigenproblem for equation (4) is

$$m_{iIjJ}\,\phi_{iI} = \lambda\,k_{iIjJ}\,\phi_{iI} \tag{5}$$

where λ is the eigenvalue and ϕ_{iI} is the eigenvector. Note the eigenvalue is the square of the frequency, $\lambda = \omega^2$.

ELEMENT-BASED STABILITY CRITERIA

For problems that do not involve contact, a conservative bound on the maximum eigenvalue of equation (5) can be obtained from the eigenproblem of the unconstrained element (e.g., Hughes [5]), that is,

$$\lambda_{e,max} \le \max_{e}\,(\lambda_e) \tag{6}$$

where λ_e is the eigenvalue for a single unconstrained element; namely

$$k^{e}_{iIjJ}\,\phi^{e}_{iI} = \lambda_e m_{iIjJ}\,\phi^{e}_{iI} \tag{7}$$

Thus, considering only the non-impacting elements, the critical time step is given by

$$\Delta t_{e,crit} = \frac{2}{\lambda^{0.5}_{e,max}} \tag{8}$$

An efficient way of obtaining an estimate of the maximum eigenvalues of an element is to employ the Rayleigh quotient

$$\lambda^{RQ}_{e,max} \geq \frac{\phi_{iI}\ \phi_{jJ}\ k^{e}_{iIjJ}}{\phi_{iI}\ \phi_{jJ}\ m^{e}_{iIjJ}} \tag{9}$$

where ϕ_{iI} is the assumed highest mode eigenvector for the element.

Because the penalty element is modelling the traction between two contacting surfaces, it does not have any mass associated with itself. Thus, the unconstrained element eigenvalues for a zero mass element are infinity and the critical time step for a mesh with contact elements cannot be obtained from equation (7). However, the assembled problem does have finite eigenvalues and can be used to obtain the critical time step.

NODE-BASED STABILITY CRITERIA

At this point, the objective is to obtain an estimate for the critical time step based upon the assembled lumped nodal mass components m_{iIjJ} and the nodal stiffness components k_{iIjJ}. To accomplish this, Gerschgorin's theorem is used to bound the maximum eigenvalue of the discretized contact problem.

Gerschgorin's circle theorem (e.g. Noble [6]) states that each eigenvalue λ_{iI} of the n x n matrix A_{iIjJ} satisfies

$$| A_{iIiI} - \lambda_{iI} | \leq \sum_{\substack{j\neq i\\ J\neq I}} | A_{iIjJ} | \tag{10}$$

So the maximum eigenvalue is given by

$$\lambda_{n,max} \leq \max_{n} (\lambda_{iI}) \tag{11}$$

A comparison between equations (6) and (11) shows that the former is based upon unassembled element information and the latter is based upon assembled nodal information.

By applying Gerschgorin's theorem, the time step Δt_{iI} for each node belonging to a contact element can be expressed as

$$\Delta t_{iI} \leq \frac{2\ (m_{iIiI})^{0.5}}{[K_{iIiI} + \sum_{\substack{j\neq i\\ J\neq I}} | K_{iIjJ} |]^{0.5}} \tag{12}$$

The maximum time step for the nodes, nc, involved in initial impact is determined from

$$\Delta t_{iI,max} = \max_{nc} (\Delta t_{iI}) \tag{13}$$

The critical time step for the problem, which consists of both impacting and non-impacting elements, is determined from

$$\Delta t_{crit} = \min (\Delta t_{e,max} , \Delta t_{iI,max}) \tag{14}$$

A significant point to note is that the row sum in equation (12) only involves the elements that are in contact at the beginning of the impact simulation. Therefore, only the stiffness for these elements is calculated. Furthermore, since only the value of the row sum is needed, they are accumulated in a column matrix, which is referred to as a partially assembled stiffness matrix. The storage requirement for this matrix is significantly less than that of a completely assembled matrix.

NUMERICAL EXAMPLES

A problem is presented to illustrate the use of the combined elemental and nodal stability criteria in application to impact.

The example is the longitudinal impact of elastic bars (Figure 1a). Dimensionless quantities were used. Both bars were assumed to have a length of 10.0 and a cross sectional area of 1.0. The density and stiffness of the bars were varied as explained below. Figure 1b shows the finite element mesh, which consists of 20 axial elements per bar with a contact element between them. The element stiffness of the left bar, k_{LB}, right bar, k_{RB}, and contact element, k_{CE}, were varied as shown in Table 1. The total mass of each element in the left bar, m_{LB}, and right bar, m_{RB}, were also varied. The values of the eigenvalue for each bar element were computed by Rayleigh's quotient, and the maximum eigenvalue, $\lambda^{RQ}_{e,max}$, is tabulated in Table 1. Next, a Rayleigh quotient for the contact element, λ^{RQ}_{ce}, was computed using the contact stiffness and the assembled nodal mass for each node at the ends of the contact element. Note, the mass contributions come only from the bar elements in contact; the contact element does not contribute to the mass. A Gerschgorin bound, $\lambda^{G}_{n,max}$, was computed only for the contact element. Finally, the exact maximum eigenvalue, λ^{E}_{max}, of the assembled problem, equation (5), was obtained with the basic Jacobi solution method. For Case 1, in which the contact element stiffness is the same as the bar element stiffness, the Rayleigh quotient for the contact element underestimates the exact value by 39 percent and thus would lead to a unstable time step; however the Gerschgorin bond overestimates by 21 percent and thus leads to a stable time step estimate. For Case 2, the contact stiffness was assumed to be ten times the bar element stiffness. Here it is seen that the Rayleigh quotient for the contact element only underestimates the exact value by 5 percent, and the Gerschgorin bound gives an overestimation of 5 percent. For the cases considered, the Gerschgorin bound always leads to a stable time step estimation, and the Rayleigh quotient always leads to an unstable time step prediction.

The impact of the two bars was simulated with the NEPTUNE code. The Young's modulus for both bars were taken to be 100, the density was 0.01. The left bar was assumed to have an initial velocity of 0.1 and the right bar was assumed to be at rest. The displacement, velocity, and force signatures during the impact event are shown in Figures 2 to 4.

CONCLUSIONS

An approach has been developed to determine the critical time step for problems involving impact using explicit time-marching integrators. The approach utilizes unassembled element information to obtain critical time steps estimates for elements outside the contact region and assembled nodal information to obtain critical time steps for nodes in the contact region. The approach is independent of element type.

ACKNOWLEDGEMENT

This work was performed under the auspices of the U. S. Department of Energy, Office of Technology Support Programs, under contract W-31-109-Eng-38.

REFERENCES

1. Plesha, M.E. Eigenvalue Estimation for Dynamic Contact Problems, Journal of Engineering Mechanics, ASCE, Vol. 113, No. 3, pp. 457-462, March 1987.

2. Isaacson, E. and Keller, H.B. Analysis of Numerical Methods, John Wiley and Sons, Inc., New York, 1966.

3. Kulak, R.F. and Fiala, C. NEPTUNE: A System of Finite Element Programs for Three-Dimensional Nonlinear Analysis, Nuclear Engineering and Design, Vol. 106, pp. 47-68, 1988.

4. Kulak, R.F. Adaptive Contact Elements for Three-Dimensional Explicit Transient Analysis, Computer Methods in Applied Mechanics and Engineering, Vol. 72, pp. 125-151, 1989.

5. Hughes, T.J.R. The Finite Element Method: Linear Static and Dynamic Finite Element Analysis, Prentice-Hall, Inc., New Jersey, 1987.

6. Noble, B. Applied Linear Algebra, Prentice-Hall, Inc., New Jersey, 1969.

Table 1. Maximum Eigenvalue Estimates

Case	$\lambda^{RQ}_{e,max}$	λ^{RQ}_{CE}	$\lambda^{G}_{n,max}$	λ^{exact}_{max}
1. $k_{LB} = 1$, $k_{RB} = 1$ $k_{CE} = 1$ $m_{LB} = 1$, $m_{RB} = 1$	4.0	4.0	8.0	6.6
2. $k_{LB} = 1$, $k_{RB} = 1$ $k_{CE} = 10$ $m_{LB} =1$, $m_{RB} = 1$	4.0	40.0	44.0	42.6
3. $k_{LB} = 1$, $k_{RB} = 1$ $k_{CE} = 1$ $m_{LB} = 1$, $m_{RB} = 10$	4.0	0.7	8.0	5.3
4. $k_{LB} = 1$, $k_{RB} = 1$ $k_{CE} = 10$ $m_{LB} = 1$, $m_{RB} = 10$	4.0	3.6	44.0	34.4

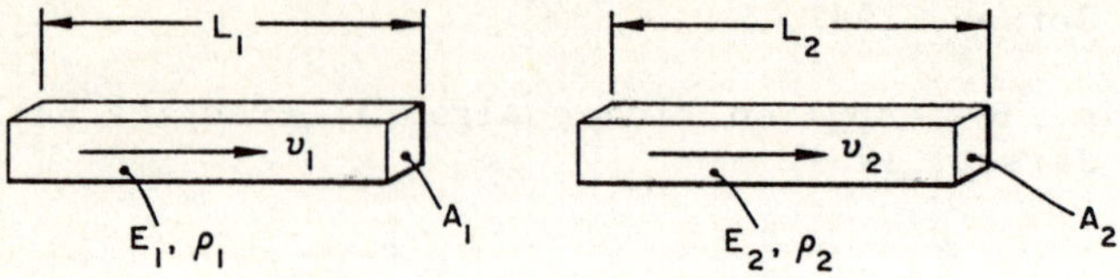

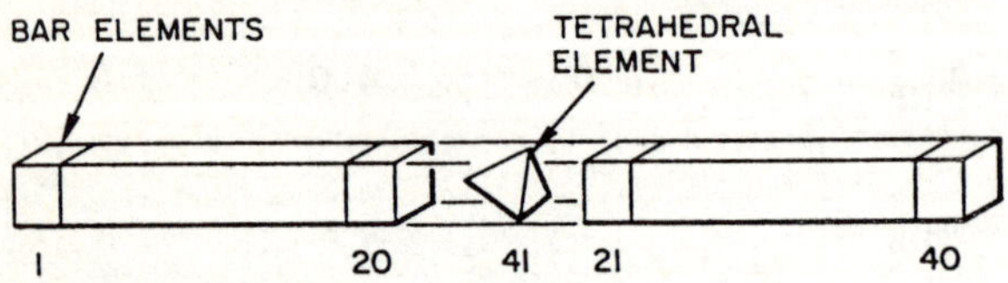

Figure 1. (a) Two impacting bars.
(b) Finite element model of impacting bars.

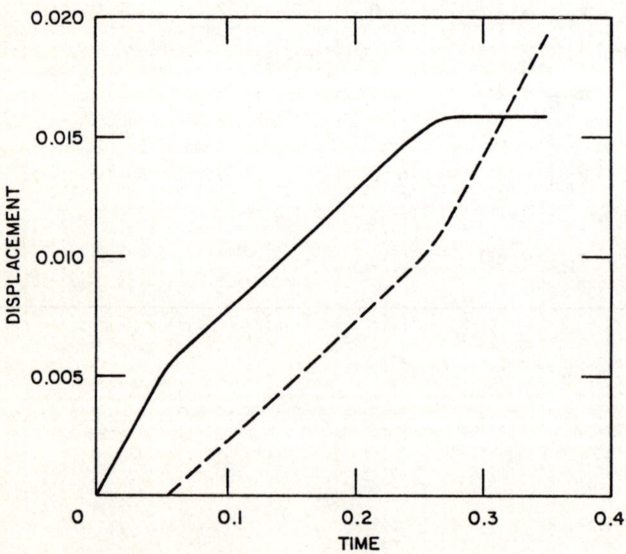

Figure 2. Displacement signature of contact interface.

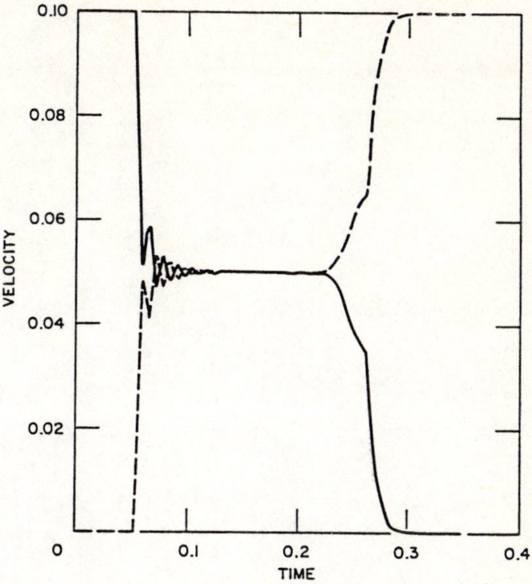

Figure 3. Velocity signature of contact interface.

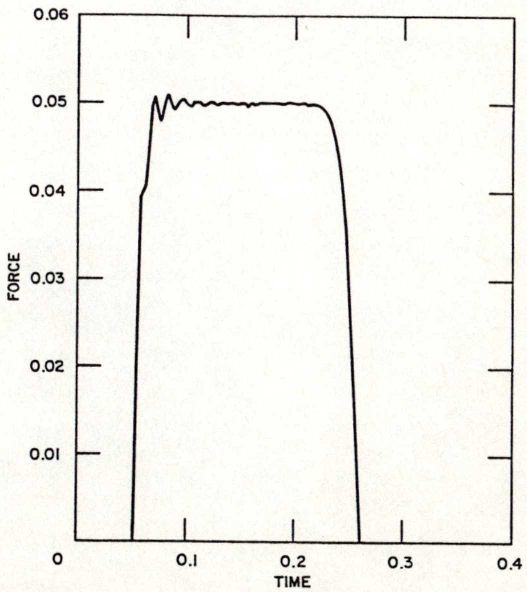

Figure 4. History of contact interface force.

A Comparison between Numerical Techniques and Analytical Models for Impact Problems

M. Elices, R. Cortés, C. Navarro and V. Sánchez-Gálvez

Department of Materials Science, Polytechnic University of Madrid, E.T.S. de Ingenieros de Caminos, Canales y Puertos, 28040 Madrid, Spain

ABSTRACT

In this paper, the problem of a projectile impacting normally a plate is analysed by using the two most common of the currently available techniques, that is, numerical methods and engineering models. The projectile, of blunt cylindrical shape, is made of a hard steel and the target plate is assumed to be of an annealed 2024-0 aluminium. An analysis of this problem is made by considering different impact velocities, for two different thicknesses of the target plate, by using the two-dimensional analytical model of Ravid and Bodner [1]. Some of the situations studied in this way, are also analysed numerically by means of a finite difference computer code developed for the dynamic analysis of two-dimensional and axisymmetric thermo-mechanical problems. The results obtained with the above procedures are then compared with the corresponding experimental results. A good agreement between analysis and experiment is observed. Also, for the particular situations studied, the analytical model of Ravid and Bodner shows some advantage over the numerical techniques, mainly in what deals with the economy of computations.

INTRODUCTION

The great interest which in the last decades has arisen, the prediction of the behaviour of real structures under situations of impact, has led to the development of a number of analytical and numerical techniques concerned with the mechanical response of materials and structures subjected to extremely high strain rates. A field of interest, for example, lies in the prediction of the material response from models which represent in detail the actual micromechanisms of damage which lead to the rupture of the materials in such a situation (Curran, Seaman and Shockey [2]). Such a field of research requires the use of large computer codes, in which the corresponding constitutive relations could be implemented. A different and more straightforward field of research relates to the development of analytical models (Anderson and Bodner [3]), which reduces the complicated two or three-dimensional problems to one-dimensional problems, by making some simplifying assumptions about the behaviour and rupture of the structure when it is subjected to impact loading. An outstanding example of this latter type of approach is related to ballistic impact.

In fact, the impact of a projectile on a plate has been studied in the past in two different ways, namely, by use of large two or three-dimensional computer codes for dynamic analysis, or by employing simple engineering models (Anderson and Bodner [3]). Although some advanced constitutive models have been presented in the literature which take into consideration the mechanisms of rupture of the material (Curran, Seaman and Shockey [2]), their application remained restricted up to now, perhaps mainly due to the experimental difficulty in obtaining the micromechanical dynamic fracture parameters required by the models. So, in the numerical analysis of impact problems some simplistic rupture models, such as those of the critical plastic strain or critical stress for instance, are frequently used in practice.

Although in ballistic impact the above numerical techniques allow for a more complete understanding of the loading histories experienced by the material, they may result in a rather expensive computational tool. In fact, when the deformation fields and the failure mode of the target can be reasonably assumed a priori, simplified engineering models become an interesting alternative. For this reason, a number of analytical models for perforation of plates have been presented in the literature (Anderson and Bodner [3]). In this paper, we will employ one of such models recently developed by Ravid and Bodner [1]. This latter model makes some detailed consideration of the two-dimensional deformation fields experienced by the target plate when it fails by plug formation. It will be here applied to some real situations in which such type of failure is actually observed. Also, for some of the situations studied with the above model, a full two-dimensional numerical analysis of the impact problem will be made, and the results obtained with both models will be compared. The results derived from the above methods of analysis will be then compared with the experimental observations.

METHODS OF ANALYSIS OF PERFORATION PROBLEMS

The penetration model by Ravid and Bodner [1] considers a rigid projectile which impacts normally a rigid viscoplastic plate. By using an energy formulation, and by assuming the deformation fields in the zone surrounding the impact region, such authors were able to derive the equations for the penetration of the target plate. According to the above model, the penetration process consists of five stages. In the first of them, several well defined plastic zones are developed in the plate. When the extension of such plastic zones reaches the rear surface of the plate, the second stage of the penetration process begins. This second stage, characterized by the lateral expansion of the bulge in the rear surface of the plate. Once the lateral dimension of the bulge reaches a maximum value determined by geometrical considerations, the third stage starts. Such a stage is accompanied by the vertical displacement of the bulge, up to the moment when a rupture condition is reached. At this moment, which is characterized by plug initiation at the free surface of the plate, the fourth stage begins. This latter stage extends up to the moment when the plug is completely formed. During the fifth stage, a friction force on the cylindrical surface of the projectile is the only resisting force, up to the instant when the projectile finally leaves the target plate.

The above model is applicable when the projectile deforms little during the impact process, and when the initial velocity is lower than about 1000 m/s. For higher impact velocities two additional stages should be considered at the beginning of the impact process, taking into account such phenomena as shock waves and projectile deformability and erosion (Anderson and Bodner [3]).

The numerical approach to solve ballistic impact problems, is based on an explicit finite difference code for the dynamic analysis of coupled thermo-mechanical two-dimensional or axisymmetric problems, which makes use of a Lagrangian formulation. The computer code allows to consider elasto-plastic or elasto-viscoplastic behaviour of the materials, including the thermal softening effect. The subdivision of the continuum is made by means of quadrilateral cells, and it allows to treat with an arbitrary topology, in a similar fashion to ordinary finite element codes. An algorithm for contact detection has been implemented in the code, in such a way that the code is suitable for dealing with problems involving contact or impact between solid bodies. Rupture conditions have also been implemented. In the present study, a rupture criterion based on a critical value of the plastic strain will be used. So, when the effective plastic strain in a given cell reaches such value, the cell is thereon assumed to retain only its capacity to bear compressive hydrostatic stresses, whereas deviatoric stresses are gradually reduced to zero from that moment. Furthermore, when the hydrostatic stresses in ruptured cells fall below a very low value, or when the distortion of the cell becomes excessive, such cell is eroded. In this way, new external faces associated to the eroded cell are created, and contacts between such new surface and other external surfaces involved in the problem, should be considered thereafter. When nodes of an eroded cell are not connected to any other non-ruptured cell, then such nodes are subsequently considered as free nodes. In this way, in a representative stage of the penetration process, the projectile/target system will be composed of active cells, ruptured cells and free nodes, which are considered as a kind of debris which may also collide with the external surfaces of the problem. Eroded cells are not considered in the calculations after erosion takes place.

APPLICATION OF RAVID AND BODNER MODEL.

In this work, some actual impact experiments performed by Goldsmith and Finnegan [4] were analysed by means of the previously described techniques. Such experiments correspond to a normal impact of a blunt cylindrical projectile made of hard steel on annealed 2024-0 aluminium plates. Plate thicknesses of 19.05 mm and 25.4 mm were considered. The projectile radius was 6.35 mm, and its length was 37.7 for the impact on the 19.05 mm thickness plate, and 36.7 mm for the impact on the 25.4 mm thickness plate. The experimental ballistic limit as determined by Goldsmith and Finnegan [4] is 305 m/s for the 19.05 mm thickness plate and 333 m/s for the 25.4 mm thickness plate. In the real experiments, the projectile was found to deform very little. Moreover, the target was found to fail by plug formation in such a way that the Ravid and Bodner model can be readily applied.

In the simulations, a projectile density of 7850 kg/m^3 was taken , a target density of 2700 kg/m^3 and a yield stress of 140 MPa. Strain rate effects were also considered in the way suggested by Ravid and Bodner [1], and taking a rate factor of C=0.013. The critical strain for plug initiation, as required by the model of Ravid and Bodner, was taken equal to 0.2. A friction coefficient between the projectile and the target of 0.05 was considered in the last stage of perforation. A typical run by using the Ravid and Bodner model required only a few minutes of CPU time.

The curves of residual velocity as a function of the initial impact velocity for the above mentioned cases are shown in Figures 1 and 2, with and without lateral friction between the projectile and the target in the last stage of penetration. It can be seen from Figure 1 that the ballistic limit predicted by the Ravid and Bodner model is 297 m/s for the 19.05 mm thickness plate, which agrees with the experimental value of 305 m/s. On the other hand, from Figure 2 we see that the ballistic limit predicted by the Ravid and Bodner model for the impact on the 25.4 mm thickness plate is 362 m/s, which also agrees with the experimental value of 333 m/s.

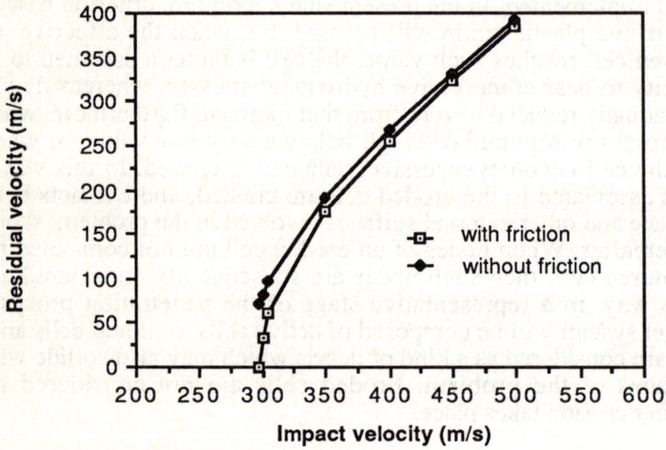

Figure 1. Residual velocity for normal impact on a 19.05 mm thickness plate, as a function of the initial impact velocity.

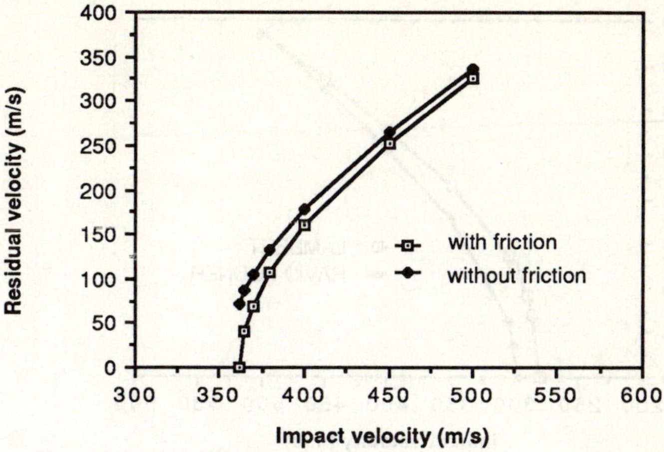

Figure 2. Residual velocity for normal impact on a 25.4 mm thickness plate, as a function of the initial impact velocity.

In the above simulations, a cylindrical plug was formed only for impacts on the 19.05 mm thickness plate at velocities of at least 450 m/s. In all the other cases, the rupture condition for plug initiation was not achieved during the analysis. Rather, a plastic instability condition leading to ductile failure was reached immediately after the third stage, thus causing the formation of an hemispherical plug of planar dimensions a little bit larger than the projectile cross section, and situated right ahead of the projectile. The above condition was equivalent to going directly from the third to the fifth stage of penetration, thus avoiding the fourth stage.

In Figures 3 and 4 the results associated to the analytical model of Ravid and Bodner for the residual velocity are compared with the empirical model of Lambert [5]. In Figure 3 the comparison is made by taking a value of the proportionality constant required by the Lambert's formula of about 700, which yields the best fit with the experimental values of the ballistic limit for both plate thicknesses. In Figure 4, the value of such a proportionality constant was selected in such a way to fit with the ballistic limit predicted by the Ravid and Bodner model for the corresponding plate thickness. It can be appreciated in such figures that the Ravid and Bodner model corresponds well with the tendency for the residual velocity predicted by Lambert's formula.

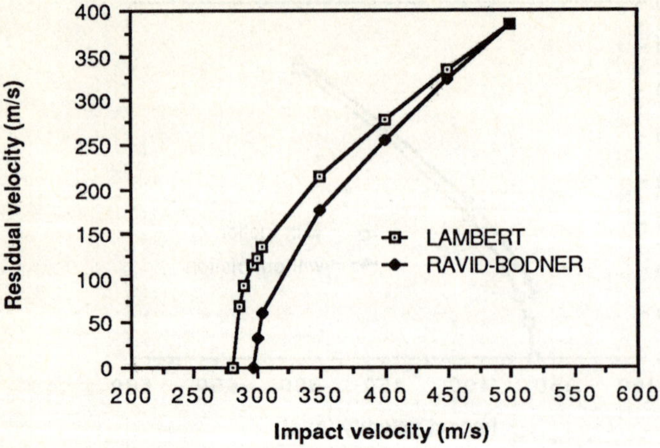

Figure 3 (a). Residual velocity curves as a function of the impact velocity for the case of a 19.05 mm thickness plate, corresponding to Ravid and Bodner model and Lambert model.

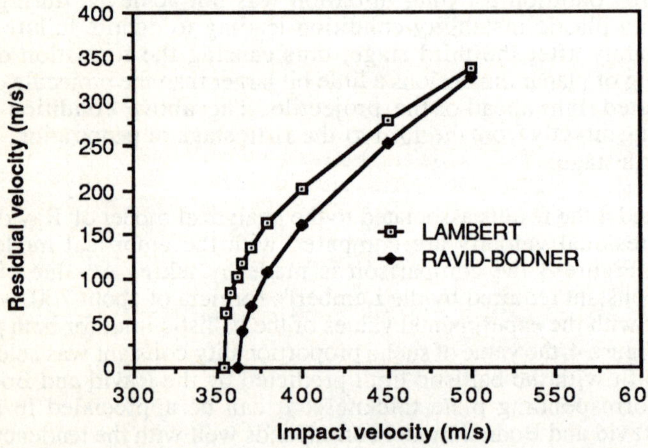

Figure 3(b). Residual velocity curves as a function of the impact velocity for the case of a 25.4 mm thickness plate, corresponding to Ravid and Bodner model and Lambert model.

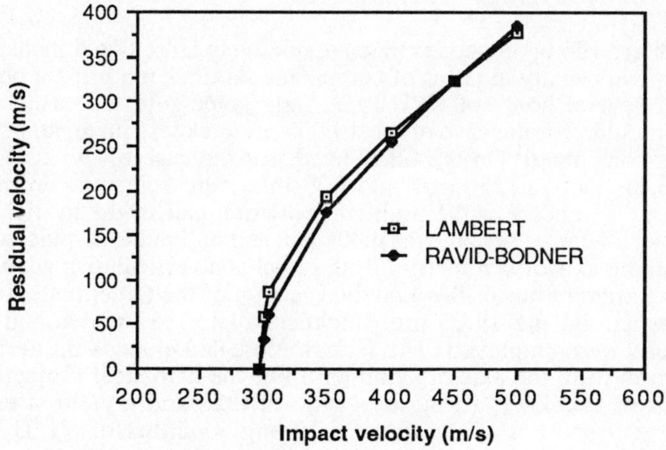

Figure 4(a). Residual velocity curves as a function of the impact velocity for the case of a 19.05 mm thickness plate, corresponding to Ravid and Bodner model and Lambert model. The ballistic limit used in this latter case corresponds to that predicted by the Ravid and Bodner model.

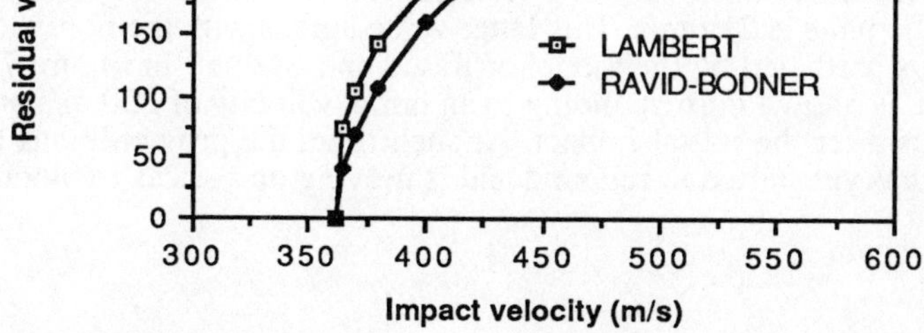

Figure 4(b). Residual velocity curves as a function of the impact velocity for the case of a 25.4 mm thickness plate, corresponding to Ravid and Bodner model and Lambert model. The ballistic limit used in this latter case corresponds to that predicted by the Ravid and Bodner model.

APPLICATION OF NUMERICAL ANALYSIS.

Since the use of large computer codes to solve relatively large two dimensional problems is relatively costly in terms of computational time, requiring a typical case of at least several hours of CPU time, only some selected cases were analysed. In particular, for the case of the 19.05 mm thickness plate, an impact at 400 m/s was considered. On the other hand, for the case of the 25.4 mm thickness plate, impacts at 250 m/s and 400 m/s were selected. From the condition of axial symmetry of the problem, only one half of the in-diameter plane of the projectile/target system was modelled, and null radial displacements were imposed at the axis of symmetry . Both radial and vertical displacements were fixed at the outer radius of the modelled section of the target plate. In the case of the impacts on the 19.05 mm thickness plate, 357 nodes and 300 quadrilateral cells were employed. The section modelled extends up to radial distance of 50 mm from the axis of symmetry. For the hard steel projectile, a Young modulus of 202 GPa, a Poisson's ratio of 0.28 and a yield stress of 1600 MPa were used. For aluminium, a Young modulus of 71 GPa, a Poisson's ratio of 0.3 and a yield stress of 140 MPa were employed. Moreover, a linear thermal softening was considered up to the melting temperature of each material, at which the yield stress was assumed to vanish both for steel and aluminium.

In Figure 5 the deformed configuration of the projectile/target system, for the impact at 400 m/s on the 19.05 mm thickness plate, is shown for the indicated times and compared with the configuration predicted by the Ravid and Bodner model. It can be appreciated that a good correspondence between both predictions exists. The residual velocity of the projectile after perforation is, according to the numerical analysis, of 282 m/s. This result agrees with the non-friction value of 268 m/s predicted by the Ravid and Bodner model.

The modelization corresponding to the impacts on the 25.4 mm thickness plate was achieved by means of 483 nodes and 420 quadrilateral cells. In Figure 6 the deformed mesh is shown corresponding to an impact velocity of 400 m/s and to a time of 90 us after the impact, when the onset of complete plug formation can be clearly observed. The residual velocity of the projectile after perforation of the plate is 219 m/s. This latter value agrees with the non-friction value of 180 m/s predicted by the model of Ravid and Bodner. In Figure 7 the deformed mesh is shown corresponding to an initial velocity of 250 m/s and to a time of 215 us after the initial impact. At such time, the projectile has been stopped, and it has yet started to rebound and is moving upward at a velocity of about 10 m/s.

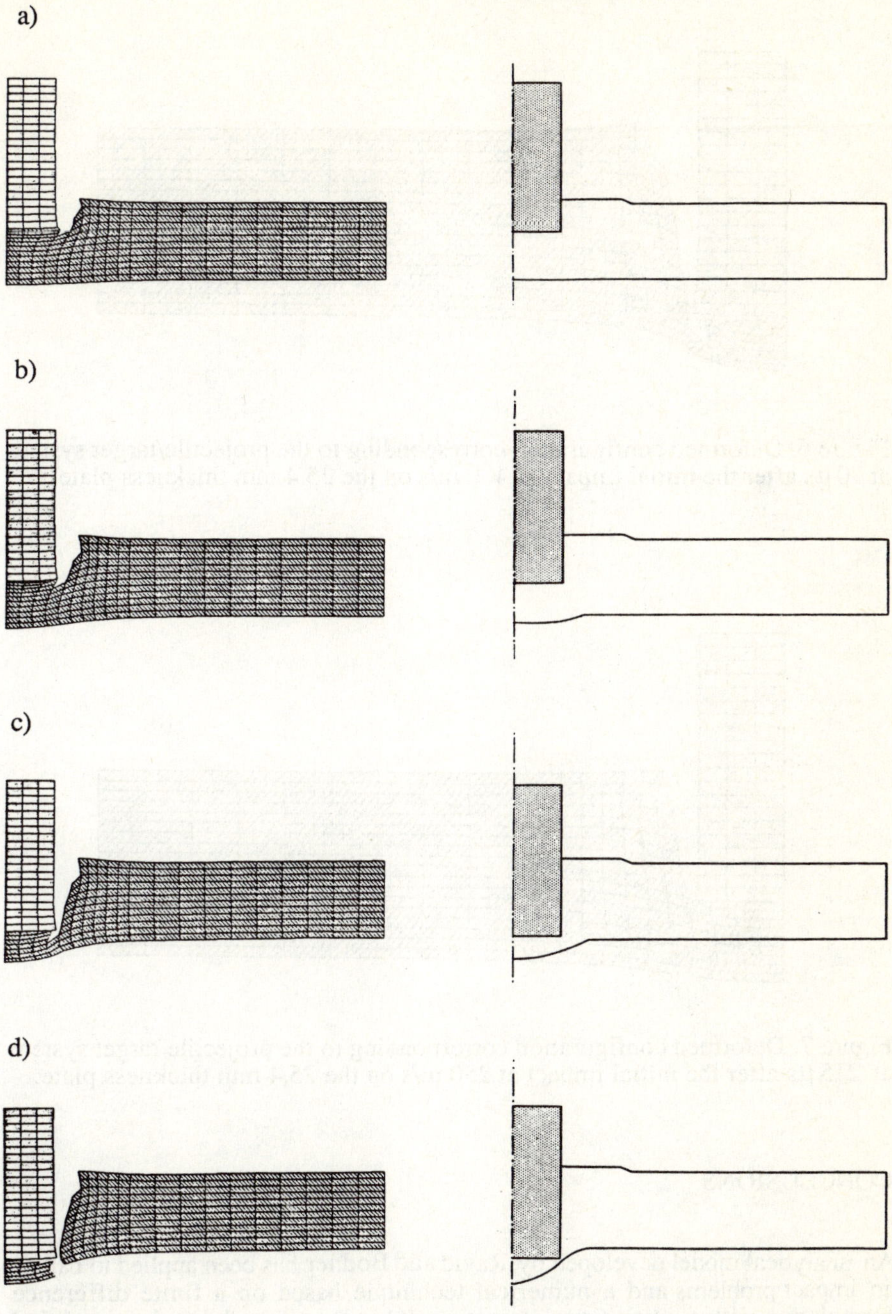

Figure 5. Comparison of the projectile/target configuration predicted by the numerical analysis (left) and the Ravid and Bodner model (right) for the following times: a) 18 μs, b) 30 μs, c) 54 μs and d) 66 μs.

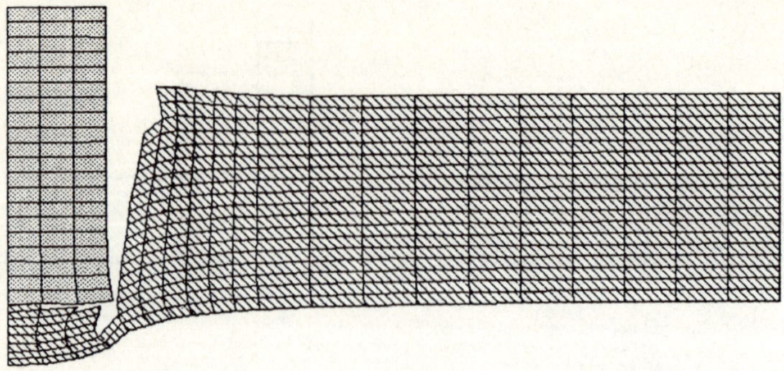

Figure 6. Deformed configuration corresponding to the projectile/target system at 90 μs after the initial impact at 400 m/s on the 25.4 mm thickness plate.

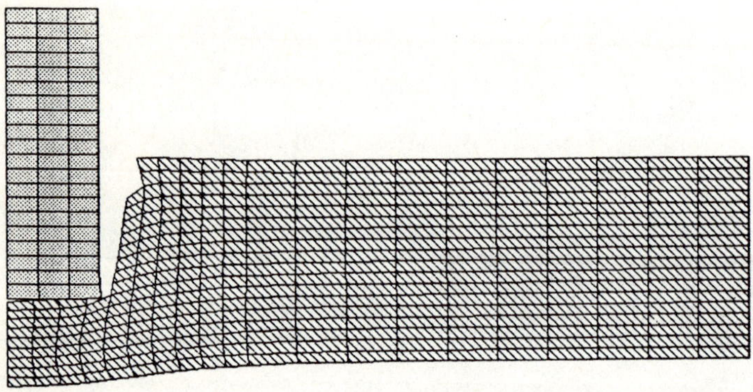

Figure 7. Deformed configuration corresponding to the projectile/target system at 215 μs after the initial impact at 250 m/s on the 25.4 mm thickness plate.

CONCLUSIONS

An analytical model developed by Ravid and Bodner has been applied to ballistic impact problems and a numerical technique based on a finite difference formulation with explicit time integration. It has been found that the simplified model of Ravid and Bodner allows to predict, with great approximation, the ballistic limit for target plates of annealed 2024-0 aluminium impacted by hard steel blunt cylindrical projectiles. A great similarity of the results obtained from the numerical analysis and by using the analytical model of Ravid and Bodner was also verified. Thus, the great economy in computations associated with the use of such analytical model makes it very advantageous with respect to numerical techniques for the perforation problems studied.

REFERENCES

1) Ravid, M. and Bodner, S. R., Dynamic Perforation of Viscoplastic Plates by Rigid Projectiles, Int. J. Engng. Sci., Vol. 21, pp. 577-591, 1983.

2) Curran, D.R, Seaman, L. and Shockey, D.A., Dynamic Failure of Solids, Physics Report, Vol. 147, pp. 254-387, 1987.

3) Anderson, C.E. and Bodner, S.R., Ballistic Impact: the Status of Analytical and Numerical Modeling, Int. J. Impact Engng., Vol. 7, pp. 9-35 , 1988.

4) Goldsmith, W. and Finnegan, S. A., Normal and Oblique Impact of Cylindro-Conical and Cylindrical Projectiles on Metallics Plates, Int. J. Impact Engng., Vol. 4, pp. 83-105, 1986.

5) Lambert , J. P., Ballistic Research Laboratory, ARBRL-TR-02066, 1978.

Boundary Element Analysis of Elastodynamic Contact Problems by the Penalty Function Method

Kouetsu Yamazaki* and Masahiro Mori**

**Department of Mechanical Systems Engineering, Kanazawa University, Kodatsuno 2-40-20, Kanazawa, 920, Japan*

***Recruit Co. Ltd., Marunouchi 1-1-2, Chiyoda-ku, Tokyo, 140, Japan*

ABSTRACT

A boundary element technique for elastodynamic contact problems is suggested by using the penalty function method. The time domain approach of higher-order boundary-time element for two-dimensional elastodynamic problems is used and some impact examples are solved numerically.

INTRODUCTION

To develop an effective method to analyze the elastodynamic contact stress fields is one of the most important subjects for mechanical engineering. Some typical methods to analyze the dynamic contact fields between arbitrarily shaped bodies have already been suggested by using the dynamic finite element method(FEM)[1-4]. Among these methods, the penalty function method for the FEM suggested by Kanto and Yagawa[3] is one of the most efficient methods, because their method does not need to increase the degree of freedom in the final equation to be solved and to modify the stiffness matrices of each body.

However, domain type of numerical methods like the FEM require a discretization of not just the surface of the domain of interest but its interior domain as well, thereby increasing the modelling requirements and the size of the problem for comparable accuracy. On the contrary, the boundary element method(BEM) discretizes only the surface of the domain of interest and decreases the size of the problem to be solved. For the contact problem, the BEM can treat easily and accurately contact conditions by taking directly the surface tractions of nodal points on the contact surface as unknown variables. Andersson and his coworkers[5] and Bezine and Fortune[6] have studied the methods to analyze the elastostatic contact problems by using the BEM, however, no method to analyze the elastodynamic contact problems is available at present because the dynamic BEM is still being developed.

In this paper, a boundary element technique for solving two-dimensional elastodynamic contact problems is suggested. When we consider an impingement between two elastic bodies, contact conditions of the displacements and tractions on an interface boundary corresponding to contact states, such as open, sticking and slipping states, have to be satisfied at each time step. A functional of the virtual work is adopted and the contact conditions are assembled to the functional by using the penalty function method. An integral representation which provides contact stress fields at each time step in terms of boundary values of displacements and tractions is derived with the aid of the method of weighted residuals. The time domain approach of higher-order boundary-time element for two-dimensional elastodynamic problems is used for discretization of time and space, and some impact problems are solved numerically.

BOUNDARY ELEMENT FORMULATION FOR DYNAMIC CONTACT ANALYSIS

Contact condition and contact state judgement

Let us consider a two-dimensional elastodynamic contact problem of two bodies A, B which occupy plane domains Ω^A and Ω^B bounded by surfaces Γ^A and Γ^B, respectively, as shown in Fig. 1. When the bodies A, B which are supported at some parts of their surfaces Γ_u^β (β=A, B) are applied tractions $\bar{p}_i^\beta$ (i=1, 2) varying with the time on the surface Γ_p^β, these bodies are contacting each other at the boundary Γ_c. The contact states on the interface boundary Γ_c can be the opening, sticking and slipping states. Let us introduce a local coordinate system ($\tilde{x}, \tilde{y}$) equally apart from each boundary for describing the contact conditions. The values with tilde denote corresponding values in the local coordinates in the following. Equilibrium equations of traction $\tilde{p}_x$, $\tilde{p}_y$, continuity condition of displacements $\tilde{u}_x$, $\tilde{u}_y$, velocities $\dot{\tilde{u}}_x$, $\dot{\tilde{u}}_y$ and acceleration $\ddot{\tilde{u}}_x$, $\ddot{\tilde{u}}_y$, and Coulomb's law of friction on the interface boundary lead to the following contact conditions for each contact state at a time t.

On a releasing surface Γ_o,

$$\tilde{p}_x = \tilde{p}_y = 0\,. \tag{1}$$

In addition, the displacements must satisfy the following condition.

$$\tilde{u}_y^A - \tilde{u}_y^B + \tilde{d}_y \geq 0\,, \tag{2}$$

in which $\tilde{d}_y$ denotes a distance between corresponding points on the interface boundary Γ_c in the $\tilde{y}$ direction, and has to be updated into a new value at each time step. Superscripts A and B denote the values which belong to the bodies Ω^A and Ω^B.

On a sticking surface Γ_{st}, on the other hand, the continuity condition of the displacements, velocities and accelerations can be written as

$$\tilde{u}_x^A - \tilde{u}_x^B + \tilde{d}_x = 0,\ \tilde{u}_y^A - \tilde{u}_y^B + \tilde{d}_y = 0, \tag{3}$$

$$\dot{\tilde{u}}_x^A = \dot{\tilde{u}}_x^B, \quad \dot{\tilde{u}}_y^A = \dot{\tilde{u}}_y^B, \tag{4}$$

$$\ddot{\tilde{u}}_x^A = \ddot{\tilde{u}}_x^B, \quad \ddot{\tilde{u}}_y^A = \ddot{\tilde{u}}_y^B, \tag{5}$$

where $\tilde{d}_x$ denotes a gap in the x direction between corresponding points on the interface boundary and is equal to zero at an initial time t_0. The resultant tangential traction must not exceed its limiting value.

$$|\tilde{p}_x| \leqq |\mu \tilde{p}_y| \ , \tag{6}$$

where μ is a dynamic coefficient of limiting friction.

On a slipping surface Γ_{sl},

$$\tilde{u}_y^A - \tilde{u}_y^B + \tilde{d}_y = 0, \tag{3'}$$

$$\dot{\tilde{u}}_y^A = \dot{\tilde{u}}_y^B, \quad \ddot{\tilde{u}}_y^A = \ddot{\tilde{u}}_y^B. \qquad (4'), (5')$$

and Coulomb's law for the slipping state leads to

$$\tilde{p}_x = \pm \mu \tilde{p}_y, \tag{7}$$

where the signs $\pm$ are opposed to a slipping direction.

The contact state at each time step will be assumed and the analysis at a time step has to be repeated to satisfy the contact conditions until the contact states judged by the analysis coincide with the assumption for all points on Γ_c. The contact state will be judged in two stages ; in the first stage the releasing state is distinguished from the contacting state by the condition (2). The sticking state and the slipping state are separated by the condition (6) in the second stage.

Boundary integral equation for contact analysis

The principle of virtual work for the elastodynamic contact problem can be expressed as

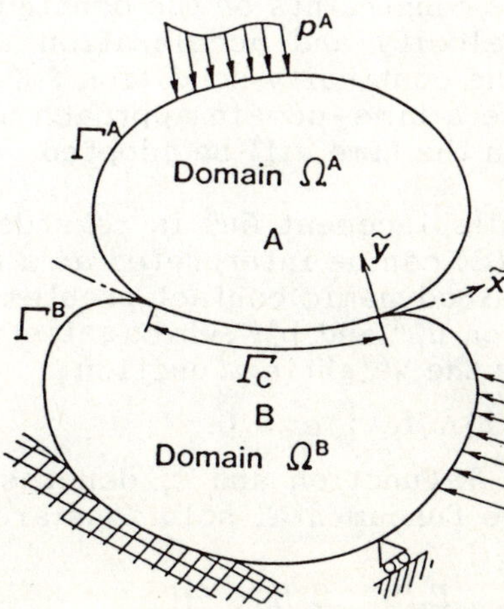

Fig.1 Elastodynamic contact model.

$$\int_{t_0}^{t_F}\{\sum_{\beta}(\delta T^{\beta}+\delta E^{\beta}-\delta W^{\beta})-\delta W_c\}\,dt = 0\,, \tag{8}$$

in which δ denotes the usual variational indicator and t_0, t_F are initial and current times. δT^{β}, δE^{β} and δW^{β} present the virtual work of inertia force, internal virtual work and external virtual work, and are described by using the usual tensor symbolism as

$$\delta T^{\beta} = \int_{\Omega^{\beta}}\rho^{\beta}\ddot{u}_i^{\beta}\delta u_i^{\beta}d\Omega\,,\quad \delta E^{\beta} = \int_{\Omega^{\beta}}\sigma_{ij}^{\beta}\delta\varepsilon_{ij}^{\beta}d\Omega\,,$$

$$\delta W^{\beta} = \int_{\Omega^{\beta}}b_i\delta u_i^{\beta}d\Omega + \int_{\Gamma_p^{\beta}}\bar{p}_i^{\beta}\delta u_i^{\beta}d\Gamma\,, \tag{9}$$

where ρ^{β} and b_i denote the specific mass and body force per unit mass. σ_{ij}^{β} and ε_{ij}^{β} denote the stress and strain tensors and u_i and p_i are the displacement and traction. On the other hand, δW_c presents an external virtual work on the contact surface Γ_c and is defined as

$$\delta W_c=\int_{\Gamma_c}\tilde{p}_x(\delta\tilde{u}_x^{A}-\delta\tilde{u}_x^{B})d\Gamma+\int_{\Gamma_c}\tilde{p}_y(\delta\tilde{u}_y^{A}-\delta\tilde{u}_y^{B})d\Gamma$$
$$(\Gamma_c=\Gamma_o+\Gamma_{st}+\Gamma_{sl}). \tag{10}$$

The penalty function method is applicable to minimize the functional (8) subjected to the constraints (3), (3') and (7). If α denotes a penalty parameter, then, the external virtual work of the contact (10) can be rewritten as

$$\delta W_C=\int_{\Gamma_{st}}\alpha(\tilde{u}_x^{A}-\tilde{u}_x^{B}+\tilde{d}_x)(\delta\tilde{u}_x^{A}-\delta\tilde{u}_x^{B})d\Gamma$$
$$\pm\int_{\Gamma_{sl}}\mu\alpha(\tilde{u}_y^{A}-\tilde{u}_y^{B}+\tilde{d}_y)(\delta\tilde{u}_x^{A}-\delta\tilde{u}_x^{B})d\Gamma$$
$$+\int_{\Gamma_{st}+\Gamma_{sl}}\alpha(\tilde{u}_y^{A}-\tilde{u}_y^{B}+\tilde{d}_y)(\delta\tilde{u}_y^{A}-\delta\tilde{u}_y^{B})d\Gamma\,. \tag{11}$$

Equation (11) coincides to (10) when α is approaching infinity. It is assumed that the constraints of the continuity (4), (5), (4') and (5') for the velocity and acceleration will be satisfied automatically when the continuity condition for the displacement is satisfied because a time-domain approach and a fundamental solution depending on the time will be adopted.

If the virtual displacement δu_i^{β} is regarded as a weighting function, equation (8) can be interpreted as a weighted residual statement for the elastodynamic contact problem. Thus, we adopt a fundamental solution $u_i^{*\beta}$ and $p_i^{*\beta}$, which satisfies the following Navier equations, as the weighting function,

$$\sigma_{ij,j}^{*}-\rho\ddot{u}_i^{*}+\Delta(\xi-x)\Delta(t_F-t)e_i = 0. \tag{12}$$

where $\Delta(\ )$ is Dirac δ-function and e_i denotes a unit direction vector. Then, if the fundamental solutions are expressed as $u_j^{*} = u_{ij}^{*}e_i$ and $p_j^{*} = p_{ij}^{*}e_i$,

$$2\pi\rho u_{ij}^{*}(\xi,x,\tau) = (\frac{B_1H_1}{A_1}-\frac{B_2H_2}{A_2})\frac{r_ir_j}{r^4}$$

$$-(A_1H_1-A_2H_2)\frac{\delta_{ij}}{r^2}+\frac{H_2\delta_{ij}}{c_2^2A_2}, \tag{13}$$

where $\tau=t_F-t$, $H_i=H(\tau-r/c_i)(i=1,2)$ when H denotes Heaviside step function, and

$$r_i=x_i-\xi_i,\ r^2=(x_i-\xi_i)(x_i-\xi_i),$$
$$A_i=(\tau^2-r^2/c_i^2)^{1/2},\ B_i=2\tau^2-r^2/c_i^2.$$

The propagation velocities of the pressure and shear waves in the body are given respectively as

$$c_1^2=(\lambda+2\mu)/\rho,\ c_2^2=\mu/\rho, \tag{14}$$

where λ and μ are the Lame constants.

The fundamental solution of the traction p^*_{ij} is also given as

$$2\pi\rho p^*_{ij}(\xi,x,\tau)=\left(\frac{H_1}{c_1^4A_1^3}-\frac{H_2}{c_2^4A_2^3}\right)\frac{R_{ij}}{r^2}$$
$$-\left(\frac{B_1\Delta_1}{c_1A_1}-\frac{B_2\Delta_2}{c_2A_2}\right)\frac{R_{ij}}{r^5}-\left(\frac{B_1H_1}{A_1}-\frac{B_2H_2}{A_2}\right)\left(\frac{S_{ij}}{r^6}-\frac{T_{ij}}{r^4}\right)$$
$$+\left(\frac{A_1\Delta_1}{c_1}-\frac{A_2\Delta_2}{c_2}\right)\frac{T_{ij}}{r^3}+\frac{H_2}{c_2^4A_2^3}T_{ij}-\frac{\Delta_2}{c_2^3A_2}\frac{T_{ij}}{r}, \tag{15}$$

and

$$R_{ij}=\lambda r^2r_in_j+2\mu r_ir_jr_kn_k,$$
$$S_{ij}=\lambda r^2r_in_j-\mu r^2(r_jn_i+2r_in_j+\delta_{ij}r_kn_k)+8\mu r_ir_jr_kn_k,$$
$$T_{ij}=\lambda r_in_j+\mu(r_jn_i+\delta_{ij}r_kn_k),\ \Delta_i=\Delta(\tau-r/c_i),$$

where n_i is a direction cosine of the unit outward normal vector to the surface.

Then, substituting from the fundamental solution (13) and (15) into (8) and taking the displacement constraints into consideration, equations (8), (9) and (11) yield

$$\int_{t_0}^{t_F}\int_{\Omega^\beta}(\sigma^\beta_{ij,j}+b^\beta_i-\rho^\beta\ddot{u}^\beta_i)u^{*\beta}_i d\Omega dt$$
$$=\int_{t_0}^{t_F}\int_{\Gamma^\beta_p}(p^\beta_i-\bar{p}^\beta_i)u^{*\beta}_i d\Gamma dt-\int_{t_0}^{t_F}\int_{\Gamma^\beta_u}(u^\beta_i-\bar{u}^\beta_i)p^{*\beta}_i d\Gamma dt$$
$$-\int_{t_0}^{t_F}\int_{\Gamma_{st}}\alpha(\tilde{u}^A_x-\tilde{u}^B_x+\tilde{d}_x)(\tilde{u}^{*A}_x-\tilde{u}^{*B}_x)d\Gamma dt$$
$$\mp\int_{t_0}^{t_F}\int_{\Gamma_{sl}}\mu\alpha(\tilde{u}^A_y-\tilde{u}^B_y+\tilde{d}_y)(\tilde{u}^{*A}_x-\tilde{u}^{*B}_x)d\Gamma dt$$
$$-\int_{t_0}^{t_F}\int_{\Gamma_{st}+\Gamma_{sl}}\alpha(\tilde{u}^A_y-\tilde{u}^B_y+\tilde{d}_y)(\tilde{u}^{*A}_y-\tilde{u}^{*B}_y)d\Gamma dt, \tag{16}$$

in which the terms with the superscript β must be taken the summation of $\beta=A,B$. By usual manner of direct BEM we can derive a boundary integral equation. It can also be separated to two equations by considering the properties that $u^{*\gamma}_i(\gamma\neq\beta)$ is zero when the point ξ is on the surface of the body β.

$$c_{ij}u_j^\beta(\xi, t_F)+\int_{t_0}^{t_F}\int_{\Gamma^\beta}p_{ij}^{*\beta}u_j^\beta d\Gamma dt=\int_{t_0}^{t_F}\int_{\Gamma^\beta}u_{ij}^{*\beta}p_j^\beta d\Gamma dt$$
$$+\int_{t_0}^{t_F}\int_{\Omega^\beta}u_{ij}^{*\beta}b_j^\beta d\Omega dt+\rho^\beta\int_{\Omega^\beta}(u_{ij}^{*\beta}\dot{u}_j^{\beta 0}-\dot{u}_{ij}^{*\beta}u_j^{\beta 0})d\Omega$$
$$+\int_{t_0}^{t_F}\int_{\Gamma_{st}}\alpha(\tilde{u}_x^\beta-\tilde{u}_x^\gamma+\tilde{d}_x)\tilde{u}_x^{*\beta}d\Gamma dt$$
$$\pm\int_{t_0}^{t_F}\int_{\Gamma_{sl}}\mu\alpha(\tilde{u}_y^\beta-\tilde{u}_y^\gamma+\tilde{d}_y)\tilde{u}_x^{*\beta}d\Gamma dt$$
$$+\int_{t_0}^{t_F}\int_{\Gamma_{st}+\Gamma_{sl}}\alpha(\tilde{u}_y^\beta-\tilde{u}_y^\gamma+\tilde{d}_y)\tilde{u}_y^{*\beta}d\Gamma dt. \quad (\beta=A, B, \gamma=B, A) \qquad (17)$$

The domain integration of the third term of (17) corresponds to the initial condition where $u_j^{\beta 0}$ and $\dot{u}_j^{\beta 0}$ denote the initial displacement and velocity at $t=t_0$. A coefficient c_{ij} is equal to $\delta_{ij}/2$ on the smooth boundary.

<u>Discretization into time-space element</u>

In order to obtain the response of current time t_F, the boundaries Γ^β will be divided into J_β quadratic isoparametric boundary elements Γ_j^β and the time t_F will also be divided into F linear time elements. Then, both displacements $\{u^\beta(x,t)\}_j$ and tractions $\{p^\beta(x,t)\}_j$ on the boundary Γ_j during a time step $t_{f-1}\leqq t\leqq t_f (f=1, \cdots; F)$ are interpolated as

$$\{u^\beta(x,t)\}_j=\varphi_1[N]\{u^\beta\}_j^{f-1}+\varphi_2[N]\{u^\beta\}_j^f,$$
$$\{p^\beta(x,t)\}_j=\varphi_1[N]\{p^\beta\}_j^{f-1}+\varphi_2[N]\{p^\beta\}_j^f, \qquad (18)$$

in which $[N]$ is a matrix of quadratic interpolating function and

$$\varphi_1=(t_f-t)/\Delta t,\ \varphi_2=(t-t_{f-1})/\Delta t,\ t_f-t_{f-1}=\Delta t.$$

$\{u^\beta\}_j^f$ and $\{p^\beta\}_j^f$ are the nodal vectors of the boundary element Γ_j^β at a time t_f and given as

$$\{u^\beta\}_j^f=\{u_1^1\ u_2^1\ u_1^2\ u_2^2\ u_1^3\ u_2^3\}^T,$$
$$\{p^\beta\}_j^f=\{p_1^1\ p_2^1\ p_1^2\ p_2^2\ p_1^3\ p_2^3\}^T, \qquad (19)$$

Substituting from (18) into (17) and neglecting the term of body force, the discretized boundary-time integration at a node ξ_l can be derived as

$$[C]_l^F\{u^\beta(\xi_l, t_F)\}$$
$$+\sum_{f=1}^{F}\sum_{j=1}^{J_\beta}(\int_{\Gamma_j^\beta}[h]^{f-1}[N]d\Gamma\{u^\beta\}_j^{f-1}+\int_{\Gamma_j^\beta}[h]^f[N]d\Gamma\{u^\beta\}_j^f)$$
$$=\sum_{f=1}^{F}\sum_{j=1}^{J_\beta}(\int_{\Gamma_j^\beta}[g]^{f-1}[N]d\Gamma\{p^\beta\}_j^{f-1}+\int_{\Gamma_j^\beta}[g]^f[N]d\Gamma\{p^\beta\}_j^f)$$
$$+\{\pi^\beta(\xi_l, t_F)\}$$
$$+\sum_{f=1}^{F-1}\sum_{j=1}^{J_c}(\int_{\Gamma_{cj}}[\tilde{g}]^{f-1}[N]d\Gamma\{\tilde{p}^\beta\}_j^{f-1}+\int_{\Gamma_{cj}}[\tilde{g}]^f[N]d\Gamma\{\tilde{p}^\beta\}_j^f)$$

$$+\sum_{j=1}^{J_c}\left(\int_{\Gamma_{cj}}[\tilde{g}]^{F-1}[S][N]\,d\Gamma(\{\tilde{u}^{\beta}\}_j^{F-1}-\{\tilde{u}^{\gamma}\}_j^{F-1}+\{\tilde{d}\}_j^{F-1})\right.$$

$$\left.+\int_{\Gamma_{cj}}[\tilde{g}]^{F}[S][N]\,d\Gamma(\{\tilde{u}^{\beta}\}_j^{F}-\{\tilde{u}^{\gamma}\}_j^{F})\right)\quad(\beta=\mathrm{A,B},\ \gamma=\mathrm{B,A})\qquad(20)$$

in which the matrix $[C]_l^F$ consists of the components c_{ij} and the vector $\{\tilde{d}\}_j^f$ denotes the gaps of displacements $\tilde{d}_x, \tilde{d}_y$ at the nodcs

$$[h]^{f-1}=\int_{t_{f-1}}^{t_f}[p^*]\varphi_1 dt\,,\ [h]^{f}=\int_{t_{f-1}}^{t_f}[p^*]\varphi_2 dt\,,$$

$$[g]^{f-1}=\int_{t_{f-1}}^{t_f}[u^*]\varphi_1 dt\,,\ [g]^{f}=\int_{t_{f-1}}^{t_f}[u^*]\varphi_2 dt\,,\qquad(21)$$

where $[p^*]$, $[u^*]$ consist of the components $p^*_{ij}(\xi_l, x, \tau)$, $u^*_{ij}(\xi_l, x, \tau)$. The time integration of (21) is done analytically by eliminating the singularity of δ-function in the fundamental solution. On the other hand, the boundary integration of (20) will be accomplished through a standard Gaussian quadrature scheme. The third term $\{\pi^{\beta}(\xi_l, t_F)\}$ of right hand side of (20) corresponds to the initial conditions. When u_j^0 and $\dot{u}_j^0$ are the constants in the region Ω^{β}, the term can be evaluated indirectly from the other terms by assuming a rigid movement.

The time integration of the terms which correspond to the external virtual work on the contact surface are separated to the previous intervals $t_0 \leqq t \leqq t_{F-1}$ and the current interval $t_{F-1} \leqq t \leqq t_F$ since the external work on the contact surface during $t_0 \leqq t \leqq t_{F-1}$ can be evaluated directly from the traction. The matrix $[S]$ to contain the contribution of contact conditions is given for each contact state as

$$[S]=\begin{cases}\begin{bmatrix}0 & 0\\ 0 & 0\end{bmatrix}\ ; & \text{on } \Gamma_o\\[2ex] \begin{bmatrix}\alpha & 0\\ 0 & \alpha\end{bmatrix}\ ; & \text{on } \Gamma_{st}\\[2ex] \begin{bmatrix}0 & \pm\mu\alpha\\ 0 & \alpha\end{bmatrix}\ ; & \text{on } \Gamma_{sl}\end{cases}\qquad(22)$$

Furthermore, the matrix $[\tilde{g}]$ is related to the corresponding values of the global coordinates by the necessary coordinate transformation.

$$[\tilde{g}]=[L]\,[g]\,[L]^{\mathrm{T}}\,,\qquad(23)$$

where the matrix $[L]$ is the coordinate transformation matrix from global coordinates to $\tilde{x}-\tilde{y}$ local coordinates.

From above formulation a computer program to analyze the elastodynamic contact problem can be prepared by adding the contribution of the last two terms of the right hand side of (20) to a general elastodynamic BEM program. This technique has no need of a global matrix expansion or decomposition for satisfying the contact conditions.

NUMERICAL EXAMPLES

The technique presented in the previous chapter was applied to analyze some collision phenomena between the slender bars and between the circular disks. The numerical analysis was implemented for the plane stress condition with keeping the double precision of word length and the value of the penalty parameter $\alpha = 10^8$.

Collision between rods

To assess an accuracy of the technique, a thin rod B of length l_B and width h fixed at one end and struck on the other free end by a thin rod A of length l_A and width h with an initial velocity $V_0 = 1$ m/s is considered ($l_B/l_A = 1.5$), as shown in Fig. 2. The materials of two rods are assumed to be same, and Young's modulus and the specific density to be $E = 206$ GPa and $\rho = 7.85 \times 10^3$ kg/m^3. The dynamic coefficient of friction between them is assumed to be $\mu = 0.2$. Two cases of slender ratio $l_B/h = 6$ and Poisson's ratio $\nu = 0$, and $l_B/h = 3$, $\nu = 0.3$ are analyzed. Because of symmetry of the model half surfaces of the rods are divided into boundary

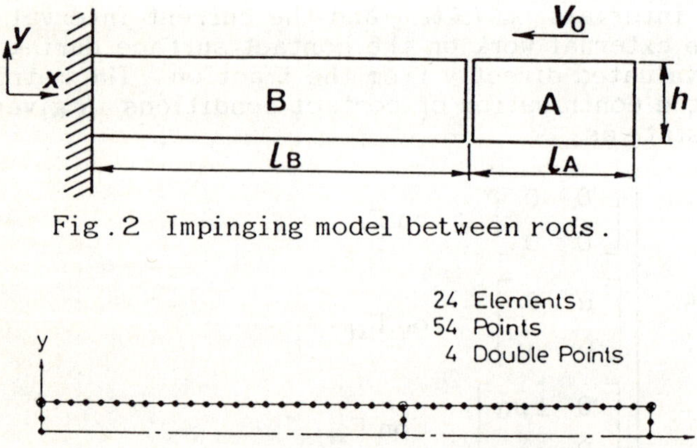

Fig. 2 Impinging model between rods.

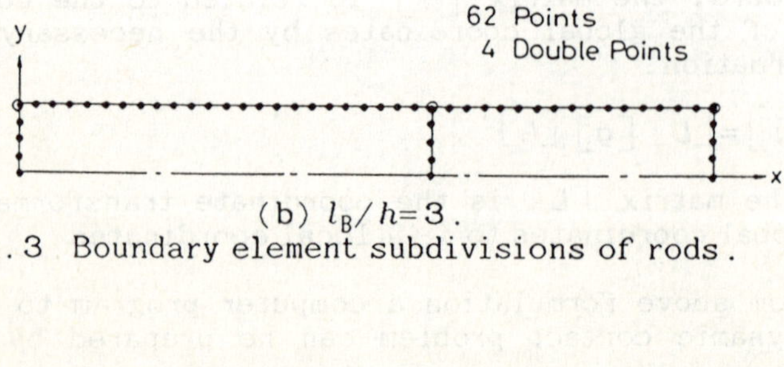

(a) $l_B/h=6$.

(b) $l_B/h=3$.

Fig. 3 Boundary element subdivisions of rods.

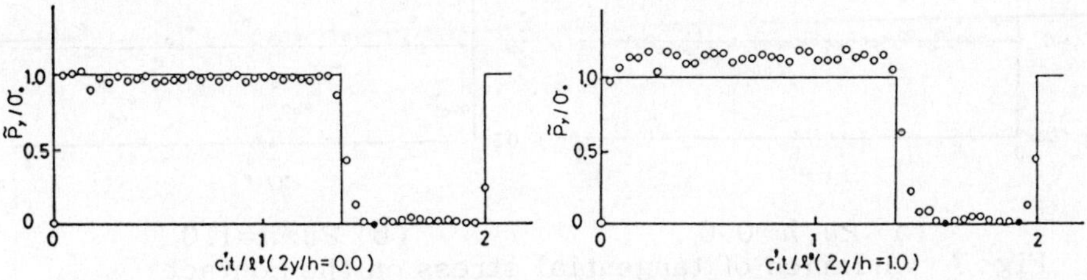

(a) $2y/h=0.0$. (b) $2y/h=1.0$.

Fig. 4 Variation of contact pressure throughout the impact($l_B/h=6$).

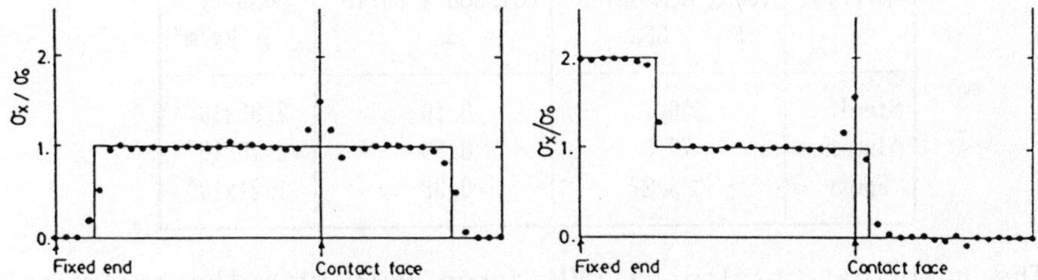

(a) $c_1't/l_B=0.90$ (b) $c_1't/l_B=1.34$

Fig. 5 Longitudinal stress distribution along the side surface($l_B/h=6$).

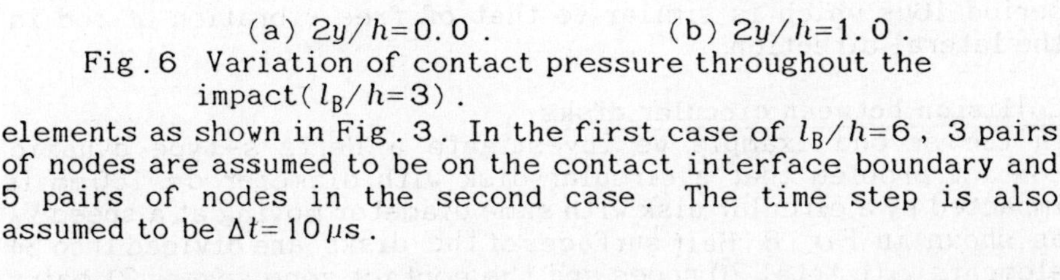

(a) $2y/h=0.0$. (b) $2y/h=1.0$.

Fig. 6 Variation of contact pressure throughout the impact($l_B/h=3$).

elements as shown in Fig. 3. In the first case of $l_B/h=6$, 3 pairs of nodes are assumed to be on the contact interface boundary and 5 pairs of nodes in the second case. The time step is also assumed to be $\Delta t=10\,\mu s$.

Figure 4 shows a variation of contact pressure throughout the impact for the first case. A solid line indicates the analytical solution $\sigma_0=\rho c_1' V_0/2$ for one-dimensional stress wave propagation, where $c_1'^2 = E/(1-\nu^2)\rho$. The symbols ○and ●plotted in the figure indicate the sticking and opening contact states.

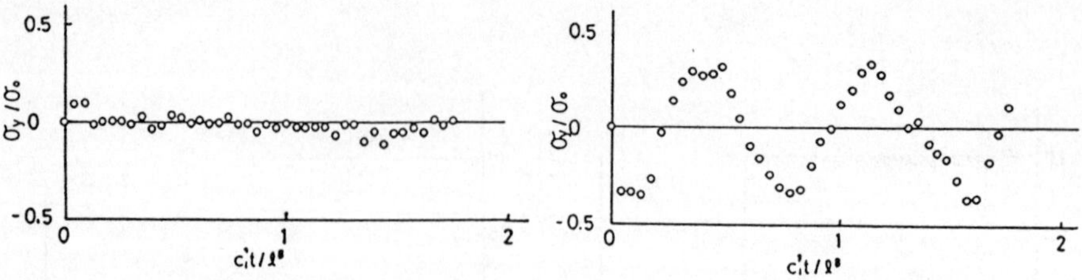

(a) $2y/h=0.0$. (b) $2y/h=1.0$.

Fig. 7 Variation of tangential stress on the contact surface($l_B/h=3$) .

TABLE 1 Elastic constants of materials.

Material	Young's Modulus E GPa	Poisson's ratio ν	Density ρ kg/m³
Steel	206	0.30	7.85×10^3
Aluminum	70.6	0.30	2.70×10^3
Epoxy	3.27	0.38	1.21×10^3

The numerical results by BEM agree well with the analytical solution at the center $2y/h=0$, but are a bit higher than the analytical ones at the corner point $2y/h=1.0$. Figure 5 shows the longitudinal stress distribution along the side surface at the time $c_l t/l_B=0.90$ and 1.34. The numerical results are in good agreement with the analytical solution indicated by the solid line except the contact interface.

Figure 6 shows the variation of contact pressure throughout the impact for the second case. The numerical result at the center is similar to that of the first case, but the result at the corner point $2y/h=1.0$ shows the oscillation of the contact pressure due to an effect of the inertia in the lateral direction. The variation of tangential stress at the same points is shown in Fig. 7. The tangential stress at the center is oscillating with a period 16μs which is similar to that of free vibration of rod in the lateral direction.

Collision between circular disks

In the second example we investigate a Hertz's-type dynamic contact problem that a circular disk with diameter $d=$ 100mm is impacted by a circular disk with same diameter moving at a speed V_0 as shown in Fig. 8. Half surfaces of the disks are divided into 34 elements with total 70 nodes and the contact zone covers 21 pairs of nodes. The material properties adopted for the analysis are tabulated in Table 1. The dynamic coefficient of the friction is assumed to be $\mu=0.2$.

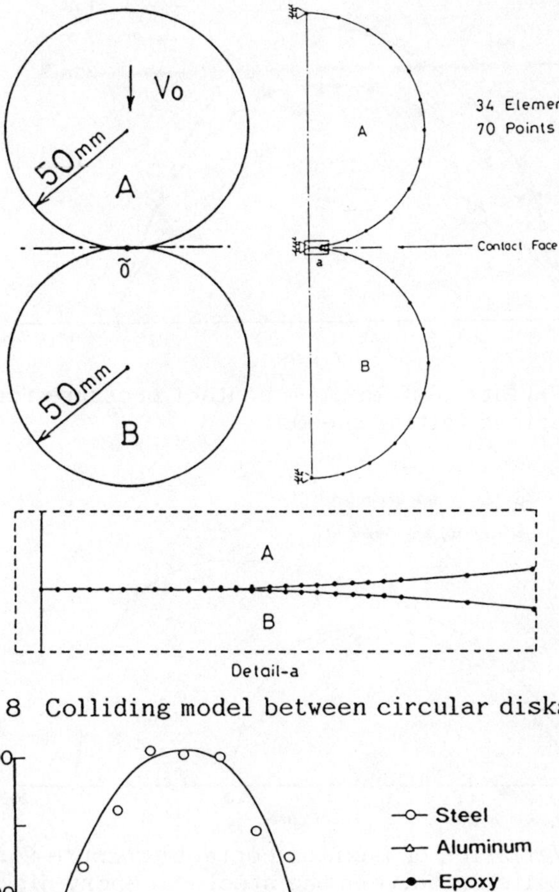

Fig. 8 Colliding model between circular disks.

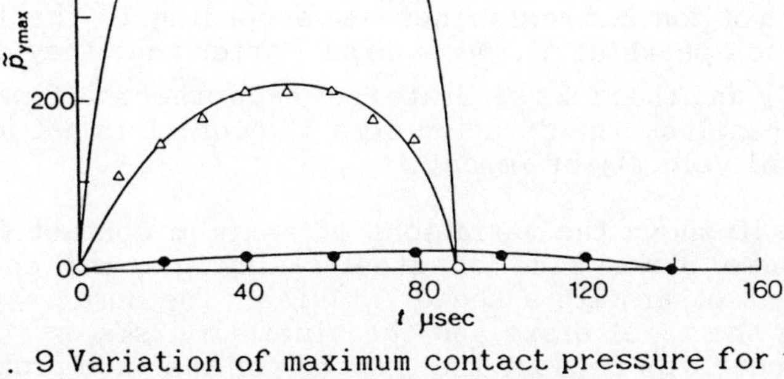

Fig. 9 Variation of maximum contact pressure for the steel, aluminum and epoxy disks.

Figure 9 shows the variations of maximum contact pressure analyzed by a time step $\Delta t = 10\,\mu s$ when a steel disk with the initial speeds $V_0 = 1, 2, 3\,\mathrm{m/s}$ collides with a steel disk. The distributions of the contact pressure at each time step are similar to that of Hertz's solution for the statically contacting

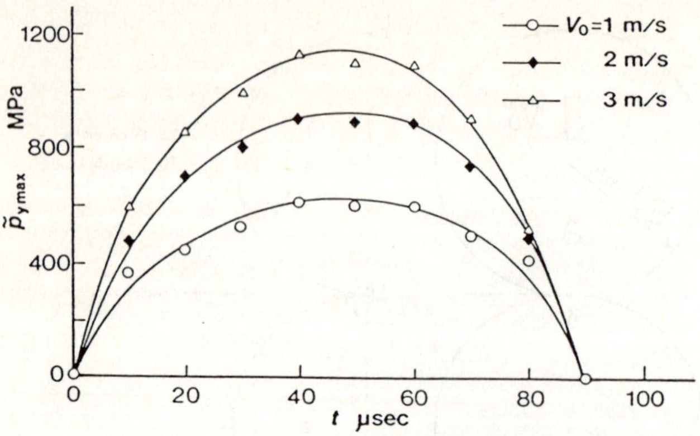

Fig. 10 Variation of maximum contact pressure for various initial speeds.

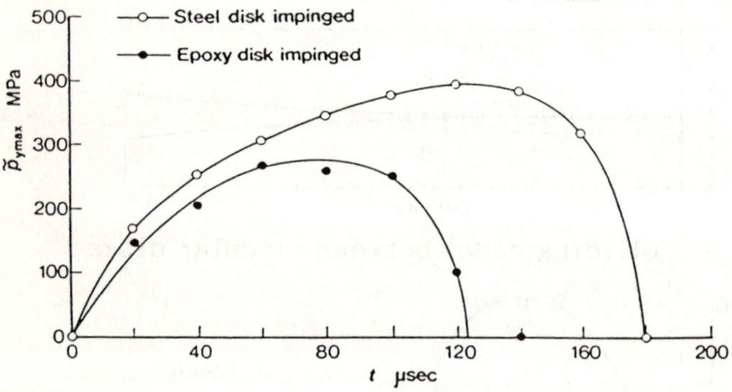

Fig. 11 Variation of maximum contact pressure for colliding between the steel and epoxy disks.

disks. After the initiation of contact the maximum value of $\tilde{p}_y$ and the width of contact region increase according to the lapse of time and reach peaks at a time $t=45\,\mu s$. After that they decrease monotonously and the disks separate from each other at a time $t=90\,\mu s$. From these results, the duration time of contact is not affected by the initial velocity of impact.

Figure 10 shows the variations of maximum contact pressure when the same disks made of steel, aluminum and epoxy are impacted each other with a speed $V_0=1$ m/s. The duration time of contact for the steel disks and the aluminum disks is the same value of about 90 μs because the speed of the wave propagation is nearly equal, but that for the epoxy disks is different from that of the other disks.

The variations of maximum contact pressure when the steel disk impacts on the epoxy disk and vice versa with a speed $V_0=1$ m/s are shown in Fig. 11. The variations are not symmetry with respect to the peak time and the duration time of contact for colliding steel disk is longer than that for the epoxy disk.

CONCLUSIONS

In this paper, an efficient boundary element technique of time domain approach for the elastodynamic contact problems is suggested by applying the penalty function method. The technique is applied to analyze the dynamic collisions between the rods and the disks. From the numerical results it is confirmed that the technique suggested here is valid and efficient.

REFERENCES

[1] Madsen, N., Numerically Efficient Procedures for Dynamic Contact Problems, Int. J. Num. Methods Engng, Vol. 20, 1-14, 1984.

[2] Asano, N., A Penalty Function Type of Virtual Work Principle for Impact Contact Problems of Two Bodies, Trans. Japan Soc. Mech. Engeers, Ser. A, Vol. 52, pp. 773-780, 1986.

[3] Kanto, Y. and Yagawa, G., Finite Element Analysis of Dynamic Buckling with Contact, Trans. Japan Soc. Mech. Engeers, Ser. A, Vol. 51, pp. 2747-2756, 1980.

[4] Pascoe, S. K. and Mottershead, J. E., Linear Elastic Problems Using Curved Elements and Including Dynamic Friction., Int. J. Num. Methods Engng., Vol. 26, 1631-1642, 1988.

[5] Andersson, T. B., Fredriksson, B. and Persson, B. G. A., The Boundary Element Method Applied to Two-Dimensional Contact Problems, Proc. 2nd Int. Conf. Recent Advances, CLM Publications, Southampton, 1980.

[6] Bezine, G. and Fortune, D., Contact Between Plates by a New Direct Boundary Integral Equation Formulation, Int. J. Solids Structures, Vol. 20, 739-746, 1984.

SECTION 3 - PENETRATION AND FRAGMENTATION

Experimental and Theoretical Approaches to the High Strain Rate Effect on the Properties of Armour Steels

Y. Meunier, R. Roux and G. Pont

Creusot-Loire Industrie, B.P. 56, 71202 Le Creusot, France

ABSTRACT

The reliable analysis of the problems associated with impact loading and the effective design of impact resistant structures require an adequate understanding of the strain rate dependence of the mechanical properties of the materials involved.

This paper emphasizes how the experimental dynamic response of armour steels, determined through shock type test equipment, over a wide range of strain rates, 10^{-3} to 10^{4} s^{-1}, can be restored by constitutive models, in terms of a mechanical equation of state relating stress to strain and strain rate for a definite structure of the material, and consistent with the well known microscopic strain mechanisms occurring at different strain rates.

1 - INTRODUCTION

The study of the dynamic plastic response of structures may consider two fairly distinct alternatives with many points of interaction :

- on the one hand, development of analytical models of the perforation process and application of relevant equations of motion and material behaviour ;

- on the other hand, full numerical solutions based on solving all the governing equations over a spatial grid at successive time increments.

The main uncertainty with these two procedures seems to be the proper characterization of inelastic material behaviour for moderate to high rates of straining as well as for large strains.

An adequate understanding of the strain rate dependence of the mechanical properties of the materials involved is required, if problems associated with impact loading are to be properly analysed, and if impact resistant structures are to be effectively designed.

In fact, the estimate of the perforating strength of armour steels, e.g. MEUNIER et al. /1 to 4/, penetrated by projectiles moving with a velocity lower than 1000 m/s, versus specific metallurgical properties of the armour, geometric and kinetic characteristics of the projectile, is founded methodically upon the understanding of (Figure 1) :

- the perforation mechanics of the armour, by analysing physically the penetration and perforation mechanisms occurring at the projectile-target impact time ;

- the dynamics of the armour, by studying the dynamic behaviour of the armour subjected to high strain rates, 10^{3} to 10^{5} s^{-1} ;

- the modelling of the armour perforation, through physical models allowing us to simulate numerically the dynamic perforation of the armour.

This dynamic behaviour then forms the middle link between the specific metallurgical properties of the armour steels and their perforating strength characteristics, in the attempt to estimate the ballistic protection by modelling the armour structural response.

The purpose of this paper is to emphasize how the experimental dynamic response of armour steels, over a wide range of strain rates, 10^{-3} to 10^{4} s^{-1}, can be restored by constitutive models, in terms of a mechanical equation of state relating stress to strain, strain rate and structure.

2 - DYNAMIC CHARACTERIZATIONS BY TENSILE AND COMPRESSIVE LOADINGS

2.1 - Studied materials The study of the mechanical behaviour under high strain rate applies to a carbon - nickel - chromium and molybdenum armour steel, 26NCD6 grade as designated by AFNOR standard. This steel is characterized for two hardness levels, 370 and 500 BHN, obtained by quenching heat treatment, followed or not by a tempering, as applicable.

2.2 - Experimental dynamic loading equipment The mechanical tests under dynamic loading were carried out by "Commissariat à l'Energie Atomique - Direction des Applications Militaires - Centre d'Etudes de BRUYERES-LE-CHATEL", using shock type test machines, each machine providing a specific range of strain rates :

. **hydraulic tensile machine**, by tensile loading, covering a strain rate range from 10^{-3} to 10 s^{-1} ;

. **cross-bow tensile machine**, by tensile loading, covering a strain rate range from 10 to 10^{3} s^{-1}. This apparatus (Figure 2) consists of two parallel guides between which a maraging steel projectile is launched by means of a rubber tape bundle. One of the two ends of the test piece is held by a long dynamometer, while the other one is secured to a sliding head. This head starts off upon the impact of the projectile, causing the test piece to be strained. The recorded values of stress- and strain-time, applied to it, allow us to plot the stress-strain curves.

. **split HOPKINSON pressure bar**, by compressive loading, covering a strain rate range from 5.10^{2} to 10^{4} s^{-1}. The principle of this apparatus uses the propagation of an elastic compressive wave across cylindrical bars, in order to load a sample into an uniaxial stress state. The sample to be squeezed (Figure 3) is held between two middle bars, input bar and output bar, on which are placed strain gages for recording data. The projectile is a striker bar, accelerated by an air gun, which impacts the output bar in order to produce the propagation of a compressive wave. A fourth bar, called the damp bar, is set behind the output bar and absorbs the kinetic energy of the projectile. The recorded values of the amplitude of the elastic compressive waves produced during the impact allow us to measure the stress applied to the sample, its strain and strain rate versus time.

. **impact test on a rigid target** or **TAYLOR's cylinder**, by impact loading, covering a strain rate range from 10^{4} to $6 \cdot 10^{4}$ s^{-1}. This experimental test (Figure 4) involves impacting a right circular cylinder against a rigid target with a firing velocity of a few hundred metres per second. If the stress level produced by the impact exceeds the yield strength of the material, the specimen buckles in the mushrooming shape, one portion of the cylinder being deformed plastically, while the other one is undeformed. The analysis of tests, performed according to the theory of WILKINS and GUINAN /5/, allows us to calculate average values of flow stress, strain and strain rate versus initial and post-impact shapes of the specimen.

2.3 - Experimental results

2.3.1 - 370 BHN armour steel The rational stress σ - strain ε curves, obtained in the strain rate range of 0.001 - 3000 s^{-1} (Figure 5), display higher stress levels as the strain rate increases. The strain rate sensitivity of the material becomes evident in a σ-log $\dot{\varepsilon}$ diagram for a given strain ratio (Figure 6), which shows two areas : a first area where sensitivity is low, and from $\dot{\varepsilon} \simeq 3000\ s^{-1}$ a second area where the stress increases significantly versus the strain rate.

2.3.2 - 500 BHN armour steel The rational stress-strain curves, obtained in the strain rate range of 0.1 - 3000 s^{-1} (Figure 7), show an obviously inconsistent behaviour as the strain rate increases, contrarily to the previous material. An assumption of explanation would be a higher sensitivity of this material to the thermal softening effect during the dynamic loading, causing a decrease of the flow stress. This armour steel also shows two areas of strain rate sensitivity (Figure 8) : a first area, $\dot{\varepsilon} < 10^3\ s^{-1}$, where the strain rate effect is low ; beyond this value, the flow stress increases slightly with the strain rate.

3 - MODELLING OF THE DYNAMIC BEHAVIOUR

3.1 - 370 BHN armour steel The results of dynamic compressive tests, obtained in the strain range of 0.02 - 0.10 where the strain is homogeneous (no shearing), allow us to model the material behaviour through empirical mathematical equations. Three constitutive models are studied :

. the power model of J.H. HOLLOMON : $\sigma = K\,\varepsilon^n$
. the linear model of P. LUDWIK : $\sigma = \sigma_0 + K\,\varepsilon^n$
. the multiplying model of D.J. STEINBERG : $\sigma = \sigma_0\,(1 + \beta\,\varepsilon)^n$.

Figure 9 exhibits an example of statistical curve fitting from these models, applied to the σ-ε diagram of the material. This example, typical of the whole of the SHPB tests, shows that the models of LUDWIK and STEINBERG are more consistent with the experimental dynamic behaviour of the material than the model of HOLLOMON.

These studied models do not take full account of the strain rate. By selecting the model of STEINBERG for σ-ε curve fitting, it is possible to obtain the strain rate effect on the progression of the σ_0 parameter, which reflects the yield strength of the material, and of the β and n parameters which reflect its hardened conditions by plotting the following graphs of equations : $\sigma_0 = f(\dot{\varepsilon})$, $\beta = g(\dot{\varepsilon})$ and $n = h(\dot{\varepsilon})$.

The statistical curve fitting analysis leads to select linear relationships for f, g and h functions, the full constitutive model being formulated as follows :

$$\sigma(\mathrm{MPa}) = (1150 + 4.5 \ln\dot{\varepsilon})\,(1 + (451 - 46 \ln\dot{\varepsilon})\,\varepsilon)^{(0.055 + 2\,.\,10^{-5}\,\dot{\varepsilon})}.$$

The function $f(\dot{\varepsilon}) = 1150 + 4.5 \ln\dot{\varepsilon}$ is consistent with the microscopic strain mecanisms and yields to an activation process for the plastic flow of the material, while the two other functions g and h are simply empirical.

As a matter of fact, for **low strain rates**, $\dot{\varepsilon} < 1\ s^{-1}$, the full modelling can be reduced to a logarithmic model

$$\sigma = \sigma_0\,(\varepsilon) + k\,(\varepsilon) \ln\dot{\varepsilon},$$

describing a thermally activated mechanism, in accordance with the general formulation of the mechanical response /6/ :

$$\tau = \tau_a + \Delta G_o/V + (kT/V) \ln \dot{\varepsilon}$$

where τ is the total applied stress, τ_a the athermal component of stress, ΔG_o the free energy of activation in the absence of stress, V the activation volume associated with the thermally activated process, k the BOLTZMANN's constant and T the absolute temperature, the logarithmic strain rate sensitivity parameter being defined as :

$$\lambda = (\partial \sigma / \partial \log \dot{\varepsilon})_{T,\varepsilon} \simeq 4\, kT/V$$

assuming, according to von MISES criterion, that $\sigma = \sqrt{3}\,\tau$.

On the other hand, for **high strain rates**, $500 < \dot{\varepsilon} < 10\,000\ s^{-1}$, the full modelling can be compared with the following function

$$\sigma = \sigma'_o(\varepsilon) + \alpha(\varepsilon)\,\dot{\varepsilon},$$

by expanding it according to the mean value theorem in the neighbourhood of $\dot{\varepsilon} = 1\,000\ s^{-1}$. For example, the following formulations obtained for :

. $\varepsilon = 0.05$: $\sigma\ (MPa) = 1346 + 2.9 \cdot 10^{-2}\,\dot{\varepsilon} + \sigma(\dot{\varepsilon}^2)$,

. $\varepsilon = 0.10$: $\sigma\ (MPa) = 1392 + 4.9 \cdot 10^{-2}\,\dot{\varepsilon} + \sigma(\dot{\varepsilon}^2)$,

are well describing a viscous friction mechanism, in accordance with the general formulation of the mechanical response /6/ :

$$\tau = \tau_a + \Delta G_o/V + \eta\,\dot{\varepsilon}$$

where η is the macroscopic viscosity.

The graph of the full constitutive model for this material is shown in Figure 10, in a σ-log $\dot{\varepsilon}$ diagram, for two strain ratios 5 and 10 %. In the studied strain range of 0-0.10, the strain rate sensitivity is actually sharp only beyond $\dot{\varepsilon} = 1000\ s^{-1}$, and seems to increase versus the strain ratio. There is a good agreement between the experiment obtained with the dynamic loading tests and the estimate made from the previous modelling.

To sum up, the mechanical behaviour of this material under dynamic loading can be modelled over a wide range of strain and strain rate. This modelling, sometimes empirical, is yet consistent with the well known microscopic strain mechanisms occurring at different strain rates.

3.2 - 500 BHN armour steel The mechanical behaviour of this steel was modelled using the three constitutive models previously studied. The graphs plotting the progression of the different model parameters versus the strain rate show a valid relationship for the σ_o parameter only, which then can be modelled as follows :

$$\sigma_o\ (MPa) = 1417 + 3.77 \ln \dot{\varepsilon},$$

which describes a thermally activated mechanism as for the previous armour steel, the logarithmic strain rate sensitivity parameters being similar for the two materials.

By selecting the model of STEINBERG for σ-ε curve fitting, and taking average values for the β and n parameters, it is possible to make a full constitutive model as follows :

$$\sigma(MPa) = (1417 + 3.77 \ln \dot{\varepsilon})\ (1 + 465\,\varepsilon)^{0.094}.$$

This approximation is all the more valid as this steel seems to be little sensitive to the strain rate (Figure 8).

The graph of the full constitutive model for this material is shown in Figure 11, in a σ-log $\dot{\varepsilon}$ diagram, for two strain ratios 3 and 10 %. The formulation of this constitutive model is suitable over a wide range of a strain rate, with yet a larger scatter of experimental results, as compared with the curve fitting, than for the previous material. This may be explained by the prevalence of either the material hardening or of the occurrencc of strain defects (plastic shearing) causing a decrease of the flow stress.

Also this formulation, which explains an activation process, can not take account of the sharp increase of the flow stress versus the strain rate for very high strain rate, $\dot{\varepsilon} \simeq 10^4\ s^{-1}$.

4 - CONCLUSIONS

In the scope of the understanding and estimate of the perforating strength of the armour steels versus their specific metallurgical characteristics, this paper has shown how the dynamic characterization of these steels, over a wide range of strain rates, 10^{-3} to $10^4\ s^{-1}$, can be determined through shock type test equipment. The dynamic response can be restored by constitutive models in terms of a mechanical equation of state, relating stress to strain and strain rate for a definite structure of the material, and consistent with the well known microscopic strain mechanisms occurring at different strain rates.

Then, it will be possible to introduce these constitutive models in computer codes in order to solve problems of dynamics under high strain rate. The accuracy of results obtained by numerical simulation of structure behaviour will be as reliable as the consistency of the constitutive models of the materials involved.

REFERENCES

1 Meunier Y., Sangoy L., Pont G., Metallurgical Aspects of Adiabatic Shear Phenomenon in Armor Steels with Perforation, in Impact'87 (Ed. Chiem C.Y., Kunze H.-D. and Meyer L.W.), pp. 711-718, Proceedings of the International Conference on Impact Loading and Dynamic Behaviour of Materials, Bremen, FRG, May 18-22, 1987, DGM Informationsgesellschaft mbH, Oberursel, 1988.

2 Meunier Y., Sangoy L., Pont G., Etude du comportement dynamique d'aciers à blindage, in Dymat 88 (Ed. Les Editions de Physique, Association Dymat), pp. C3/109-C3/114, Proceedings of the International Conference on Mechanical and Physical Behaviour of Materials under Dynamic Loading, Ajaccio, Corsica, September 19-23, 1988, les Editions de Physique, Les Ulis, 1988.

3 Meunier Y., Sangoy L., Pont G., Dynamic Behaviour of Armoured Steels, in International Conference on Ballistics (Ed. East China Institute of Technology), pp. IV/74-IV/81, Proceedings of the International Conference on Ballistics, Nanjing, China, October 25-28, 1988, ECIT, Nanjing, 1988.

4 Meunier Y., Sangoy L., Pont G., Steels for Ballistic Protection, in the Fourth Israel Materials Engineering Conference (Baram J. and Eliezer D.), Proceedings to be published, Beer-Sheva, Israel, December 7-8, 1988.

5 Wilkins M.L., Guinan M.W., Impact of Cylinders on a Rigid Boundary, J. Appl. Phys., Vol. 44, No.3, pp. 1200-1206, March 1973.

6 Harding J., Material Behaviour at High Rates of Strain, in Impact'87, ibid. 1, pp. 23-42.

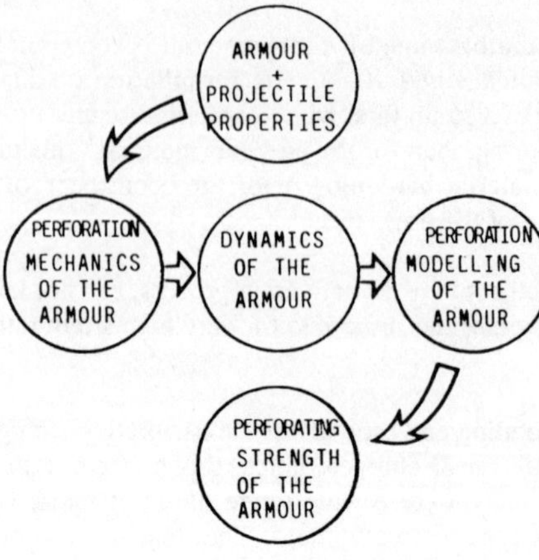

FIG. 1 - Estimate of the perforating strength of armour steels.

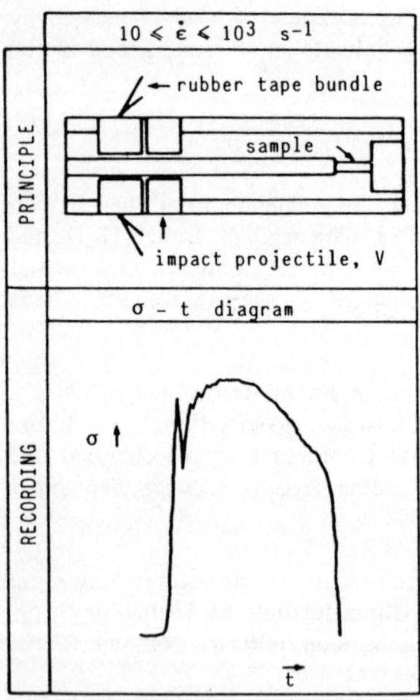

FIG. 2 - Tensile loading equipment : crow-bow machine.

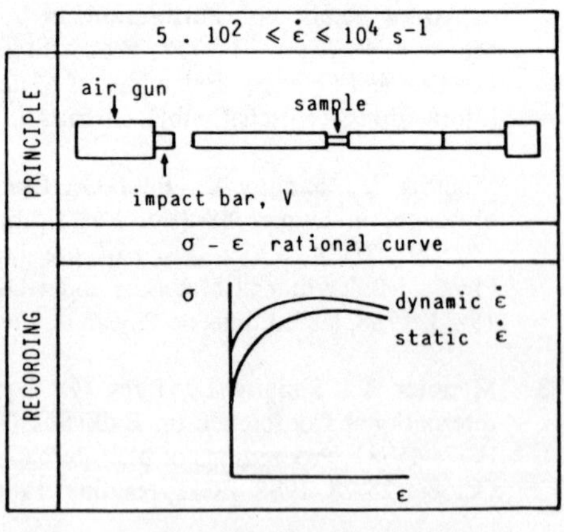

FIG. 3 - Compressive loading equipment : split HOPKINSON pressure bar.

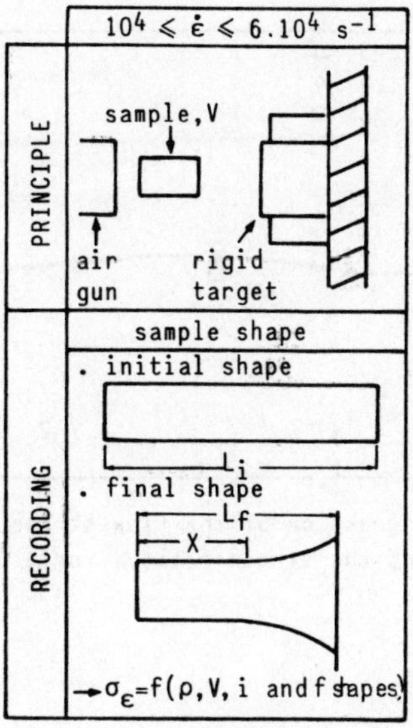

FIG. 4 - Impact loading equipment : TAYLOR's cylinder on a rigid target.

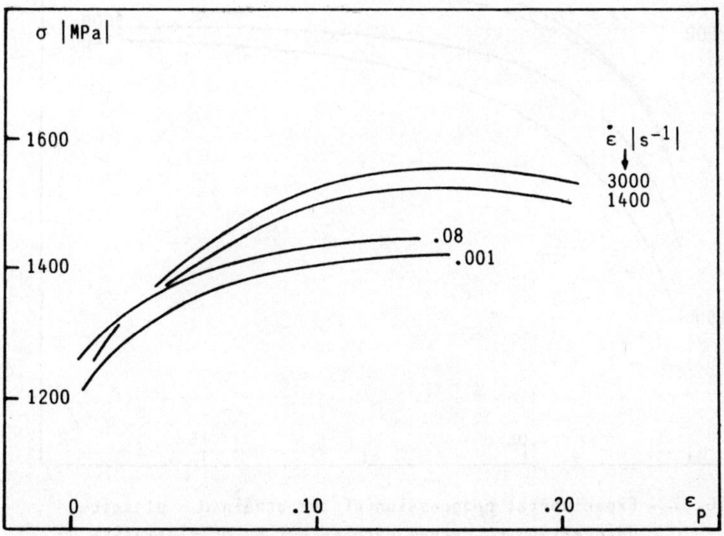

FIG. 5 - Experimental progression of the stress σ-plastic deformation ε_p curves carried out by quasi-static and compressive loadings for a 370 BHN armour steel.

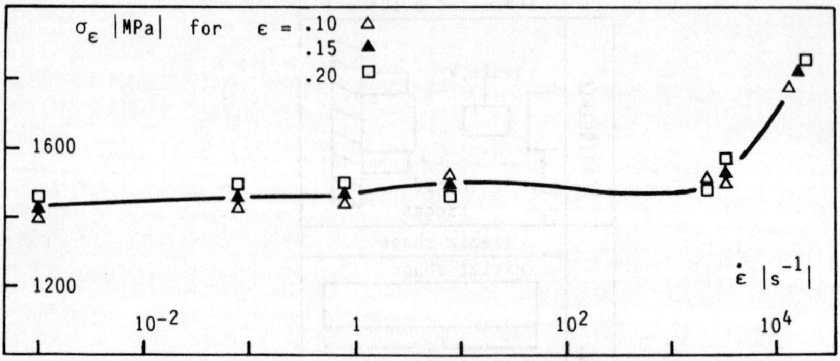

FIG. 6 - Experimental progression of the flow stress σ_ε versus the strain rate $\dot{\varepsilon}$ and the strain ratio ε for a 370 BHN armour steel.

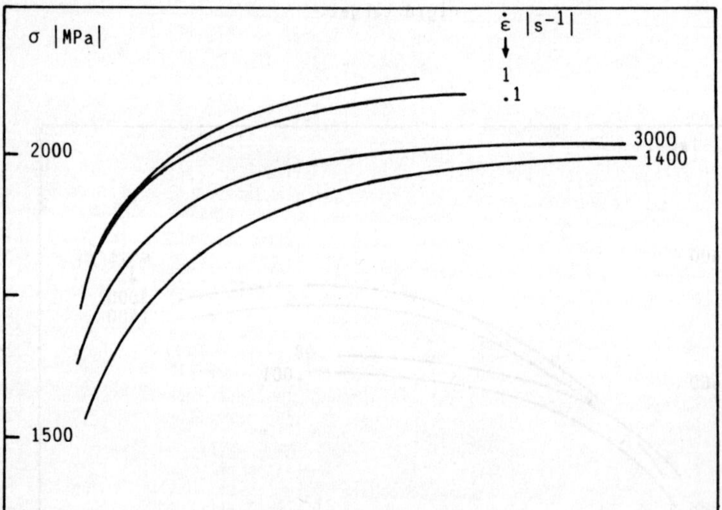

FIG. 7 - Experimental progression of the strain σ - plastic deformation ε_p curves carried out by quasi-static and compressive loadings for a 500 BHN armour steel.

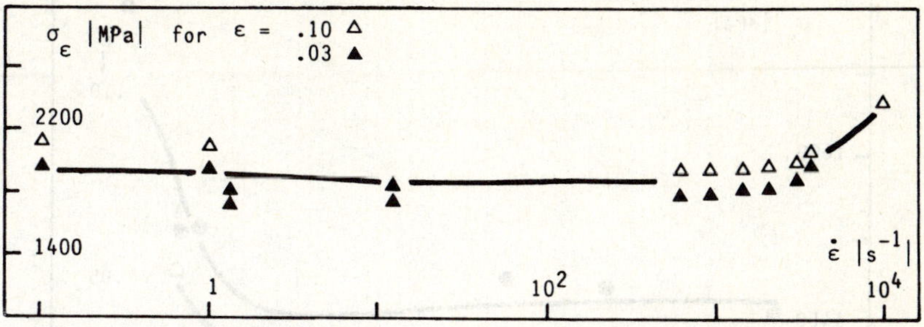

FIG. 8 - Experimental progression of the flow stress σ_ε versus the strain rate $\dot{\varepsilon}$ and the strain ratio ε for a 500 BHN armour steel.

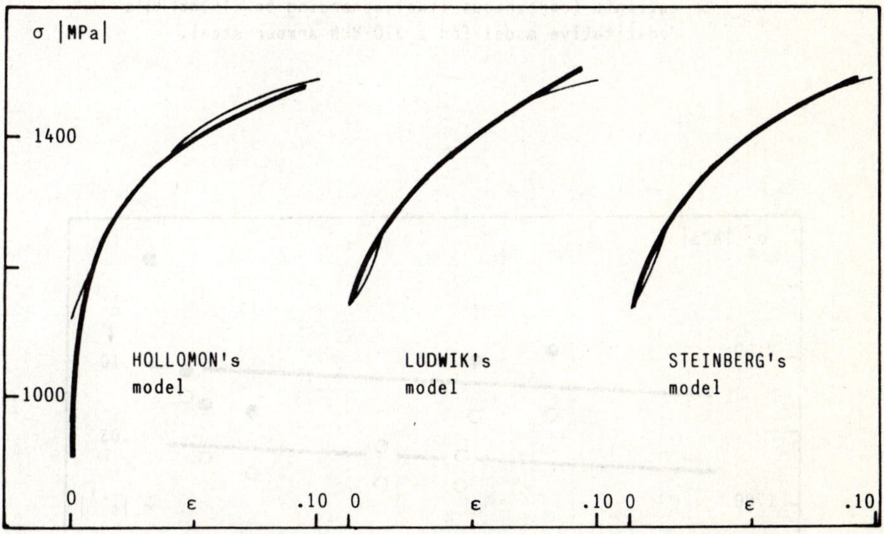

FIG. 9 - Comparison of the strain ratio dependence of the flow stress for $\dot{\varepsilon}$ = 1380 s^{-1} between experiment (thin line) and estimate (heavy line) according to three constitutive models for a 370 BHN armour steel.

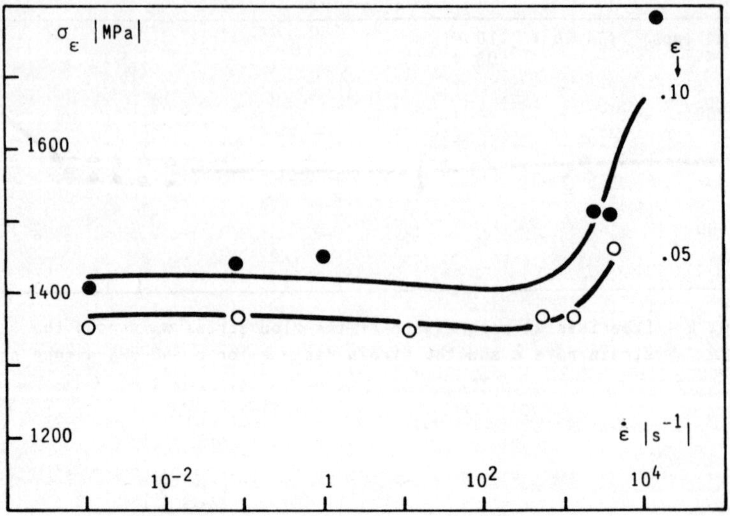

FIG. 10 - Comparison of the strain rate dependence of the flow stress between experiment (conventional signs) and estimate (continuous line) according to STEINBERG's constitutive model for a 370 BHN armour steel.

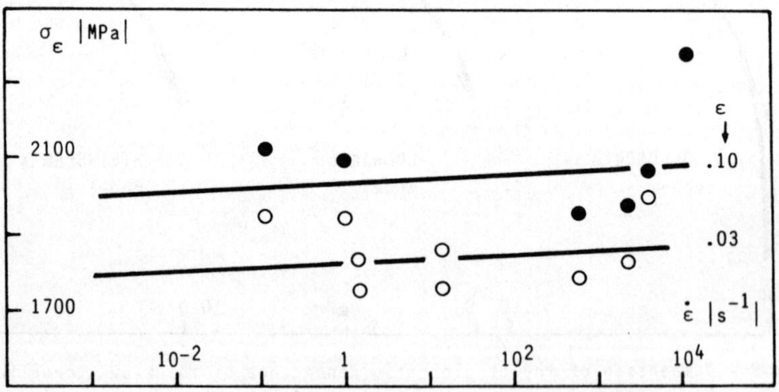

FIG. 11 - Comparison of the strain rate dependence of the flow stress between experiment (conventional signs) and estimate (continuous line) according to STEINBERG's constitutive model for a 500 BHN armour steel.

Mathematical Models in Penetration Mechanics

J. Dehn

USA Ballistic Research Laboratory, Aberdeen Proving Ground, MD 21005, USA

ABSTRACT

Three types of mathematical models in use are described. These range from simple, empirical models suitable as subroutines in larger codes to complicated wave codes requiring many hours of cpu time on state-of-the-art machines.

SIMPLE MODELS

The Ballistic Research Laboratory uses many models both experimental and mathematical for different purposes. Here we will describe three types of mathematical models which are used to simulate the terminal effects of projectile impact on a target. Since each has its advantages and limitations different ones are used for different jobs.

The simplest model used is a pair of empirical equations called the Thor equations which summarize a large body of experimental plate perforation data. The residual speed of a compact projectile after perforation is represented by

$$V = V_o - 10^a M_o^b V_o^c (\sec\theta_o)^d T^e \tag{1}$$

where M_o, V_o and θ_o are the striking mass, speed and obliquity of the projectile while T is the target plate thickness and the base 10 is arbitrary. The parameters a,b... are determined for a particular projectile/target material combination by curve fitting. A similar equation for residual mass begins $M = M_o - \ldots$ with a second set of five fitting exponents on the same factors. If these equations are used outside the range of the experimental data they can give erroneous or even impossible results like negative masses. Other limitations and advantages of such models have been discussed [1].

Why is such a model still used in our high-tech age? The answer is simple. Large computer codes which analyze the vulnerability of complicated aircraft or ground vehicles often consider many attack angles for every mechanical, electrical, chemical, optical, electronic and human component. Little time is available to determine penetration by primary and secondary (spall) fragments. When millions of penetration subroutine calls are made, only microseconds can be spent on each. Penetrations by rods or jets require more complicated models and take more time, but in no case can we afford to use wave codes which take hours of cpu time on the fastest computers. Exterior ballisticians use simplified models to describe mile long trajectories and employ wave codes only in the immediate vicinity of an artillery shell. Terminal ballisticians must do analogous things.

INTERMEDIATE MODELS

The more complicated models referred to above can also be useful in initial design studies of armors and penetrators. Such models should be based on physical laws and contain very few adjustable parameters. Because of this they can be qualitatively correct well beyond conditions already tested and can be used to predict qualitative trends when geometric and material changes are made. Judicious use of such models can shorten design tasks and reduce the number of experiments needed. Many models of this type exist and everyone has their favorite. Here is one example.

Assume that the remaining mass M of a projectile depends on its initial mass M_o and the depth of penetration P:

$$M = M_o - \mu_o P \qquad (2)$$

Here μ_o is an adjustable constant for each projectile/target material combination but is independent of speed and other factors. P is determined by solving the force equation

$$M \ddot{P} = -(a_o + c_o \dot{P}^2) \qquad (3)$$

where a_o and c_o are combinations of material parameters. Since $\dot{M} = -\mu_o \dot{P}$ from Eq (2), we can include an $\dot{M}\dot{P}$ term in the $\dot{P}^2$ term on the right. The only other assumed parameter is a constant flattening factor, (A/A_o), where A is the projectile cross-sectional area, initially A_o.

Consider the example of a steel rod with aspect ratio length/diameter = 22mm/2mm =11 and mass 7.8 g striking a semi-∞ magnesium target, a case for which time-dependent experimental data is available [2]. The solution to equation (3) for time $0 \leq t \leq 40$ μs can be written as

$$P = 64(\text{mm}) \; (1 - [1 + (4 - t/10)^2]/17) \tag{4}$$

where the numbers are combinations of known parameters [3]. From this we can find M(t) in equation (3) as well as $\dot{P}$. The remaining length of the rod is $L=M/(\rho_o A_o)$ where ρ_o is the constant rod density. A library of μ_o and (A/A_o) values can be built up from a number of basic experiments and rules developed for estimating them when experimental data is lacking. Fig 1 shows experimental points determined by flash x-rays (triangles) together with curves calculated using Eq (4). Such close agreement is possible only in cases used to determine the two adjustable parameters. However qualitative trends can be predicted correctly in quite different cases.

CODE MODELS

Sometimes experiments are difficult or impossible to carry out for a variety of availability, cost or safety reasons. For example, foreign military equipment may not be available. If trends are not good enough and best estimates of performance are needed, wave code calculations may be desirable. Many codes of varying complexity are available for this purpose and everyone has their favorites. Here I will give an example of an armor design which is too difficult to simulate with an intermediate model. It is a generic case not associated with any particular equipment and is only meant to show an example which is too complicated for intermediate models. The calculations were done using the HULL code [4] on a Cray computer.

Fig 2 shows a tungsten alloy rod 20 μs after striking a block of alumina ceramic at 1,700 m/s. Some pressure waves have already reflected from the sides of this relatively small target. Fig 3 shows that cracking has begun 70 μs after impact. A smaller alumina block cracks completely after only 50 μs as we see in Fig. 4. If we surround the block with 2 cm of armor steel first, ceramic cracking is still severe as shown in Fig 5.

Fig 6 shows a wider ceramic block surrounded by a layer of plastic which isolates it from an outer aluminum confinement box. It is about to be struck by a steel rod moving at 1,700 m/s. Some 20 μs after impact the rod nose slows dramatically while its erosion rate increases. Fig 7 shows how much the rod has eroded without really penetrating the ceramic much 40 μs after impact. By 50 μs its speed increases significantly as it begins to penetrate the ceramic and by 150 μs about half the rod emerges from the target moving at about 80% of its striking speed. Fig 8 shows complete target perforation, while Fig 9 shows the position histories of the nose and rear of the rod. Clearly the target is not an adequate defense against this attack. Improved designs can be achieved and the total task time reduced if gun range experiments and computer simulations are done interactively. Still more complicated designs are very difficult to simulate with intermediate models.

Since we do the experiments anyway, why do we bother with computational simulations too? Part of the answer is the fact that many ballistics experiments are not well instrumented. Modern targets and projectiles can be very complicated and even with two or three (superimposed) flash x-ray pictures it is often difficult to determine the sequence of events or even the major factors at work because of large quantities of target and projectile debris. If cine x-radiography becomes a common tool, the experimental situation will be much improved although the cost will rise. Ideally, we would like to do time-resolved experiments and computations interactively with the cost of the latter expected to decrease. Geometric and material variations are easier to do using computational physics, but we will always need experimental trial-by-fire. Better experiments and better calculations done interactively should challenge each other and shorten the overall effort.

We have briefly examined three types of mathematical models used in penetration mechanics. Purely empirical models still have a place as subroutines in large vulnerability codes, while intermediate models can be helpful in understanding the performance of individual armor/anti-armor components, thus aiding in the design of more complicated systems. Code calculations can be helpful in the design of ever more complicated modern arrays and may even be the best tool available when experiments are difficult or impossible for one reason or another. The interested reader can find further information on the use of mathematical models in penetration mechanics by consulting references [5] and [6].

[1] J.Dehn, BRL TR 2188 (1979).

[2] E. Perez, Science et Techniques de l'Armement 56 (1), 11 (1982).

[3] J. Dehn, BRL TR 2770 (1986).

[4] D. A. Matuska and R. E. Durrett, AFATL TR 78-125, (1978).

[5] J. Zukas et al., Impact Dynamics, (New York, Wiley, 1982).

[6] C. E. Anderson, Jr. and S. R. Bodner, "Ballistic Impact: the Status of Analytical and Numerical Modeling", Int. J. Impact Engng 7 (1) 9-35 (1988).

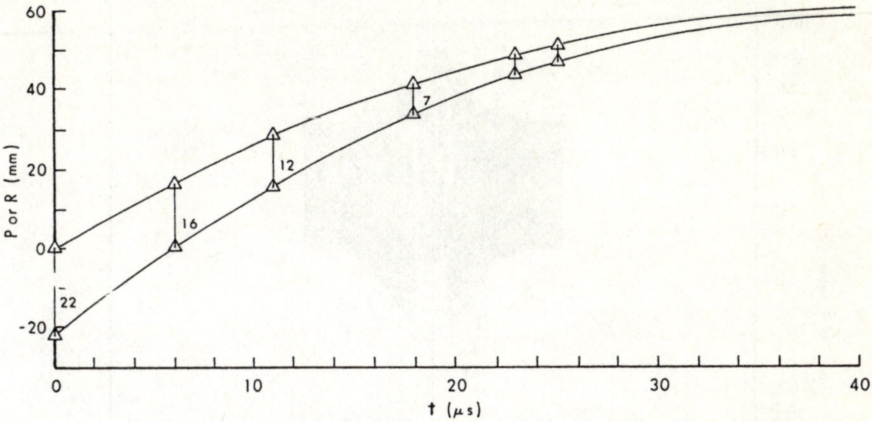

Fig. 1. Front and rear positions versus time for a steel rod impacting a magnesium target at 3,270 m/s. Numbers on vertical lines indicate decreasing length in mm.

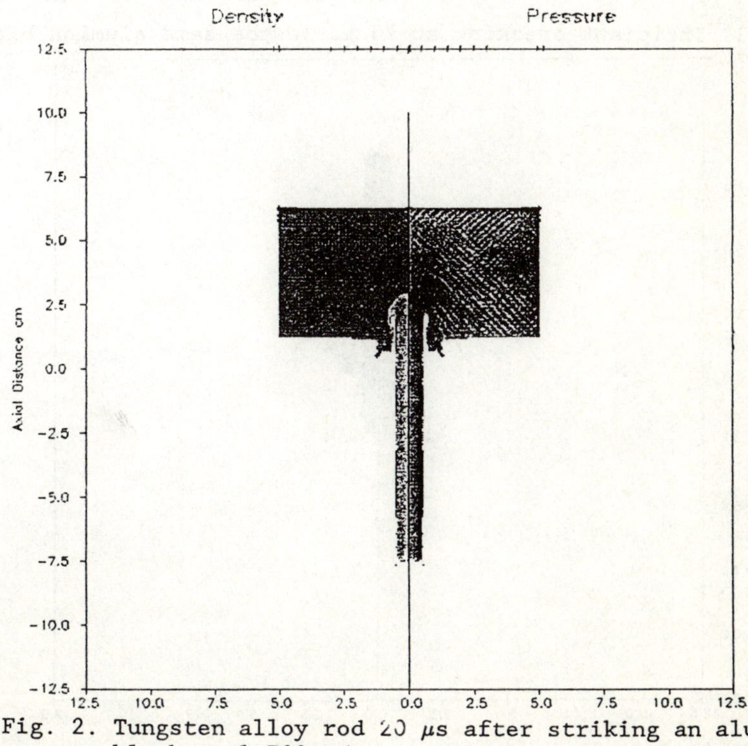

Fig. 2. Tungsten alloy rod 20 μs after striking an alumina block at 1,700 m/s.

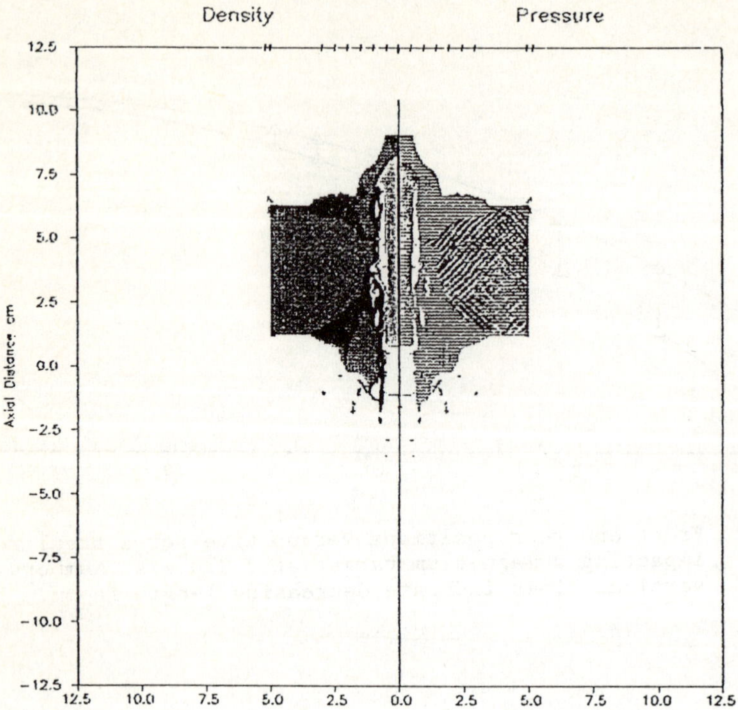

Fig. 3. Incipient cracking at 70 μs in the same alumina block.

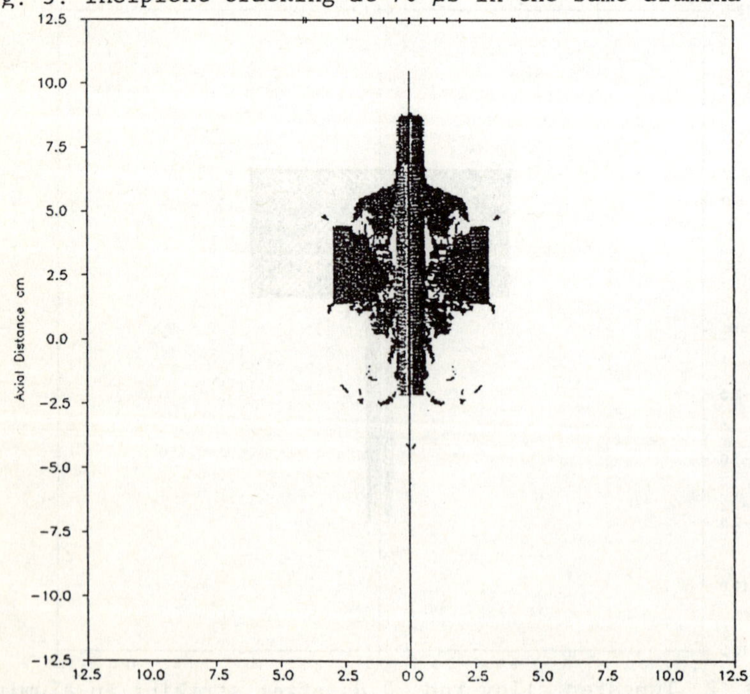

Fig. 4. A smaller alumina block cracked after 50 μs.

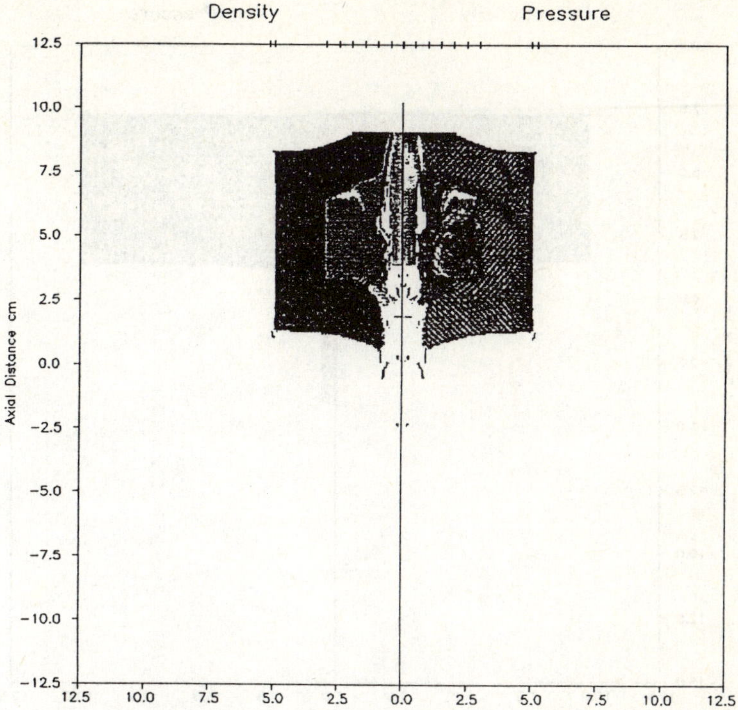

Fig. 5. The smaller alumina block in a steel surround 90 μs after impact.

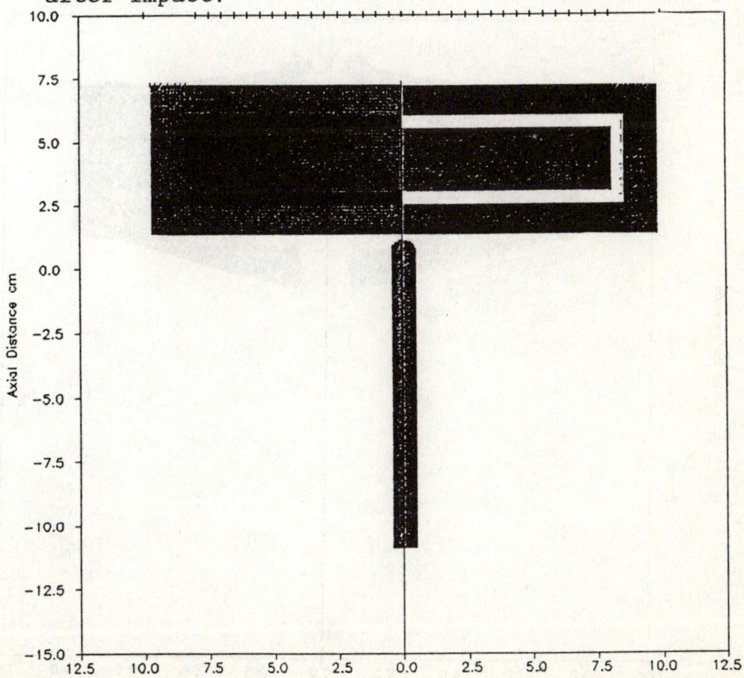

Fig. 6. Wider alumina block isolated by plastic in an aluminum surround about to be struck by by a steel rod at 1,700 m/s.

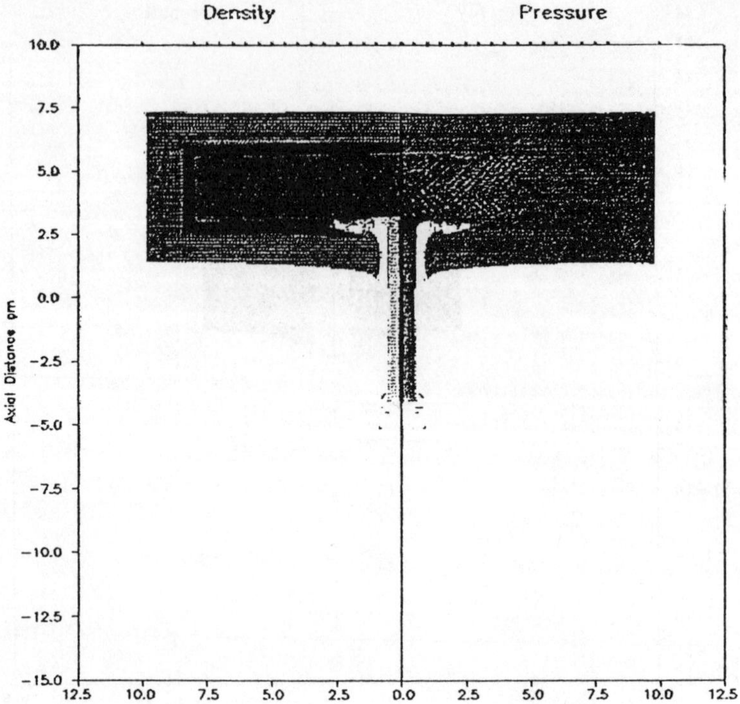

Fig. 7. Case of Fig 6 40 μs after impact showing penetration delay and rod erosion.

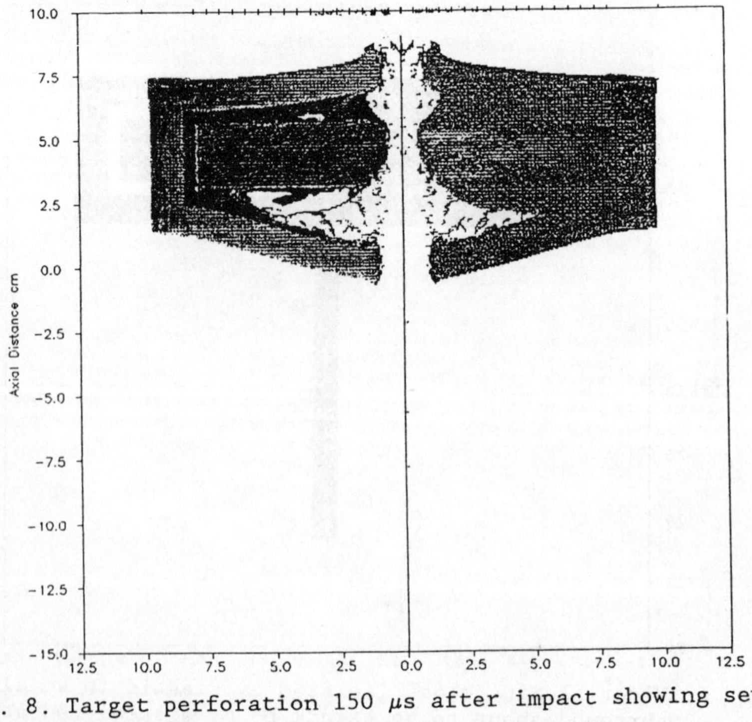

Fig. 8. Target perforation 150 μs after impact showing severe distortion.

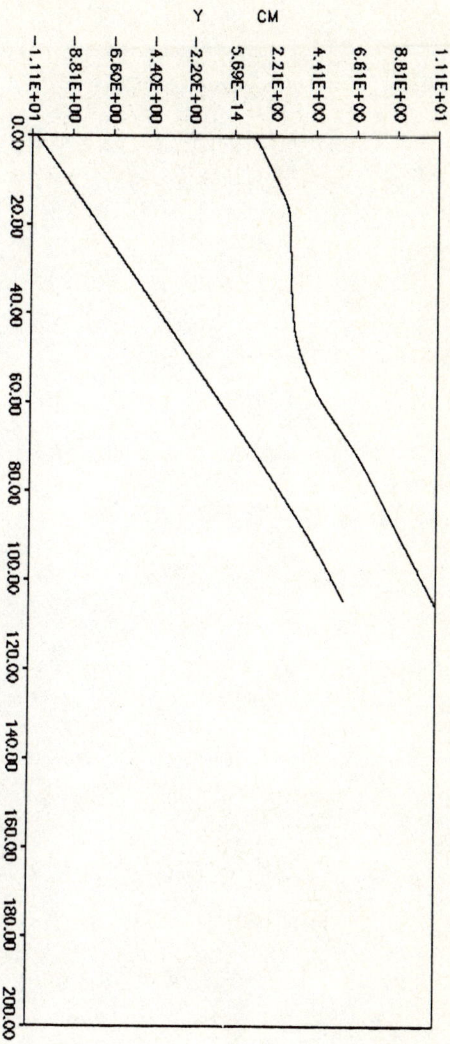

Fig. 9. Displacement of rod nose and rear showing velocity decrease when passing through the ceramic.

Hypervelocity Fragmentation and Terminal Effectiveness

Ian G. Cullis
Royal Armament Research and Development Establishment, Fort Halstead, Sevenoaks, Kent, United Kingdom

ABSTRACT

The application of two and three dimensional eulerian hydrocode evaluations to the study of hypervelocity fragment impact and penetration of thin target plates is discussed and representative results presented.

INTRODUCTION

The response of a structure to hypervelocity fragment impact is of interest to a large number of varied branches of science and engineering.

This response is quite complex since it can cover the whole range of material behaviour (Schulz [1], Kinslow [2]) from dynamic failure to vaporisation of both fragment and target, the resultant cloud of debris being dispersed over a significantly greater area than the initial impact. The vast experimental database generated during the 1960s is fully reported in the Proceedings of the Symposia on Hypervelocity Impact.

Empirical Models

In the field of ordnance applications, of interest in this paper, the empirically observed sensitivity of fragment breakup and target response to impact velocity, orientation, materials properties and relative fragment/target dimensions have been used to define a series of analytic models based on dimensional analysis (eg Yatteau [4], Naz [5]]. Although these models essentially average out detailed target response effects they nevertheless provide an invaluable tool for lethality modelling studies. As performance requirements for future fragmenting warheads become more demanding, however, the detailed physics of the fragment target interaction (and the importance of material properties) needs to be understood. Much of this detailed understanding, however, cannot be gained through experimentation alone, since flash x-ray or velocity probe data is difficult to interpret and cannot supply information on many of the required physical field variables. This is particularly true when determining the timing of fragment and target failure.

Hydrocode Modelling

Hydrocode modelling studies, when carefully integrated into an experimental programme can supply much of this missing information. At RARDE we attempt to use theoretical modelling to predict the first experiment and the experiment to modify the model. Very often the lack of material data or the inadequacy of material algorithms and numerical methodology implies that the numerical simulations fail to provide the necessary degree of agreement over the complete timescale of the experiment.

Nevertheless, numerical simulations do provide a sound understanding of the physics involved and likely sensitive material properties or impact situations. Equally important, however, because a calculation can simulate an ideal scenario with precise material behaviour, the dominant and sensitive physical processes can be identified.

Most early calculations (Kinslow [2]) addressed impact situations at extremely high velocity where melting and vaporisation of fragment and target are important. Good agreement between experiment and calculations can be obtained in such cases since the problem is essentially hydrodynamic. For example the eulerian code HULL used at RARDE can predict quite well (Figure 1) the debris cloud from the impact of a 20g lead sphere impacting a 6.3mm lead plate at 6.6km/s (Burkett [3]). The calculated debris envelope is in very good agreement with the experiment, although detailed agreement with the axial density profile is less good at highlighting inadequacies in the equation of state and the numerical treatment of the thermodynamics.

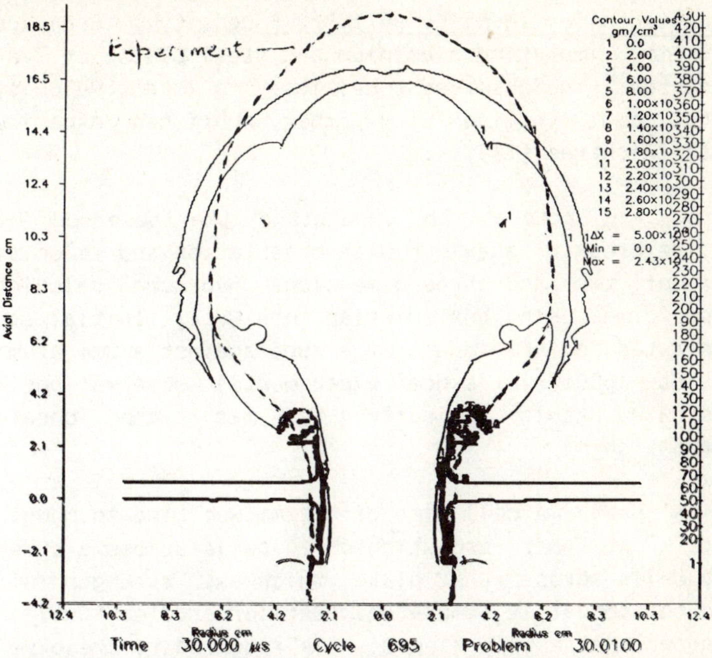

Figure 1: Lead Sphere Impact of 6.6km/s, time 30 usec

Similarly, recent studies (Schulz [1], Naz [5]) have qualitatively addressed the importance of fragment/target dimensions and impact velocity on debris behaviour at lower impact velocities where strength effects dominate. An understanding of the magnitude and form of the initial shock waves generated on impact is vital since their subsequent development and release defines the form and nature of the fragment failure, the behind target debris cloud, and the condition of the fragment prior to impacting subsequent target plates. Equally the ability of current hydrocode methodologies to model these problems needs to be assessed.

MODELLING STUDIES

Preformed fragmenting warheads employ fragments that are either cubes to maximise fragment packing number, or spheres for aerodynamic stability. Both fragment types have been studied theoretically and experimentally at RARDE.

CUBE IMPACT STUDIES

Experiments at RARDE, using the 2-stage gas gun (Richards [6]) have identified various fragment breakup modes in tungsten alloy (W-Ni-Cu 90/7.5/2.5 density = 18.0) cubes, side 4.5mm, impacting aluminium and steel plates of various thicknesses. The observed sensitivity to impact velocity and orientation is in line with other published experimental results (Richards [7]).

In an attempt to understand the observed highly sensitive breakup behaviour with orientation and velocity, a number of two and three dimensional hydrocode calculations were performed using the eulerian code HULL. Initial studies concentrated on the impact of a cube against a 3mm aluminium plate at 1500m/s, since experimental observations had demonstrated this velocity to mark the onset of fragmentation.

Numerical Mesh - A cell size of 0.22mm was used to model the impact. At this resolution of 20 cells across a cube side and 10 cells across the plate thickness strength effects should be adequately simulated. Earlier studies (Cullis [8]) have shown that a very fine mesh is required to resolve the formation and development of fracture surfaces in a HULL calculation. In a 3 dimensional calculation this requirement imposes memory requirements beyond the capability of our CRAY1S. A fracture algorithm was therefore not included in the solution methodology. As we are interested in the early evolution of the penetration process this does not represent a significant restriction on our numerical simulations.

Flat Face Impact - The normal impact of a cube face against a 3mm aluminium plate generates 200kbar shocks that propagate into the cube, at 4.3mm/ usec, and plate, at 7mm/ usec. Radial expansion coupled with the lateral release waves from the cube sides reduces the peak shock pressure to 180kbar at 0.5 usec (Figure 2) as well as imposing a curvature on its profile. The grazing incidence of the shock with the vertical cube sides, however, generates a rarefaction fan rather than the intense tensile pulse associated with normal incidence shock reflection at a free surface responsible for spalling. The rarefaction fan is thus unable to generate a tensile pulse as it propagates back into the cube (Figure 2). As it does so it imposes the symmetry of the cube on the incident compressive shock.

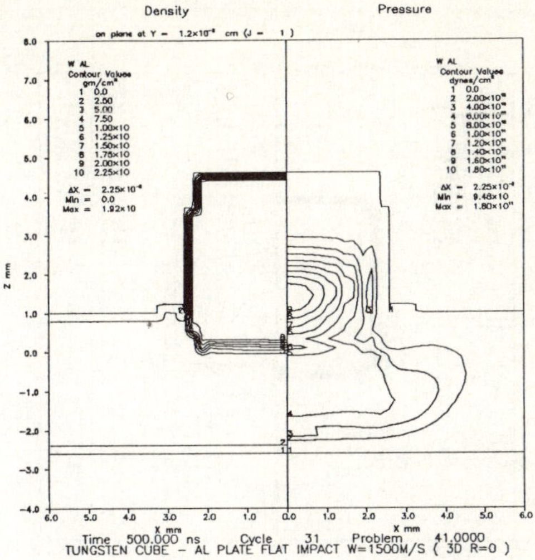

Figure 2: Tungsten Cube Flat Face Impact 0.5 usec

The plate thickness to fragment diameter ratio (T/D) at 0.67 is such that, given the relative shock and sound speeds for tungsten and aluminium, the reflected tensile wave from the rear of the plate arrives back (~0.8 sec) at the cube-plate interface before the shock in the cube has reflected from its top surface. Thus for this impact scenario the release wave from the plate is the dominant mechanism in the development of the penetration process.

By 1- usec the cube plate interface has begun to separate and a strong tensile pulse has developed along the axis of the cube (Figure 3). The aluminium plate thus acts as a momentum trap and explains the experimental flash x-ray observation (Yatteau [4]) of aluminium plate material being projected ahead of the fragment after penetration . The magnitude of the tensile pulse generated in the cube is 80kbar, of the same order, as the spall strength determined by Anderson [9] working with W-Ni-Fe alloys.

The interaction of the release waves generated by the reflection of the original incident shock with this tensile region is such that, with the exception of the corners, the cube is in tension (Figure 4). By 1.5 usecs, however, further reverberation generates tensile enhancements along the upper diagonals of the cube and over a plane 0.5mm thick at the impacted face - Figure 3.

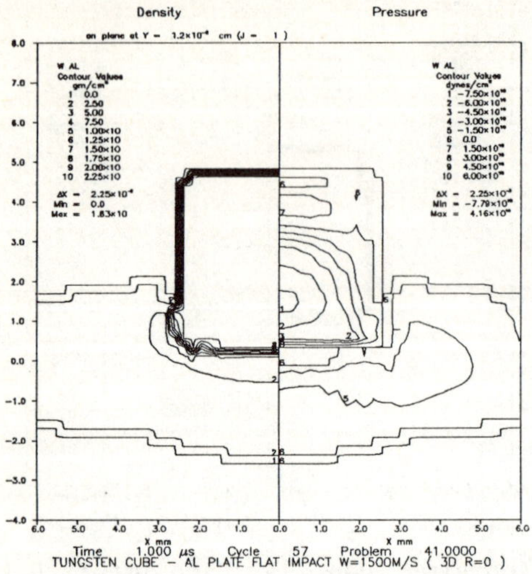

Figure 3: Tungsten Cube Flat Face Impact 1 usec

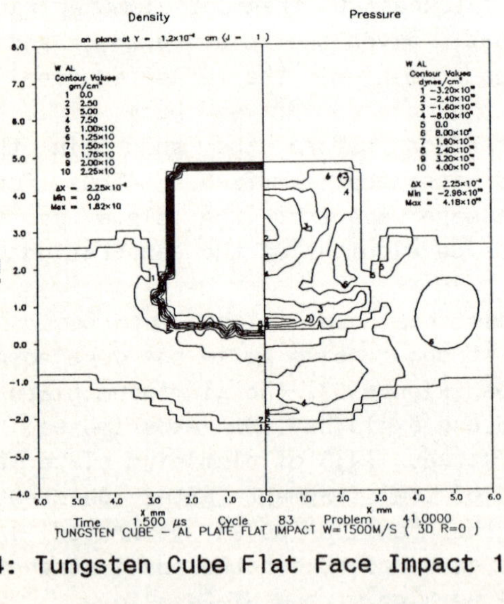

Figure 4: Tungsten Cube Flat Face Impact 1.5 usec

Edge Impact - In the case of an edge impact the cube/plate impact surface is reduced to a line. The impact shocks therefore decay much more rapidly as they propagate, until at 0.5 usecs the shock pressure in the cube is 50% of the pressure in the flat impact orientation (Figure 5). The geometry of the impact is such that the intersection velocity is supersonic. Release waves can only originate, therefore, from the end free surfaces, as evidenced by the pressure plot (Figure 5), and the horizontal diagonal once the cube has penetrated to that depth. The tensile release from the rear of the plate, because of its reduced velocity does not reach the cube plate interface until 1.3 useconds. The interface does not now separate to any significant degree and the tensile pulse transmitted into the cube is 20% of the peak level calculated for the flat impact. For this orientation, therefore, both the plate and cube release waves control the breakup mechanism.

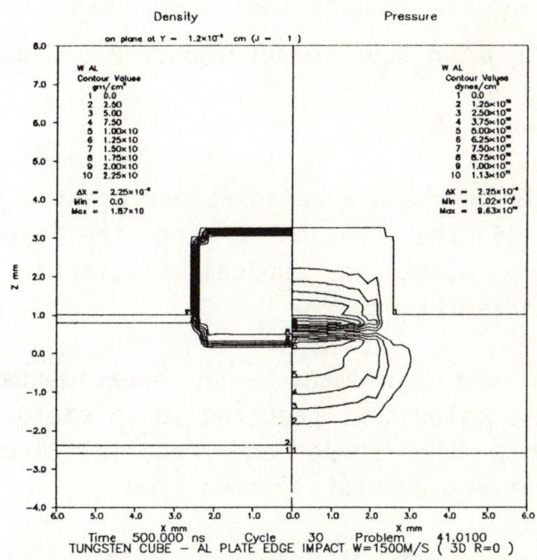

Figure 5: Tungsten Cube Edge Impact 0.5 usec

<u>Corner Impact</u> - For a cube corner impact the impact surface degenerates into a point. The impact shocks therefore decay even more rapidly than the other 2 orientations, until by 0.5 useconds the peak shock pressure is 42kbar, 25% of the flat impact pressure. The intersection velocity for this orientation and velocity is just subsonic and release waves can propagate into the cube and plate, shaping the shock pressure profile accordingly (Figure 6). Once again both the release waves generated by the impact shock in the cube and the plate control any breakup mechanisms.

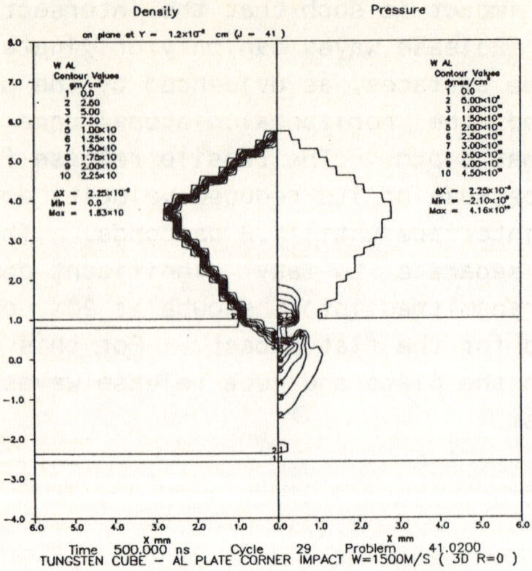

Figure 6: Tungsten Cube Corner Impact 0.5 usec

Cube Impact Summary

The analysis of these 3 calculations clearly allows the identification of the physics behind the experimentally observed breakup modes in cubical fragments and their sensitivity to orientation.

In general the agreement with experimental data for impact velocities below that required to initiate breakup is excellent. Both the shape and material erosion of the emerging fragment are accurately predicted.

The flat face impact is clearly more severe in the initiation of breakup once the critical impact velocity has been reached. Other orientations need higher impact velocities to initiate breakup, in good agreement with experiment.

The combination of fragment and plate density, sonic and shock properties and relative dimensions (T/D) is perhaps even more sensitive than previously realised.

For thin plates the plate release wave controls the mode and timing of cube breakup. For very thin plates, and implied short pulse durations, it can even overtake the shock within the cube, implying an increase in breakup velocity.

For thick plates the cube release waves control the breakup mode and its timing. As the cube dimension increases, the release waves, able to travel further and having broader pulse shapes before interfering, are more likely to initiate breakup; ie the breakup velocity will be lower.

For intermediate thickness plates, both mechanisms will be important. In this regime the breakup velocity will be lower, since although neither mechanism on its own may be able to initiate breakup; their combined effect can.

The calculations have, therefore, qualitatively explained the experimentally observed 'V' form of breakup velocity with T/D and the influence of cube size reported by Yatteau [4] for steel cubes impacting aluminium plates.

Sphere Impact Studies

The importance of fragment shape on the symmetry of the shock and release waves during penetration has been demonstrated to have an important influence on breakup mechanisms. To study this further a calculation investigated the impact of an equal mass tungsten alloy sphere (diameter = 5.6mm) against a 3mm aluminium plate.

The impact characteristics of an equal mass sphere show many similarities with a cube. The impact surface is now a circle which expands at the intersection velocity. Initially this is supersonic preventing the propagation of release waves into the interior of the fragment until much later in the penetration process (Figure 7). Once again the spherical expansion of the shock greatly reduces its magnitude compared with the flat face cube impact. For this velocity the sphere and plate remain in close contact throughout the penetration process. The arrival of the plate release wave propagates a tensile pulse into the sphere which interacts with the sphere release wave to generate a tensile disc at its centre). At 1500m/s the magnitude of the tensile pressure at 25kbar is probably unable to initiate significant breakup consistent with the experimentally measured breakup velocity of 1720m/s, and the published results of Naz [6], and Finnegan [10].

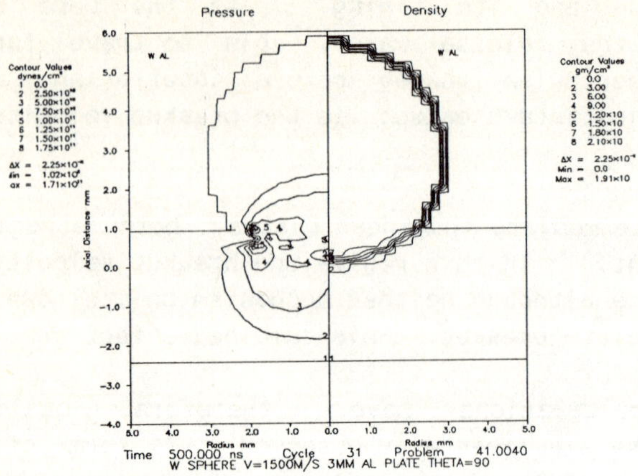

Figure 7: Tungsten Sphere Impact 0.5 usec

For spherical fragments, therefore, symmetry effects can be expected, through radial convergence, to play a more important role in the initiation of breakup.

FRAGMENT BREAKUP

The calculations have identified a number of possible failure modes in cubes ranging from loss of corners (Figure 2), axial rupture (Figure 2), "apple coring" (Figure 3), reported by Yatteau [3], and surface spall (Figure 3). For spherical fragments the calculations suggest spallation will be the primary failure mode. They also predict that a given failure mode will initiate at a specific time after impact.

The mode of fragment breakup is highly dependent on its dynamic fracture properties and its deformation history. Brittle fracture, characteristic of sintered tungsten alloys, is governed by the magnitude of the tensile stress or strain. Ductile fracture, characteristic of mild steel, is governed by the triaxiality of the stress field. At high rates of strain and elevated temperature, however, the dynamic material properties can undergo marked behavioural changes such that a ductile material can appear brittle and a brittle material more ductile.

The degree of fragment deformation during penetration can also influence the breakup mode, since a high energy deformation state in a steel projectile can be relived by a tensile pulse without initiating failure (Finnegan [10], Figure 2). To complicate matters further the duration of a tensile pulse is a deciding factor in whether failure is initiated and if so whether the voids thus created coalesce to form cracks.

Future Experimentation

Future experimentation clearly needs to investigate the timing and pattern of break up with velocity and orientation to determine the governing failure mechanisms and estimate the fragments dynamic fracture properties as described by Anderson [9]. With this improved level of experimental data new fracture models can be developed that are capable of predicting fragment breakup and resolving some of the issues outlined above. This in turn will demand improved numerical methodologies and material algorithms.

CONCLUSIONS

This paper has described the results of a series of two and three dimensional eulerian hydrocode calculations simulating the impact of cubical and spherical tungsten projectiles against a thin aluminium plate. By concentrating on the physics of the penetration process during the first 3 microseconds the calculations have identified the important features that subsequently influence the behaviour of the behind plate debris and fragment breakup.

Whilst the lack of a failure algorithm in the solution limits the ability of a calculation in predicting late time behaviour, the calculations have still qualitatively described the observed breakup modes and their sensitivity to plate thickness.

The confidence gained from these studies performed at RARDE will enable future theoretical work to address the inclusion of failure models to confirm the next series of experiments. In the longer term this will enable the development of failure models capable of supplying accurate data for improved fragment and performance algorithms in lethality codes.

Acknowledgement

The author would like to thank Mr M Burkett for supplying details of the Los Alamos experiments with lead projectiles.

REFERENCES

1. Schulz J, Heimdahl E, Finnegan S. Computer Characterisation of Debris Clouds Resulting from Hypervelocity Impact. Int J Impact Engineering, 5: 577-584 1987.

2. Kinslow R. High Velocity Impact Phenomena. Chap 4, Academic Press 1970.

3. Private Communication.

4. Yatteau J, Recht R. High Speed Penetration of Spaced Plates by Computer Fragments. Ninth International Symposium on Ballistics 2:365-374, Shrivenham 1970.

5. Naz P. Spaced Plates Penetration by Spherical High Density Fragments at High Velocity. Tenth International Symposium on Ballistics 2: San Diego 1987.

6. Richards A, Bowden A. The Development of a Projectile Velocity Measuring System for the RARDE No 2 Hypervelocity Launcher. Int J Impact Engineering, 5:519-523, 1987.

7. Richards A. Unpublished MOD Report 1988.

8. Cullis I, Nash M. The Use of Fracture Models in the Theoretical Study of Explosive-Metal Interactions. Ninth International Symposium on Ballistics 2: 285-292, Shrivenham 1986.

9. Anderson C, O'Donoghue P, Bodner S. Fracture of Cubical Tungsten Specimens under Explosive Loading. Int J Impact Engineering 7 No 4: 371-377, 1988.

10. Finnegan S, Schulz J, Heimdahl O. Fragmentation Characterisation Techniques for Ordnance Speed Impacts. Tenth International Symposium on Ballistics 2, San Diego 1987.

Numerical Analysis of Spall Fracture in Ductile Metals

R. Cortés*, M. Elices* and Z. Bílek**

*Department of Materials Science, Polytechnic Universty of Madrid, E.T.S. Ingenieros de Caminos, Canales y Puertos, 28040 Madrid, Spain

**Institute of Metallurgy, CAS, 616 62 Brno, Zizkova 22, Czechoslovakia

ABSTRACT

In this paper, the results of the numerical simulation of spall experiments in ductile metals are presented and compared with the corresponding experimental observations. The particular situation under study is that of cylindrical specimens made of aluminium, copper and steel, subjected near their axis of symmetry to the action of an explosive loading. The specimens were found to fail by spall fracture near the free surface, due to the reflection of the initial pressure wave as a tensile wave at such surface. The above referred explosive load experiments were modelled numerically by employing a Lagrangian finite difference code for dynamic 2-D and axisymmetric problems allowing the complete thermo-mechanical analysis of the shock process experienced by the specimens. The analysis was performed by considering an elasto-plastic behaviour of the material including strain-rate and temperature effects on flow stress. Rupture criteria such as those of the critical tensile stress and of the Tuler and Butcher type were used for predicting spall fracture of the specimens. Finally, the degree of of approximation to reality obtained was discussed, and recommendations for further work on spall fracture modelling were performed.

INTRODUCTION

Spall damage is a type of fracture produced when large tensile stresses are developed in material bodies as a result of the interaction of stress waves. Since the duration of the loading pulse is short, damage develops independently at different points of the material, and interacts in a complicated manner with the stress wave. Dynamic loads which may produce spall fracture result from high velocity impact, detonation of high intensity explosives and short term radiation deposition. Damage levels ranging from a few small cracks or voids to complete fracture of regions of the material may be produced under varying loading conditions. Various aspects of the spall problem have been reviewed (Meyers and Aimone [1]; Curran, Seaman and Shockey [2]; Buchar, Bílek and Dusek [3]), and several attempts have been made to develop a predictive capability for spall damage (Davison and Stevens [4]; Davison, Stevens and Kipp [5]; Johnson [6]; Seaman, Curran and Shockey [7]).

This paper describes the results of the study of the spall damage and fracture development in aluminium, copper and steel by applying computational and experimental procedures. The particular situation under study is that of cylindrical specimens subjected to an axial pressure pulse near their axis of symmetry. As it will be seen, in real experiments such specimens were found to fail by spall fracture near the free surface, due to the reflection of the initial pressure wave as a tensile wave at such surface.

The experiments were modelled numerically by use of a Lagrangian finite difference code for dynamic 2-D and axisymmetric problems allowing the complete thermomechanical analysis of the impact loading experienced by the specimens. Various spall fracture criteria were implemented in the computer code, and the predicted final damage distributions were compared with the experimental findings. The good agreement of both supports the validity of the theoretical approach adopted in this paper.

EXPERIMENTAL WORK

From the given materials the cylindrical specimens of 100 mm in diameter and 20 mm in height were prepared. These specimens were loaded by means of an explosive located on the upper face of the specimen, around the axis of symmetry, as shown in Figure 1. The properties of the explosives used in the experiments are shown in Table I. The action of such explosive was concentrated within a circular region of 30 mm in diameter. The cover plate shown in Figure 1, and which is located between the specimen and the explosive, has the effect of modifying the maximum loading pressure exerted on the specimen by simply changing its thickness. In this manner, with the same value of the charge height, different values of the maximum loading pressure can be obtained. The loading pressure histories within the loaded region were measured with accuracy by means of manganin foil gauges up to the maximum values p_m of the pressure exerted by the explosive. For larger times, the load was expected to decrease very rapidly, in a fairly exponential form. The materials employed in this study were aluminium, copper and a steel very close to Armco iron. The chemical compositions of such materials are indicated in Table II, where materials A, B and C correspond to aluminium, copper and steel, respectively . Aluminium also contained 0.002% Mg, and in steel was also present 0.026% S and 0.019% P. On the other hand, the values of the maximum loading pressures reached in each experiment are indicated in Table III.

Table I. Properties of the explosives used in the experiments

Explosive	Heat of Explosion kJ/kg	Detonation Velocity m/s	Density kg/m^3
Semtex S35	2050	2750	1100
Semtex 2	1420	5500	2200
Semtex 1A	4970	7400	1000

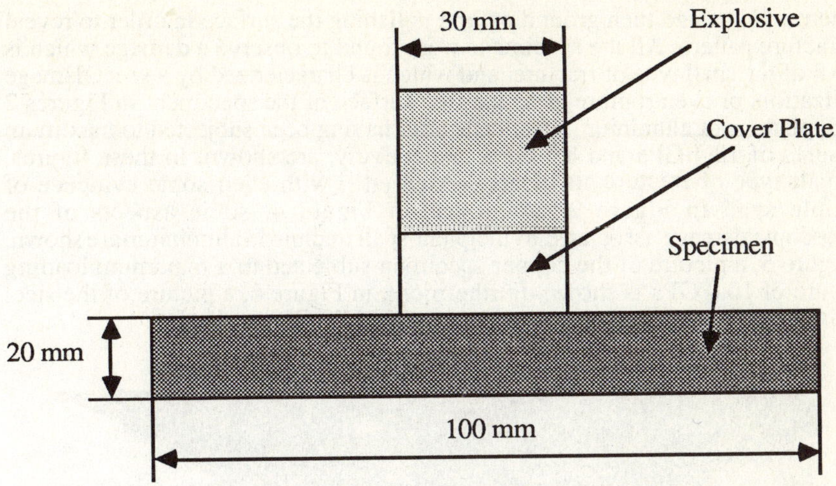

Figure 1. Experimental arrangement for the investigation of spall fracture due to explosive loading.

Table II. Chemical composition of the materials used in the experiments.

Material	Content in %							
	Al	Cu	Fe	Mn	C	Zn	Ni	Cr
A	99.84		0.14	0.02		0.002		
B		99.61	0.006			0.38		
C			99.56	0.23	0.082		0.05	0.03

Table III. Values of the maximum loading pressure reached in the experiments.

Material	Pressure (GPa)
Aluminium	8.4
Aluminium	10.2
Aluminium	10.8
Aluminium	12.4
Copper	10.0
Steel	14.9

The fracture damage experienced by the specimens as a result of the explosive loading, was studied by sectioning the recovered specimens by a diametral plane, and then grounding and polishing the surface in order to reveal the fracture pattern. All the specimens were found to observe a damage which is typical of the spall type of fracture, and which is characterized by a great damage localization, or even rupture, near the free surface of the specimen. In Figures 2 and 3, pictures of aluminium specimens after having been subjected to maximum pressures of 12.4 GPa and 8.4 GPa, respectively, are shown. In these figures, the spall type of fracture is clearly observed , with even some evidence of multiple spall in Figure 2. Moreover, in Figure 4, some aspects of the microscopic damage associated to incipient spall fracture in aluminium are shown. In Figure 5, a picture of the copper specimen subjected to a maximum loading pressure of 10.0 GPa is shown. Furthermore, in Figure 6, a picture of the steel specimen subjected to a maximum pressure of 14.9 GPa is shown.

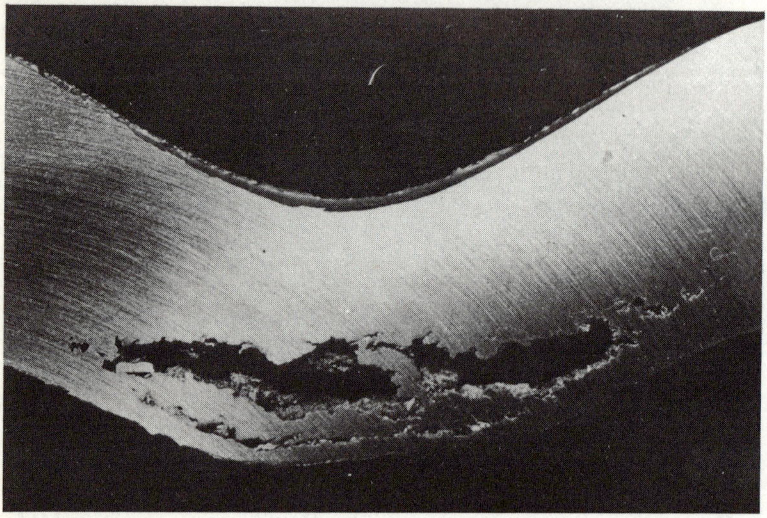

Figure 2. An aspect of an aluminium specimen after having been subjected to a maximum loading pressure of 12.4 GPa (2x; Amplification: 12%).

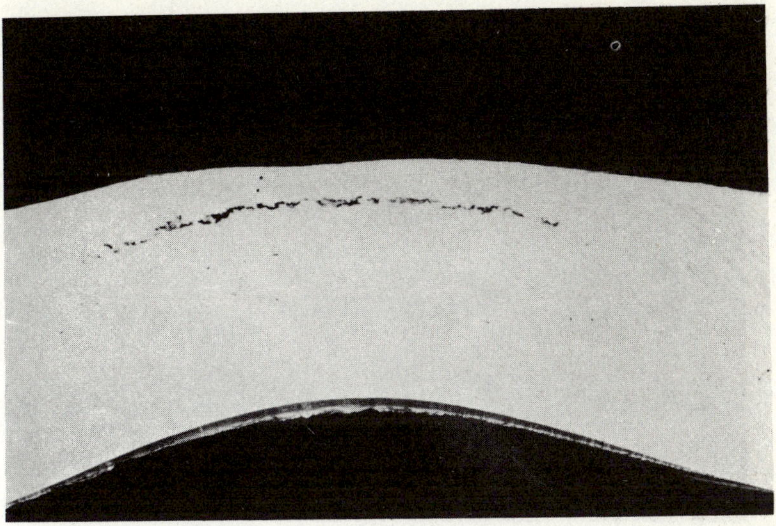

Figure 3. An aspect of an aluminium specimen after having been subjected to a maximum loading pressure of 8.4 GPa (2x; Amplification: 12%).

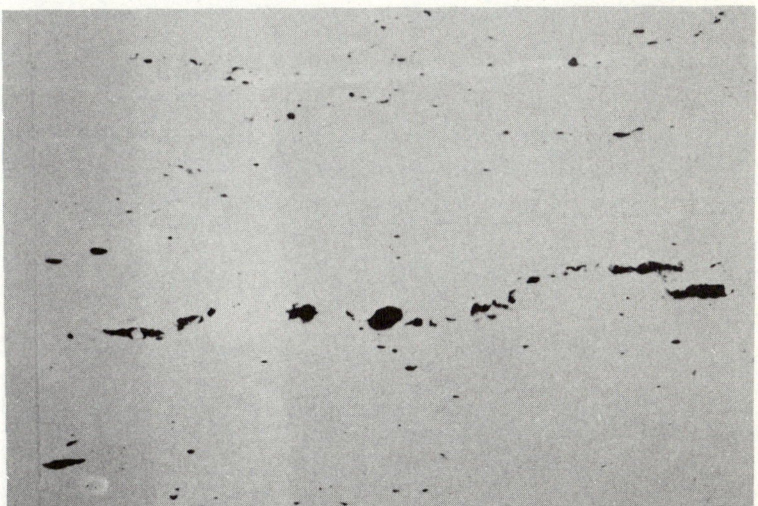

Figure 4. An aspect of incipient spall in an aluminium specimen after the experience (40.8x; Amplification: 7%).

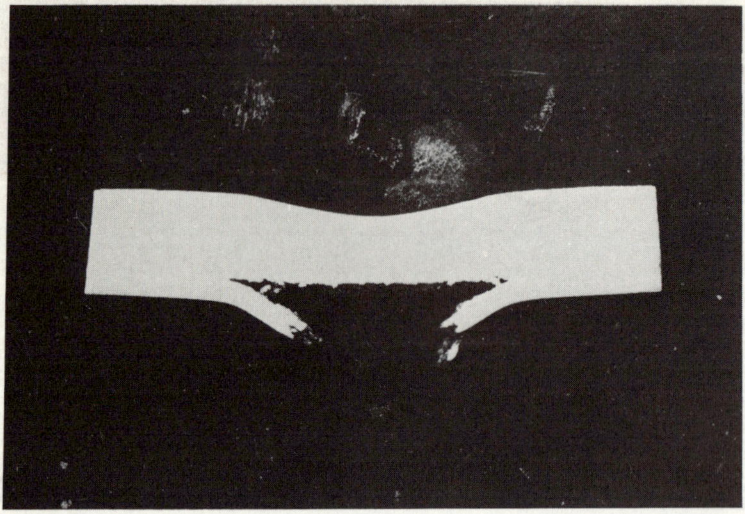

Figure 5. An aspect of a copper specimen after having been subjected to a maximum loading pressure of 10.0 GPa (0.75x; Amplification: 17%).

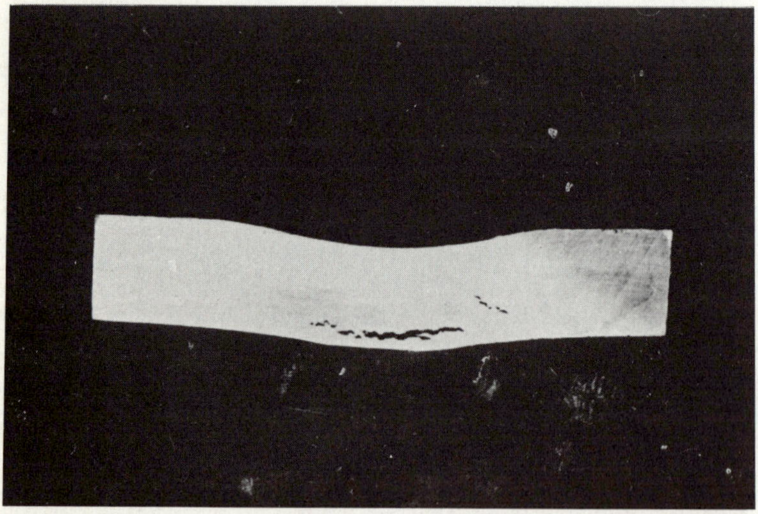

Figure 6. An aspect of a steel specimen after having been subjected to a maximum loading pressure of 14.9 GPa (0.75x; Amplification: 16%).

NUMERICAL ANALISIS

The explosive load experiments described in the previous section were then analysed numerically, by means of an explicit finite difference code for the dynamic analysis of coupled thermo-mechanical two-dimensional or axisymmetric problems, which make use of a Lagrangian formulation. The computer code allows to consider elasto-plastic or elasto-viscoplastic behaviour of the materials, including the thermal softening effect. The subdivision of the continuum is made by means of quadrilateral cells, and it allows to treat with an arbitrary topology, in a similar fashion to ordinary finite element codes. From the conditions of axial symmetry only one half of the in-diameter plane of the specimen was modelled, and null radial displacements were imposed at the axis of symmetry. At the outer radius of the modeled section, a simple support condition of the specimen was assumed. In the modelisation, 986 nodes and 924 quadrilateral cells were employed. A fully coupled thermo-mechanical analysis of the problem was performed. For aluminium a Young modulus of 71 GPa was used, a Poisson ratio of 0.35, a yield stress of 300 MPa, a density of 2700 kg/m^3, a thermal expansion coefficient of 2.3×10^{-5} K^{-1}, a thermal conductivity of 249 $Wm^{-1}K^{-1}$ and a heat capacity of 902 W/kg K. For copper, a Young modulus of 119 GPa, a Poisson ratio of 0.33, a yield stress of 900 GPa, a density of 8930 kg/m^3, a thermal expansion coefficient of 1.77×10^{-5} K^{-1}, a thermal conductivity of 399 $Wm^{-1}K^{-1}$, and a specific heat of 385 W/kg K were used in the calculations. Finally, for steel a Young modulus of 210 GPa, a Poisson ratio of 0.28, a yield stress of 1067 GPa, a density of 7850 kg/m^3, a thermal expansion coefficient of 1.1×10^{-5} K^{-1}, a thermal conductivity of 73 $Wm^{-1}K^{-1}$, and a specific heat of 452 W/kg K were used in the calculations. Strain rate effects on the yield stress were also taken into account. Linear thermal softening was also considered for the materials, up to the corresponding melting temperature, when the yield stress of the materials was assumed to vanish.

From the experimental measurements of the pressure pulse, reliable values of the applied pressure histories at different distances from the axis of symmetry of the specimen were estimated, which were introduced as an input in the computer runs.The analyses were made up to 12 μs after the instant of load initiation. For aluminium, a critical stress criterion for rupture, defined by the condition:

$$\sigma = \sigma_c \tag{1}$$

was used, an the critical stress for spall fracture σ_c was taken equal to 2.7 GPa. For steel, the same rupture criterion was also used, the critical stress for spall fracture being equal to 1.4 GPa. For copper, the above criteria was also used, and the critical stress for rupture was taken equal to 2.5 GPa. Furthermore, a damage accumulation criterion of the Tuler and Butcher type, defined as:

$$K = \int_0^t (\sigma - \sigma_0)^n dt \qquad (2)$$

was used for this latter material, the critical damage for rupture K being equal to 1.9 x 10^{-7} GPa s, the threshold stress for damage initiation σ_0 being equal to -0.75 GPa and the damage exponent n being equal to 2.

In Figure 7, the deformed mesh for aluminium subjected to a maximum pressure of 8.4 GPa at a time of 8 μs is shown. The ruptured zones of the material are shown in dark. It can be seen that a good correspondence with the experimental rupture pattern shown in Figure 3 has been obtained. Moreover, in Figure 8 the deformed mesh for the case of the copper specimen subjected to a maximum pressure of 10.0 GPa at a time of 8 μs is shown, when the Tuler and Butcher rupture criterion was used. In Figure 9, the same situation is illustrated when the critical stress rupture criterion was used for copper. Furthermore, in Figure 10 the deformed mesh for steel at a time of 8 μs, for the case of a maximum pressure of 14.9 GPa is shown. In general, it can be seen that the development of the ruptured zones predicted by the numerical analysis agrees relatively well with the experimental observations. Moreover, in the case of steel, the numerical analysis has correctly predicted the appearance of multiple spall, as can be seen by comparing figures 6 and 10.

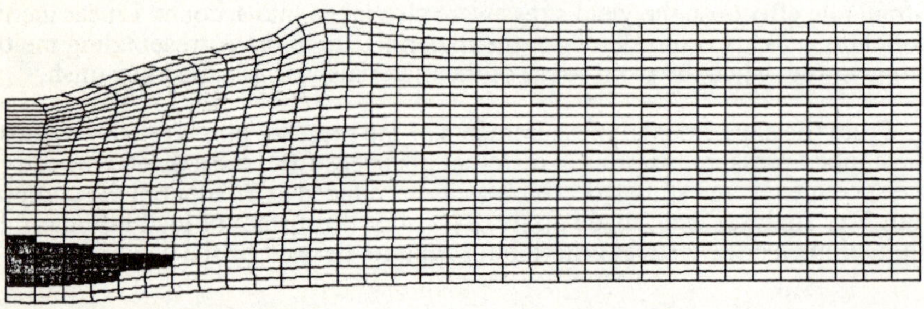

Figure 7. Deformed mesh for an aluminium specimen subjected to a maximum pressure of 8.4 GPa, corresponding to a time of 8 μs after the initiation of loading. Ruptured zones are shown as dark.

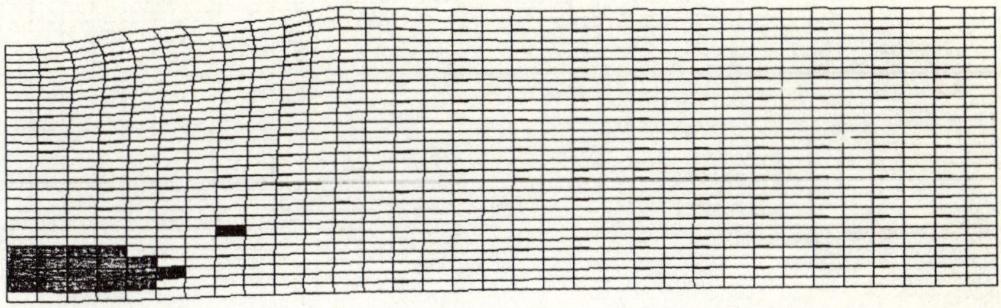

Figure 8. Deformed mesh for a copper specimen subjected to a maximum pressure of 10.0 GPa, corresponding to a time of 8 μs after the initiation of loading. A Tuler and Butcher type rupture criterion has been used in the analysis. Ruptured zones are shown as dark.

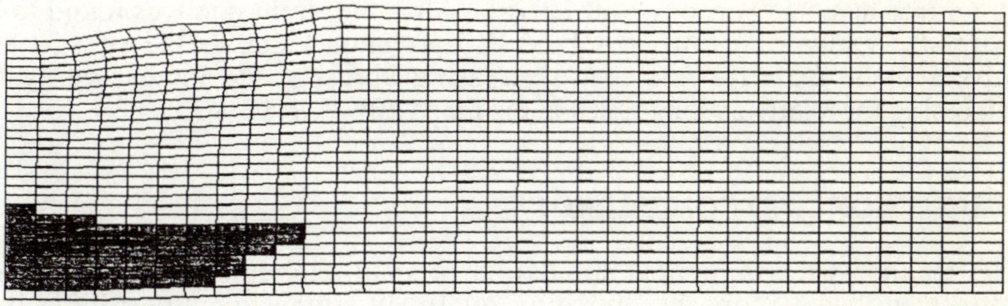

Figure 9. Deformed mesh for a copper specimen subjected to a maximum pressure of 10.0 GPa, corresponding to a time of 8 μs after the initiation of loading. A critical stress rupture criterion has been used in the analysis. Ruptured zones are shown as dark.

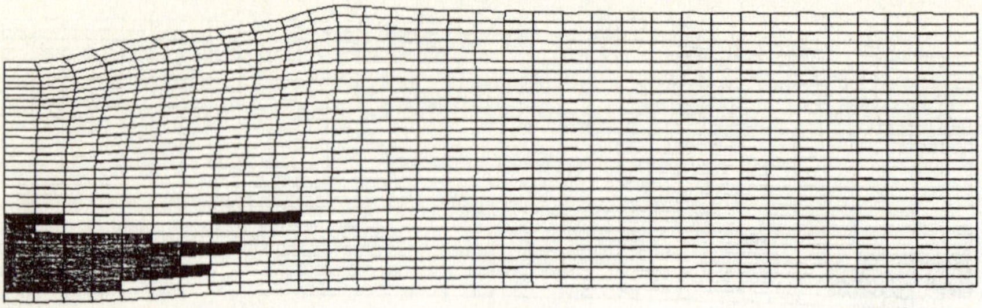

Figure 10. Deformed mesh for a steel specimen subjected to a maximum pressure of 14.9 GPa, corresponding to a time of 8 μs after the initiation of loading. Ruptured zones are shown as dark.

An examination of the strain-rate histories experienced by the material near the axis of symmetry of the specimen, showed that in all cases the maximum strain rate detected was of about 10^5 s^{-1} . Also, the strain rate was found to be markedly variable in time, reaching its maximum value as the stress wave traversed the material, and then decreasing abruptly and increasing again, remaining for relatively long times in the range 10^4 s^{-1} to 10^5 s^{-1}.

DISCUSSION AND CONCLUSIONS

The results of the numerical simulation of spall experiments in aluminium, copper and steel show that by using relatively simple fracture criteria it is possible to predict relatively well the main features of spall fracture in ductile metals. However, it should be kept in mind that the validity of the calculations is mainly limited by the validity of the values of the parameters included in the rupture criteria. In general, rupture criteria such as those of the critical stress or that of Tuler and Butcher, are closely tied to the experimental conditions at which they have been obtained, in such a manner that the degree of aproximation to reality is poorer as the conditions at which those criteria are applied depart from the experimental conditions. For this reason, in spite of some success achieved with some simple criteria, it is more convenient to implement some dynamic rupture models which take into account in some detail the micromechanisms of rupture. In the case of aluminum and copper, such models should consider phenomena such as void nucleation and growth. In the case of steel, crack nucleation and growth should be the dominant phenomena. In this way, it should be possible to obtain a more general capability for predicting dynamic fracture of metals. Furthermore, it is desirable to develop some rupture criteria which take explicitly into account the influence of the strain rate in the rupture condition, since this parameter may influence the dynamic fracture processes.

REFERENCES

1) Meyers, M.A. and Aimone,C.T., Dynamic Fracture (Spalling) of Metals, Progress in Mat. Sci., Vol. 28, pp.1-96, 1983.

2) Curran, D.R, Seaman, L. and Shockey, D.A., Dynamic Failure of Solids, Physics Reports, Vol. 147, pp.253-388, 1987.

3) Buchar, J., Bílek, Z. and Dusek,F., Mechanical Behaviour of Materials at Extremely High Strain Rates, Trans Tech Publications, Switzerland, 1986.

4) Davison, L. and Stevens, A.L. (1973), Thermomechanical Constitution of Spalling Elastic Bodies, J. Appl. Phys., Vol. 44, pp.668-674, 1973.

5) Davison, L., Stevens, A.L. and Kipp, M.E., Theory of Spall Damage Accumulation in Ductile Metals, J. Mech. Phys. Solids, Vol. 25, pp.11-28, 1977.

6) Johnson, J.N, Dynamic Fracture and Spallation in Ductile Solids, J. Appl. Phys., Vol. 52, pp. 2812-2825, 1981.

7) Seaman, L., Curran, D.R. and Shockey, D.A., Computational Models for Ductile and Brittle Fracture, J. Appl. Phys., Vol. 47, pp.4814-4826, 1976.

The Apparent Synergism in Combined Blast and Fragment Loadings

K.A. Marchand and P.A. Cox

Southwest Research Institute, PO Drawer 28510, San Antonio, Texas 78284, USA

ABSTRACT

Recent tests and analyses concerning the response of standard construction components subjected to close-in combined blast and fragment loadings are discussed. The apparent synergism of the effects of the explosive shock applied concurrent with the fragment impact is described.

INTRODUCTION

In the design and analysis of protective structures designed to resist the attack of modern conventional weaponry, many methods are available for the prediction of dynamic response of the structure of concern [1],[2],[3],[4]. Additionally, several methods for the prediction of projectile and fragment penetration are available for these same structural varieties. What is missing, and is typically not addressed in a design or simplified analysis reference, is the effect of combined load types such as simultaneously applied impact and non-impact, yet impulsive, loadings.

The effect of simultaneous loadings has been observed, however, and the differences in structure responses as compared to separately applied loads of the same type are severe. In some early Waterways Experiment Station work [5] concerning the testing of field fortifications used in the Vietnam War years, it was noted that the integrity of reinforced concrete walls loaded with artillery shells detonated in close proximity to the fortifications (5-15 ft.) was dependent on the use of sandbags placed in front of the walls. The sandbags in these tests provided a measure of protection from the impact loading generated by the fragments. Recent work at Southwest Research Institute (SwRI) [6] has shown that the damage caused by the combined impact and blast loading of a steel flying metal plate is significantly more damaging to concrete barrier panels than the loading generated by approximately three times the weight of uncased explosive. Other work reported by Coltharp, et al. [7] has provided additional information concerning the necessity of fragment protection.

It can be shown, then, that simultaneous impact and blast loads applied to a structure can cause responses that in some cases are more severe than the sum of the damage generated in the structure through the independent application of the loads. Hence, this simultaneous loading is considered to be synergistic in the sense that the simultaneous damage is greater than the sum of the impact and non-impact loadings.

A rationale, both analytically and empirically supported, is needed by the hardened structure designer and vulnerability analyst to define the response modes and degree of response expected from synergistic blast and fragment loadings. Significant test data and applicable analytic tools are available which could allow design or analysis information to be generated which would adequately predict and account for these kinds of structural responses.

BACKGROUND

A background discussion of the definition of synergistic blast and fragment loadings can best be presented by considering the following:

(1) the practical hardened structure vulnerability or design problems at hand which would drive any research effort in this area, and
(2) the specific concerns of response mode definition and the effect of threat type, material type, and nonuniformity of load application on the response mode.

Structure Types and Applications

Consider first the types of protective structures defined by function or purpose of the structure. A variety of protective structure geometries and configurations exist which make use of various materials and structural element types. Large cast in place conventional reinforced concrete structures used as command and control facilities are common protective designs. Requirements for shorter erection times for facilities in the modern combat arena have driven some designs toward modular, prefabricated and portable structures. This type of structure, while obviously more expedient and often more economical, is much more sensitive the type of loading applied. Localized, close-in blast or fragment loadings on these types of structures, where loads are concentrated, and response modes are non-uniform and localized, will require detailed panel or element design to defeat these types of loadings but may not require special consideration in connection design. Large uniform and longer duration airblast or groundshock loads may tear typical connections at the supports.

Cast-in-place and modular pre-cast structures are two extremes of structures likely to be considered in design. In actuality, some combination of these types of structures and some combination of conventional or unconventional materials such as slurry infiltrated fiber concrete (SIFCON) or polymer infiltrated fiber concrete (PIFCON) may exist. Additional design criteria which could be imposed on the structure designer or analyst includes penetration resistance of the barrier to small or medium caliber projectiles. The structural details necessary for projectile defeat (material toughness or surface hardness) may not be complimentary to the requirements for structural integrity under other types of loads. Another threat type might include terrorist weapons. It can be seen, then, that a complete understanding of threat constituents, weapon standoff and construction material and configuration interaction must exist prior to making estimates of response mode and severity.

Response Mode Definition; Material, Threat and Standoff Interaction

Parameters of Interest for Design

The response modes generated through the application of localized blast and fragment loadings will include both flexural modes and local shear or breaching modes. The structure designer, however, must select materials and must detail the design based on his understanding of a primary mode of response; yet, the sets of requirements of the two modes described above can be mutually exclusive. Factors important for design of structural members adequate for flexural response include:

- continuous reinforcement for increased ductility
- matrix toughness for structural integrity at large flexural deformations

Factors important for design of members for local shear or breaching response include:

- low stiffness in system for load transfer to surrounding member responding in flexure
- high fracture strength of matrix material, not necessarily ductile

Thus, an understanding of the responses generated, and the material and structure interaction must be gained in order to predict and design hardened structures for these environments.

Recent Tests and Experience for Response Mode Definition

Recent work by Ross [8], Baylot et al. [9] and Marchand et al. [10] has described and defined the occurrence of changes in response mode with varying load uniformity. Ross described four response mechanisms which could be postulated for reinforced concrete when loaded with a time and spatially varying load. Figure 1 shows schematics of these mechanisms. This work also describes analytic techniques for predicting time and spatial distributions of the loading from cylindrical explosive charges. Baylot presents some test and analysis results of slabs tested with buried explosive charges which illustrate the occurrence of an elliptical yield line pattern due to the orthotropy in the slab design and due to the highly localized groundshock loads. The analytic procedure presented in that reference also includes predictions of time and spatial load distributions for cylindrical charges of varying orientation with respect to the loaded slab.

Ross presents a method for determining a critical impulse at which a local conventional reinforced concrete section fails in shear. This procedure is based on a method described for edge shear response of ductile metals[11]. The equation presented for the critical impulse is as follows:

$$I_{cr} = \rho h v_{cr} \tag{1}$$

where:

ρ = slab density
h = slab thickness

and

v_{cr} = critical velocity, defined as:

$$v_{cr} = \frac{2\sqrt{2}}{3}\sqrt{\frac{\sigma_{ut}}{\rho}} \tag{2}$$

where

σ_{ut} = ultimate tensile stress of the slab material

The tensile strength of the material can be determined from the calculated maximum moment for a beam as:

$$\sigma_{ut} = \frac{M_{max}c}{I} \tag{3}$$

where

I = planar moment of inertia, in.^{2}

c = maximum fiber distance, in.

The maximum moment for a conventional reinforced concrete beam can be expressed as:

$$M_p = 0.9bd^2f'_c q(1-0.59q) \tag{4}$$

where

b = beam width, in.
d = distance from bottom of beam to centerline of steel, in
f'_c = compressive strength of concrete, psi

$q = \rho(f_y/f'_c)$

ρ = tensile reinforcement ratio
f_y = yield strength of reinforcement, psi

The above equations allow the sensitivity of the various parameters of interest in structural components designed to resist localized loadings to be checked. Figures 2 and 3 show the calculated critical impulse for varying beam thickness and steel percentage. Other parameters such as concrete strength and density in conventional reinforced concrete beams can be evaluated as well.

Quantitative Assessment of Synergistic Effects

The above described tests and analyses are illustrative of the understanding that presently exists for deformation mode changes in conventional reinforced concrete structural systems generated by time and spatially varying blast loads. This understanding is the foundation for the extrapolation of response predictions inclusive of fragment impact effects. Local response in conventional structural members will be significantly more severe when localized impact is superimposed on the blast load. Thus, a synergism, or enhancement of damage can be postulated.

Recent tests by Marchand et al. have compared the response of several barrier types subjected to airblast from an uncased explosive to their response when subjected to the same explosive cased in heavy metal. The differences in response are dramatic. Figures 4 and 5 show a conventional reinforced concrete barrier wall subjected to both loadings at the same standoff. The significant feature of the wall loaded with the cased explosive is the extent of the spall zone on the back face of the panel, and along the free edges of the panel. The shock generated in the wall by the application of the fragments generates peak pressures according to the following relationship:

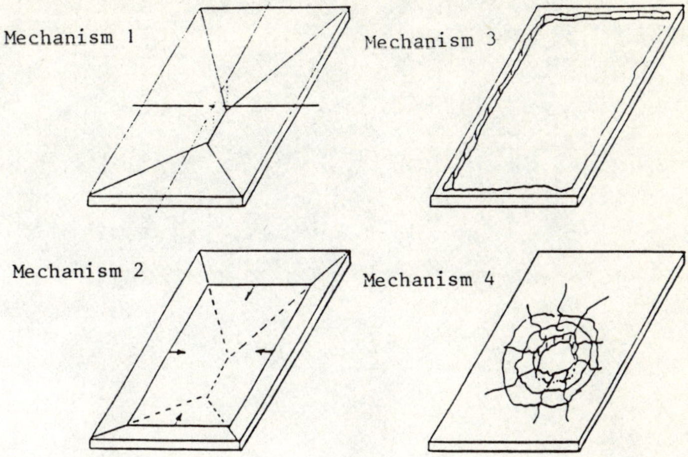

Figure 1. Schematics of Response Mechanisms for Blast Loaded Reinforced Concrete.

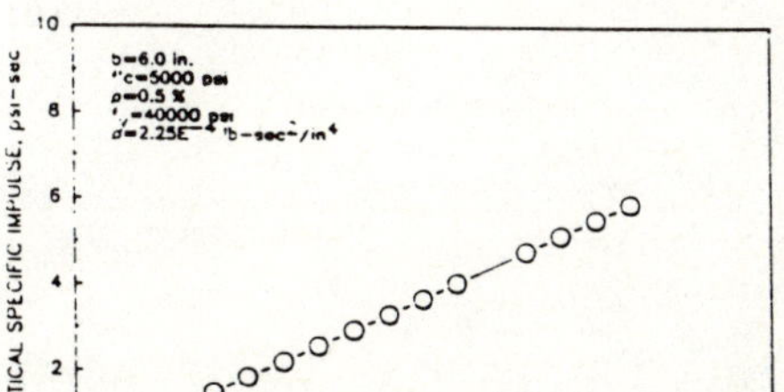

Figure 2. Critical Specific Impulse Versus Varying Beam Thickness.

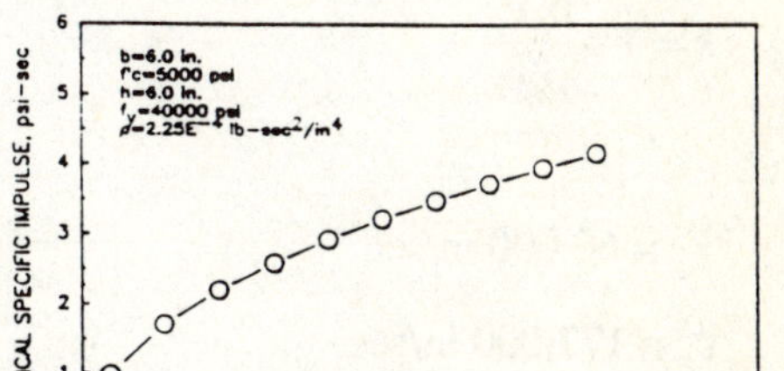

Figure 3. Critical Specific Impulse Versus Varying Steel Percentage.

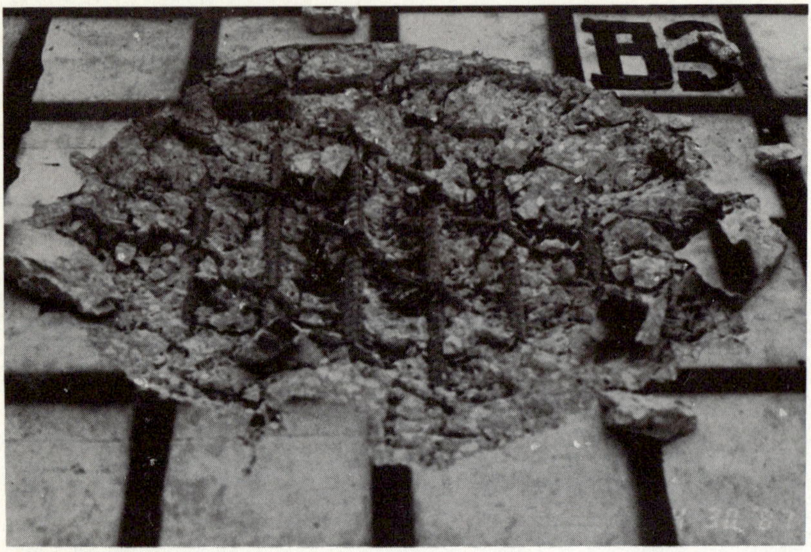

Figure 4. Reinforced Concrete Barrier Panel Loaded by Uncased Explosive.

$$P_c = \rho_c c_c V_p / k \tag{5}$$

where:

$$k = 1 + \frac{\rho_p c_p}{\rho_c c_c} \tag{6}$$

and:

$V_p = 60{,}000$ in/sec

$c_p = 177{,}000$ in/sec

$\rho_p = .00164$ lb-sec^2/in^4

$\rho_c = .000225$ lb-sec^2/in^4

$c_c = 133{,}000$ in/sec

then:

$k = 10.7$

and:

$$P_c = 167{,}800 \text{ psi}$$

This impact stress is much higher than the compressive strength of the concrete matrix, which suggests that some material will be failed in compression in the impact region of the wall. Additionally, after the stress wave is propagated through the thickness of the barrier wall and out to the free edges, it is reflected back, causing tensile failures in the matrix at these locations. Conversely, for the scaled distance of approximately 0.44, the reflected pressure expected at the blast wave and wall interface is approximately 26,000 psi, or about 6.5 times less than the impact stress induced by the fragments. This suggests that the local material damage will be significantly less in the blast only case than in the fragment load case. Obviously, any fragment momentum trapped by the wall will also add to the overall response of the barrier. The main point here, though, is that the fragment impact generates material failure and degradation locally, which is not considered in the response calculation through simple momentum addition.

Figure 5. Reinforced Concrete Barrier Panel Loaded by Cased Explosive.

Additional tests performed as reported by Marchand considered steel plate barriers filled with sand. The local material phenomena described above is not of concern for this structural configuration, as the material differences do not allow the propagation of the same magnitude stress wave to the rear or sides of the barrier. Thus, the expected response of the barrier subjected to fragment impact and blast should be different from the blast case only through additional trapped fragment momentum. Figures 6 and 7 show the sand filled steel barrier after testing with the blast only and blast and fragment loadings. It can be seen that the response mode is very similar in each case, but with larger deformation in the fragment and blast test due to the addition of fragment impulse. Thus, it is important to recognize that synergism is a function of material types and properties as well as structural configuration.

Figure 6. Sand Filled Barrier Panel Subjected to Uncased Charge.

Other recent tests performed at SwRI have shown that the effects of blast and fragment synergism can be defined and adequately considered in design [6]. Tests were performed on composite reinforced concrete panels subjected to flying metal plate attack. The loading on the panel consisted of the airblast from a flying metal plate explosive package detonated nearby coupled with the impact of the high velocity plate. Local material failure occurred as the concrete matrix was microfractured under the high impact stresses. The tests showed, however, that the flying metal plate was defeated by the simple addition of a rear panel membrane plate, which trapped the momentum of the displaced concrete. The plate thickness selected to allow plastic deformation in a membrane mode. Analysis was performed for these tests using the equations above for local shear calculations. These predictions of flexural and local response compared quite favorably with the observed results.

Additional testing was accomplished to observe the response of composite cementitious materials when subjected to blast only and combined loadings. Figures 8 and 9 show the rear face response of slurry infiltrated fiber concrete (SIFCON) panels when subjected to the same loadings and configuration as described above for the concrete and sand panels. It is obvious from these photos that the SIFCON does not exhibit the same quantity of local spall or rear face tensile failure as does the concrete as shown in Figures 4 and 5. The properties of the material (compressive strength of 12,000 psi and high compressive ductility as well as tensile strengths three

Figure 7. Sand Filled Barrier Panel Subjected to Cased Charge.

times that of the concrete) allow it to respond exclusively in flexure--without any severe local failure. In its flexural response, however, low bending ductility is exhibited as indicated by the formation of a centerline flexural crack through the cross section as indicated by the note on Figure 10.

Alteration of Blast Characteristics

Another consideration in synergistic effects assessment is threat definition in terms of fragment and blast characteristics. Much work has been done in the prediction and verification of case breakup, fragment velocity and mass distribution for bombs and warheads. Marchand showed that carefully modeled bombs can be made to generate a representative velocity and mass distribution of fragments. Blast effects from cased munitions, however, are less well understood. Data was gathered for blast loadings on barriers for both cased and uncased munitions. Although this data is preliminary, it indicates that the arrival of the blast pressure wave at the barrier position is delayed due to the expansion of the heavy case surrounding the explosive. This expansion of the case, when compared with data from other tests, can account for approximately 30% of the delay. The other 70% of the delay time observed can be accounted for if some reduction in the amplitude of the pressure wave (and impulse) is assumed. Conversations with other researchers have indicated that reductions in impulse for heavily cased munitions of up to 50% of equivalent charge weight have been observed in tests. For the tests reported by Marchand, a 25% charge weight decrease (and proportional impulse decrease) can account for the reduction in wave front velocity and a corresponding later time wave front arrival [12].

Figure 8. Rear Face of SIFCON Panel After Bare Charge Loading (Ref. 10).

Figure 9. Rear Face of SIFCON Panel After Cased Charge Loading (Ref. 10).

Figure 10. Flexural Crack Indicating the Onset of Collapse for SIFCON Panel After Cased Charge Loading (Ref. 10).

CURRENT WORK

Current studies performed over recent months have included the following studies. Additionally, new work is planned for the next year.

Composite Wall Studies

Tests and analysis were performed by Marchand, et al.[13]for the US Naval Civil Engineering Laboratory to allow the prediction of flying metal plate penetration in sand, earth, steel, reinforced concrete and targets made of contiguous layers of these materials. Additionally, tests were required to validate this analytic model using model headwall designs and model headwall retrofit designs with the following specifically stated objectives:

The analysis of the penetration of flying plates through single and multiple thicknesses of construction materials was a complex task, when the physics of the interaction of the plate and the walls or barriers is considered on a first principles basis. The approach in the work reported was to first put existing test data for flying plate penetration of four types of headwall systems into non-dimensional form using similitude analysis. This data, then, when plotted non-dimensionally, could be compared with predicted values of penetration when the input and results of that analysis were also put in non-dimensional form.

An analytic model was developed using a local impulse failure model for concrete coupled with energy relationships associated with the impact event which predicts "critical energy" or the energy input necessary to breach a wall or retrofit/wall combination. The applied energy of the flying plate is then compared with this critical energy and breach is assumed to occur when the applied energy is greater than the critical value for a given set of plate and wall parameters.

The four cases used to validate the analytic model were:

Case "a") Flying metal plate attack of single reinforced concrete panels - for prediction of baseline penetration into headwall designs.

Case "b") Flying metal plate attack of single reinforced concrete panels (headwall designs) with earth or sand backfill - for prediction of plate penetration into headwall designs with earth or sand berms or backfill placed as a retrofit to an existing structure.

Case "c") Flying metal plate attack of a single reinforced concrete headwall design with a retrofit reinforced concrete panel placed at some standoff - for plate penetration prediction of spaced reinforced concrete panels with no backfill.

Case "d") Flying metal plate attack of a single reinforced concrete headwall design with a retrofit reinforced concrete or SIFCON panel placed at a standoff with earth, sand or rock backfill - for prediction of plate penetration in spaced panels with earth, sand or rock backfill.

Similitude analysis was used to establish non-dimensional equations made up of "pi" terms for each of the above "cases". The complete variable list used to make up these terms and equations is presented in the reference. Also presented is the author's estimation of the range of applicability of the variable values. The basis for these limits was determined from the range of values tested in previous work. Figure 11 shows the resulting non-dimensional presentation of the data for case "a" described above.

Planned Research

The proposed effort of investigating the synergistic effects of blast and fragments can be summarized in three parts. The three Phases should include the following activities as a minimum:

PHASE I:

(1) Comprehensive literature search
(2) Testing of threats and materials to quantify the synergistic effects as they vary with material type, threat type, and munition proximity
(3) Evaluation of currently available response models concurrent with testing

PHASE II:

(1) Laboratory tests and numerical analysis to define fragment and blast shock front interaction and material failure in a loaded specimen
(2) Characterization of response mode change with respect to standoff based on fragment density and shock intensity calculations

PHASE III:

(1) Analytic model development/refinement to allow the prediction and definition of synergistic effects, response modes, and response magnitude
(2) Selection of optimum materials and structural configurations for combined blast and fragment loading scenarios

The efforts described above are just beginning and are being sponsored by the US Air Force Engineering and Services Center at Tyndall Air Force Base, Florida. Phase I results will be published in the Spring of 1990.

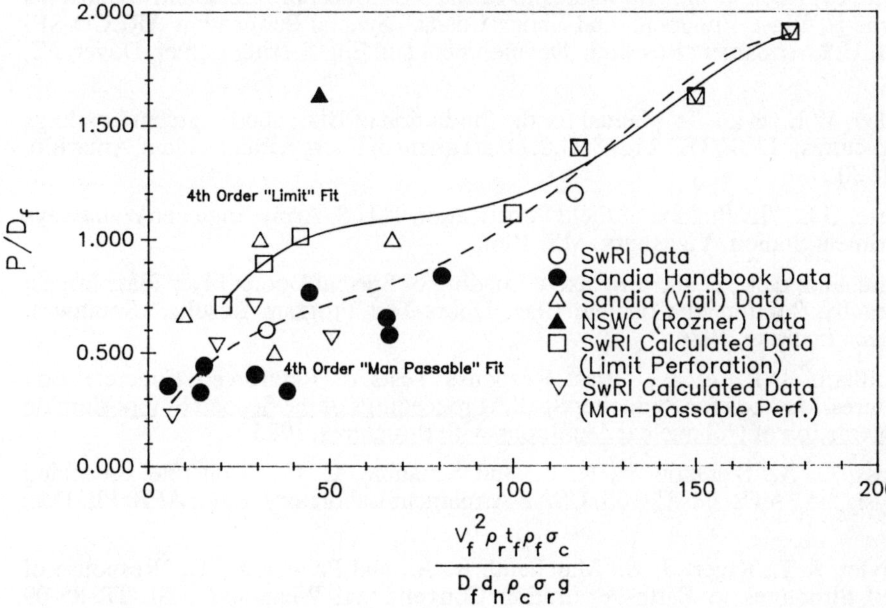

Figure 11. Non-dimensional Data for Case "a" Tests and Analysis (Ref. 14).

CONCLUSIONS

A brief overview of the apparent synergism in blast and fragment loadings on common construction materials has been presented and illustrates the research needs in this regime of dynamic loading. Tests are planned to address these needs and analysis will be performed and analytical and numerical techniques developed to address the observed response.

REFERENCES

1. Baylot, J.T., et al., "Fundamentals of Protective Design for Conventional Weapons," Structures Laboratory, U.S. Army Engineer Waterways Experiment Station, Vicksburg, MS, 1985.

2. Crawford, R.E., et al., "Protection from Nonnuclear Weapons," Technical Report No. AFWL-TR-70-127, Air Force Weapons Laboratory, Kirtland AFB, NM, 1971.

3. Whitney, M.G., et al., "Structures to Resist the Effects of Accidental Explosions Volume II, Blast, Fragment, and Shock Loads," Special Publication ARLCD-SP-84001, U.S. Armament Research, Development and Engineering Center, Dover, NJ, 1986.

4. Baker, W.E., et al., "A Manual for the Prediction of Blast and Fragment Loadings on Structures," DOE/TIC-11268, U.S. Department of Energy, Pantex Plant, Amarillo, TX, 1980.

5. Carre, G.L., "Evaluation of Field Fortifications," U.S. Army Engineer Waterways Experiment Station, Vicksburg, MS, 1968.

6. Marchand, K.A., et al., "Impulsive Loading of Special Doors: Flyer Plate Impact of Heavily Reinforced Concrete Blast Doors-Test Program Results," Southwest Research Institute, 1986.

7. Coltharp, D.R., et al., "Blast Response Tests of Reinforced Concrete Box Structures-Methods for Reducing Spall," Proceedings of the Second Symposium on the Interaction of Nonnuclear Munitions with Structures, 1985.

8. Ross, C. A., Sierakowski, R. L., and Schauble, C. C., "Concrete Breaching Analysis," AFATL-TR-81-105, USAF Armament Laboratory, Eglin AFB, FL, Dec. 1981.

9. Baylot, J. T., Kiger, S. A., Marchand, K. A., and Painter, K. T., "Response of Buried Structures to Earth-Penetrating Conventional Weapons," ESL-TR-85-09 Engineering & Services Laboratory, Tyndall AFB, FL, Jan 1985.

10. Marchand, K.A., et al., "Development of an Alternate Munition Storage Barrier System, Phase II Report-Tests of the Barrier Concepts," Contract Report No. DACA88-86-D-0017 Delivery Order Nos. 002, 003, 004 and 010, U.S. Army Construction Engineering Research Laboratory, 1988.

11. Jones, N., "Plastic Failure of Ductile Beams Loaded Dynamically," ASME Journal of Engineering for Industry, February 1976.

12. Marchand, K.A. and Vargas, L.M., "The Effects of Heavy Tungsten Casings on the Airblast Characteristics of a Cylindrical Charge," Proceedings of the 59th Shock and Vibration Symposium, Albuquerque, New Mexico, October 1988.

13. Marchand, K.A. and Garza, L.R., "Countering Explosive Threats--Analysis Report," SwRI Project 06-1473-090, Tecolote Research, Inc., Santa Barbara, CA Subcontract TRI87-107, NCEL Prime Contract No. N00123-86-D-0299, August, 1988.

Societal Risks from Detonations in Explosive Storage Structures

K.A.J. Williams and C.P. Ellinas

Advanced Mechanics & Engineering Ltd., Surrey Research Park, Guildford, Surrey, UK

ABSTRACT

Explosives either for industrial or military use are usually stored in well protected concrete structures, designed in the UK to ESTC standards. Such structures, however, are usually designed to ensure that an explosion in one storage unit will not cause sympathetic detonation in adjacent units, and as a result they do not necessarily fully contain the effects of an explosion.

The paper reviews available information on the mechanics and effects of such explosions and presents an approach for the assessment of risks to people outside explosive storage installations.

INTRODUCTION

Wherever explosives, chemicals, fuels, gases and other potentially explosive materials are manufactured, handled or stored there is an inherent risk of accidental explosion.

It is not usually possible to design structures to contain the effects of an explosion and the approach taken is to ensure that damage to adjacent buildings is limited to within acceptable limits. This is embodied in a number of regulations [1-3] and especially those concerned with the storage of explosives. In this latter case the design of adjacent buildings and their spacing are chosen to minimise the risk of a sympathetic detonation and the consequent propagation of an explosion.

The effects on people outside the industrial or military installation, where an explosion has occurred, are associated mainly with blast overpressure and missiles, which can cause injury and often death. Large explosions cause fragmentation of the storage structure, with fragments travelling at high velocities to considerable distances endangering people not associated with the particular installation.

In the UK, igloos, which are reinforced concrete structures with earth cover specifically designed to resist blast effects, are the preferred means of storing explosives. This type of construction, illustrated in Figure 1, has been found to be most effective in protecting stored explosives and in resisting the effects of adjacent explosions.

In potentially hazardous installations (PHIs) it is normally required to ensure that the manufacture, storage and handling of potentially explosive substances do not expose personnel within the PHIs, but especially the general public, to an unacceptable level of risk.

This paper briefly reviews the subject, identifies potential hazards and, following analysis, it illustrates how quantification of levels of risk can be achieved. As a result it recommends procedures on how societal risks from detonations in explosive storage structures can be estimated and now their acceptability can be assessed.

HAZARD ASSESSMENT

Hazard assessment and analysis of new installations is usually carried out at three distinct, consecutive, stages:

- Feasibility study: Its purpose is to identify major hazards in order to make fundamental planning decisions, such as site suitability, process type, population at risk and so on.

- Detailed design stage: By this stage the site layout has been decided and the plant has been designed. This is the most appropriate time to carry out risk analysis, to identify and implement appropriate protection measures.

- Pre-commissioning study: By this stage construction is substantially complete, staff responsibilities have been defined and the preliminary operation procedures have been prepared. It is advisable at this stage to reassess hazards and risks and take appropriate measures to improve safety.

This paper is predominantly concerned with the first two stages.

Major hazards may be broadly identified as either internal or external to the installation. External hazards are obviously site dependent, and for example include:

- earthquakes
- aircraft crash
- sabotage.

Historically available information indicates that the probability of such hazards occurring is usually small, but can be assessed when required in special circumstances. The probability of occurrence of internal hazards is by comparison considerably higher. These include:

- release of poisonous gases
- fire and radiant heat
- explosions.

Of these, the last poses the greater risk because of its far-reaching consequences. Reference [4] indicates that an explosion initially produces a fire ball at an extremely high pressure and temperature. However, cooling is very rapid and the effect of fire and radiant heat tends to be highly localised. The major hazards arise from blast effects due to high overpressure and from fragments and debris travelling at very high velocities.

HISTORICAL DATA

The probabilities of explosions occurring during manufacture, handling and storage of explosive substances are not the same. A review of UK accidents involving explosives [1] over the period 1945 - 1969 indicated that approximately 85% of such explosions occur during manufacture/processing, 10% during transportation and only 5% during storage.

The probability of an explosion occurring at the first place can be estimated on the basis of available historical data relating known numbers of explosions to the manufacturing experience of the particular explosive substance under investigation.

ANALYSIS

The magnitude of an explosion is established by the amount of energy released. This is normally expressed in relative terms and the generally accepted standard is to compare the energy of an explosion to that released by the explosion of an equivalent amount of TNT.

An explosion does not give its targets any advance warning of impending destruction because explosive blasts travel faster than the speed of sound. The very first mechanical effect of explosive blast is a forceful blow from the instantaneous pressure jump in its shock front. This is followed immediately by the crushing effect of blast overpressure and a high velocity blast wind. Distance offers the best protection against damage from explosions.

The ESTC regulations [1] require that the various structures within a PHI are separated from each other to prevent propagation of an accident from one structure to another. Similarly, PHIs should be situated away from population centres and uses so as to ensure that the public is not exposed to unacceptably high hazards. Safe distances are determined with regard to:

- quantity of explosives manufactured, handled or stored;
- structure construction;
- nature of adjacent public usage;
- effects of an explosion; overpressure, radiant heat, primary and secondary fragmentation.

In the UK safe distances for determining the internal layout of PHIs and their separation from societal uses are recommended in the ESTC regulations [1]. Their choice can also be made through probabilistic analysis and proper evaluation of individual and societal risks.

The maximum distances to which fragments and debris can be propelled outwards from an explosion depend upon their initial velocity, and are related to the explosion magnitude [1]. In Reference [5] the maximum radial horizontal distance, r_m, in metres from an explosion centre is given by:

$$r_m = 45W^{1/3} \tag{1}$$

where W is the equivalent amount of TNT, expressed in kilogrammes.

From the expected maximum distance relationship it is possible to estimate the maximum velocity of fragments. Assuming that the maximum distance occurs when the initial angle of the fragment trajectory is 45°, the maximum fragment velocity, v, may be obtained from:

$$v = \left(g r_m\right)^{1/2} \tag{2}$$

where g is the acceleration due to gravity.

For the type of igloo shown in Figure 1 it is reasonable to assume that that the greatest proportion of debris is ejected at approximately the same velocity. It may, therefore, be conservatively assumed in risk analysis that all debris in such an explosion travels at the same velocity as given by equation (2). A more accurate prediction of initial fragment velocity distribution may be obtained using a method presented in Reference [3], provided that the structural details of potential explosive site are known.

The range of fragment ejection angles is also very dependent upon the structural details of the storage unit. For the type of igloo shown in Figure 1, with $b/h \approx 2$, ejection angles would be expected to be about ±45° to the vertical, given that the igloo roof is the direction of least resistance. The statistical distribution of ejection angles may then be easily calculated by assuming that 99% of all fragments are ejected within this range and are characterised by a normal distribution.

The greatest threat would be posed by fragmentation of the igloo roof. Observations from large explosions, reported in Reference [6], indicate that the number, n, and mass, m, of fragments are related exponentially, so that:

$$n = Ae^{-Bm} \quad (3)$$

A and B are constants defined as follows:

The total mass ejected, M, ignoring soil cover and, for example, considering the igloo roof only, is given by:

$$M = \int n \, dm = \frac{A}{B} \quad (4)$$

Assuming also that 99% of all fragments are smaller than 10kg, leads to:

$$Ae^{-10B} = 0.01 \quad (5)$$

Solution of equations (4) and (5) enables estimation of constants A and B appearing in equation (3). Given the fragment velocity, Reference [7] gives guidance on the smaller fragment size which on impact has a 50% probability of lethal consequences. This, together with equation (3), can be used to estimate the total number of fragments from an explosion which have sufficient energy to cause death. The relationship between r and the ejection angle, α, is given by:

$$r = \frac{v^2}{g} \sin 2\alpha \quad (6)$$

Having established these relationships, it is then possible through probabilistic analysis to estimate the probability of a lethal fragment landing at any distance r from the explosion site, and the probability of a person at this distance being killed.

If the details of population centres, public transportation, and generally the presence of people that may be exposed to fragments, are known, it is possible to estimate the probable number of deaths per explosion and per year, given knowledge of the probability of occurrence of an explosion.

ACCEPTANCE CRITERIA

The risks to population surrounding PHIs are usually considered at two levels:

- individual risk;
- societal risk.

Individual risk, $I(h,r,\theta)$, refers to the risk of death to an individual at any given location (r,θ), see Figure 2, outside the PHI, given a hazard h (in this case h = explosion), and it is defined by:

$$I(h,r,\theta) = F(h) \times P[h,r,\theta] \quad (7)$$

where $F(h)$ = frequency of hazard h at the PHI, per year
$P[h,r,\theta]$ = probability that an individual at location (r,θ) is killed, given that hazard h has occurred.

An acceptable maximum limit for individual risk appears to be 10^{-5} per year [8].

Societal risk is defined as the possibility of a number of people dying as a result of accidents at a PHI, and may be calculated as follows.

The number of deaths, $N_i(h)$, given an accident scenario, i, may be obtained from:

$$N_i(h) = \int_r \int_\theta P_i\,[h,r,\theta]\; dr d\theta \quad (8)$$

Assuming k accident scenarios, so that each of these is associated with $N_i(h)$, i = 1, .., k, number of deaths, then the probability of $N_i(h)$ number of deaths occurring is the same as the probability of occurrence $P_i\left[j=N_i(h)\right]$ of accident scenario i. The cumulative probability, $P[j \geq N(h)]$, of N(h) or more deaths occurring, given hazard h, may be obtained from the following summation:

$$P\,[j \geq N(h)] = \sum_{j=}^{j=N_{max}} P_i\left[j = N_i(h)\right] \quad (9)$$

where N_{max} is the maximum number of deaths for any one of the k accident scenarios.

The frequency, $F(j \geq N(h))$, with which N(h) or more deaths occur per year, given hazard h, is then:

$$F(j \geq N(h)) = F(h) \times P[j \geq N(h)] \tag{10}$$

Acceptability of societal risk may be examined with regards to the example shown in Figure 3. Thus, for example, if the frequency of one death as a result of hazard h is less than 10^{-3} per year, the societal risk may be considered to be acceptable.

CONCLUSIONS

The mechanics and effects of explosions resulting from detonations in explosive storage structures have been examined and a risk analysis method has been proposed. This enables the estimation of both individual and societal risks and examples of appropriate acceptance criteria have been presented. These can be used to assess whether the frequency and consequences of explosions in such installations containing explosive storage units are acceptable and, therefore, make decisions on safety and the provision of additional protection measures to ensure adequate safety.

REFERENCES

1. UK Explosives Storage and Transport Committee (ESTC), Regulation Leaflets 4, 6 and 7, 1979.

2. UK Ministry of Defence, "Explosives Regulations," Procurement Executive.

3. US Army, Navy and Air Force, "Structures to Resist the Effects of Accidental Explosions," TM5-1-1300/NAVFACP-397/AFM88-22, US Government Printing Office, Washington, 1969.

4. Marshall, V.C., "Major Chemical Hazards," Ellis Horwood Ltd, 1987.

5. Kinney, G., and Graham, K.J., "Explosive Shocks in Air," Springer-Verlag, New York, 1985.

6. Christopherson, D.G., "Structural Defence," Ministry of Home Security, Research and Experiments Department, UK, 1945.

7. Baker, W.E., et al., "Explosion Hazards and Evaluation," Elsevier, New York, 1983.

8. Institution of Chemical Engineers, "The Assessment and Control of Major Hazards," Symposium at the University of Manchester Institute for Science and Technology, 1985.

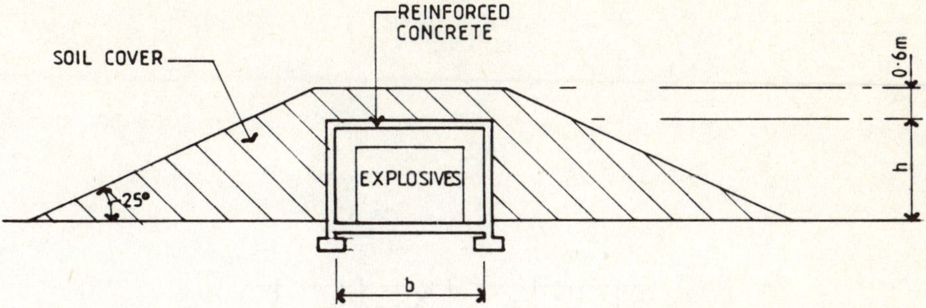

Fig. 1 Typical igloo construction

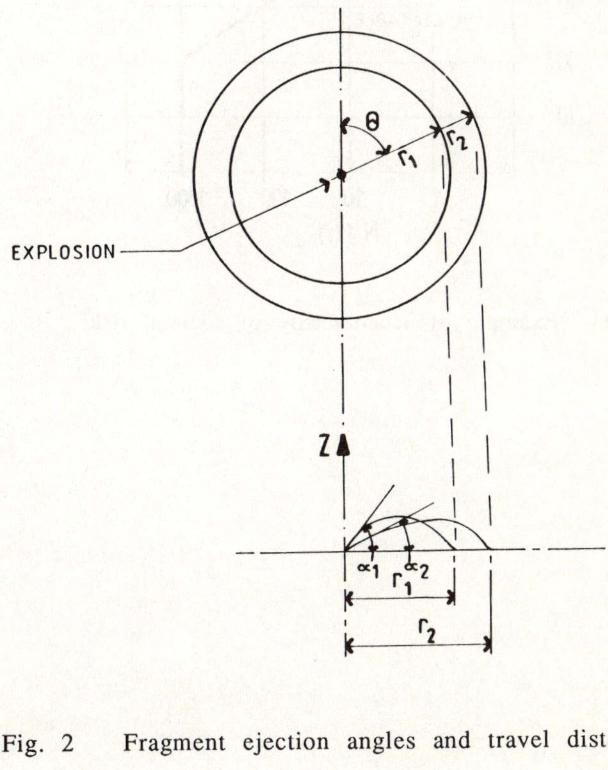

Fig. 2 Fragment ejection angles and travel distances

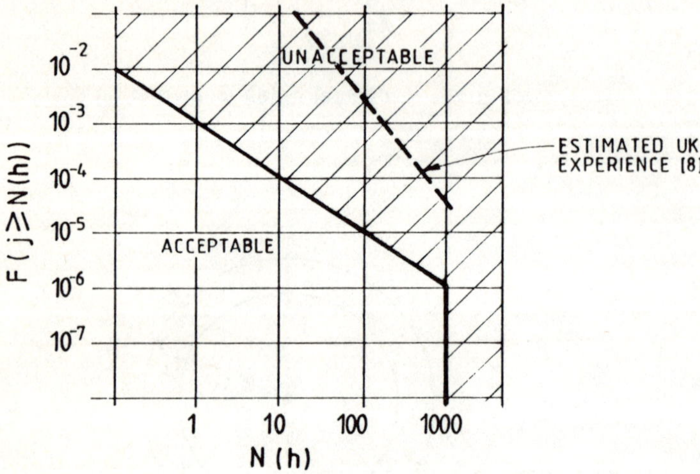

Fig. 3 Example of acceptability of societal risk

SECTION 4 - COLLISIONS

Risk Modelling of Vessel Impacts on a Multispan Bridge

Peter Bein

N.D. Lea International Ltd., 1455 West Georgia Street, Vancouver, Canada, V6G 2T3

ABSTRACT

Existing methods are not suitable to analyse the risk of vessel impacts on long, multispan bridges located in marine environment. Sea waves, tidal and storm surge currents, and severe ice pack conditions in winter increase the risk of vessel aberrance, for which local statistics may be limited. Foreign statistics of vessel aberrance are not appropriate because of regional bias, obsolescence and assumptions. A model was developed to determine the energies and associated return periods of vessel impacts on both the main and the side spans and piers of a long bridge. The model considers local marine environment and vessel traffic conditions, including winter navigation, and allows testing of the sensitivity of impact energy return periods to vessel traffic and bridge geometry assumptions. The application of the risk model is illustrated using preliminary data collected during planning of the Northumberland Strait Crossing in Canada.

INTRODUCTION

A number of assumptions must be made and documented in risk analyses. Available knowledge and data, and even scarce information should be utilized and provided for consideration to the decision maker, who combines the data with value judgments. A fixed value is assigned to human life in bridge risk assessments. Decisions based on these analyses inappropriately mix loss of human life with economic efficiencies (Rosenblueth[1]). In an alternative approach avoiding explicit evaluation, vessel impacts have been compared to earthquake loads, and the impact forces which induce critical limit states in the structure were selected as design criteria (Bein[2]). The critical limit states do not admit structural failures that may lead to loss of human life.

Usually each hazard is analyzed separately, then the highest load from among all hazards is selected for bridge design. Return periods for vessel impacts should be chosen on the basis of acceptable risk, in conjunction with other threats such as earthquake or extreme ice conditions. If this procedure is not followed, the sum of all threat probabilities is greater than the assumed acceptable probability of each individual threat, and the return period of a load from any of the threats is shorter than accepted for individual threats. This point has been missed by many bridge risk analyses, while other structures are designed to more appropriate philosophy. Systems and decision perspective should be maintained to avoid such pitfalls. Because non-monetary values (human life, habitat) are involved, economic efficiency should not be the only decision criterion.

This paper is limited to the methodology and modelling of vessel impacts on a multispan bridge in marine environment. An example from preliminary planning of the Northumberland Strait Crossing in Canada is used for illustration. In a comprehensive planning process, the vessel impact risk analysis should be integrated with all the other components of bridge design and operation decisions.

VESSEL AND BRIDGE CASUALTY DATA

Casualty Statistics and Limitations

In this paper vessel casualties (collision, grounding, flooding, capsizing, foundering, sinking, fire, explosion, ice damage or other) and near-casualties are "incidents". "Collision" is contact between two or more vessels, and "striking" is a contact between vessel and stationary object. In North America the law requires reporting of all incidents but near-casualties would not always be reported. Casualty reports and investigations may miss causative factors and terminology across countries is open to misinterpretation.

Vessel and bridge casualty statistics from the U.S.A, Europe and Japan are commonly used in bridge risk analyses, even though conditions may differ. The statistics reflect local and historical conditions. Advances have been made in ship and navigational aids technology. Cultural and temperamental differences between the navigators in foreign waters and at a study site may give rise to different aberrance frequencies. This is a dilemma because strikings of bridges are rare events. The lack of local data also encourages the investigation of vessel casualties involving objects other than bridges.

Visibility decreased by fog or snow was regarded the second leading factor after the human factor in increasing navigational risk. Other factors were traffic density, fishing vessels and pleasure craft in shipping lanes, and winds (Canadian Coast Guard[3]). European and Japanese studies demonstrated a dramatic increase in the risk of stranding in reduced visibility (Cockcroft[4], Fuji and Yamanouchi[5]). Foreign ships mismaneuver more often than local ships, and there is large variation between sites (Fuji et al[6]). All probabilities derived from these statistics are expressed as "number of casualties per one vessel passing by a locality". The statistics are used in many studies for lack of better data. It will be demonstrated in this paper that risk analysis of long bridges requires a different unit for the vessel aberrance parameter.

Computational methods based on mechanistic models miss the human error. Statistical analyses of historical data do include it, but they imply that every ship movement has the same probability of casualty. Vessel casualty data banks, such as those operated by the marine insurance companies, or the Canadian or the United States Coast Guards, provide a sound basis to establish statistics reflecting local factors.

Bridge Striking Data

Human factor was the main cause in major Canadian bridge strikings (Canadian Coast Guard[7]) Mechanical failure, tide and current, and wind were less frequent contributors . The U.S. (Marine Board[8]) found that contrary to conventional design practice ships choose to ram the approach piers and spans more often than the main piers. Only six of 19 serious worldwide strikings were at main piers, while 13 involved side piers (Frandsen [9]). Bridges over waterways may create or aggravate difficult areas of navigation. Most of the incidents in the U.S.A. occurred within one mile of a bridge, near or in a bend or turn of the channel, and many were rammings of bridge piers. Most of the encounters with bridges and other fixed objects involved tugs, towed vessels and barges. The prime cause of strikings was personal fault (United States Coast Guard[10]). Sabotage or vandalism could be of serious consequences if it involved a heavy vessel such as a loaded barge set adrift in high seas and currents.

Sea Ice Factor in Vessel and Bridge Casualties

In the Baltic Sea a large share of all tanker casualties occur in winter due to sea ice, short daylight time, freezing spray, storms and snowstorms, which affect the performance of both vessels and helmsmen (Kostilainen and Hyvarinen[11]). Ice conditions were ranked as medium hazard (Canadian Coast Guard[5]). Presence of bridges was consistently ranked at the bottom of the list, and natural underwater hazards ranked less risky than ice conditions. In Canada, ice was the primary cause in three major bridge casualties (Canadian Coast Guard[7]). The 1980 - 1985 casualty records from Northumberland Strait indicate that the risk in winter is about seven times greater per transit than in ice-free season.

IMPACT ENERGY AND RISK FACTORS

Relevant Variables

The damage of vessels and bridge elements depends on the kinetic energy and its dissipation during impact, and thus on the mechanical properties and hydrodynamic behavior of the vessel, the type of bridge construction and foundations and their structural properties. Impact energy is proportional to the mass of the moving vessel times its instantaneous velocity squared. The mass at impact includes an added hydrodynamic mass effect created by a water volume surrounding the hull. When a large ship is beset and moving with the pack ice, the mass is that of a portion of the pack plus the displacement of the ship.

Risk Factors For multispan bridges located in marine environment the risk of striking by vessels is due to: (a) mixed use involving fishing, recreation and commercial vessels; (b) large proportion of tows and ballasted vessels; (c) tides, currents, winds and ice; and (d) a large number of spans and piers. Low-clearance side spans pose an obstacle to a transit of an aberrant large vessel. Evasive action may be taken before the bridge is reached, but if it fails the vessel strikes a pier or a span with a probability defined by both the vessel and the bridge geometry. Small craft will not govern the design of piers because one can expect larger forces from pack ice.

Impact Velocity The impact velocity of vessels is determined by the operating speed at the time an aberrance occurs, the evasive actions taken, and the modifying effects of wind, wave drift and current forces. For larger vessels, the horizontal component of drift velocity far exceeds the vertical component even if waves are present. Large passenger cruisers, laden freighters and deep-draft tankers can produce impacts of the order 300 MNm for impact speeds up to 5 m/s. A berm around a main pier would be able to prevent a direct impact, but ballasted tankers and cargo ships can ride up the slope. The berms surrounding piers should be designed against the ride-up, but the cost of berms in deep water would be a major economic factor for multispan bridges.

Vessel Behaviour Barges, deep draft vessels and ballasted vessels are risky because they are difficult to steer, especially if currents are present. The maximum drifting velocity of such vessels would be in following wind and current, and the heaviest vessel impacts would be broadside. Smaller vessels experience instantaneous accelerations in all directions, but only surge has an effect on bow-on or stern-on impacts, sway on impacts broadside, and yaw on other impacts. A disabled vessel drifts beam on to the weather, but vessels with superstructures well forward will weathervane and drift stern against the sea faster than broadside. The distribution of drift velocities of disabled supply vessels in the North Sea without the effect of current was insensitive to vessel size ranging from 820 tonnes displacement to 4,400 tonnes. In Sea States 3 and 4 (the maximum probable conditions in Northumberland Strait), the average drifting velocities were 0.4 - 0.6 m/s broadside, and 0.6 - 0.7 m/s head-on (Standing and Brendling[12]).

Environmental Factors
Water depths, currents, winds, sea ice and meteorological conditions at a bridge site should be examined for a comprehensive risk analysis. The horizontal water movements are intensified, and current speeds and macro-turbulence are typically increased in passages and narrows, where bridges are most often located. The near-surface horizontal water flows may be due to tidal, wind-stress, and storm surge currents, which are independent and linearly additive. The resultant flow depends on the hydrography and bathymetry and on the meteorological conditions in a bridge site area. Currents determine drift velocities of disabled vessels and ice floes, and vessel course-keeping on approaches to the bridge. The relevant current components and other environment factors will be discussed next.

Tidal and Storm-Surge Currents The strength of the tidal currents does not necessarily depend on the occurrences of high and low water, but currents are stronger when tides are higher than normal (spring tides). Currents due to meteorological conditions prolong flood and ebb and increase the current flow if directions coincide. Spatial variations also occur. Minor oscillations in sea level in adjacent large bodies of water due to meteorological conditions create temporary storm surges through the smaller body of water, over which a bridge may be located.

Wind-Stress Currents Cross-winds pile up water against the coast, setting up longshore currents parallel to the coast, and rip currents which feed the return flow against the incoming wind-driven flow. The rip currents give rise to cross-flows which are difficult to predict. For both cross- and through-wind orientations the flow of water also depends on the actual bathymetry and the shape of the coastline. The wind-stress currents are stronger at higher water levels, and in approximate proportion to the increase in water level. A graph (US Naval Oceanographic Office[13]) can be used to predict wind-stress currents. Relatively slow currents are expected for cross winds, but the returning rip currents may be swifter than the wind-drift currents. The velocity vector field will be further randomized by the longshore currents, and by eddies forming in the adjacent bays. The rip currents can be a major factor in loss of control of smaller and towed vessels, and of all vessels under ballast.

Winds and Waves Wind direction, speed and duration statistics are usually obtained from adjacent land stations and require adjustment for the actual bridge site conditions. Reliable wind data is crucial since in most cases it is the only basis for prediction of the wave regime. Long-crested waves cause free-floating bodies to surge back-and-forth, to heave, and to drift at the same time regardless of orientation (Sibul[14]). A ship drifts in the direction of the waves if they are short relative to ship length, stays still in the longer waves, and may occasionally move against the direction of wave travel in very long waves. A model representative of a modern barge or scow gave the highest velocities. Studies for recent offshore developments confirm these findings. The conclusion for bridge risk studies is that the surge component adds to the impact velocity, and that runaway or disabled vessels can drift towards the bridge against wind and waves in some cases. The effect of wave surge on the motions of vessels attending platforms in the North Sea was a 20 to 50% increase of the impact speed (Standing and Brendling[12]).

Sea Ice Both ice-capable vessels and ships under icebreaker escort are unable to move in very dense ice pack concentrations pressing against the hull (Bradford[15]). If this happens simultaneously with ice pack drift, the beset ship will threaten the bridge. The piers are usually designed to break the impeding ice cover in flexure rather than resist the kinetic energy of the pack. They may thus suffer damage if subjected to lateral loads delivered by sea ice through a hull of an aberrant ship. Damage will also occur if the beset ship's sail height exceeds the clear height of side spans.

Winds, currents and tidal streams may cause the opening of coastal leads, which then can be utilized by ships for easier navigation on the far approaches to the bridge. This would increase the risk to the side flank of the bridge. Through-winds over narrows create a bottleneck effect on the flow of ice. The actual situation at a bridge site will be modified by currents, tides, and possibly ice jams. Ship damage due to ice includes hull puncture and damage to propeller and rudder. The possible loss of control over the vessel's course may eventually lead to a collision with the bridge.

Other Conditions Visibility and freezing spray data is generally not available for low traffic volume waterways. There may be considerable differences in visibility between mid-channel and locations closer to shore where the data is easier to collect. Long-lasting strong breezess in the early winter will pick up spray from the wave crests, which will freeze and may grow heavy enough to affect vessel's stability. The presence of bridge piers and protective structures alters the original flows of water and air in the immediate vicinity of a bridge. This has an effect on the less steerable vessels such as towed barges and ships under ballast.

PROBABILISTIC MODEL AND EXAMPLE

Assume all side spans are equal length and the navigational channel is centered about the middle of the bridge. Each pier is represented by a circle of a diameter equal the width of the pier base plus a characteristic average vessel dimension. All errant paths of a vessel are equiprobable straight lines fanning out from the point where loss of control started (Figure 1). When the fan's outside just glances the inside of the main pier circles, there are no strikings. When the outside of the fan intersects the bridge approaches at the shoreline the upper limit of the number of potential strikings is reached. Moving the fan farther away will strand the vessel. Intuitively, between the two limits there are strikings of the bridge.

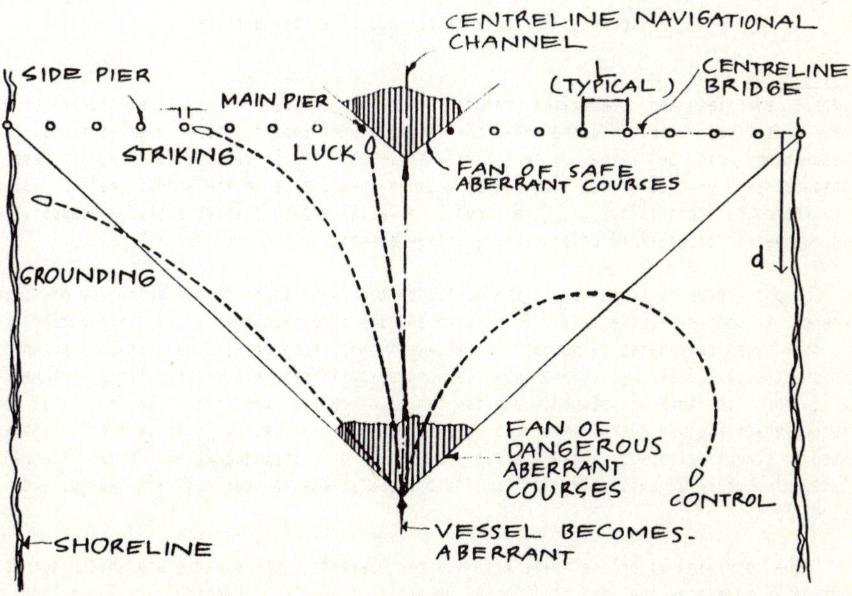

Figure 1. Situation graph for probabilistic model.

The geometric likelihood of striking the piers is shown in Figure 2. For aberrance occurring up to about five span lengths from the bridge, main piers are prime targets. As the start point of aberrance moves away the likelihood of hitting one of the side piers increases. Similar results can be shown for striking the spans. The tendency to hit side piers and spans will be countered by an increasing chance for successive evasive action as exemplified by the "lucky" ship in Figure 1 top centre. The effects of evasive actions on-board are included in aberrance statistics, but there would be no dedicated stand-by resources at a bridge site. Since the distance between the bridge and the farthest point of aberrance is large, aberrance probabilities will be conditional on the distance. For a navigational channel located against a shore there will be some reduction of frequency of striking the side elements.

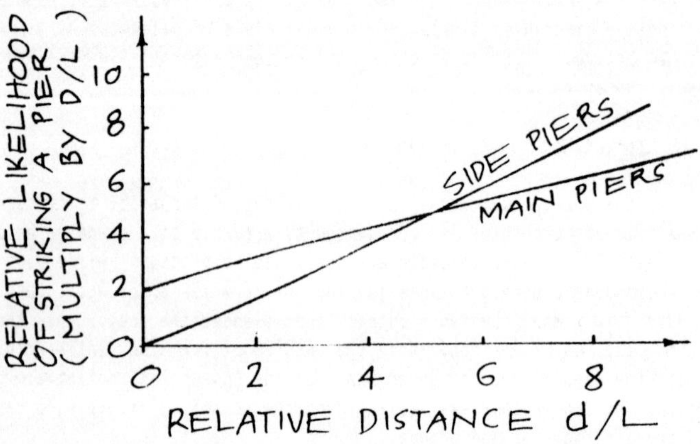

Figure 2. Relative likelihood of striking piers.

Methodology

Impact energies and associated return periods are calculated by the model assuming representative vessels, added mass coefficients and impact speed distributions. The assumptions can be tested by sensitivity analyses. The probability of impact mass is computed as a product of (a) fraction of a given vessel type in the annual traffic volume, (b) aberrance probability, (c) likelihood of a vessel's path against a bridge element, and (d) geometric probability of striking a bridge element.

Impact speed distribution is the most uncertain data item and the magnitude of impact energy is most sensitive to it. The speed and its marginal probability distribution will in most cases be adopted by judgment of site and navigation conditions. It is recommended to use a simple distribution, for example a three-point discrete distribution. The results will then be more trustworhty and can be smoothed if required. Environmental data representative of an average year are required in the model since extreme events (notably used in structural design) have a small probability of coinciding with a vessel aberrance. Data are usually available for winds and tides at a site, but not for waves and ice conditions.

The striking of bridge spans with slender elements such as masts and antennas, would not be as severe as the impact of "hard" elements of a ship. Length overall and width of a vessel enter geometric probability of striking a pier. Windage and underwater areas and drag coefficients are necessary for calculating wind and current forces acting on a drifting ship. Vessels should be grouped into representative types and sizes.

Displacement and dimensions of representative laden and light vessels can be taken from ship specifications. The added mass coefficient can be judged from the type of vessel, loading condition, orientation at impact and the draft. Air draft is estimated to the top of the highest element of each ship. The ratio of a span's vertical clearance to vessel's air draft laden and ballasted is required by the model for both the side spans and the main span.

Geometric Probabilities
The geometric probability of striking a pier is PG = (D + S)/LS, where PG = geometric probability, D = pier base (or island) diameter, S = ship effective width, and LS = span length. The island diameter shall be taken at the vessel draft depth, and will thus depend on both the tide elevation and vessel draft distribution. The ship effective width, S, is the width projected on the longitudinal axis of the bridge. It ranges from ship breadth for bow-on and stern-on collisions to ship length for collisions broadside. The effective width can be assumed S = (L + B)/3 for ships impacting oblique (ballasted freighter, ballasted tankers and all barges), and S = (L + B)/4 for other ships, where L = vessel length overall, and B = vessel breadth. The geometric probability of striking a span is 1.0 if the "hard" elements can reach the bridge structure and 0.0 otherwise.

Example
The Abegweit Passage of Northumberland Strait between the Prince Edward Island and mainland Canada is 12 km wide. In preliminary planning a number of alternative bridge designs and a tunnel were considered. This example refers to a design with some 40 spans each 300 m in length. The side spans have a low clear height of 14 m, but two spans over the navigational channel are elevated to permit transit of large ships.

Two ebbs and two floods move through the passage each day. The flood rates range from 0.4 m/s in fall and spring to 1 m/s in winter and fall. Surge currents caused by storms in the Gulf of St. Lawrence occur on five days per month on the average. The surges can reach 1 m/s and last long if driven by major storms. Rip currents caused by wind stress can be a major factor in loss of control of ballasted or towed vessels at the site. Winds exceeding 50 km/h blow infrequently and their durations are not sufficient to develop Sea States 5 to 6. Ice conditions are among the most difficult of all approaches to Canadian ports, and last 10.5 weeks on the average. Even the most powerful ships may then experience difficulty, while lower-powered vessels become beset by the moving sea ice.

Vessel Traffic The vessels in the navigation channel range from occasional passenger cruisers and large cargo ships and tankers, to fishing trawlers and barge tows. The base case traffic volume is 260 vessel movements per year, of which 10% occur in ice conditions. Approximately 40% of all movements of cargo and tankships are under ballast. Barges constitute 20% of the total traffic. The cargo vessels were assumed fully laden in 65% of movements, and ballasted in 35%, and tankers and barges fully laden in 50% of movements. The base case assumes a 3-point distribution with 2, 4, and 6 m/s impact speed and probability mass 25%, 50% and 25%, respectively. For barges and other vessels, the three points are 1, 2, and 3 m/s. The low speed represents impacts in head currents, while the high speed represents impacts in following currents.

Aberrance frequencies based on "vessel-km" rather than "passing by a geographic point" could not be calculated for lack of traffic volume data, and were assumed from world statistics. To account for the large length of the bridge, a "path probability" was introduced. It refers to the likelihood of an aberrant vessel transit within the main channel and piers. Other paths would lead to a collision with side piers and spans. The two probabilities sum up to 1.0. In the base case 50% of aberrant courses are within the main channel, except ballasted freighters, ballasted tankers, and all barges will have this probability reduced to 30% owing to their steering difficulties.

Analysis The analysis was carried out for main piers, side piers and side spans, using a microcomputer spreadsheet. None of the representative vessels can reach the underside of the main span even at high tide. The return periods were computed from cumulative probability mass functions of impact energies and the results were smoothed out. The sensitivity of results was tested relative to impact speeds, traffic, aberrance frequencies, path probability, and bridge geometry. The results were graphed with the return period in years as a function of the impact energy in MNm. A graph can be entered with a predetermined return period and the corresponding impact energy can be found. The graphs can be used as a guide to the magnitude of return periods and impact energies.

Results The most severe loads can be expected on side spans, since the return period-impact energy curve has the smallest slope (Figures 3 and 4). A 500 year energy is 400 MNm on side spans, while it is only 140 MNm on main piers and 60 MNm on side piers in the base case. This agrees with intuition, since the side spans "catch" all vessels that cannot clear the height of the side spans above water level. Side piers receive less energy than main piers for given return period owing to higher main channel path probability assigned to the larger vessels, and higher geometric probability of striking main pier island than side piers. Changing the range of speeds from low to high, doubles or triples the impact energy for a fixed return period (Figure 3). Changing the skewness of speed distribution from left to right has a similar effect, since right skew of the distribution implies more frequent occurrences of higher speeds.

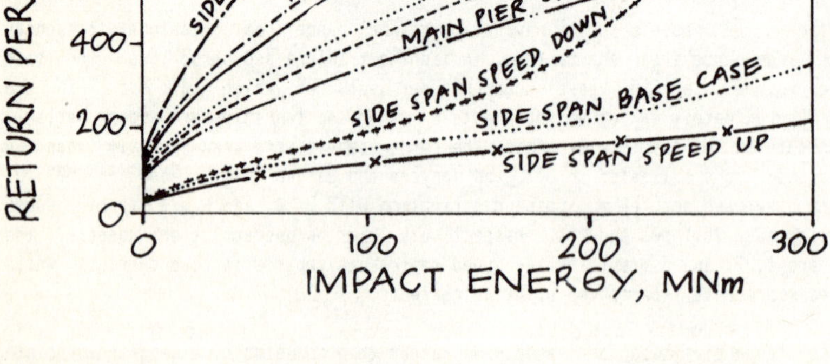

Figure 3. Effects of changing impact speed by 1 m/s.

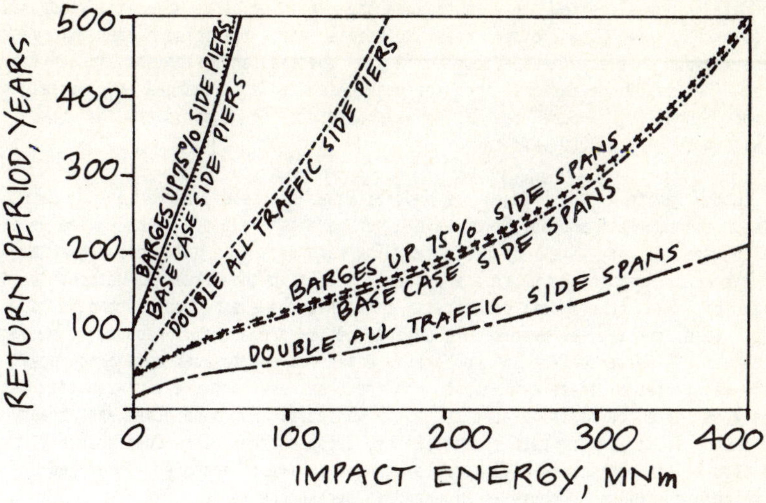

Figure 4. Effects on side piers and spans.

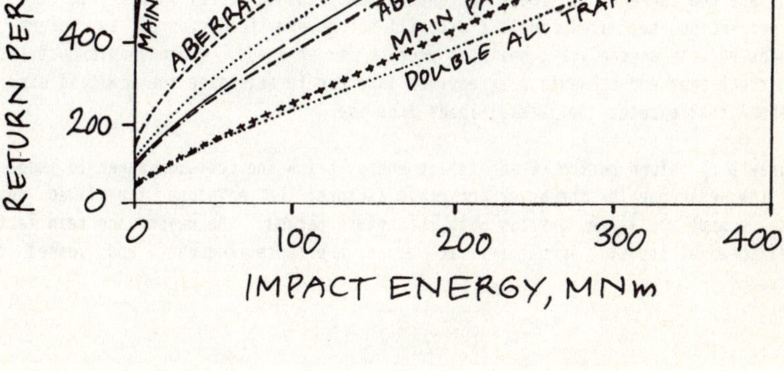

Figure 5. Effects on main piers.

The effects of doubling the annual traffic to 520 movements and increasing barge traffic by 75% are shown in Figures 4 and 5. The barge traffic adjustment has a negligible effect for all bridge elements. This is because the number of barge movements increases by cutting down the number of the other vessel movements, while the total number of vessels remains unchanged. This simulates a mode shift from the smaller freighters and tankers to the barge. The small deviations from the base case reflect changes in vessel mass distribution. Doubling annual traffic, however, triples or quadruples the base case energies within the 200 to 500 year range.

Due to the length of the bridge, the probabilities of vessel aberrance and path are crucial assumptions. Path probabilities have a strong effect but the example uses extreme cases which are not very likely in practice. The path effect is not comparable to the effect of aberrance frequencies which are probably closer to the actual situation. A high probability of main channel path drastically increases the energy corresponding to 200 years, but less for the 500 year return period. A low probability reduces the impact energies to nil. A weaker effect in the same direction is due to aberrance frequency. The approximate magnitudes of these deviations from the base case vary greatly with the return period because the slope of the graphs is highly variable. For side piers and spans the opposite trend is true for path probabilities, because they are complementary. The aberrance probability deviations give rise to changes in impact energies (for fixed return periods) which are similar to those in Figure 5 in percentage terms.

The effect on side pier of increasing side span length by 10% and of doubling pier base diameter was found minor. An interesting result, however, was obtained for the deviations from side span vertical clearance. Decreasing the 14 m clearance to 10 m (saving of over 160 vertical metres of side pier structures) does not practically change the risk of collisions with the "hard" elements of vessels.

CONCLUSIONS

The risk of bridge strikings by vessels should be viewed from a systems perspective which includes other threats such as earthquakes and extreme sea ice conditions. Multispan bridges pose a unique problem, owing to bridge length, large number of piers and the trade-offs between capital cost and risk involved in low-clearance side spans. Marine environment tends to increase the risk and the velocity of uncontrolled vessel movements. No published appropriate prototype model exists, and foreign data on vessel casualties should be used with caution.

Rough seas and currents can add a subtantial component to vessel impact velocity. In waters experiencing sea ice, small craft will not govern the design of any piers for lateral loads because severe ice conditions produce larger loads. Hydrodynamic analyses of vessels in rough seas and currents are required in order to determine the limiting size and type of vessel that exceeds the lateral loads from ice.

The graphs of return period versus impact energy allow the decision maker to judge the width of the envelope for the major uncertain factors. The envelope for fixed return periods can be wide and varies greatly with the return period. The major uncertain factors are vessel speed at impact , annual traffic volume, aberrance frequency, and vessel path probability.

ACKNOWLEDGEMENT

The author would like to thank Public Works and the Canadian Coast Guard for providing the data for the example. The opinions expressed herein are those of the author.

REFERENCES

1. Rosenblueth, "What Should We Do with Structural Reliabilities", Proceedings ICASP 5 Reliability and Risk Analysis in Civil Engineering, Vancouver, Canada, Vol.1, p.24, 1987.
2. Bein, "Probabilistic Model of Barge Impact Energy on the Annacis East Bridge Piers", Proceedings 1st International Conference Modelling Under Uncertainty, Fulmer Research Institute, England, 1986.
3. Canadian Coast Guard, "Study Vessel Traffic Services: Final Report", Transport Canada, TP5965-1, 1984.
4. Cockcroft, "The Circumstances of Sea Collisions", Journal of Navigation, Vol.35, p.100, 1982.
5. Fuji and Yamanouchi, "Visual Range and the Degree of Risk", Journal of Navigation, Vol.27, p.248, 1974.
6. Fuji, Yamanouchi and Mizuki, "Frequency of Accidents in Marine Traffic", Journal of Navigation, Vol.27, p.239, 1974.
7. Canadian Coast Guard, "Vulnerability of Bridges in Canadian Waters", Transport Canada, TP3446, 1982.
8. Marine Board, "Ship Collisions with Bridges: The Nature of the Accidents, Their Prevention and Mitigation", Commission on Engineering and Technical Systems, National Research Council, Washington D.C, 1983.
9. Frandsen, "Accidents Involving Bridges", IABSE Colloquium on Ship Collision with Bridges and Offshore Structures - Copenhagen, Proceedings IABSE, Vol.41, p.11, 1982.
10. United States Coast Guard, "Statistics of Casualties", Proceedings of the Marine Safety Council, 1980, 1981, 1982, and 1983.
11. Kostilainen and Hyvarinen, "Ship Casualties in the Baltic, Gulf of Finland and Gulf of Bothnia in 1971 - 1972", Journal of Navigation, Vol.27, p.220, 1974.
12. Standing and Brendling, "Collisions of Attendant Vessels with Offshore Installations", Part 1 and 2, National Maritime Institute Ltd., London, England, 1985.
13. United States Naval Oceanographic Office, "Ocean Thermal Structure Forecasting", U.S. Navy Manual Series, Vol.5, SP-105, 1966.
14. Sibul, "Laboratory Studies of the Motion of Freely Floating Bodies in Non-Uniform and Uniform Long Crested Waves", Proceedings of the First Conference on Ships and Waves, Hoboken, New Jersey, 1954.
15. Bradford, "A Preliminary Report of the Observation of Sea Ice Pressure and Its Effect on Merchant Vessels Under Icebreaker Escort", Proceedings of International Conference on Sea Ice, Reykjavik, Iceland, 1971.

The Dynamic Load Arising from a Crashing Military Combat Aircraft

K. Kinsella and J. Jowett

Safety & Reliability Directorate, Wigshaw Lane, Culcheth, Warrington, WA3 4NE, UK

ABSTRACT

In the UK military combat aircraft pose the major aircraft crash hazard to nuclear facilities. To make a probabilistic assessment of the structural failure frequency arising from such impacts a methodology must account for the wide spread of impact characteristics which can be realised in an accident. For example, aircraft within the military combat category may have impact masses between 5 and 50 Te and impact velocities between 50 and 300 m/sec. The paper describes a methodology for assessing the time dependent loading created by an aircraft with impact characteristics anywhere within the above ranges.

The numerical method presented is applicable to aircraft impacting normally against a rigid structure. It is based on previously published techniques used to determine the load-time history of a crashing Phantom F4 type aircraft but uses linear scaling laws to extend the calculation procedure to aircraft of differing size.

In the methodology, the crash event is treated as a combination of the separate time dependent impacts of the aircrafts airframe and engines. The airframe is classed as a soft missile which will suffer considerable deformation and a finite difference method of calculation is employed to describe its perfectly plastic impact. The engines, which are considered separately, are known to constitute a much harder missile which will undergo little deformation. In this case the results of a previous lumped mass - spring model calculation have been used to describe the impact behaviour.

Using the method, calculations have been performed to obtain total load-time histories for military combat aircraft in the mass and velocity ranges of interest. These have then been used to develop a simple equation for the average dynamic load, a parameter which may be used to assess the impact response of concrete structures.

INTRODUCTION

In the UK military combat aircraft (MCA) pose the major aircraft crash hazard to nuclear facilities. As such, the determination of both crash frequency and consequences for this group of aircraft is of primary importance in assessing the associated risk.

As part of a method for properly assessing the consequences of an MCA impact, it is necessary in the first instance, to determine the loading that would be delivered to the target. In the past, this has been computed for individual aircraft [1,2,3] with the Phantom type MCA (due to its specification in FRG design codes for nuclear containment structures) receiving particular attention. However military combat aircraft have a wide spread of impact characteristics (mass 5-50 Te, velocity 50-300 m/sec). Hence if a fully probabilistic evaluation of crash consequences is to be performed a simple method of calculating impact forces for a range of aircraft masses and velocities is needed.

This paper presents a methodology suitable for this purpose, applicable to aircraft impacting normally against a rigid structure. It is based on previously published techniques used to determine the load-time history of a crashing Phantom F4 type aircraft [2,3] but uses linear scaling laws to extend the calculation procedure to MCA of differing size.

METHOD

Linear Scaling

A cursory examination of military combat aircraft most common to UK air-space suggests that they may be related, as a first approximation, by linear scaling, in that their length varies approximately in proportion to their mass$^{1/3}$.

More particularly, as most MCA are built with fuselage mounted engines, are of similar shape and are constructed of the same materials, the approximation may be good enough to justify the assumption that all possess relatively similar structural strength and mass characteristics. If this is the case, it will be possible to develop a general method for calculating the loading during a MCA impact by applying appropriate scaling criteria to mass and crushing strength data obtained for a reference aircraft.

To investigate the existence of linear scaling between different types of MCA, 8 of the most common aircraft types flying over the UK have been selected (mass range 5 to 45 te) in addition to a Vulcan bomber (mass approximately 80 te). A list of the aircraft with corresponding mass, length and cross section data is given in Table 1. If linear scaling exists then aircraft length should vary with cross section area $^{1/2}$ and with mass$^{1/3}$.

Measurements of length and cross section for the 9 aircraft have been obtained from scaled drawings [4]. (The cross sectional area was calculated as the mean of seven ellipsoidal representations of the fuselage plus engines, taken at equally spaced intervals along the aircraft length.) By plotting the measured length against both cross section and maximum take off mass (Figures 1 and 2 respectively) it is found that length varies with cross section$^{0.47}$ and with mass$^{0.38}$. Furthermore, a plot of total engine mass against aircraft mass (Figure 3) shows that the mass of the engines increases in direct proportion to that of the aircraft.

These results provide sufficient evidence to show that approximate linear scaling does exist between the types of aircraft selected.

Table 1

Mass, length and cross-section data for MCA common to the UK

Aircraft type	Max take-off mass (te)	Total engine mass (te)	Overall length (m)	Average fuselage cross section (inc engines) (m^2)
Jet Provost	5.2	0.34	10.3	1.19
Hawk	7.7	0.72	11.2	1.19
Harrier	11.3	1.40	13.9	1.83
F16	15.0	1.34	14.5	2.08
Jaguar	15.5	1.47	16.8	1.92
MRCA (Tornado)	26.3	2.16	16.7	2.28
Phantom	28.0	3.33	19.2	2.42
F111	45.0	3.80	23.0	3.78
Vulcan	81.7	-	30.5	11.28

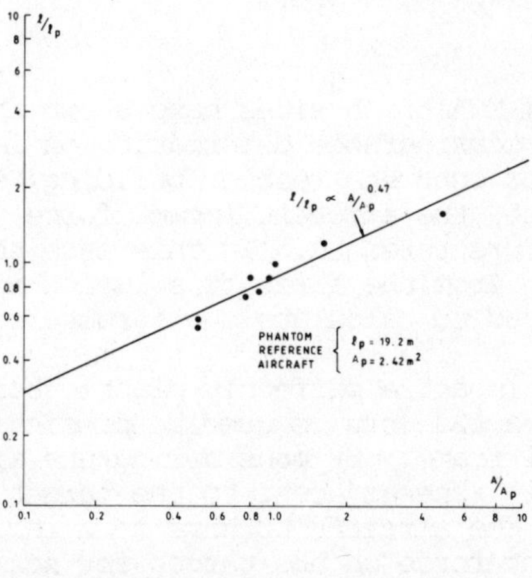

FIGURE 1 RELATIONSHIP BETWEEN AIRCRAFT LENGTH AND AVERAGE CROSS - SECTION

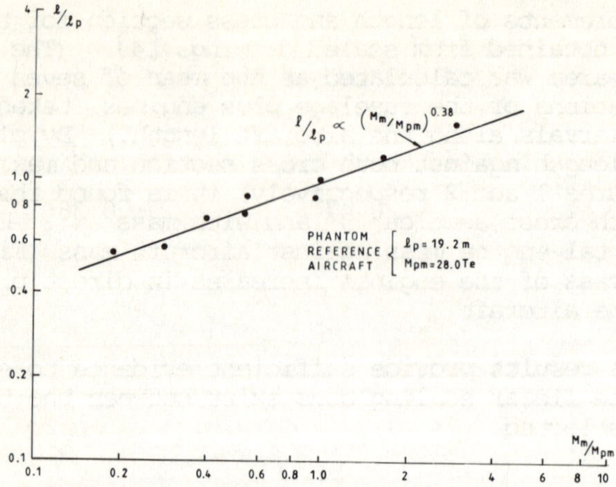

FIGURE 2 RELATIONSHIP BETWEEN AIRCRAFT LENGTH AND MAXIMUM TAKE - OFF MASS

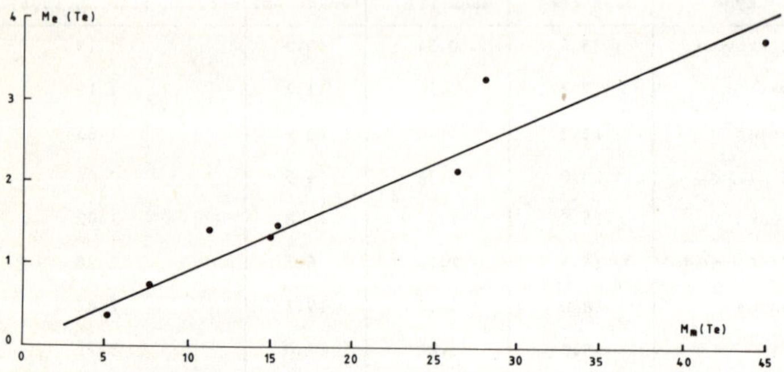

FIGURE 3 RELATIONSHIP BETWEEN ENGINE MASS (M_e) AND MAXIMUM TAKE - OFF MASS (M_m)

Air Frame Impact

Observations from MCA crash sites have shown that the aircraft airframe suffers considerable deformation on impact. For a typical reinforced concrete reactor building, when an aircraft fails to perforate, the aircraft impact force is independent of the target structure strength. In this case the force-time history resulting from the aircraft's impact against a rigid surface may be used to calculate the structures response.

For such an impact a perfectly plastic soft missile formula developed by Riera [2] can be used to predict the loading exerted by the airframe. By considering the airframe to deform only within a thin zone adjacent to the target and assuming the deformed part of the aircraft to have zero residual velocity, the time dependent force on the target for normal impacts is given by:

$$F(t) = P_c\,[x(t)] + \mu\,[x(t)]\,V^2(t) \qquad (1)$$

where:

$F(t)$ is the force on the target at time t, (N)

$P_c\,[x(t)]$ is the crushing strength of the airframe at a distance x from the aircraft nose, (N)

$\mu\,[x(t)]$ is the mass per unit length of the aircraft at a distance x from the aircraft nose, (kg/m)

$V(t)$ is the velocity of the undamaged portion of the airframe, at time t after impact, (m/s)

and $\mu\,[x(t)]\,V^2(t)$ is the component of force at time t created by the reduction in momentum as mass enters the deformation zone and is brought to rest.

During the impact, the force, P_c, required to crush the airframe, will act against the uncrushed portion of the aircraft such that its instantaneous deceleration a(t) is given by:

$$a(t) = -P_c(t)/m(t) \qquad (2)$$

where m(t) is the mass of the uncrushed portion of the aircraft at time t measured in kg.

In order that this model can be generalised for use with the range of MCA specified in Table 1, it is necessary to specify crushing strength (P_c) and mass/unit length (μ) distributions for a reference aircraft. Ideally the reference aircraft would have a mass which lies close to the centre of the MCA mass range considered, so that any degree of inaccuracy inherent in applying the scaling approximations is kept to a minimum. An aircraft which satisfies this criterion and for which the required data is readily available is the Phanton F-4.

Due to the specification of this aircraft in German design codes for nuclear power plant containments, data on the crushing strength and mass distributions has already been published [3]. This is shown in Figure 4. (Note: The mass of the engines shown in Figure 4(a) is not included in the mass distribution used for the airframe model as engine impacts are considered separately in the next sub-section.)

Equations (1) and (2) can now be written in the more general form:

$$F(t) = Q(P_{cp}[x_p(t)/Q^{1/2}] + \mu_p\,[x_p(t)/Q^{1/2}]\,V^2\,(t)) \qquad (3)$$

and $$a(t) = -Q\,P_{cp}[x_p(t)/Q^{1/2}]/m(t) \qquad (4)$$

where $P_{cp}[x_p(t)]$ and $\mu_p[x_p(t)]$ are the crushing strength and mass/unit length distributions of the Phantom aircraft; with $x_p(t)$ the length of the reference aircraft deformed at time, t. Q is the scaling factor for the airframe cross section given by:

$$Q = \left(\frac{M}{M_p}\right)^{2/3} \quad (5)$$

The Phantom mass, M_p, used throughout is 22 tonnes (including the mass of the engines) and is mid-way between the maximum take-off mass (28 te) and the lower end of crash mass estimates, ~ 16 te [3].

Using a numerical time integration method to solve equations (3) and (4) it is possible to generate an approximate load-time history for any MCA within the mass range considered (5-80 te). To perform this calculation only the mass and velocity of the aircraft at the time of impact are required.

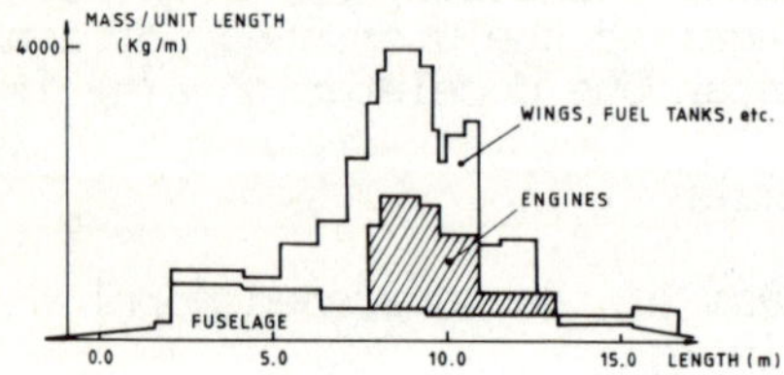

FIGURE 4 (a) MASS DISTRIBUTION OF PHANTOM F4

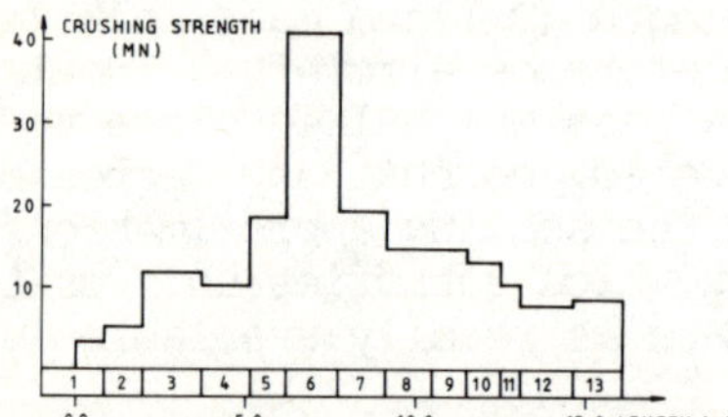

FIGURE 4 (b) CRUSHING STRENGTH DISTRIBUTION OF PHANTOM F4 FUSELAGE

Engine Impact

The extreme decelerations experienced by MCA during the initial stages of impact are such that the connections between the fuselage and engines can be expected to fail (typically 1-2mS after impact) resulting in the engines detaching from the fuselage and travelling at near constant velocity until they reach the target. Observations from crash sites of the Phantom aircraft [5] show that the engines undergo little deformation on impact and that the impact location is close to that of the fuselage. The degree of closeness is such that when considering the loading to a structure it will be necessary to combine the independent effects of the fuselage and engine impact into a single loading event. To achieve this a model is required to describe the impact of the engines.

A suitable model for describing the behaviour of a single Phantom engine impacting against a rigid target has been developed by Riera & Zorn [6]. In this, the structural characteristics of the engine are represented by a series of one dimensional lumped masses connected by elasto plastic springs, and the reaction is obtained as the time derivative of the total momentum. An encouraging indication of the validity of this method is that its predictions of deformation and rebound velocity have been found to correspond closely with crash site observations [6].

By correlating Riera & Zorn's results in terms of impact velocity it can be shown that the loading function for a normal engine impact can be adequately described by the following parameters illustrated in Figure 5:

Peak Force, $F_1 = \mathrm{Exp}\,((V_i+979.4)/302.3)$ (MN) (6)

Time $t_1 = 1.1$ (ms) (7)

Time, $t_2 = 0.082\, V_i^{0.862}$ (ms) (8)

Mid-impact Force/Time gradient,

$F_1-F_2/t_2-t_1 = \mathrm{Exp}\,((V_i-114.2)/81.7)$ (MN/ms) (9)

Impact Duration, $t_3 = 4.58 \ln V_i - 15.0$ (ms) (10)

where V_i is the impact velocity of the engine in m/s (taken as equal to the impact velocity of the aircraft). These equations are applicable in the velocity range 70-210 m/s.

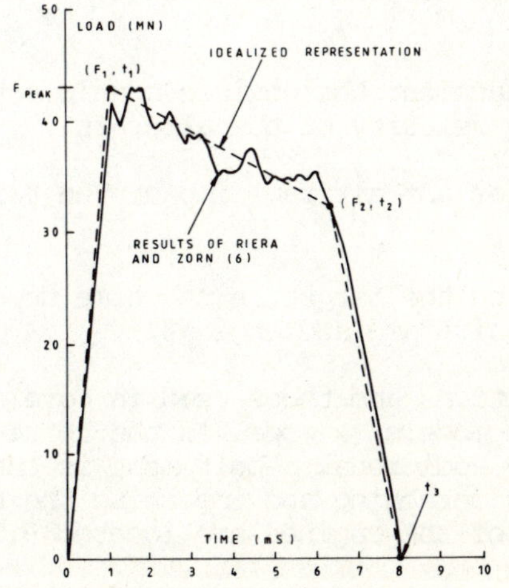

FIGURE 5 IDEALIZED REPRESENTATION OF PHANTOM F4 ENGINE LOAD HISTORY

For the majority of aircraft crashes the assumption of a perpendicular engine impact will be conservative. In practice, the small misalignment of engines and fuselage included in the aircraft design, and any rotation of the engine after detachment will result in the engine impacting the target obliquely.

The existence of approximate linear scaling between the type of MCA considered is assumed here to extend to the mass, location and total cross sectional area of the engines. The fact that some MCA have one engine and others two will not effect this assumption providing their total cross-section and mass vary accordingly. In this way it is possible to generate approximate load-time histories for the engines of aircraft within the mass range of those listed in Table 1, by simply applying the laws of linear scaling to equations (6) through (10), such that:

$$F_1 = 2\ Q\ \mathrm{Exp}(\ (V_i + 979.4)/302.3) \qquad (\mathrm{MN}) \quad (11)$$

$$t_1 = 1.1\ Q^{1/2} \qquad (\mathrm{ms}) \quad (12)$$

$$t_2 = 0.082\ Q^{1/2}\ V_i^{0.862} \qquad (\mathrm{ms}) \quad (13)$$

$$F_1 - F_2/t_2 - t_1 = 2\ Q^{1/2}\ \mathrm{Exp}(\ (V_i - 114.2)/81.7) \qquad (\mathrm{MN/ms}) \quad (14)$$

and $$t_3 = Q^{1/2}\ (4.58\ \ln V_i - 15.0) \qquad (\mathrm{ms}) \quad (15)$$

The factor of 2 introduced in equations (11) and (14) derives from the Phantom F-4 having two engines.

Total Impact

In arriving at a model of the total aircraft impact, the following additional assumptions are made:

i The engines detach from the airframe immediately on impact.

ii After detachment the engines continue to travel at the initial impact velocity of the aircraft.

iii The engines and airframe impact the target at the same location.

iv The load on the target is the time dependent sum of that from the airframe and engines.

Taking these assumptions and those used in developing the airframe and engine models, a model of the total impact can be obtained. The only additional requirement is the time delay between the engines detaching and impacting the target. For the Phantom, the front of the engines are located 9.7m from the nose of the aircraft. Hence the time taken from detachment to target, T_E, is given by:

$$T_E = 9700/V_i \qquad \text{(ms)} \quad (16)$$

where V_i is the impact velocity of the aircraft in m/s.

Extending this to other aircraft, assuming linear scaling gives:

$$T_E = 9700\ Q^{1/2}/V_i \qquad \text{(ms)} \quad (17)$$

For a given aircraft mass and impact velocity, the total load-time history can therefore be computed by summing the results from the airframe and engine models with the initiation of the impacts separated by the time interval, T_E.

RESULTS

Using a numerical time integration scheme, the impact model described above has been used to compute the loading on a rigid target due to normal impacts of MCA with mass and velocity in the ranges 5-80 te and 70-300 m/s respectively.

Figure 6 shows an example calculation for a Phantom F-4 type aircraft with mass = 22 tonnes and velocity = 210 m/s. Also shown in this figure are the effects of including the engines into the mass distribution of the airframe and of considering the impact of the airframe only.

Further load-time histories demonstrating the effects of different masses and velocity are given in Figure 7. These show how the peak load increases and impact duration reduces with increasing impact velocity and how peak load and duration both reduce with reduced aircraft mass.

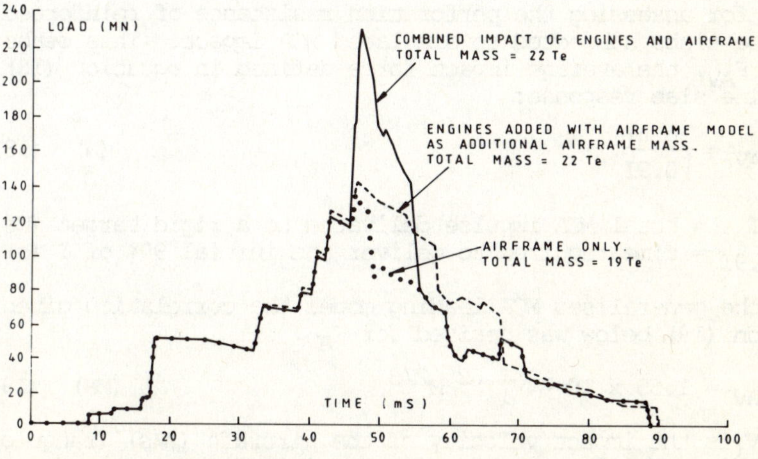

FIGURE 6 PREDICTED LOAD-TIME HISTORY FOR A PHANTOM F4 IMPACTING A RIGID TARGET AT AN IMPACT VELOCITY OF 210 m/S

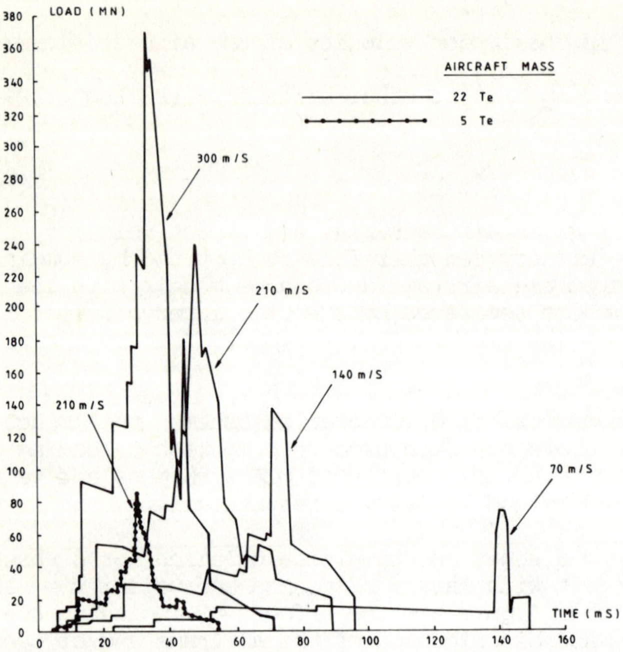

FIGURE 7 LOAD-TIME HISTORIES CALCULATED FOR VARYING IMPACT VELOCITY & MASS

Another paper in this conference [7] describes a simplified method for assessing the perforation resistance of reinforced concrete slabs subjected to simulated MCA impact. This method uses, F_{AV}, the average impact force defined in equation (18) to determine slab response:

$$F_{AV} = \frac{0.9I}{t_{0.9I}} \qquad \text{(N)} \qquad (18)$$

where I = total MCA impulse delivered to a rigid target (Ns) and $t_{0.9I}$ = time required to deliver the initial 90% of I (s).

Using the generalised MCA loading model the correlation given in equation (19) below was derived for F_{AV}:

$$F_{AV} = 1.55 \times 10^{-3}\ V_i^{1.62}\ M^{2/3} \qquad \text{(MN)} \qquad (19)$$

where V_i is the impact velocity of the aircraft (m/s) (range of application 70-300 m/s) and M is the aircraft mass (te) (range of application 5-80 te).

CONCLUSIONS

i Approximate linear scaling has been shown to exist between a variety of military combat aircraft common to the UK in the mass range 5-80 te.

ii Using the linear scaling approximation, a general and simple methodology has been developed for predicting the loading characteristics of a military combat aircraft impacting normally against a rigid target. The model can be used for MCA in the mass and velocity ranges 5-80 te and 70-300 m/s respectively.

iii Results from the model have been correlated into a simple equation for predicting an average loading condition suitable for assessing the impact response of civil structures. The equation relates the average loading to the mass and velocity of the aircraft at the time of impact.

REFERENCES

1. Drittler, K. and Gruner, P. The Force Resulting from Impact of Fast-Flying Military Aircraft upon a Rigid Wall, Nuclear Engineering & Design, Vol. 37, pp. 245-248, 1976.

2. Riera, J.D. A Critical Reappraisal of Nuclear Power Plant Safety against Accidental Aircraft Impact, Vol. J, paper J9/2, 1979.

3. Zorn, N.F. and Schueller, G.I. On the Failure Probability of the Containment under Accidental Aircraft Impact, Nuclear Engineering & Design, Vol. 91, pp. 277-286, 1986.

4. Swanborough, G. Military Aircraft of the World, Ian Allan publicn, 1981.

5. Zorn, N.F. Statischtische Unkrsuchung du Phantomabsture der Bundeswchs. Internal Working report RS-201 C-16, Munich University, 1981.

6. Riera, J.D. and Zorn, N.F. Risk Due to Accidental Aircraft Crash against Nuclear Power Plants: On the Definition of the Excitation due to Engine Impact. Report CT-28/81, University Rio Grande do Sul, Brazil, 1981.

7. Jowett, J. and Kinsella, K. Soft Missile Perforation Analysis of Small and Large Scale Concrete Slabs. This Conference.

NOMENCLATURE

A	Average cross sectional area of aircraft fuselage (m^2)
F(t)	The force exerted on the target at time t from the point of impact (N)
F_{AV}	Average impact force. defined by equation (18 or 19) (N or MN)
F_1, F_2	The force levels used to define the aircraft engine load (see Fig 5) (MN)
I	Total aircraft impulse delivered to a rigid target (Ns)
ℓ	Length of aircraft (m)
m(t)	The mass of the uncrushed portion of the airframe (excluding engines) at time t (kg)
M	Total mass of aircraft including engines (te)
M_E	Mass of aircraft engine (te)
$P_C[x(t)]$	Crushing strength of the airframe at a distance x from the aircraft nose (N)
Q	Mass scaling parameter defined by equation (5)
t_1,t_2,t_3	The time values used to define the aircraft engine load (see Fig 4) (ms)
T_E	Time between instant of aircraft impact and impact of engines (ms)
$t_{0.9I}$	Time required to deliver initial 90% of I (s)
V(t)	Instantaneous velocity of uncrushed portion of airframe. (m/s).
V_i	Impact velocity of aircraft or aircraft engines. (m/s)
x(t)	Length of fuselage crushed at time t (m)
$\mu[x(t)]$	Mass per unit length of fuselage (kg/m)

<u>Subscripts</u>

p	Refers to phantom reference aircraft.
m	Maximum Take-off value.

Crashworthiness Analysis of Shell Structures using the Finite Element Method

Z.H. Zhong
Department of Mechanical Engineering, Linköping University, S-581 83 Linköping, Sweden

ABSTRACT

An overview is made on crashworthiness analysis of shell structures by using the finite element method. Special considerations are then given to the treatment of general contact-impact interfaces with friction involved in a crash process of shell structures. New algorithms for crashworthiness analysis are discussed. Numerical examples solved by using the new algorithms are presented.

INTRODUCTION

The fast developments of both numerical methods and computing technologies have now been giving increasingly strong influences on many engineering areas. Computer simulations of various physical phenomena are becoming common practice both in scientific researches and in industry applications. One remarkable achievement in computer simulation is perhaps the crash simulation of engineering structures, like passenger cars and aircrafts. In many engineering applications, a majority of structures can be considered of shell type. The shell structure has thus been studied extensively in different contexts. A great deal of both experimental and numerical work has been reported on this, e.g. Jones and Wierzbicki [1] to Hughes [11]. Since experiments are often time-consuming and expensive, computer simulations, e.g. Jones and Wierzbicki [1] to Hazet and Guiraud [5], are taking the place of them, though not all of them.

Due to the extreme complexity involved in a crash process of shell structures, however, a lot has to be done to make a large scale and complicated simulation possible and sufficiently reliable and efficient. Among a huge number of colleages, the author is also working with the crashworthiness analysis. In this paper, part of the work done is presented. In the following sections, an overview is first made on crashworthiness analysis. Special considerations are then given to the contact-impact interfaces involved in the crashworthiness analysis of shell structures. In this study, the DYNA3D program, Hallquist [8], has been used as a base code for the implementation of new algorithms and for the finite element analysis.

MECHANICAL BEHAVIOURS OF SHELL STRUCTURES UNDER CRASH LOADING

Talking about crash, one often thinks of "death". Indeed, crash often means destruction in one way or another. That is due to the nature of a crash process, i.e. high rate heavy loading on the structure. Such loading can often produce large dynamic deformations, strains and stresses as waves travelling around in the structure and can be very destructive. Depending on the stress level developed in the structure, the deformation of the structure can be considered elastic or elasto-plastic. If the loading rate is sufficiently high, rate dependence of the material can become dominant, rendering a visco-plastic behaviour of the structure. With shell structures, buckling will almost always be produced under crash loading. Due to large displacements and plastic deformations, contact-impact interfaces can often be presented within a single component of a structure. Actually, contact-impact interfaces can be involved at any stage of the loading process if it is contained in the undeformed structure. With the contact-impact interface, friction will be brought into effect , which further complicates the structural response. If defects and cracks or other facts are presented in the structure, simulation difficulties will be upgraded significantly. That is, however, not concerned in this study.

METHODS FOR CRASHWORTHINESS ANALYSIS

There are several important and difficult aspects which account for the main work in crashworthiness analysis. Using the finite element method to discretize the structure, accurate and efficient element formulations are required. Although various shell element formulations are available, the lower order elements are mostly recommended and used in the crashworthiness analysis. Constitutive formulations must be able to model the behaviour of the concerned materials under severe conditions, like large deformations and large strains etc. Numerical methods are also required for the integration in the time domain. For dynamic problems with high frequency loading, the explicit time integration method is proven to be more preferable than the implicit method. Alternative contact searching algorithms have been proposed to detect the a priori unknown contacting boundaries. In most previous applications, the penalty method has been used as a contact constraint method to impose the zero-penetration condition. With a sophisticated interface algorithm, it is now also possible to use the Lagrange multiplier method as the contact constraint method. When friction effects are to be considered, a friction model has to be selected and implemented.

GENERAL FORMULATION OF NONLINEAR DYNAMIC PROBLEMS WITH CONTACT-IMPACT INTERFACES

Consider a dynamic system of two bodies with volumes ${}^{t}V^{1}$ and ${}^{t}V^{2}$ at time t. With the principle of virtual work applied and the explicit time integration method employed in the updated Lagrangian formulation, the governing equations of the system can be found as, see Zhong [15],

$$\int_{{}^{t}V^{\alpha}} {}^{t}\tau_{ij}\, \delta\, {}_{t}e_{ij}\, {}^{t}dV = {}^{t}r_{o}^{\alpha} + {}^{t}r_{I}^{\alpha} + {}^{t}r_{c}^{\alpha} \qquad \alpha = 1, 2 \qquad (1)$$

where α denotes the number of the body the equation pertains to, ${}^{t}\tau_{ij}$ and ${}_{t}e_{ij}$ are the stress and strain components at time t, ${}^{t}r_{o}^{\alpha}$ the virtual work of the external forces on the bodies excluding the contact interaction forces on the contact boundaries, ${}^{t}r_{I}^{\alpha}$ are the virtual work of the inertia force within the bodies, and ${}^{t}r_{c}^{\alpha}$ are the virtual work of the contact interaction forces. The last term on the right-hand side of Eq. (1) is evaluated as follows:

$${}^{t}r_{c}^{\alpha} = \int_{{}^{t}\Gamma_{c}^{\alpha}} {}^{t}f_{n}^{\alpha}\, \delta\, u_{n}^{\alpha}\, {}^{t}dS + \int_{{}^{t}\Gamma_{c}^{\alpha}} {}^{t}f_{t}^{\alpha}\, \delta\, u_{t}^{\alpha}\, {}^{t}dS \qquad \alpha = 1, 2 \qquad (2)$$

where ${}^{t}\Gamma_{c}^{\alpha}$ denote the contact boundaries, ${}^{t}f_{n}^{\alpha}$ and ${}^{t}f_{t}^{\alpha}$ are the normal contact-impact forces and the friction forces, respectively, and u_{n}^{α} and u_{t}^{α} the normal and tangential components of the displacement on the contact boundaries, respectively.

In the above, the evaluation of ${}^{t}\tau_{ij}$ is based on specific element formulations and constitutive formulations, ${}^{t}f_{n}^{\alpha}$ are to be evaluated under the zero-penetration condition and ${}^{t}f_{t}^{\alpha}$ under the friction law.

ELEMENT FORMULATION AND CONSTITUTIVE FORMULATION

Since shell structures take extremely important places in engineering, shell element formulations receive central attention in the finite element method. Numerous literatures and papers on shell theories and shell elements are available. Various shell elements have been proposed and examined. For explicit analysis, however, the lower order elements are mostly recommended. Two well known elements due to Belytschko [10] and Hughes [11] are widely studied and implemented for large scale nonlinear analyses. Both elements were developed with emphasis on computing efficiency, with the first much faster, and have been efficiently implemented in the DYNA3D program. As well known, sophisticated element formulations are not sufficient for good solutions. Well designed finite element mesh is also essential in crashworthiness analysis. In a large scale analysis with severe conditions, mesh generation takes quite a lot of human work. How a mesh should be generated is largely based on convenience, experience and judgement . In practice, mesh refinements and regenerations are often required for a possible and usable solution.

To accurately predict the structural behaviour, it is essential to have reliable constitutive formulations. Considerable improvements in both the accuracy and efficiency of the constitutive models have been made over the last few years, Hughes [12]. Constitutive laws for elastic and elasto-plastic materials under normal conditions are now well established. There are, however, situations where material modelling presents difficulties. For example, rate dependence and temperature dependence is more difficult to handle in numerical simulations although much research work has been done on this. It should also be observed that a good constitutive formulation does not guarantee a valid solution. Correct choice of constitutive formulations in accordance with different applications is important.

CONTACT SEARCHING ALGORITHM

General

To detect the a priori unknown contacting boundaries, reliable and efficient contact searching algorithms are required. An earlier work on the contact searching algorithm can be found in Hallquist [13], in which the master-slave algorithm is described. After that, a single surface algorithm is developed for better efficiency, Benson and Hallquist [14]. Recently, a reliable and more efficient contact searching algorithm, i.e. the HIerarchy-Territory Algorithm (HITA), for general contact-impact systems has been developed, Zhong [15]. In the HITA algorithm, a contact system is decomposed into contact hierarchies. A territory is then defined for each contact hierarchy to obtain an efficient indication of the position of the contact hierarchy. With the introduction of the contact hierarchy and the contact territory, quite efficient and reliable contact searching algorithms have been developed and validated, Zhong [15] to [18].

Further developments of the HITA algorithm

Further developments of the HITA algorithm are described in detail in Zhong [22]. The main developments are shortly reviewed here. For three dimensional analysis, a new contact hierarchy, the contact edge, has been added. From topological point of view, this is natural. It has also been found that it is very beneficial to define an edge as a contact hierarchy both for searching contacting nodes and for imposing contact constraints. A contact segment is now defined by contact edges instead of contact nodes. With the introduction of contact edges, both storage and operations required for processing the contact-impact interface are reduced.

For the convenience of description, the contact hierarchies are classified into two categories, i.e. the primitive contact hierarchy and the composite contact hierarchy. By definition, contact nodes, contact edges and contact segments are primitive contact hierarchies and contact bodies and contact surfaces are composite ones. Furthermore, the concept of contact frontiers is also introduced. A contact frontier is defined, for each contact hierarchy, as the boundary of its contact territory, in which the thickness of the shell has to be included when shell structures are encountered.

To handle more general cases, the concept of contact pairs is used. A contact pair always consists of a hitting node and a target hierarchy, which can be any one of the primitive contact hierarchies. Both the hitting node and the target hierarchy may appear in more than one contact pair. With the definition of the contact pairs, the situations shown in Fig. 1 can then be handled efficiently. In Fig. 1, node 5 can be matched with both segment 12 and segment 34. It is now very important to observe that the contact force can no longer be associated with a contact node but a contact pair.

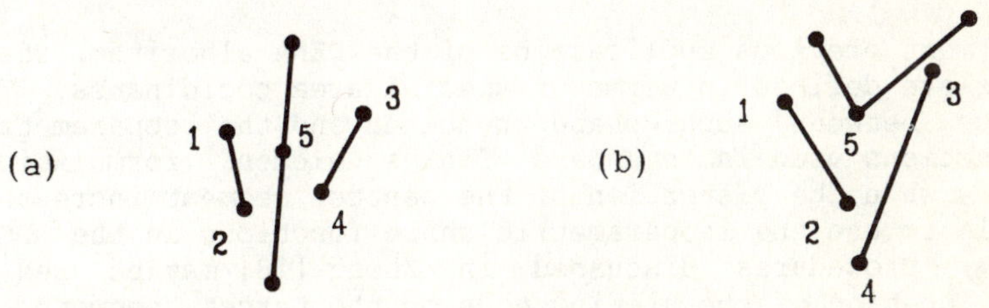

Fig. 1 Situations where the definition of contact pairs is desirable.

CONTACT-IMPACT INTERFACE ALGORITHM

General

In the explicit environment, the penalty method was the only constraint method to impose the zero-penetration condition and to find the contact-impact force until the DEfence Node Algorithm (DENA) was developed, Zhong [15]. In the DENA algorithm, a defence node is introduced to prevent a hitting node from penetrating through its target segment. The Lagrange multiplier method can then be used as the contact constraint method. The advantages of the Lagrange multiplier method over the penalty method have been shown. Both the HITA and the DENA algorithms have been implemented in the DYNA3D program and examined in various applications with satisfactory results.

Refinements of the DENA algorithm

In the earlier version of the DENA algorithm, each contact pair is considered only once within one time step. When one primitive contact hierarchy appears in more than one contact pair, however, all the contact pairs containing this contact hierarchy will be coupled and iterations may be desirable for better accuracy. The iteration can be done as follows.

```
10   CONTINUE
     DO 100 IP=1,NP  (Total number of contact pairs
                        to be processed)
        (Process contact pair IP to find
```

```
    the contact-impact force)
100 CONTINUE
    IF (Contact force increments small enough) THEN
      RETURN
    ELSE
      GO TO 10
    END IF
```

It should be pointed out that this iteration process will be quite efficient. It is the first iteration which takes time. Most computed results in the first iteration, like velocities and accelerations etc., which will not be updated within this iteration process, can be used in the subsequent iterations. It is also found that the convergence of the iteration is very fast.

In most previous applications of the DENA algorithm, shape functions are defined in terms of special area coordinates. The difference between such shape functions and the isoparametric shape functions used in standard finite element formulations will grow when the distortion of the contact segment increases. To be able to use the isoparametric shape functions in the DENA algorithm, procedures discussed in Zhong [18] may be used to find the location of the hitting node on the target segment in terms of the isoparametric coordinates. From the isoparametric coordinates, the normal vector can also be calculated accurately. Again the additional computation work is negligible.

Friction evaluation

When contact-impact interfaces are involved, friction effects will come to play their rolls, Chaudhary and Bathe [19] to Oden [21]. There are two main difficult aspects in modelling the friction effects. First, the friction law itself has now been under discussion . Secondly, the implementation of a given friction law demands even bigger efforts than imposing the zero-penetration condition in the contact-impact interface. Although friction coefficient can be very small in lubricated contact-impact surfaces, friction effects should, in general, not be neglected. It can affect the solution and is an active factor causing solution bifurcations.

When friction is not to be included in the contact-impact interface, the defence node is defined on the midsurface of the shell element in the DENA algorithm. If friction effects are to be considered, however, the defence node can no longer be defined on the midsurface of the shell element. Instead, the defence node should be defined on the contact face as shown in Fig. 2(a). The construction of the defence node is the same as in the frictionless case. The contact-impact forces, including the friction force, on the defence node will also be distributed to the target nodes in the same way. It is, however, important to observe that depending on the shell thickness, moments will be produced by the friction force and should also be distributed into the target nodes. One way to consider this is to introduce the so called "secondary target nodes" as shown in Fig. 2(b).

Now all the contact-impact force components on the defence node will be first distributed to the secondary target nodes and then transferred into general force components of the target nodes. The moment due to friction effects on a target node will be defined by the friction force distributed to its corresponding secondary target node and the shell thickness.

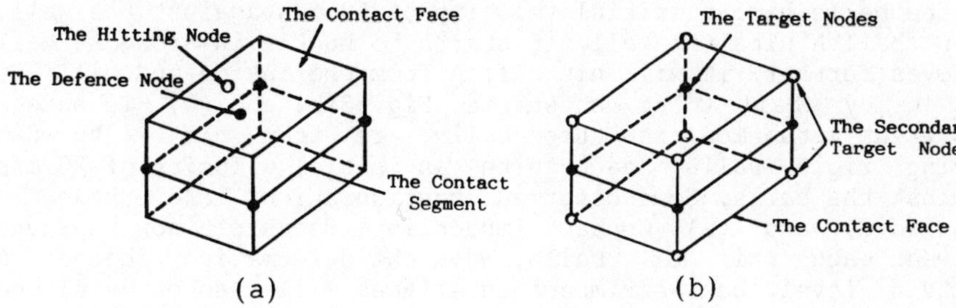

Fig. 2 (a) The defence node on the contact face.
(b) The secondary target nodes.

Although a few alternative friction laws are available, the classical Coloumb's friction law is mostly used, especially for general crash analysis. With the Coloumb's friction law, a pair of hitting nodes and defence nodes will be either in pure sticking state or in pure sliding state. The friction force under pure sliding condition is calculated by simply multiplying the current normal contact-impact force by the friction coefficient.

To calculate the friction force under the pure sticking assumption, it is necessary to understand the assumption and tranform it to equivalent constraint conditions. Let T_X and T_Y be two orthogonal unit vectors in the tangent plane defined at the defence node. Kinematic analysis shows that if the sliding displacement increments in both the tx and the ty directions are zero, no sliding is possible between the pair of nodes. To impose the pure sticking condition, we thus require that the sliding displacement increments in both the T_X and the T_Y direction are zero. This zero-sliding condition is completely the same as the zero-penetration condition in the calculation of the normal contact-impact force. Therefore, the DENA algorithm can be used to calculate the friction force increment. In the calculation of the normal contact-impact force, the DENA algorithm is applied in the normal direction. In the calculation of the friction forces, however, the DENA algorithm has to be applied in both the T_X direction and the T_Y direction.

PROGRAM IMPLEMENTATION AND NUMERICAL EXAMPLES

In the original version of the DYNA3D program, efficient shell elements and material models are available. Several interface algorithms were also implemented. In this study, the new contact-impact algorithms with the recent developments have been implemented in the program on different computers from work stations to supercomputers. Details about the implementation can be found in Zhong [15], [22] and Oldenburg [17]. The implementation has been examined in different applications. Two examples are presented here.

Thin-walled balls crashing onto rigid walls

In this example, the three balls crashing onto a rigid wall, see Fig. 3(a), are first studied, excluding friction effects. Each of the balls has an initial velocity of 50 m/s against the wall. When ball A hits the wall, it starts to buckle inwards. As ball B moves further, it will hit ball A from the inside and will be hit by ball C as shown in Fig. 3(b) and (c). To have a different situation, the three balls are then hit by two moving rigid walls, each with an initial velocity of 20 m/s against the balls. Some deformed configurations are shown in Fig. 3(d) and (e). Contact-impact interfaces can now be found between each pair of balls. When the deformation reaches a certain level, contact-impact interfaces will also be developed within both ball B and ball C. In Fig. 3(e), contact-impact interface within ball C is developed.

It can be seen that large deformations are presented in all the balls. Whether the deformation is correctly modelled depends on the element formulation, the constitutive formulation and the contact-impact interface algorithm. By using the new contact-impact algorithms, the contact-impact interfaces are well taken care of automatically without tedious input definitions for the algorithms.

A ball impacting two plates

Next, a ball impacting two plates is considered, see Fig. 4. In the first case, the ball and the plates are totally free to move, and the ball impacts the upper plate with an initial velocity (-50,0,-50) m/s. The deformed configurations with and without friction effects are shown in Fig. 5(a) to (d). In the second case, the boundary conditions are changed by fixing the plates at their ends further from the ball. The deformed configurations with and without friction effects are shown in Fig. 5(e) to (h).

As expected, the deformations with and without friction effects are quite different. In the first case, friction prevents large relative sliding between the two plates. In both cases, friction tends to keep the two plates closer. It is very interesting to examine the energy transformation. The kinetic energy and the total energy of the whole system for the first test case, with and without friction, are plotted in Fig. 6(a) to (d). It can be seen that the total energy decreases more when friction is included. The kinetic energy, however, decreases less with friction. One reason is the following. With friction, the two plates tend to stick together under crash loading from the ball. Consequently, the plates become stiffer and the total deformation of the plates is smaller. Thus, the kinetic energy loss is less. On the other hand, energy is lost due to friction effects. In this case, the total energy loss is bigger with friction. It is, however, not necessarily true in general.

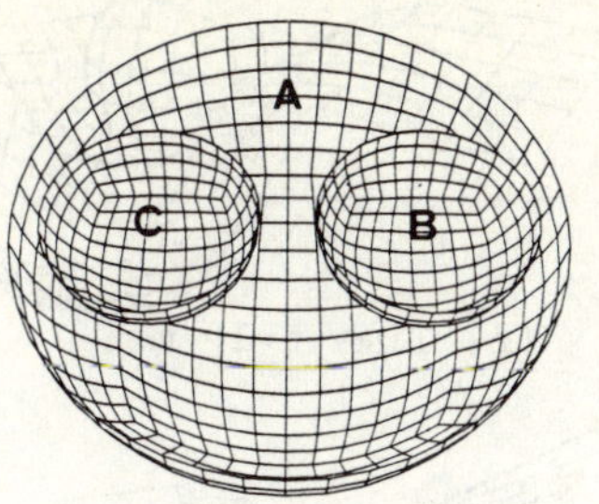

Material data:

Young's modulus	E	$= 0.21\times10^{12}\ N/m^2$
Hardening modulus	E_t	$= 0.63\times10^{9}\ N/m^2$
Poisson's ratio	ν	$= 0.3$
Yielding stress	σ_y	$= 0.20\times10^{9}\ N/m^2$
Density	ρ	$= 7840\ Kg/m^3$
Thickness	t	$= 0.0015\ m$

(a) Time t=0 ms.

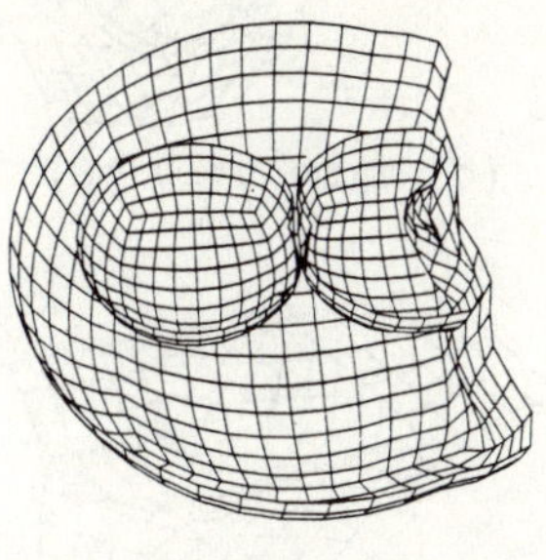

(b) Time t=1 ms.

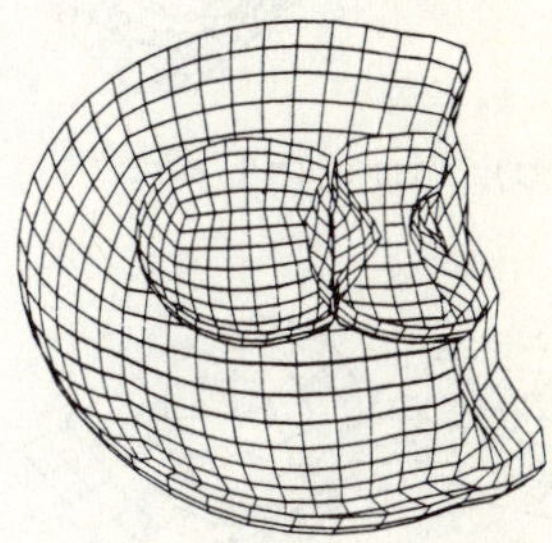

(c) Time t=2 ms.

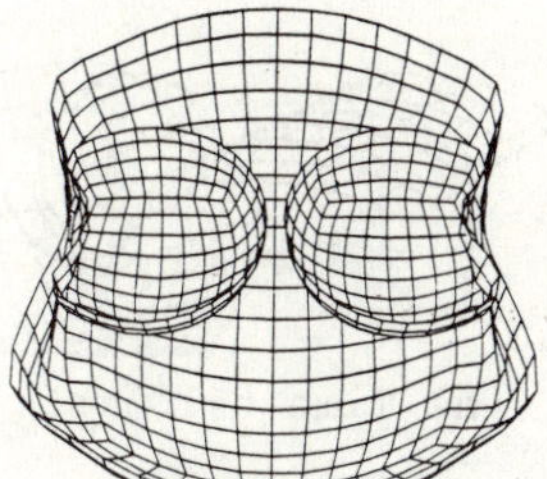

(d) Time t=1 ms.

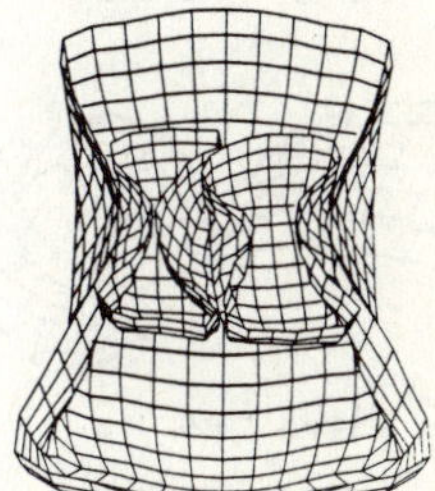

(e) Time t=2 ms.

Fig. 3 Thin-walled balls crashing onto rigid walls
(a) Geometry and material data.
(b),(c) Deformed configurations from test case one.
(d),(e) Deformed configurations from test case two.

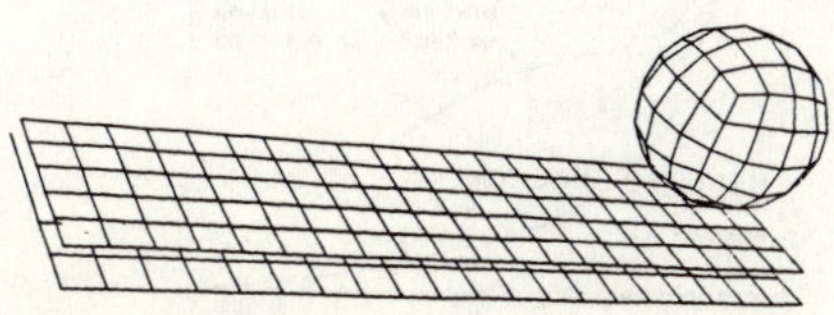

Material data:

Young's modulus	E	$= 0.21\times10^{12}\ N/m^2$
Hardening modulus	E_t	$= 0.63\times10^{9}\ N/m^2$
Poisson's ratio	ν	$= 0.3$
Yielding stress	σ_y	$= 0.20\times10^{9}\ N/m^2$
Density	ρ	$= 7840\ Kg/m^3$
Thickness	t	$= 0.0015\ m$

Fig. 4 A ball impacting two plates. Geometry and material data.

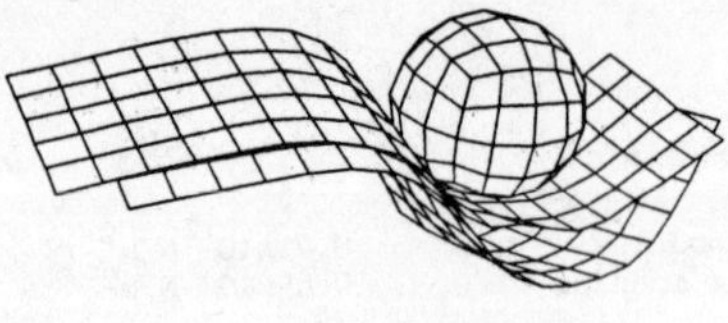

(a) Time t=4 ms.

(b) Time t=10 ms.

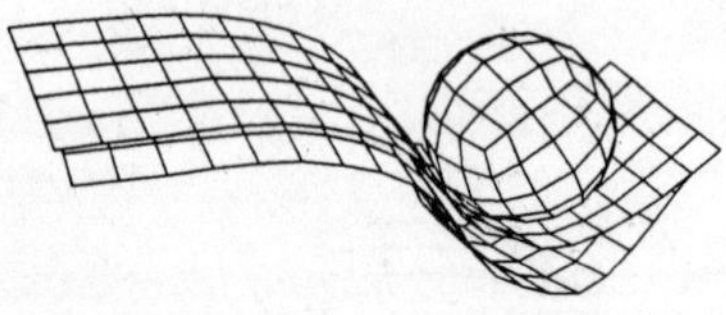

(c) Time t=4 ms.

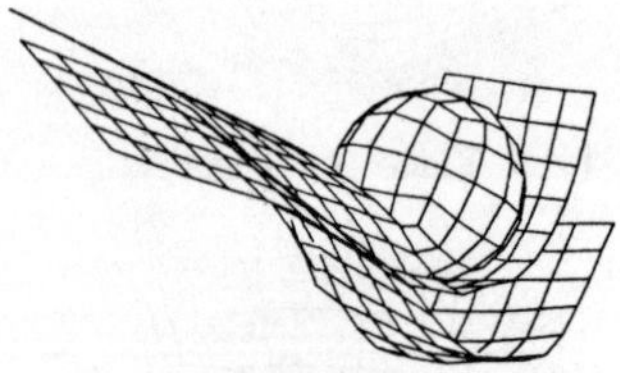

(d) Time t=10 ms.

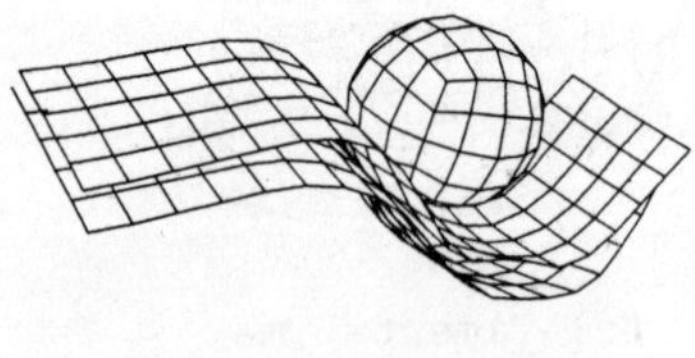

(e) Time t=4 ms.

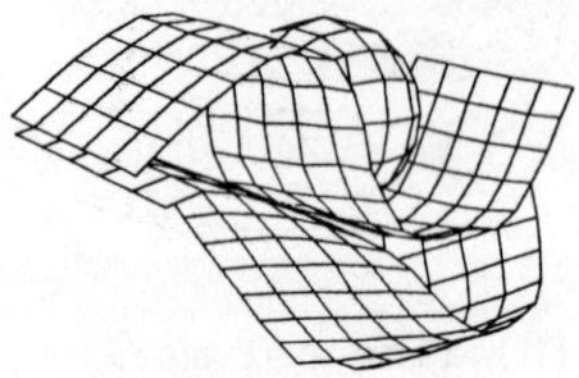

(f) Time t=10 ms.

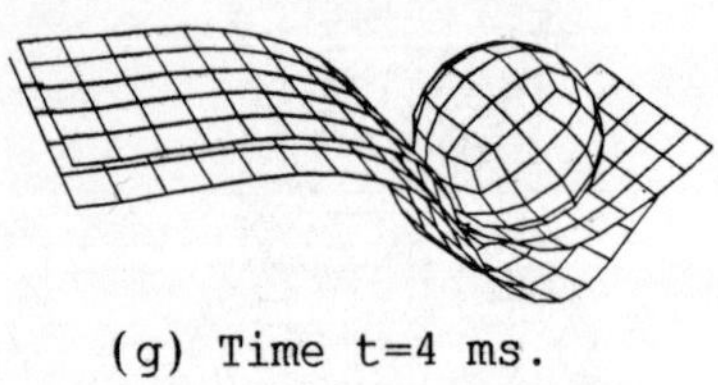

(g) Time t=4 ms.

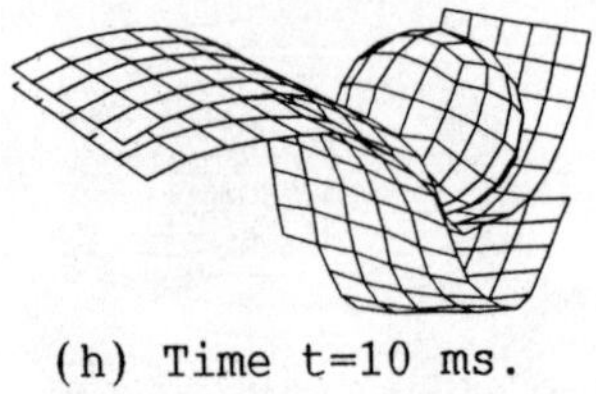

(h) Time t=10 ms.

Fig. 5 Deformed configurations of the ball and plates with friction coefficient equal to (a), (b), 0.0; (c), (d), 0.5; (e), (f), 0.0; (g), (h), 0.5.

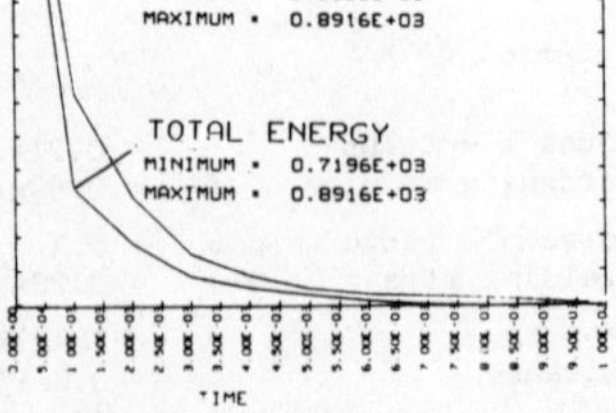

(a) Friction case.

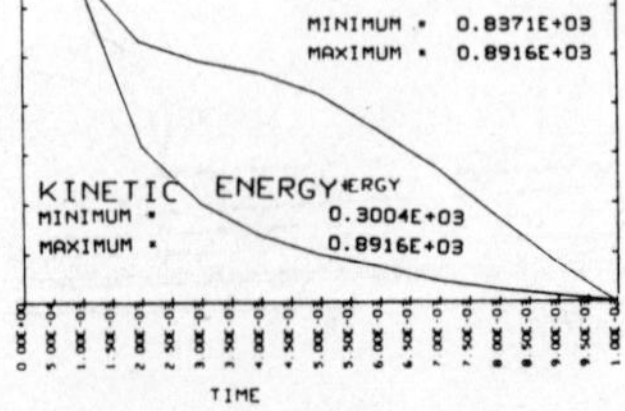

(b) Frictionless case.

Fig. 6 The energy curves for the ball and plates problem.

CONCLUDING REMARKS

It can be seen that beside the element formulation and the constitutive formulation, the treatment of general contact-impact interfaces with friction demands quite some work in the crashworthiness analysis of shell structures. If contact-impact interfaces are presented in a crash process, reliable contact-impact algorithms are essential for a valid solution. Despite all the achievements, further work is largely demanded in all aspects of crashworthiness analysis to make numerical results competitive with experimental ones in general applications.

ACKNOWLEDGEMENT

The author is grateful to Professor Larsgunnar Nilsson for his support and encouragement during the course of this study.

REFERENCES

1. Jones, N. and Wierzbicki, T. Structural Crashworthiness, Butterworths, 1983.

2. Davies, G.A.O. Structural impact and crashworthiness, Elsevier, 1984.

3. Nilsson L and Larsson M, Saab "Crash" Program: A route to customer safety, the SAAB-SCANIA Griffin, 1987.

4. Stanger J et al, A numerical simulation of the proposed European side impact test procedure, Proceedings of the Second International Conference on Supercomputing Applications in the Automobile Industry, Seville, Spain, 1988.

5. Hazet B and Guiraud M, Some applications of crash simulation in car manufacturing, Proceedings of the Second International Conference on Supercomputing Applications in the Automobile Industry, Seville, Spain, 1988.

6. Brush, Don O. and Almnth, Bo O. Buckling of Bars, Plates and Shells, McGRAW-HILL Book Company, 1975.

7. Bushnell, D. Computerized buckling analysis of shells, Martinus Nijhoff Publisher, 1985.

8. Hallquist, J.O. DYNA3D User's Manual, Rev. 4, Lawrence Livermore Laboratory, Livermore, 1988.

9. Bathe, K.J. Finite Element Procedures in Engineering Analysis, Prentice-Hall, Englewood Cliffs, New Jersey, 1982.

10. Belytschko, T., Lin, J.I. and Tsay, C.S. Explicit algorithms for the nonlinear dynamics of shells, Computer Methods in Applied Mechanics and Engineering 42, 1984.

11. Hughes, T.J.R. and Liu, W.K. Nonlinear finite element analysis of shells: Part I - three-dimension shells, Computer Methods in Applied Mechanics and Engineering 26, 1981.

12. Hughes, T.J.R. Numerical implementation of constitutive models: Rate independent deviatoric plasticity, Theoretical Foundation for Large-scale Computations of Nonlinear Material Behaviour, Martinus-Nijhof, Dordrecht, The Netherlands, 1984.

13. Hallquist, J.O. A numerical treatment of sliding interfaces and impact, University of California, Lawrence Livermore Laboratory, Livermore, 1978.

14. Benson, D.J. and Hallquist, J.O. A Single Surface Contact Algorithm for the Postbuckling Analysis of Shell Structures, University of California, San Diego La Jolla, 1987.

15. Zhong, Z.H. On Contact-Impact Problems, Dissertations No. 178, Linköping University, Linköping, 1988.

16. Zhong, Z.H. and Nilsson, L. A contact searching algorithm for general contact problems, accepted for publication in Computers & Structures, 1988.

17. Oldenburg, M. Finite Element Analysis of Thin-walled Structures Subjected to Impact Loading, Doctoral thesis 1988:69D, Luleå University of Technology, Luleå, 1988.

18. Zhong, Z.H. New algorithms for numerical treatments of general contact-impact interfaces, Proceedings of the fourth SAS-World Conference FEMCAD, Vol. 1, 1988.

19. Chaudhary, A.B. and Bathe, K.J. A solution method for static and dynamic analysis of three-dimensional contact problems with friction, Computers and Structures 24, 1986.

20. Zhong, Z.H. Evaluation of friction in general contact-impact interfaces, submitted to Communications in Applied Numerical Methods, 1989.

21. Oden, J.T. and Martins, J.A.C. Models and computational methods for dynamic friction phenomena, Computer Methods in Applied Mechanics and Engineering 52, 1985.

22. Zhong, Z.H. Recent Developments of the HITA and DENA contact-impact algorithms, Publication LiTH.IKP.R.547, Linköping University, Linköping, 1989.

The Use of Vehicle Accelerometer Data to Estimate the Kinematics of Automobile Occupants during Collisions

M.H. Ray and J.F. Carney III

Department of Civil and Environmental Engineering, Vanderbilt University Box 1708, Station B Nashville, Tennessee 37235, USA

ABSTRACT

The primary goal of highway safety research is to reduce the occurrence of severe and fatal injuries to motor vehicle occupants. This goal has been pursued from a variety of perspectives: automotive design, human factors engineering and the design of more forgiving roadsides. This paper is concerned with improvements to an evaluation criteria used by most researchers and engineers in the roadside safety appurtenance field.

It is widely accepted that the severity of occupant injury in a motor vehicle collision is related to the velocity with which the occupant impacts the vehicle interior. *National Cooperative Highway Research Program Report 230* [1] specifies a procedure for estimating the occupant impact velocity based on a simple integration of the data obtained from accelerometers and rate gyros located on the test vehicle. Fundamental to the proper calculation of occupant response is a correct interpretation of the response of the impacting vehicle.

This paper demonstrates that the simple method advocated in *Report 230* for determining the vehicle and occupant responses neglects important effects. These effects can account for exceptionally large errors in predicting the vehicle and occupant kinematics. Since the vehicle accelerometers and rate gyros are attached to a moving frame of reference, the components of post-impact vehicular velocity are coupled. These effects are accounted for in the proposed technique and the accuracy of this improved method is demonstrated.

INTRODUCTION

In the United States, roadside safety appurtenances like crash cushions, guardrails, and breakaway sign supports are accepted for use as operational devices based on their performance in tests described in *Report 230* [1]. A matrix of tests is defined in this document for each type of roadside appurtenance along with specifications for performing these tests, analyzing the resulting data and evaluating the performance of the device. This ensures that some standard of performance is maintained throughout the appurtenance design and testing community.

The subject of this paper is the technique recommended for calculating the hypothetical occupant impact velocity with the vehicle interior during the collision. Known as the flail space method, it was first introduced by Michie in 1980 [2] as a means of unifying a variety of test evaluation parameters such as 50-msec average accelerations, change in momentum, peak accelerations, and change in vehicle velocity. The flail space method assumes that the occupant can be modeled as a mass which is free to translate, or flail, within the vehicle passenger compartment. If the distance between the occupant and the interior surface is known the time when the occupant strikes the interior can be found and hence the relative velocity between the occupant and the interior can be calculated. After the occupant has collided with the interior the important evaluation parameter becomes the highest 10-msec acceleration experienced by the occupant; this is defined as the ridedown acceleration.

The occupant risk criteria has proven to be a very useful tool for comparing otherwise different types of impact scenarios. One can, for example, compare the relative severities of an impact with a crash cushion and a guardrail terminal even though these two types of devices serve very different functions.

Report 230 recommends the following procedure for calculating the hypothetical occupant impact velocities. First, the acceleration records from an actual test are numerically integrated to determine when the occupant has contacted the compartment interior:

$$r = \int_{t_o}^{t_r} \int_{t_o}^{t_r} a_r dt\, dt \tag{1a}$$

$$s = \int_{t_o}^{t_s} \int_{t_o}^{t_s} a_s dt\, dt \tag{1b}$$

where

r = displacement in the vehicle longitudinal direction,
s = displacement in the vehicle lateral direction,
a_r = acceleration in the vehicle longitudinal direction,
a_s = acceleration in the vehicle lateral direction,
t_o = time of vehicle impact with appurtenance,
t_r = time when the occupant has traveled the longitudinal flail distance,
t_s = time when the occupant has traveled the lateral flail distance.

The time histories of a_r and a_s are measured using accelerometers during the full-scale crash test. The values of r and s, the longitudinal and lateral flail distances, are generally assumed to be 2 and 1 feet respectively. Once the times of impact, t_r and t_s, are known the relative velocity of the occupant with respect to the vehicle can be found using the following equations:

$$v_{occ-r} = \int_{t_o}^{t_r} a_r\, dt \tag{2a}$$

$$v_{occ-s} = \int_{t_o}^{t_s} a_s \, dt \tag{2b}$$

where

v_{occ-r} = occupant impact velocity in the vehicle longitudinal direction,
v_{occ-s} = occupant impact velocity in the vehicle lateral direction,

The forms of equations 1 and 2 contain several implicit assumptions which are problematic when considering impacts in general. Since vehicular rotation rates are neglected, the equations are uncoupled. No lateral acceleration or velocity terms appear in the expression for v_{occ-r} and no longitudinal terms appear in the expression for v_{occ-s}. While equations 1 and 2 yield reasonable results when the yaw rotation rate and time of impact are small in comparison with the total event time, they do not provide adequate results when the vehicle or occupant response late in the impact event is desired. The following sections illustrate a technique which eliminates the need for the assumptions listed above and provides a more robust technique for calculating the vehicle and occupant response.

COUPLED EQUATIONS OF MOTION

In order to determine a time record of an impact event, accelerometers and rate gyros are usually mounted on the vehicle. Accelerometers provide instantaneous values of the three perpendicular acceleration components at the mounted location. The rate gyros yield the roll, pitch, and yaw rate histories of the vehicle.

Consider the two reference frames shown in figure 1. The global $x-y-z$ frame is attached to the earth and assumed to be an inertial system. The non-newtonian frame of reference, $r-s-t$, is attached to the vehicle.

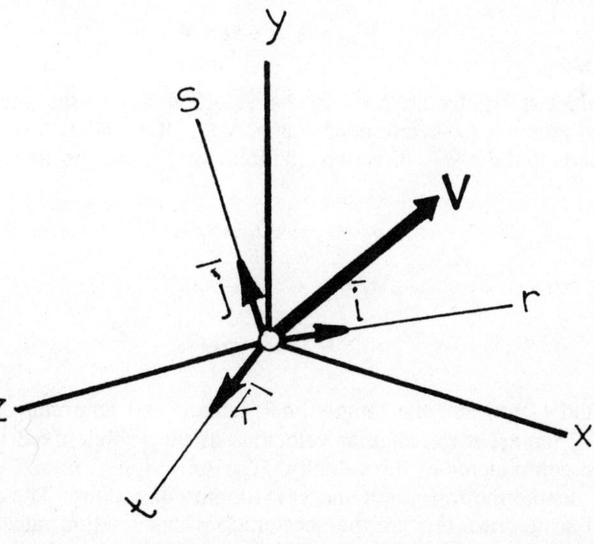

Figure 1. Vector Defined Two Coordinate Reference Frames.

During a collision event the vector representing the vehicle velocity is both changing its magnitude and its direction with respect to the global reference frame, x–y–z. The accelerometers are rigidly fixed in the vehicle local reference frame, r–s–t. If **i**, **j**, and **k** are unit vectors in the vehicle-fixed reference frame, r–s–t, we can express the velocity vector, **V**, as:

$$\mathbf{V} = V_u\mathbf{i} + V_v\mathbf{j} + V_w\mathbf{k} \tag{3}$$

Differentiating **V** with respect to time yields the rate of change of **V** in the r–s–t frame.

$$\dot{\mathbf{V}}_{rst} = \dot{V}_r\mathbf{i} + \dot{V}_s\mathbf{j} + \dot{V}_t\mathbf{k} \tag{4}$$

The analyst, however, is concerned with the vehicle kinematics in the fixed reference frame. Thus, to find the rate of change of **V** in the global x–y–z reference frame we must account for the fact that **i**, **j**, and **k** are changing with respect to time. Differentiating **V** in the global x–y–z reference frame yields the following expression:

$$\dot{\mathbf{V}}_{xyz} = \dot{V}_r\mathbf{i} + \dot{V}_s\mathbf{j} + \dot{V}_t\mathbf{k} + V_r\frac{\partial\mathbf{i}}{\partial t} + V_s\frac{\partial\mathbf{j}}{\partial t} + V_t\frac{\partial\mathbf{k}}{\partial t} \tag{5}$$

The first three terms of equation 5 are simply the rate of change with respect to the vehicle-fixed r–s–t reference frame. The partial derivatives represent the rotation rate of the vehicle-fixed coordinate axes. If we define the vector $\boldsymbol{\Omega}$ to be the angular velocity of the vehicle-fixed r–s–t reference frame we can rewrite equation 5 as follows:

$$\dot{\mathbf{V}}_{xyz} = \dot{\mathbf{V}}_{rst} + \boldsymbol{\Omega}\times\mathbf{V} \tag{6}$$

The accelerometers rigidly attached to the vehicle measure the acceleration of the vehicle in the global x–y–z reference frame, $\dot{\mathbf{V}}_{xyz}$. If equation 6 is resolved into its component parts in the r–s–t directions the following equations are obtained:

$$a_r = \dot{v}_r + v_t\dot{\theta}_p - v_s\dot{\theta}_y \tag{7a}$$

$$a_s = \dot{v}_s + v_r\dot{\theta}_y - v_t\dot{\theta}_r \tag{7b}$$

$$a_t = \dot{v}_t + v_s\dot{\theta}_r - v_r\dot{\theta}_p \tag{7c}$$

where r, s, and t represent the longitudinal, lateral, and binormal directions on the vehicle; the $\dot{\theta}_i$ represent the angular velocities of the vehicle-fixed reference frame; and v_i are the components of the velocity. The subscripts r, p, and y on the angular velocities, θ, denote the roll, pitch and yaw rotation directions. The quantities measured in a full scale crash test are the accelerations and rotation rates. Solving equations 7 for the unknown rates of change of velocity yields:

$$\dot{v}_r = a_r - v_t\dot{\theta}_p + v_s\dot{\theta}_y \tag{8a}$$

$$\dot{v}_s = a_s - v_r\dot{\theta}_y + v_t\dot{\theta}_r \tag{8b}$$

$$\dot{v}_t = a_t - v_s\dot{\theta}_r + v_r\dot{\theta}_p \tag{8c}$$

If one assumes that all rotation rates and the vertical velocity are negligibly small in comparison to the longitudinal and lateral velocities, equations 8 reduce to $\dot{v}_r = a_r$ and $\dot{v}_s = a_s$ and the *Report 230* method described by equations 1 and 2 is valid. In general, however, equations 8 demonstrate that the equations of motion are coupled.

When a_r, a_s, and a_t are known in closed form, it might be possible to obtain the exact solution of this coupled system of equations. In a typical vehicular crash test scenario, however, the character of the accelerometer outputs precludes this possibility. The authors have demonstrated [3] that the fourth-order Runge-Kutta method is very well suited to this type of analysis because the method proceeds in a step-wise manner using only the state of the system during the previous time step and the measurements for the current step. This allows the data reduction to proceed from the beginning of the impact event through the end with no *a priori* knowledge about the exit conditions of the vehicle.

If the experimenter employs a combination of three rotation rate gyros and three accelerometers, determining the velocity history follows directly from equations 8. Other experimenters, however, prefer to use six accelerometers located at known positions on the vehicle.

Suppose that the values of a_r, a_s, and a_t are measured at two different locations, 1 and 2, on the vehicle. If a position vector relating the portions of point 2 to point 1 has components r, s, and t, then the following system of equations can be written:

$$a_{r_1} = \dot{v}_{r_1} + v_{t_1}\dot{\theta}_p - v_{s_1}\dot{\theta}_y \tag{9a}$$

$$a_{s_1} = \dot{v}_{s_1} + v_{r_1}\dot{\theta}_y - v_{t_1}\dot{\theta}_r \tag{9b}$$

$$a_{t_1} = \dot{v}_{t_1} + v_{s_1}\dot{\theta}_r - v_{r_1}\dot{\theta}_p \tag{9c}$$

$$a_{r_2} = \dot{v}_{s_2} + v_{t_2}\dot{\theta}_p - v_{s_2}\dot{\theta}_y - r(\dot{\theta}_p^2 + \dot{\theta}_y^2) + s(\dot{\theta}_r\dot{\theta}_p - \ddot{\theta}_y) + t(\dot{\theta}_r\dot{\theta}_y + \ddot{\theta}_p) \tag{10a}$$

$$a_{s_2} = \dot{v}_{s_2} + v_{r_2}\dot{\theta}_y - v_{t_2}\dot{\theta}_r + r(\dot{\theta}_r\dot{\theta}_p + \ddot{\theta}_y) - s(\dot{\theta}_r^2 + \dot{\theta}_y^2) + t(\dot{\theta}_p\dot{\theta}_y - \ddot{\theta}_r) \tag{10b}$$

$$a_{t_2} = \dot{v}_{t_2} + v_{s_2}\dot{\theta}_r - v_{r_2}\dot{\theta}_p + r(\dot{\theta}_r\dot{\theta}_y - \ddot{\theta}_p) + s(\dot{\theta}_p\dot{\theta}_y + \ddot{\theta}_r) - t(\dot{\theta}_r^2 + \dot{\theta}_p^2) \tag{10c}$$

The unknown quantities in equations 9 and 10 are the translational velocities, v_i, and the rotational velocities, $\dot{\theta}_i$. This coupled system of equations can be numerically solved for the unknowns using a fourth-order Runge-Kutta technique as discussed earlier.

Furthermore, the velocity at point 1 can easily be related to the velocity of point 2 through the relation:

$$\vec{v}_1 = \vec{v}_2 + \vec{v}_{1/2} \tag{11}$$

$$\text{where } \vec{v}_{1/2} = \begin{bmatrix} \vec{i} & \vec{j} & \vec{k} \\ r & s & t \\ \theta_r & \theta_p & \theta_y \end{bmatrix} \tag{12}$$

It follows, therefore, that the components of velocity at points one and two and the three components of angular velocity can be determined for each time step.

Magnitude of Errors in Typical Full-Scale Crash Tests

The assumption that all rotation rates and the vertical velocity are negligibly small is valid for some types of collision scenarios, such as an impact with a small sign support where no rotation occurs. If the effect of the coupled terms in equations 8 and 9 are small for most realistic impact scenarios then using the simpler approach of equations 1 and 2 would be justified. In fact, neglecting the coupled terms in equations 8 and 9 causes large errors for several typical tests recommended in *Report 230*. In addition, some roadside safety researchers are becoming concerned about the performance of roadside safety devices in non-traditional types of impact scenarios such as side impacts and non-tracking approach trajectories [4]. By definition, non-tracking approach trajectories are characterized by significant yaw rotation rates, $\dot{\theta}_y$, prior to impact. For these non-traditional types of collisions the effect of the coupled terms cannot be ignored.

As an illustration of the magnitude of the errors produced by ignoring the coupled nature of the kinematic relationships, consider the following typical redirectional crash test [5]. In this test, a 4490-lb automobile impacts a standard blocked-out W-beam steel-post guardrail at 86.2 fps at an impact angle of 25.5 degrees. The accelerometer and rate gyro outputs were employed to determine the vehicular velocity at the instant it ceased to be in contact with the rail. The coupled and uncoupled formulation predictions are presented in table 1. The coupled predictions, which are in agreement with the results of the high speed film analysis, differ significantly from those of the uncoupled formulation. The uncoupled lateral velocity prediction is in error by 209.4 percent. This fact has important ramifications when developing techniques for tracking the occupant and vehicle interior positions during an impact event. In a redirectional impact situation, an unbelted occupant could impact several locations in the vehicle interior. Since the flail space associated with later occupant impacts could be in the order of 3-4 feet, it is quite possible that the serious injury producing impact could be the second or third occupant impact event. It may be necessary to track both the occupant and the vehicle interior for several seconds after initial impact, and the coupled formulation is therefore essential. Consider the vehicular velocity versus displacement results shown in figures 2 and 3 for the 86.2-fps, 25.5-degree crash test. These figures compare the velocity-displacement history predictions of the impacting vehicle using the coupled and uncoupled equations of motion. The coupled equations of motion accurately predict the post-impact response of the vehicle. Both figures demonstrate the increasing inaccuracies associated with the uncoupled formulation as the impact event proceeds.

Impact Velocity	Impact Angle		Vehicle Wt.
86.2 fps	25.5 degrees		4490 lb.
Parameter	**Coupled**	**Uncoupled**	**Error (%)**
Velocity at Loss of Contact with Rail			
Longitudinal (fps)	55.3	61.5	11.2
Lateral (fps)	17.5	54.2	209.4

Table 1. Velocity Predictions at Loss of Contact with Rail

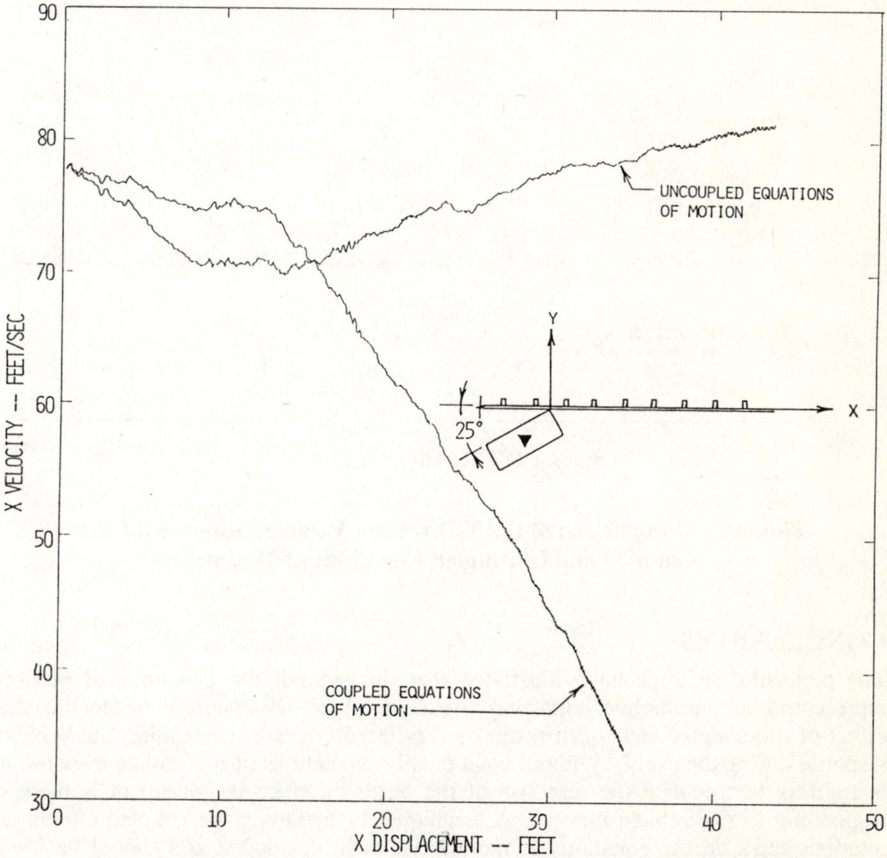

Figure 2. Comparison of the X-Direction Vehicle Response for the Coupled and Uncoupled Equations of Motion.

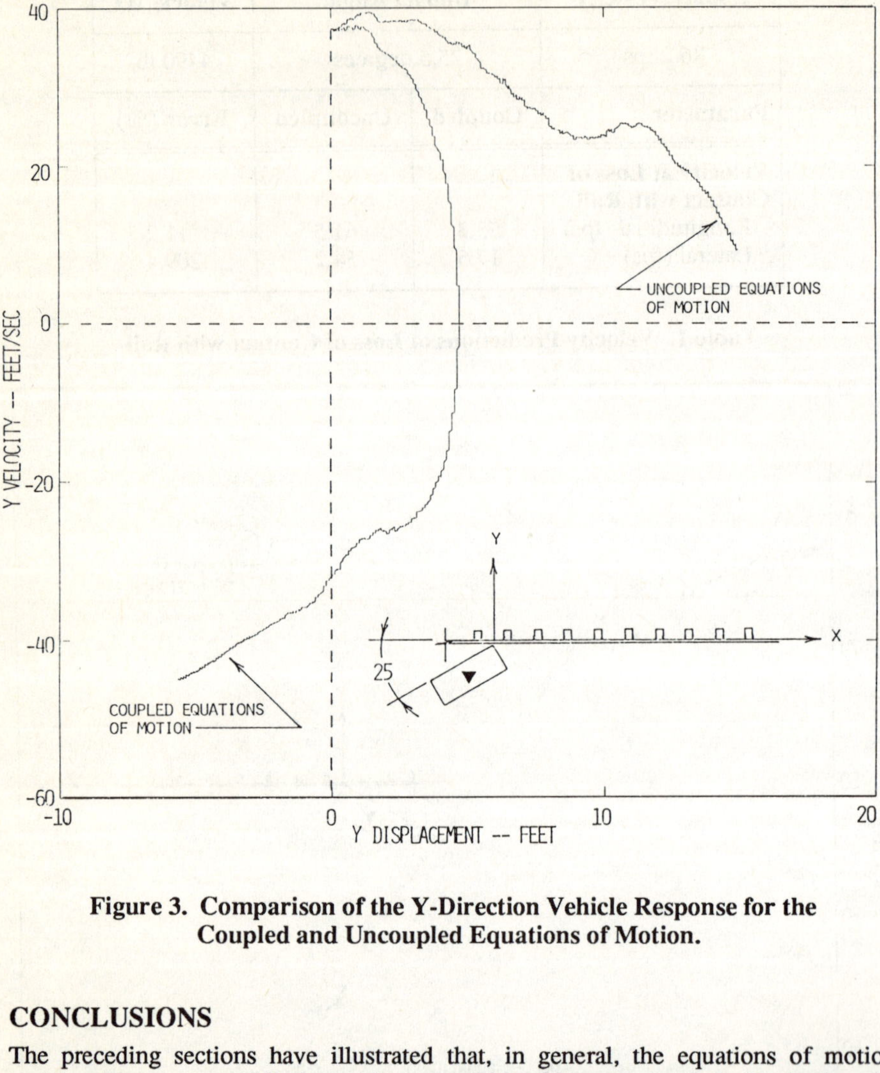

Figure 3. Comparison of the Y-Direction Vehicle Response for the Coupled and Uncoupled Equations of Motion.

CONCLUSIONS

The preceding sections have illustrated that, in general, the equations of motion representing an automobile impacting a roadside object are coupled. Neglecting the effect of the coupled terms can result in very large errors in calculating the vehicle response during the event. Without good quality predictions of the vehicle response it is fruitless to calculate the response of the occupant since the occupant is merely responding to the vehicle motion. A technique for including the coupled effects of rotations rates on the equations of motion has been developed and should be used when calculating the response of a vehicle in an impact situation.

REFERENCES

1. J. D. Michie, ''Recommended Procedures for the Safety Performance Evaluation of Highway Appurtenances,'' *NCHRP Report 230*, National Cooperative Highway Research Program, (1981).

2. J.D. Michie, ''Collision Risk Assessment Based on Occupant Flail-Space Model,'' *Transportation Research Record No. 796*, Transportation Research Board, (1981).

3. M. H. Ray and J. F. Carney III, *An Improved Method for Determining Vehicle and Occupant Kinematics in Full-scale Crash Tests,* Transportation Research Board, Washington D.C. (1989).

4. M. W. Hargrave, A. G. Hansen, and J. A. Hinch, *A Summary of Recent Side Impact Research Conducted by the Federal Highway Administration,* Society of Automotive Engineers, Warrendale, PA (February 1989). SAE Technical Paper No. 890377

5. M. H. Ray, J. B. Mayer, and J. D. Michie, ''Replacing the 4500-lb Passenger Sedan in Report 230 Tests,'' Evaluation of Design Analysis Procedures and Acceptance Criteria for Roadside Hardware, Federal Highway Administration, Washington, D.C. (September 1987). Final Report

Numerical Simulation of Airbag Behaviour

A.M.A. van der Hiejden, A.J. Buijk and P.H.L. Groenenboom

PISCES International B.V., Groningenweg 6, 2803 PV Gouda, The Netherlands

1. INTRODUCTION

One of the occupant safety systems in automobiles is the airbag. Though not common use yet, in the near future the airbag will be part of safety regulations in automobiles. The behaviour of an impacting object (the driver or passenger) on a deploying airbag has so far mainly been studied experimentally. These experiments are very costly and do not yield detailed information. Airbag behaviour can be studied using computer programs which contain both the Eulerian and the Lagrangean or Shell characterization and the full coupling between them. Results of numerical studies will be presented, and in some cases comparison with experimental data will be shown.

2. AIRBAG MODELS IN CVS PROGRAMS

Two- and three-dimensional models for the simulation of airbag restraint systems have been available in Crash Victim Simulation (CVS) programs for over a decade. In their current status [1, 2, 3] these models are useful tools to forecast airbag system performance in a very limited range around validated data. This data may be obtained from experiments or from numerical simulations as described in this paper. These models are based on the assumption that the pressure and the temperature in the airbag are uniformly distributed during the inflation. Furthermore it is assumed, that there is no transient overpressurization of the passenger compartment as a result of the inflating airbag. The final shape of the airbag is an input parameter, and is usually a simple shape (e.g. cylindrical, ellipsoidal). Bag stretch is either not taken into account, or in an approximate way. Inertia effects of the bag are ignored both during the deployment and the impact phase.

The inflator itself is usually not included in the simulation, but is represented by prescribing the mass flow rate history and the temperature history together with the thermodynamic properties of the gas. Since the temperature history is hard to measure due to the very short duration of the discharge process it is usually determined in an approximate way [2].

3. TANK TEST ANALYSIS

A common way of testing the behaviour of an inflator is the tank test. In this test an inflator discharges into a closed fixed volume tank and the pressure-time history is measured during the discharge process. Due to the rapidity of the discharge process the dynamic temperature of the inflator gas is very hard to measure. A model calculation in rotational symmetry has been performed using the PISCES-2DELK code. The inflator mass flux, the pressure time history and the temperature-time history of the gas at the inflow opening were prescribed. It will be clear that the temperature time history can only be a very rough estimate. More detailed information about the temperature time history may be obtained from inflator simulations as discussed in section 4. The gas is injected under an angle of 70 ° with respect to the axis of symmetry. The flow field in the tank after 15 resp. 35 ms is shown in the figures 1 and 2. Notice that at 35 ms there is a tendency to form a double vortex. At later times the vortex disappears.

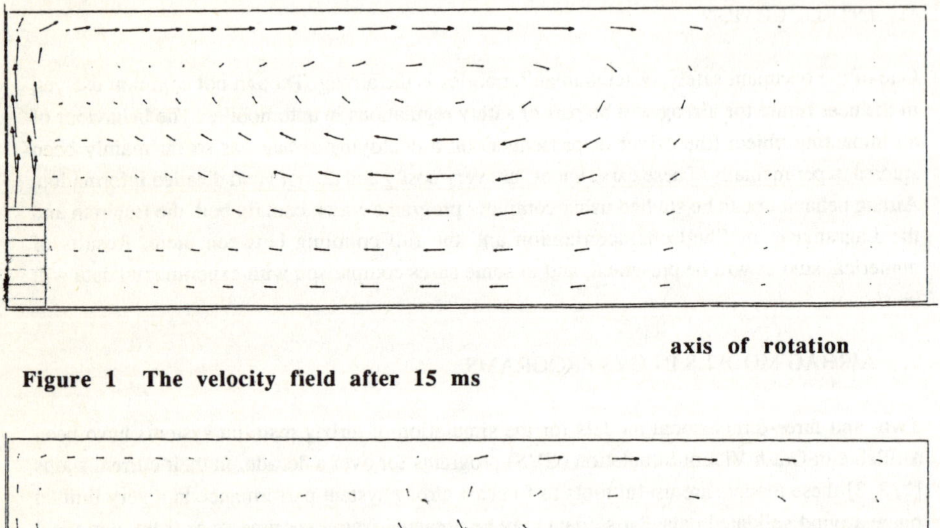

axis of rotation

Figure 1 The velocity field after 15 ms

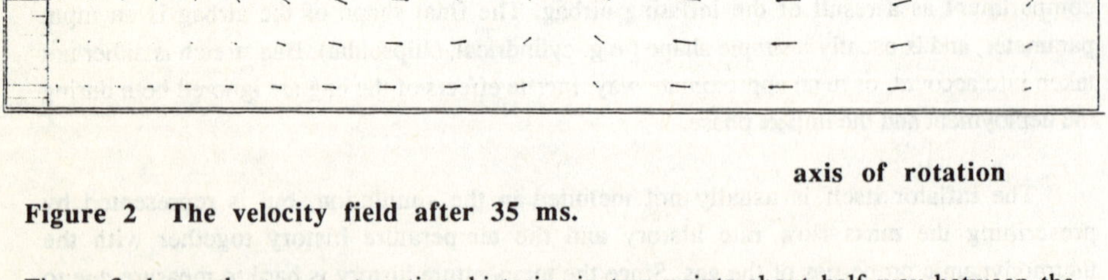

axis of rotation

Figure 2 The velocity field after 35 ms.

Figure 3 shows the temperature-time histories in two points about half way between the inflator exhaust openings and the cylindrical wall of the tank.

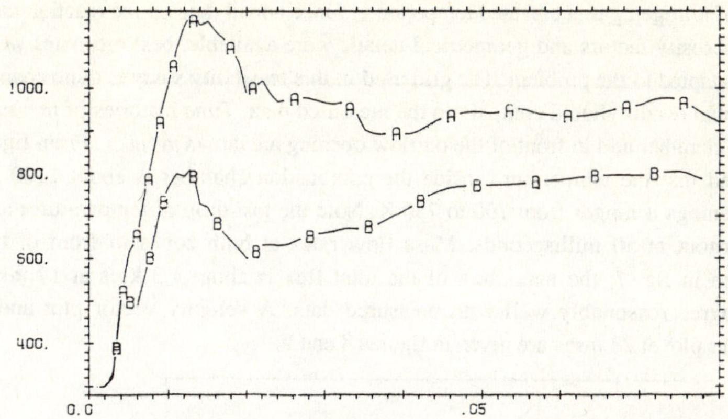

Figure 3 Temperature-time histories in two neighbour zones.

This figure shows that the temperature distribution in the tank is not uniform at all, which also has implications for the temperature distribution within airbags.

4. INFLATOR SIMULATION

The purpose of this section is to demonstrate the capability of a special version of the PISCES-2DELK code to simulate the chemical reaction and nitrogen flow for an actual inflator. This GRAINS code describes the dynamic two-phase flow of a reactive mixture of burning particles and a compressible fluid and has been applied successfully to numerical simulation of the interior ballistics of propellant guns [4]. An important asset of this computer program is the flexibility to include user-written routines to define models for the basic physico-chemical processes.The pyrotechnic inflator unit has been modelled as a rigid circular cylinder of 104 mm diameter and a height of 40 mm (fig. 4). The circumferential openings with the filter/cooling assembly are replaced by a circular region of low porosity in this model. Internally, the inflator consists of an outer ring connected by a small opening to the combustion chamber. In this combustion chamber, propellant discs are present, surrounded by porous material. The propellant is ignited by the central initiator.

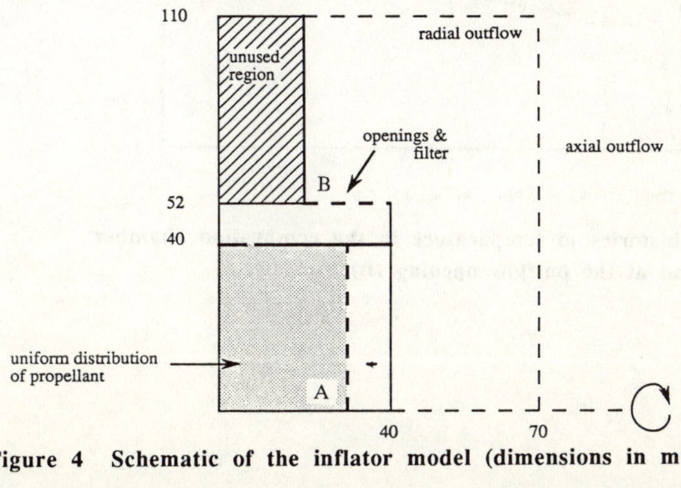

Figure 4 Schematic of the inflator model (dimensions in mm)

The inflator contains 85 gram of a mixture of sodium-azide and copper-oxide which will produce 34 gram of nitrogen gas at elevated temperature. Since not all data on the reaction rate, heat production, porosity factors and geometrical details were available, best estimates were used and slightly adapted to the problem. The grid used in this feasibility study is quite coarse, but at least the global results should compare to the measured data. Time histories of pressure in the combustion chamber and in front of the outflow opening are shown in fig. 5 . From fig. 6 it may be observed that the temperature inside the combustion chamber is about 1100 K, whereas at the openings it ranges from 700 to 750 K. Note the fast drop of temperatures and pressure after burnout at 50 milliseconds. Mass flow rates at both zones in front of the openings are given in fig. 7; the maximum of the total flux is about 1.3 kg/s at 17 msec which seems to agree reasonably well with measured data. A velocity vector plot and a temperature contour plot at 28 msec are given in figures 8 and 9.

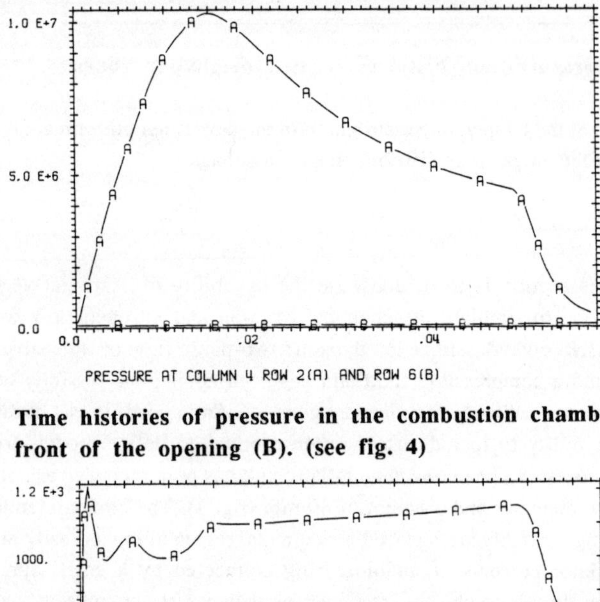

Figure 5 **Time histories of pressure in the combustion chamber (A) and in front of the opening (B). (see fig. 4)**

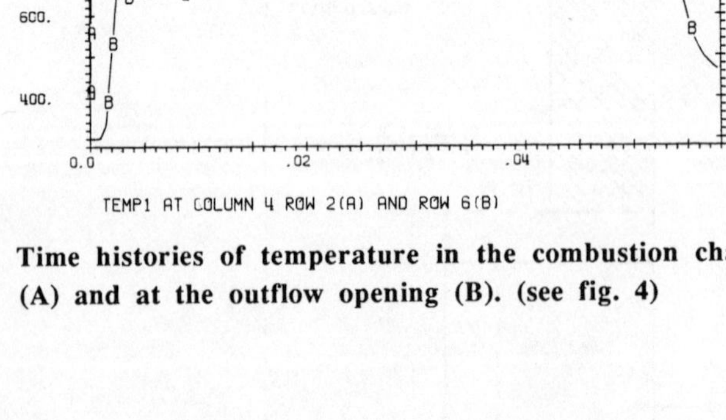

Figure 6 **Time histories of temperature in the combustion chamber (A) and at the outflow opening (B). (see fig. 4)**

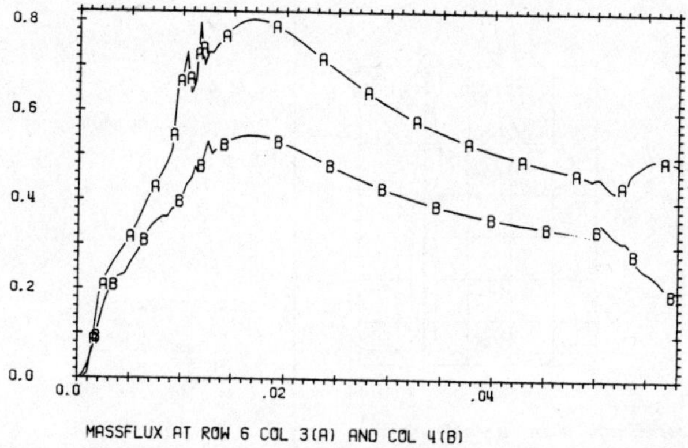

Figure 7 Mass flow rates in the zones in front of the outflow openings (A and B) (see fig. 8)

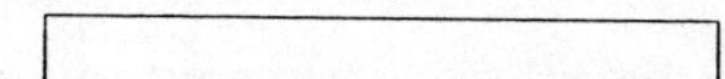

Figure 8 The velocity field after 28 ms

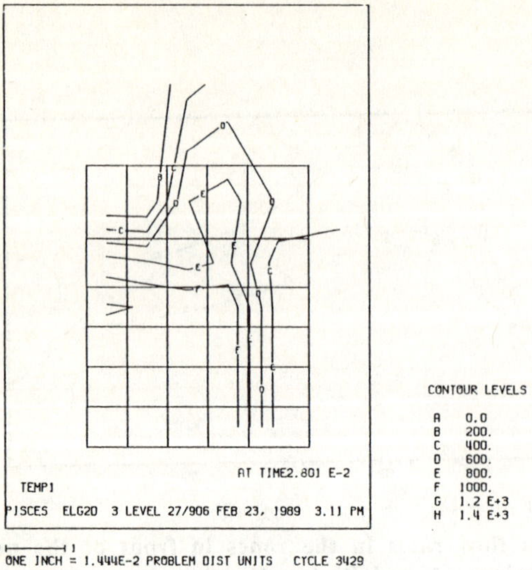

Figure 9 The temperature contours after 28 ms

Detailed analyses with finer meshes and with more sophisticated models including fluid-structure interaction on inflators are possible and are expected to be performed in the future.

5. 2D-AIRBAG MODELS

Aspects of airbag behaviour such as unfolding and wrinkling, and non-orthogonal or non-central body impact are fully three-dimensional phenomena. However, parameter studies on airbag inflation, including effects of straps and holes and on orthogonal, central impact in axial symmetry (2D) are very useful and can be performed economically and the results may be compared to most experiments. In this section we shall discuss two analyses with PISCES-2DELK; the first one describes the so-called suffocating airbag during inflation, the second one deals with the axial-symmetric body impact.

The inflation of an axial symmetric airbag with two small holes close to the inflator and with elastic straps has been computed as a coupled Euler-shell model in PISCES-2DELK. The bag (fig. 10) is modelled by membrane elements of which the left side is fixed to the steering column. Additional constraints are posed by the attached straps and by the steering wheel. The initial shape is that of a slightly curved, flat pancake with the original surface area. At the outside atmospheric pressure is assumed. The Eulerian mesh is also shown in fig. 10. Since the part outside the bag is covered, gas is only allowed inside the bag. The gas enters from the inflator with specified density, temperature and mass flux. The holes in the zone next to the inflator opening are modelled as a narrow circumferential slit in the polygon that allows transfer from air/nitrogen through the bag; the structure of the bag itself remains intact. The computed airbag shape, flow field and material distribution after 11 and 23 milliseconds are shown in figure 11. The small regions of different material inside the bag demonstrate that air from outside the bag has flowed into the bag due to the pressure drop in the flow field in front of the inflator openings. After 35 to 40 milliseconds the inflow of gas becomes very small and the bag has reached its final shape. Damping is provided by the presence of gas. The straps limit the deflection of the bag.

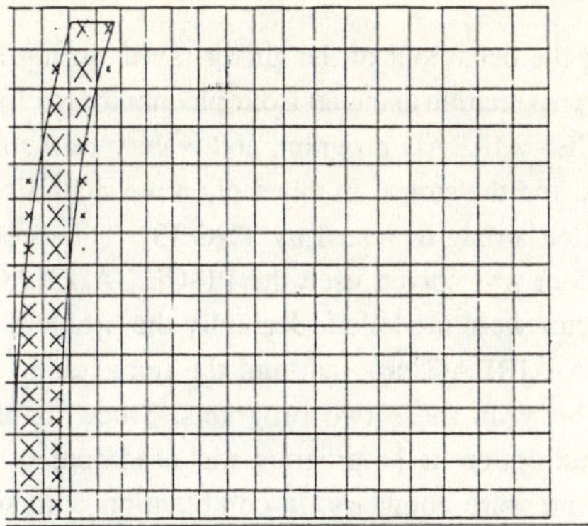

Figure 10 Initial shape of the bag and the Euler grid

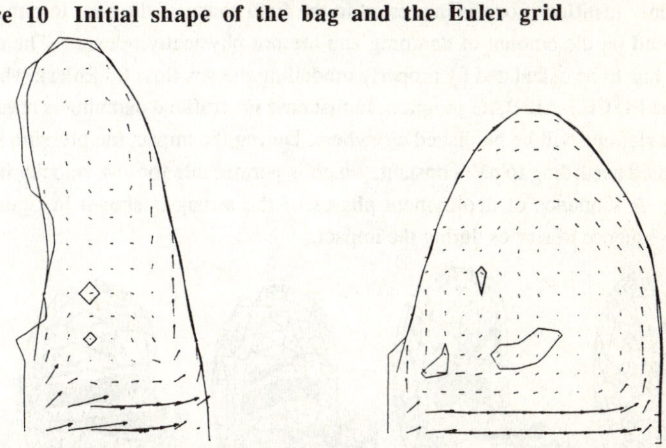

Figure 11 Situation after 11 ms (left) and 23 ms (right)

The orthogonal, central impact of a sphere with a mass of 5.25 kg and a velocity of 5.835 m/s on an axial symmetric airbag has also been computed as a coupled Euler-shell model in PISCES-2DELK. The configuration of the problem is shown in fig. 12. The sphere is modelled as a rigid. The displacement of the sphere during contact with the bag is given in fig. 15, where a comparison is made with an experiment done by TNO [3] and where also the results of a calculation with the PISCES-AIRBAG (see next section) are shown.

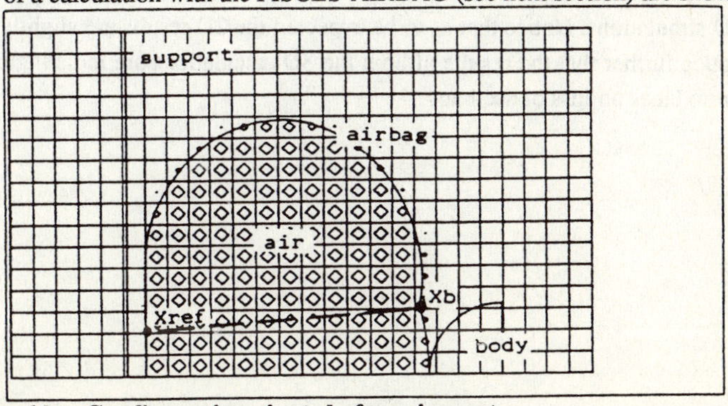

Figure 12 Configuration just before impact.

6. 3D-AIRBAG MODELS

As mentioned earlier the behaviour of the airbag fabric during inflation and impact may be modelled to a good approximation using an isotropic membrane formulation. Such an element is available in the PISCES-AIRBAG program, and is very well suited for realistic modelling of both the airbag fabric and the straps. In this section we shall discuss the impact of a spherical impactor on an inflated airbag as tested by TNO [3]. This problem has been solved in two ways. First the problem was solved used the PISCES-AIRBAG program in which both the airbag and the impactor were modelled. Secondly the problem was solved by modelling the airbag in the PISCES-AIRBAG program and the impactor in the MADYMO program and using the coupling between these two programs. Since for this problem gas flow is not important, the simplest option in the program was used namely the deployment of the bag by means of a constant pressure boundary, in combination with some nodal velocity damping. This procedure is only justified if one is interested in the final shape of the bag. Intermediate configurations depend on the amount of damping and are not physically relevant. The actual inflation behaviour has to be calculated by properly modelling the gas flow (in Euler), which is another option in the PISCES-AIRBAG program. In that case no artificial damping is required. Results of such calculations will be published elsewhere. During the impact the pressure inside the airbag is calculated according to pV=constant, which is permissible for low velocity impact on a closed airbag. A sequence of deployment phases of the airbag is shown in figure 13. Figure 14 shows a sequence of shapes during the impact.

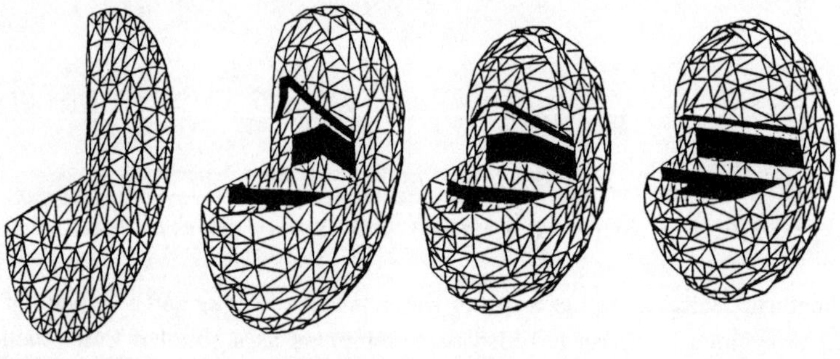

Figure 13 Deployment of 3D airbag with straps

Figure 15 shows the displacement history of the body (sphere) as obtained from the experiment and the 2D and 3D simulations. Notice that as to be expected the 2D results are slightly below the 3D results. Notice further that the results of both the 3D calculations are indistinguishable i.e. the coupling introduces no loss of accuracy.

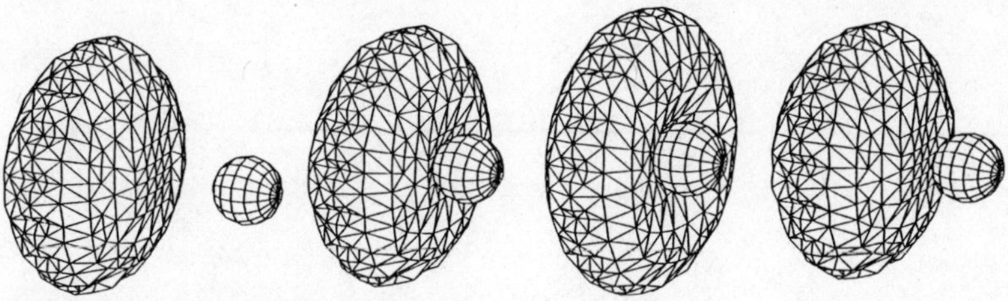

Figure 14 Airbag during contact

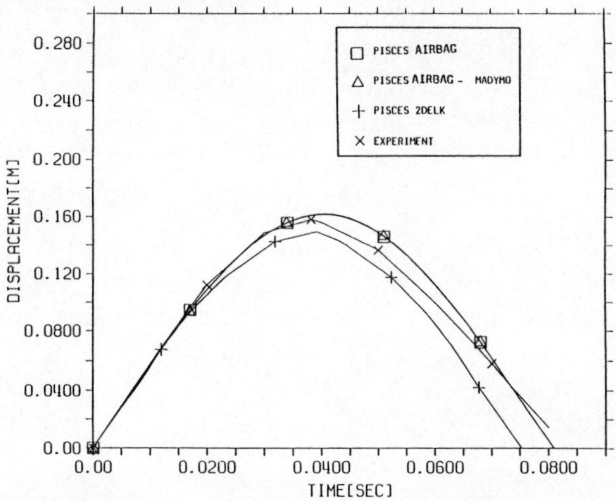

Figure 15 Displacement of sphere during contact

CONCLUSIONS

The results presented in this paper show that various aspects which are important for both the design of airbag devices and the proper understanding of the airbag and airbag-occupant behaviour can be modelled accurately in both 2D and 3D numerical simulations. Numerical simulations yield a number of results which cannot easily be measured like e.g. the temperature distribution in the airbag. Other aspects like the influence of the deployment of an airbag on the pressure in a compartment and vice versa will be discussed elsewhere.

REFERENCES

1. Nieboer J.J., Wismans J. and Fraterman E., Status of the Madymo 2D Airbag model.

2. Wang J.T. and Nefske D.J., A new CAL3D Airbag Inflation model
SAE 880654, International Congres and Exposition, Detroit, 1988.

3. de Coo P.J.A., Wismans J., Nieboer J.J. and van der Heijden A.M.A., Integrierte Kollisionssimulation mit FE- und Insassenmodellen.
VDI Berichte Nr. 699, 1988, pp 179-195.

4. Groenenboom P.H.L. and Thomson P., Two phase flow for a 40 mm granular propellant gun in comparison with experimental results.
11th International Symposium on Ballistics, May 1989, Brussels.

SECTION 5 - PLATES, PANELS AND SHELLS

Response of Panels Subject to Simultaneous In-Plane Compressive Loading and Transverse Blast Loading

C.C. Chong*, J.G. Hetherington**, T.S. Lok** and P.D. Smith**

**CDC Construction and Development Pte. Ltd., Singapore*

***Civil Engineering Group, Royal Military College of Science, Shrivenham, UK*

ABSTRACT

Any structure consists of elements which support a range of static loads. This paper considers those elements which are subjected to in-service compressive stresses and seeks to assess the response of such elements when transversely loaded by a blast wave produced by the detonation of a condensed high explosive charge.

The following four approaches are presented which are all directed towards assessment of the inter-relation between compressive load, blast load and element deformation: an energy solution incorporating the effect of membrane stresses; a single degree of freedom approximation of the system; a finite-difference numerical simulation using the PR2D code and a set of experimental results. In all cases the structural elements are loaded by in-plane static compressive forces and transversely by blast pulses derived from small spherical charges of plastic explosive.

This latter experimental procedure involved the compressive pre-loading of clamped-ended mild steel plates and the measurement of blast overpressures using a piezoelectric pressure transducer together with a simple method of measuring the maximum central deflection of the plates.

Results from the four approaches are presented and implications of the findings discussed.

INTRODUCTION

The response of structures loaded by explosively generated blast waves has been the subject of considerable research over a number of years. In such investigations the behaviour of complete military and civilian 'target' structures at both full and model scale has been assessed together with a sizeable number of projects concerned with the response of structural elements. Baker et al [1] presents a comprehensive overview of the experimental techniques, numerical simulations and methods of analysis that are used in defining structure loads, assessing response and designing to withstand such inputs.

In the specific area of structural element response there have been recent developments in the assessment of the dynamic elastoplastic deformation of plates both analytically as presented, for example, by Yankelevsky [2] and experimentally by, for example, Gürke and Bücking [3]. The evaluation of resistance functions for masonry structures has received recent attention from Hetherington [4] who included precompressive loads in that analysis.

Such studies have tended to concentrate on the response of the element in isolation with little or no regard to the location of the element within the complete structure. However, recent work by Forsén [5] has given consideration to the way in-plane compressive forces due to inertia in wall panels subjected to air blast loading change and Holasut and Ruiz [6] have analysed the effect of a pressure pulse on the stability of a statically compressed structural element.

This present investigation is an attempt to assess the response of a simple structural component - a one way spanning plate - which is subjected to an in-plane compressive static loading which should be maintained while the plate is laterally loaded by a blast wave generated by the detonation of a small charge of condensed high explosive. The aim of the investigation was to establish how the maximum elastic deformation generated by the lateral loading varied for blast waves of different impulse and for different in-plane (here described as axial) loading. The effect of different boundary conditions on plate behaviour was also studied.

Four approaches were employed in this study - experimentation, numerical simulation, analysis using an energy method and a single degree of freedom approximation. On the experimental side, a robust loading apparatus was designed to hold one way spanning steel plates, their ends being held in clamps to prevent rotation. In one series of tests the clamped ends were also fixed against in-plane translation (notation F on graphs) while in a second series the ends were allowed to slide relative to each other (notation S of graphs). Numerical simulation of the experiment was conducted using the finite difference program PR2D made available to the authors by Principia Mechanica Ltd of London. The input to the program was a simulation of the reflected overpressure signal measured in the experimental phase described above. Finally, by equating the kinetic energy imparted to the structural element by the blast wave, characterised in terms of reflected overpressure impulse, to the elastic strain energy of deformation of the element an estimate of maximum central displacement of the plate was derived. This analysis was performed in two ways. In the first, the structure was analysed as a simple, equivalent, single degree of freedom, system and in the second elastic membrane stresses were included in the model.

EXPERIMENTAL PROCEDURE

To facilitate the study of the response of the plates, a loading frame was designed and manufactured from mild steel components. The components of the frame were designed such that failure would not occur under the combined loads. In addition, adequate stiffness was provided to ensure that the response of the plates was not affected by excessive flexibility of the components.

A diagram of the loading frame is shown in Figure 1. The frame permits an in-plane compressive force to be applied to the plates while the charge is detonated at a prescribed stand-off distance. Compressive force is applied through a system of springs connected in line with the longitudinal axis of the plate. The three channels are held in position by two galvanised pipes. One of the channels is permitted to slide freely. To reduce friction at the pipe/channel interface, brass bushes were fitted and were kept well lubricated throughout the experiments. Rotation of the plates at their supports was prevented by rigid clamping blocks. Care was taken during manufacture to ensure that no eccentricity was introduced into the loading frame. The entire frame rests on a mild steel, rigidly weided supporting structure.

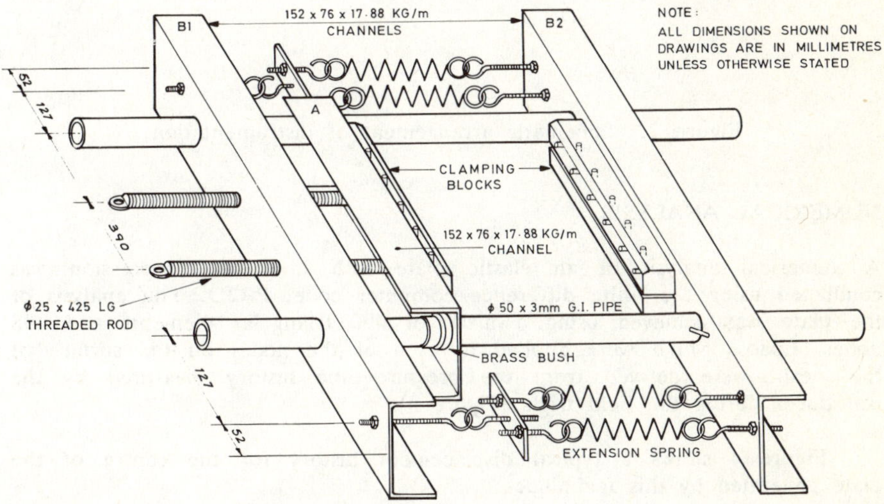

Figure 1. Loading Frame

Prior to the dynamic tests, the loading frame was calibrated on an Instron Hydraulic Testing Machine. The load was measured on a proving ring attached to the crosshead. Dial gauges and electrical strain gauges were used to calibrate the lateral deflection and the corresponding strains on the plates which were 300 mm wide, 298 mm span and of thickness 1.2 mm. These static tests were necessary as they provided data which were then used in the field tests described below. Some results of the static tests are given later in the paper.

The field tests were conducted in the Explosive Range Demonstration Area at RMCS. The measuring equipment used in the field test is shown diagrammatically in Figure 2. The loading frame was placed horizontally on top of the supporting frame and levelled before each test. Spherical charges of plastic explosive (PE4) of known mass were suspended from a tripod and the position of a pressure transducer was established at the same radius from the charge as from the centre of the plate. Further, a plasticine cone was placed directly under the centre of the plate to measure the maximum lateral deflection. The results of all the tests may be found in Chong [7].

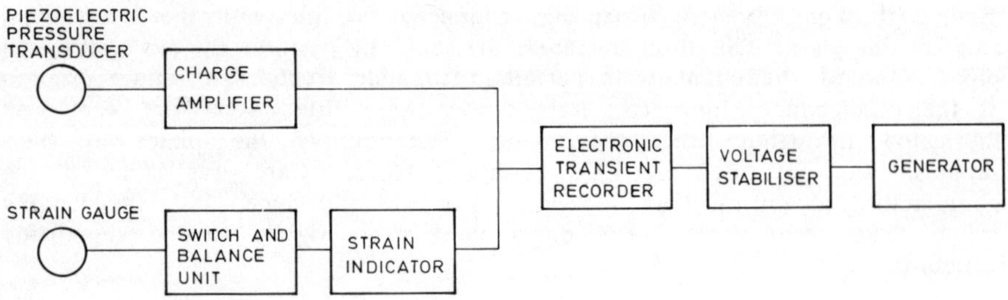

Figure 2. Schematic arrangement of instrumentation

NUMERICAL ANALYSIS

A numerical analysis of an elastic plate with in-plane compression was conducted using the finite difference computer code, PR2D. The analysis of the plate was achieved using a mesh of 496 triangular elements and 498 nodes. Loads, which were applied to each of the nodes on the surface of the mesh, were derived from the pressure-time history measured by the transducer during the field trials (Figure 3).

Figure 4 shows a typical displacement history for the centre of the plate generated by this technique.

Results generated in this way permitted an examination of the dependence of peak central plate deflection on reflected impulse, i_r, axial load P_{ax}, and end fixity. Figure 5 shows the relationship between peak central deflection and reflected impulse, i_r for the fixed and sliding boundary conditions. The results of this analysis conform with the prediction that the system would respond impulsively since its period of oscillation is long in comparison with the positive phase duration of the loading pulse.

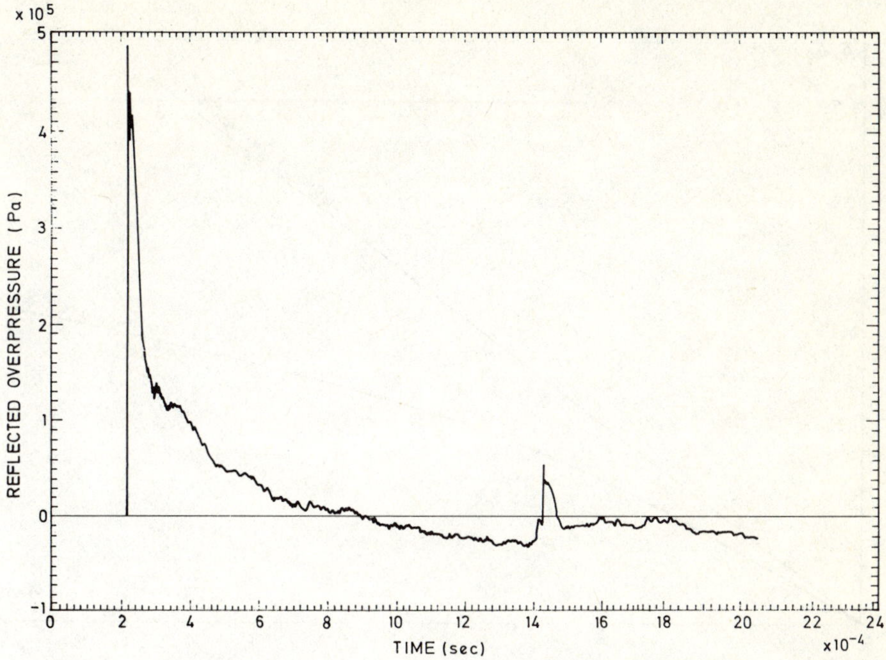

Figure 3. Reflected overpressure – time history from experiment

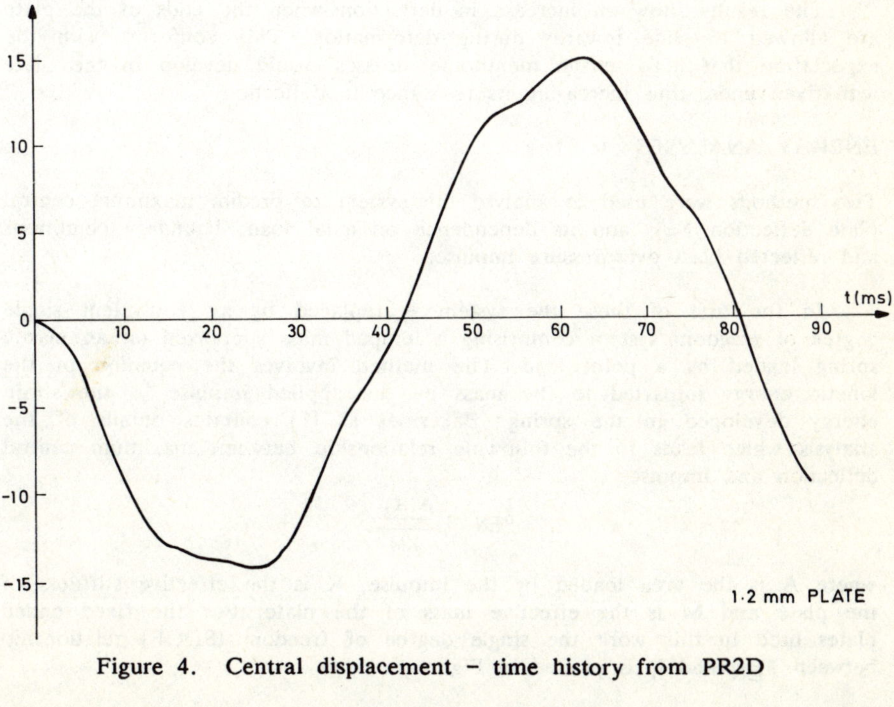

Figure 4. Central displacement – time history from PR2D

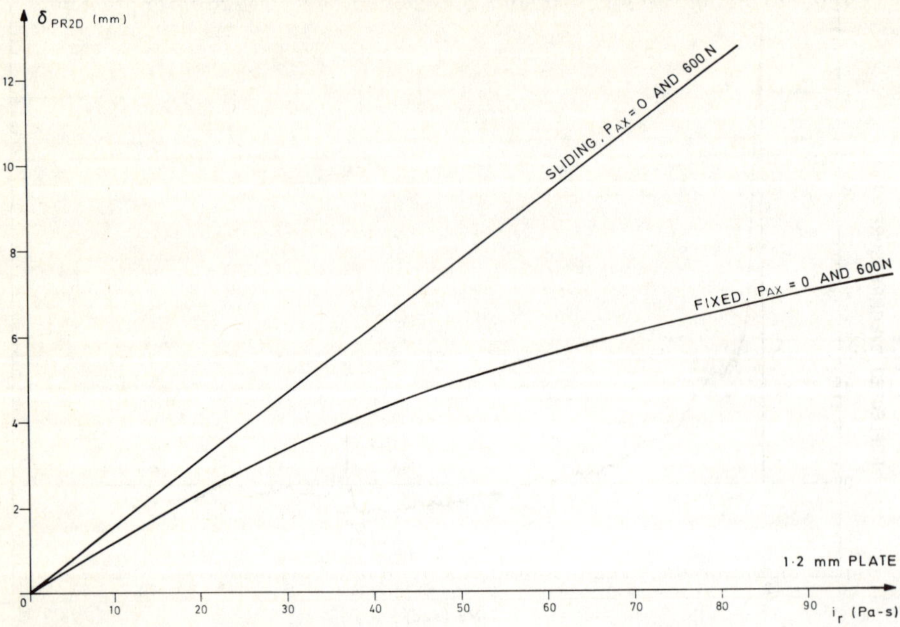

Figure 5. Central displacement versus reflected impulse from PR2D

The results show an increase in deflection when the ends of the plate are allowed to slide inwards during deformation. This conforms with the expectation that high tensile membrane stresses would develop in the plate with fixed ends, thus increasing its resistance to deflection.

ENERGY ANALYSES

Two methods were used to analyse the system to predict maximum central plate deflection δ_{EN} and its dependence on axial load, boundary conditions and reflected blast overpressure impulse.

In the first of these the system is replaced by an equivalent single degree of freedom system comprising a lumped mass supported on an elastic spring loaded by a point load. The method involves the equation of the kinetic energy imparted to the mass by the applied impulse to the strain energy developed in the spring. Baker et al [1] contains details of the analysis which leads to the following relationship between maximum central deflection and impulse

$$\delta_{EN} = \frac{A\ i_r}{\sqrt{KM}}$$

where A is the area loaded by the impulse, K is the effective stiffness of the plate and M is the effective mass of the plate. For the fixed ended plates used in this work the single degree of freedom (SDOF) relationship between δ_{EN} and i_r is plotted in Figure 6.

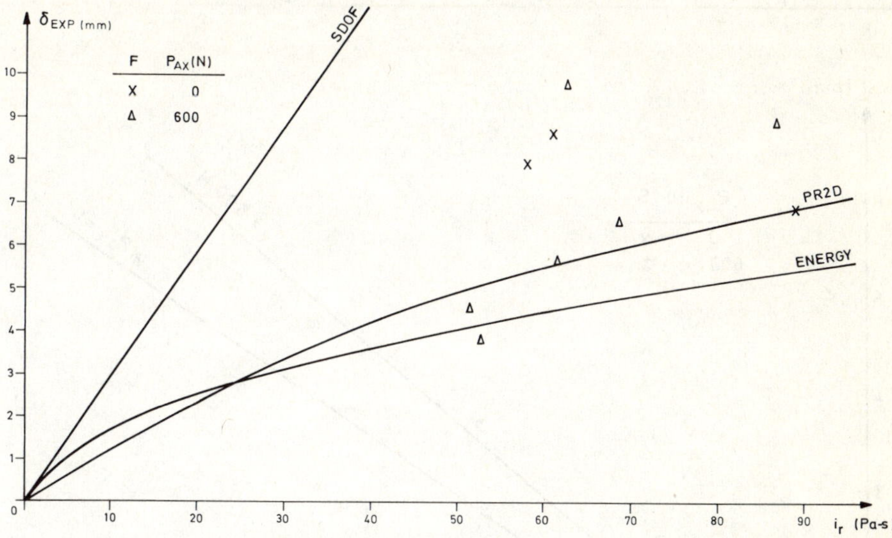

Figure 6. Comparison of energy method and PR2D predictions with experimental results

This simplified analysis is based solely on consideration of bending of the plates which neglects the effects of membrane stresses. These effects can most easily be accommodated in the following energy analysis.

A deformed shape of the plates was assumed of the form

$$y = \frac{16}{L^4} \delta_{EN} x^2 (L-x)^2$$

where y is the displacement at distance x from the end of a one-way spanning plate of length L. An energy balance between the kinetic energy imparted by the blast wave impulse, the strain energy in bending and the strain energy associated with the generation of plate membrane stresses may be formulated as

$$\frac{i_r{}^2 bL}{2\rho h} \approx \frac{Ebh}{8} \int_0^L \left[\frac{dy}{dx}\right]^4 dx + \frac{512EI \ \delta_{EN}{}^2}{5L^3 \ (1-\nu^2)}$$

where b, h, I, ρ, E and ν are plate breadth, depth, second moment of area, density, elastic modulus and Poissons ratio respectively.

For the plates used here the line labelled ENERGY in Figure 6 results. As expected the maximum deformation for a given impulse is less than that derived by consideration of bending alone.

EXPERIMENTAL RESULTS

The results of static tests on plates, with ends fixed against rotation but with in-plane translation permitted, are shown in Figure 7.

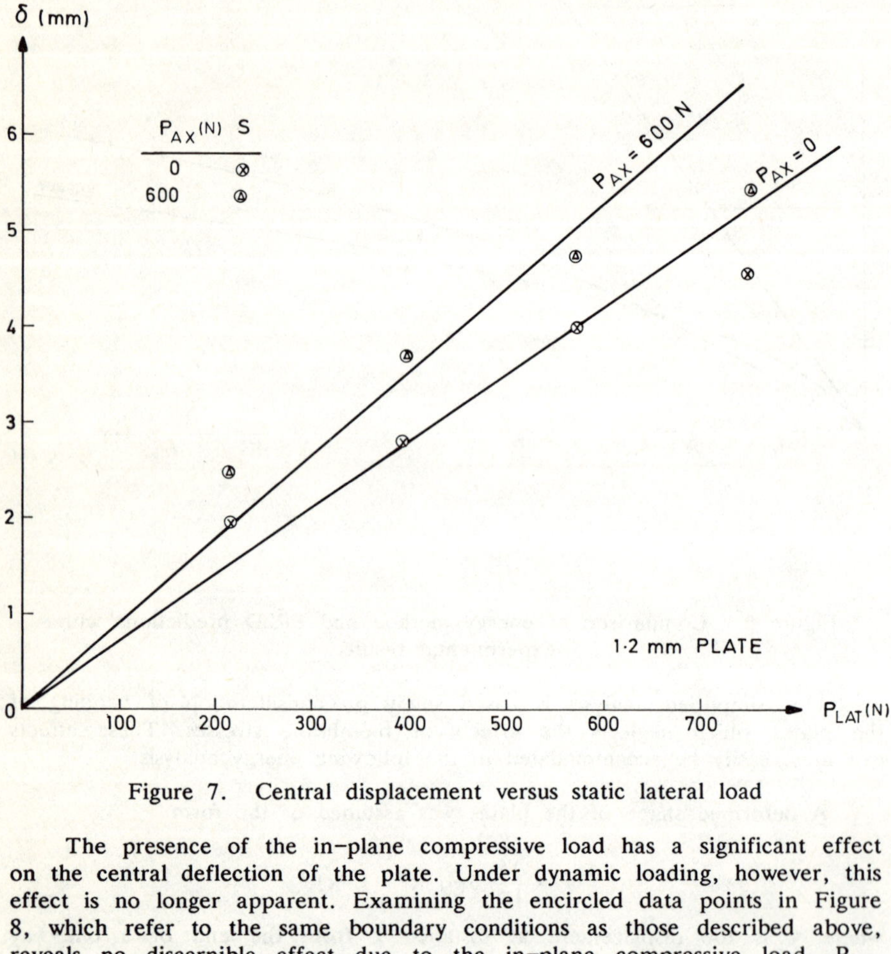

Figure 7. Central displacement versus static lateral load

The presence of the in-plane compressive load has a significant effect on the central deflection of the plate. Under dynamic loading, however, this effect is no longer apparent. Examining the encircled data points in Figure 8, which refer to the same boundary conditions as those described above, reveals no discernible effect due to the in-plane compressive load, P_{ax}. This is due to the high inertia of the loading system which is simply left behind when the plate begins to deform. This results in the axial precompression being quickly eroded by developing tensile membrane stresses.

The validity of this explanation of events is confirmed by the similarity in response between fixed and sliding supports in the same figure. The inertia of the sliding support means that its position is effectively fixed and that its behaviour is indistinguishable from that of a fixed support. Figure 9 shows the correlation between displacement derived from PR2D and those measured experimentally.

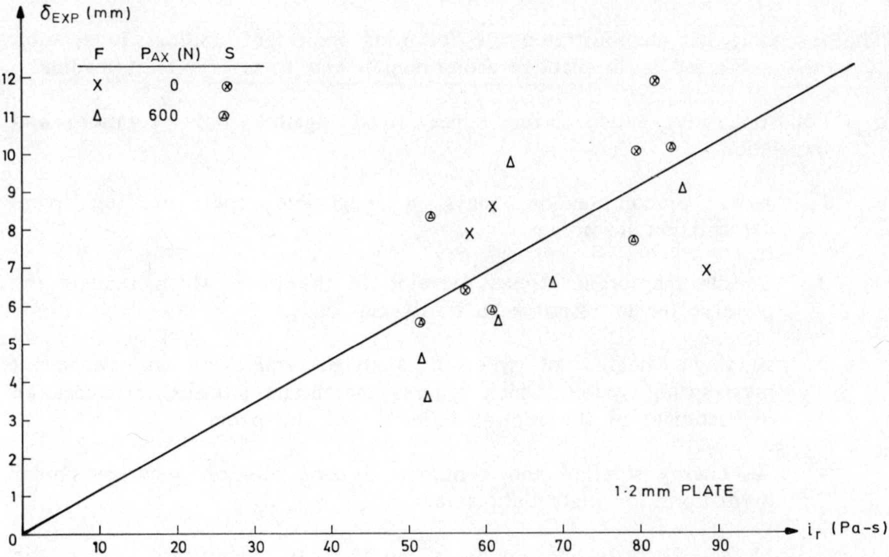

Figure 8. Central displacement versus impulse for different axial loads and boundary conditions

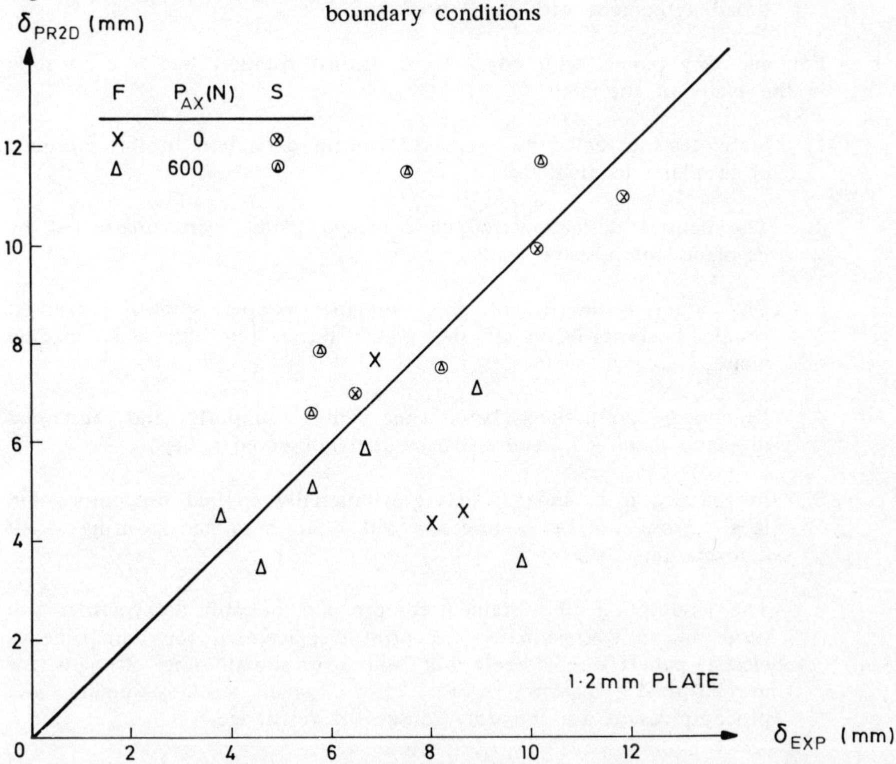

Figure 9. Comparison of numerically derived and experimentally measured displacement

DISCUSSION

The foregoing has demonstrated the following important findings in relation to panels subjected to in-plane precompression and transverse blast loading:

a. For one-way panels with edges fixed against both rotation and translation

1. Axial precompression plays a negligible role in the plate deformation history.

2. Tensile membrane stresses develop in the plate which renders the plate stiffer in response to transverse loads.

3. A single degree of freedom analysis, employing an equivalent mass-spring system which ignores membrane stresses, produces an overestimate of the central deflection of the panel.

4. An energy solution and a numerical code solution generate similar predictions for plate deflection.

5. Measured plate response is subject to significant errors in measuring the impulse delivered to the plates and the subsequent displacement history. However, the measured deflections are in broad agreement with those predicted.

b. For one way panels with edges fixed against rotation but free to slide in the plane of the plate

1. Static tests revealed an increased central deflection in the presence of in-plane loading.

2. The central deflections of blast loaded plates were unaffected by in-plane compressive loads.

3. The inherent inertia of the in-plane loading system prevented sustained compression of the plates throughout the blast loading phase.

4. Numerical predictions based on sliding supports and sustained in-plane loading exceed experimentally observed values.

5. In practice it is unlikely that gravitationally applied precompression (e.g. from higher storeys) will be sustained during wall deformation.

6. The assumption of sustained compression in wall deformation will result in an overestimate of central displacement for homogeneous elastic panels (e.g. steel) but will overestimate the strength for unreinforced masonry walls, for which self weight and precompression are the only sources of resistance.

CONCLUSIONS

1. Energy solutions and a numerical code successfully predict deflection of one way panels with edges fixed against rotation and translation.

2. In–plane compression is not sustained during blast loading for plates with fixed or sliding edge supports.

3. The assumption of sustained compression results in conservative design for steel panels, but unsafe design for masonry walls.

REFERENCES

1. Baker WE, Cox PA, Westine PS, Kulesz JJ, Strehlow RA,Explosion Hazards and Evaluation, Elsevier, Amsterdam, Oxford and New York, 1983.

2. Yankelevsky DZ, Elasto–plastic Blast Response of Rectangular Plates, International Journal of Impact Engineering, Vol 3, No. 2, pp 107–119, 1985.

3. Gürke G, Bücking P et al, Elasto–Plastic Response of Steel Plates to Blast Loads, pp 382–394, Proceeding of the Tenth International Symposium on Military Application of Blast Simulation, Bad Reichenhall, Federal Republic of Germany, 1987.

4. Hetherington JG, Arching Theory for Masonry Walls, Miscellaneous Paper, Structures Laboratory, Waterways Experimental Station Vicksburg MS, 1987.

5. Forsén R, Increase of In–plane Compressive Forces due to Inertia in Wall Panels Subjected to Air–Blast Loading, pp 369–380, Proceedings of the Tenth International Symposium on Military Applications of Blast Simulation, Bad Reichenhall, Federal Republic of Germany, 1987.

6. Holasut S, Ruiz C, Effect of an Impulsively Disturbing Load on the Stability of a Statically Loaded Structure, International Journal of Impact Engineering Vol 3, No. 1, pp 57–73, 1985.

7. Chong CC, Response of Panels Subject to Simultaneous In–Plane Compressive Loading and Transverse Blast Loading, No. 2 Blast Course MSc Project Report, RMCS, Shrivenham, 1988.

Dynamic Failure of Sandwich Structures

M.A. Rollins and Y.R. Xia

Department of Engineering Science, Oxford University, UK

Abstract

Experimental and numerical methods are used to study the modes of deformation and failure of sandwich panels under impact. Two types of panels are considered. The first type is constructed from two face sheets separated by a corrugated core, the second type from several layers bonded by an adhesive. In the case of the corrugated structure different failure modes are found to be dependent on the core geometry and the thickness of the faces. An optimal design may be reached to obtain the maximum impact resistance while keeping the stiffness-to-weight ratio at a maximum. For the adhesively bonded layered panels, the adhesive is found to have an important effect on load transmission across layers, on the energy absorption capability and on the mode of failure. This may switch between fracture of the layers, and shear failure of the adhesive allowing large interlayer slip.

1 Introduction

Sandwich structures are recognised to be weight saving with high stiffness, and therefore have become widely used in aircraft engineering. Various types of sandwich panels are used, from simple structures of two faces with a thick low density continuum core, to corrugated and honeycomb cores. Layered structures with adhesive bonding present another type of sandwich structure with further interesting properties.

Corrugated core sandwich structures, made of high strength alloys, offer significant advantages over conventional honeycomb sandwich structures in many aerospace applications. Structurally they offer improved integrity under impact and resistance to corrosion due to the larger area of bonding between the core and faces.

The failure modes of the corrugated structure under transverse impact are mainly determined by the material characterisation of the core. Conventional design requires that the core must be stiff enough in the direction perpendicular to the faces to ensure that the faces remain the correct distance apart. Also the core must be stiff enough in shear to ensure that when the panel is bent the faces do not slide over each other acting in effect as independent plates. Under high intensity impact, the panels are bound to fail. The question is how the core geometry affects the whole structural response and whether the conventional design guides still hold. The first part of this paper begins to answer these questions by discussing qualitatively the results of some impact tests.

The potential advantages of adhesives over mechanical fasteners for joining have long been recognised. However, industrial acceptance of adhesives for structural application has been low. This is due to lack of adequate information about reliability and poor understanding of the failure mechanism of adhesive bonds.

Adhesively bonded layered plates have a unique failure mechanism in the ability to delaminate. This failure mode may be favourable when it comes to absorbing energy, for example, if it is desired to prevent penetration. However, if a measure of toughness is the compression strength or buckling load in the plane of the layers after an impact, then delamination is probably the most detrimental failure mode. The second part of this paper outlines some experimental and numerical studies of the effect of adhesive properties on the preferred failure mode of the structure.

2 Deformation of the corrugated structure

Progress in the development of impact resistant panels in the past has been hindered by lack of understanding of the mechanisms of impact damage. For lightweight corrugated sandwich structures, see Figure 1, the works are limited. Reports on the maximisation of the shear modulus of the core and on the analysis of least-weight core configurations can be seen in [1]. Effective elastic constants such as extensional stiffness, bending stiffness and shear stiffness in the plane normal to the axis of corrugations, were obtained in [2], providing a database for design in the elastic regime.

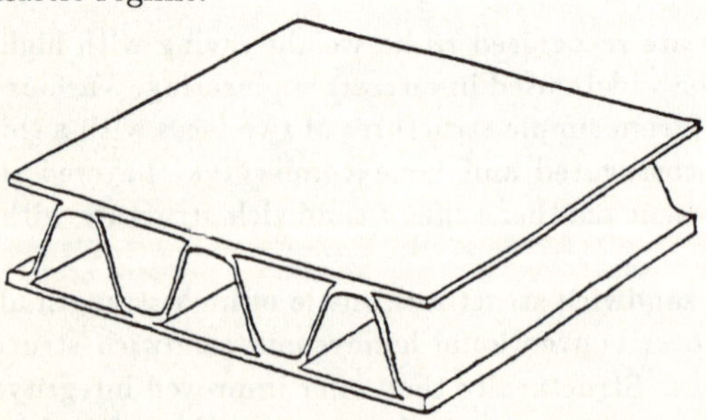

Figure 1: Corrugated core sandwich panel

Impact tests were conducted on coupons of the corrugated core sandwich structure using an instrumented Hopkinson bar apparatus. This consists of a

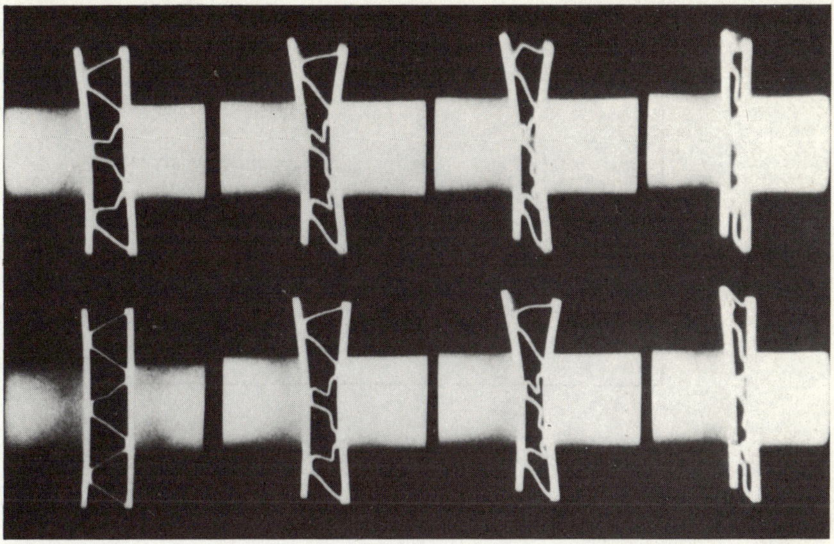

Figure 2: Compression tests of the panels

70mm bore gas gun which accelerates a 1 metre long, 25.4 mm diameter bar (either titanium or high strength steel) to a suitable velocity. This bar impacts another identical bar which has one end in contact with the specimen. The bars have strain gauges which measure the stress pulse generated by the impact. Details of the equipment and instrumentation can be found in [3,4]. Two types of tests are considered here. These are direct compression and bending. High speed photography was used to observe the deformation of the specimens wherever possible. The photographs in Figure 2 clearly illustrate the transient deformation of a specimen of panel in a typical compression test. At initial impact the core undergoes elastic deformation in the transverse direction as if the core were a continuous medium. However, the core structure is a series of plate elements and as such admits local instabilities. As soon as a critical global stress is exceeded, the core structure buckles in one or more of its elements and this buckling quickly develops into a plastic collapse mechanism as the local yield stress of the core material is reached. From the plot of incident stress against time shown in Figure 3, it can be observed that this process of core crushing is marked by a constant level of resistance given by the plateau stress. The rapid rise in stress accompanies the total crushing of the core structure and contact of the two face sheets. Clearly the plateau stress is an important design parameter as it represents the direct stress required to crush the panel. From a series of compression tests, the characteristics of many core configuration can be obtained and the effect of geometry changes, such as the leg angle, quantified and used in design.

Numerical modelling of this type of deformation is reported in [5] where the

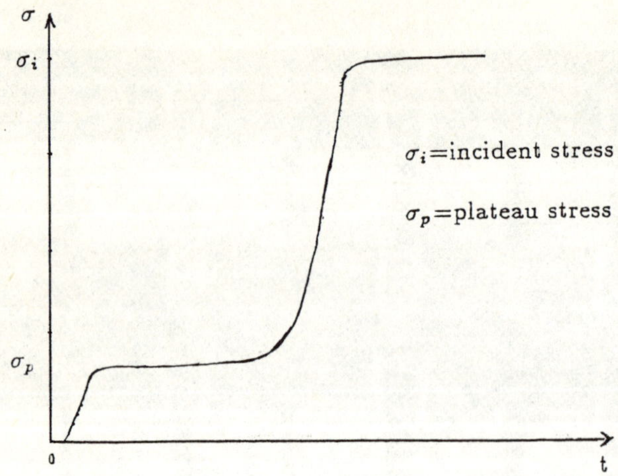

Figure 3: Stress history

finite element code ABAQUS is used.

3 Failure modes of the panels

In a realistic situation, a panel of corrugated material will be in effect simply supported and combined bending and compressive stresses will be encountered. This is the load configuration that was investigated in the bending tests. There are two basic modes of failure, one being the failure of the core and the second entailing the failure of the faces while the core integrity is maintained. These modes can be further split into different categories.

The failure of the core is termed a mode I failure. Mode Ia is symmetric plastic collapse of the core structure. Mode Ib is asymmetric plastic collapse resulting in large interfacial slip. These mode I failures occur when the core structure is relatively easy to crush, for example, large corrugation pitch and a thin core membrane.

The failure of the faces is a mode II failure that occurs when the initial transverse incident stress is insufficient to cause mode I. If the faces are maintained the correct distance apart throughout the bending, very large membrane stresses are generated in the them. The type of failure is then governed by their thickness. Mode IIa is the local buckling of the compressive face caused by the axial stress in a local region of the inner face exceeding the buckling strength. This requires that the inner face sheet be relatively thin. This can happen in several elements at the same time and the result is wrinkling in the face. Mode IIb can occur if the relative thickness of the faces is such that the ultimate tensile strength of the outer face is reached before mode IIa occurs. This results in a tear occuring in the outer face.

Mode I and mode II are mutually exclusive since the mechanism of mode II requires the skins to develop very high axial stresses. This can only occur at moderate curvatures provided the skins remain a large distance from the neutral

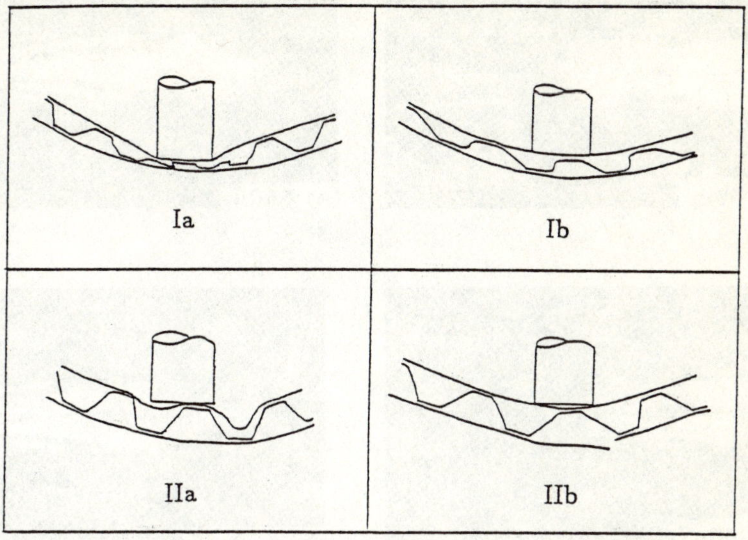

Figure 4: Failure modes of the corrugated structure

axis of the panel. The switch between the two modes can be of a subtle nature. Therefore, it can be seen that a detailed understanding of the core geometry and localised deformation effects is necessary to design for a preferred mode of failure in the case of corrugated sandwich structures.

4 Stress waves in layered structures

For adhesively bonded sandwich beams under pulse loading, the wavelength is comparable to the characteristic dimension of the configuration. The boundaries or interfaces cause wave reflection and refraction. In addition to these, the layered geometry generates inhomogeneous waves due to the interfaces and possibly to the anisotropic properties of each constituent if the material itself is a composite material. The interfaces govern the load transmission across the layers. Therefore, the strength of bonding becomes the main factor in this respect. The thin interlayer of bond presents additional material discontinuities which complicate the wave pattern. When stress waves spreading out in a solid encounter a geometric or material discontinuity, high levels of local stress are caused.

Since attention is focused on the wave transmission through the interface, the adhesive material effect, and the impedance mismatch of the bonded layers, the material properties of both the bonded layers and the adhesive serve as parameters for study.

'Araldite' models of one to four-layer cantilever specimens with different ad-

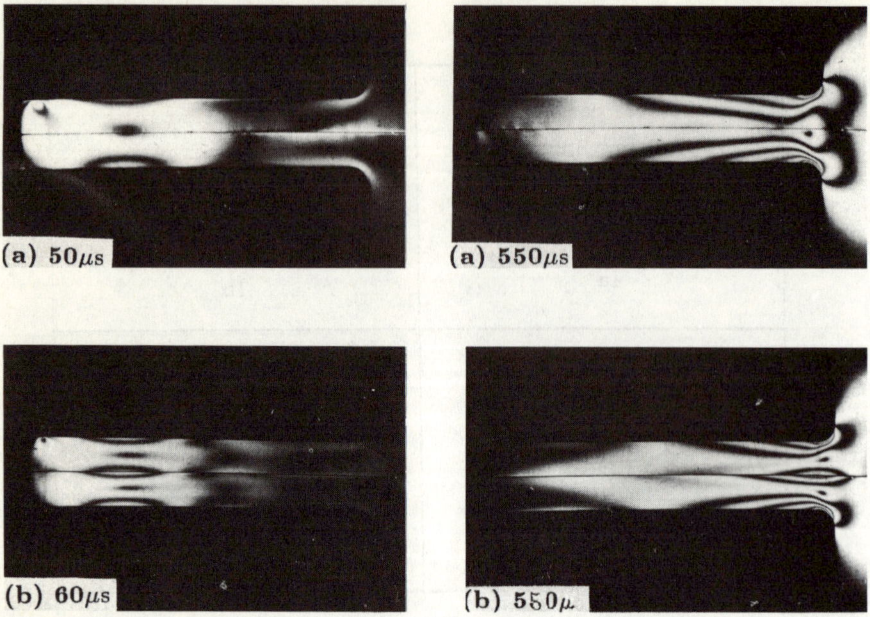

Figure 5: Dynamic fringes in layered beams (a) superglue bonding (b) flexible bonding

hesives or without bonding, were subjected to impact loading by a falling weight. The beam was 100 mm long and overall thickness of 20 mm. Dynamic photoelastic tests were undertaken to observe the wave motion in the various specimens. Figure 5 shows some selected photographs for specimens of two layers with flexible bonding and with very strong bonding.

In the case of a rigid adhesive, the wave initiated at the tip travels both along and across the beam. The wave pattern is almost the same as for the case of a single layer beam. The zero order fringe travels along the interface of the two layers. In the case of a flexible adhesive, there is a time delay when the wave is transmitted from the upper layer to the lower layer. The two layers tend to behave independently. For the multilayer case, the adhesive effect on the wave transmission is even more pronounced. Away from the impact area load transfers are largely in the form of shear, although complex wave diffractions occur along the bond line. This implies large shear deformation of the adhesive which significantly affects the wave transmission, and absorbs impact energy. The adhesive may become a mixed region of bonded and frictional slip areas due to the shear failure.

4.1 Dynamic effects

Under static loading, shear stresses in the adhesive result from the need to satisfy the equilibrium conditions. These stresses, in the case of a cantilever beam, are

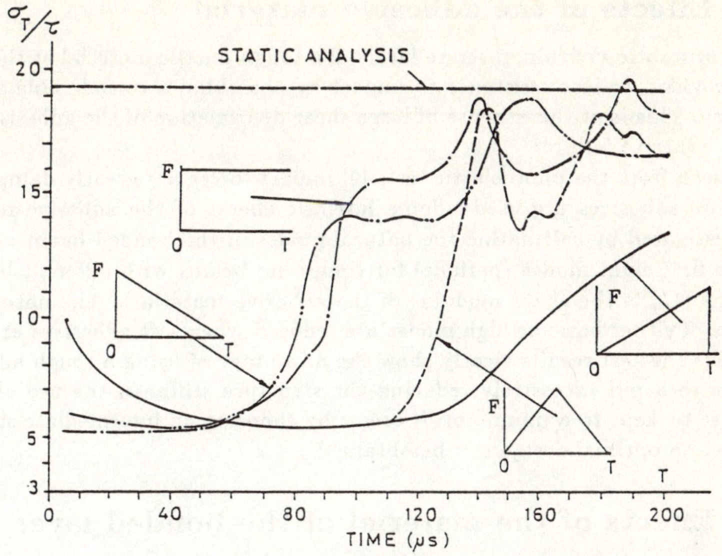

Figure 6: Pulse effect on the failure modes

maximum along the neutral axis. For example, under concentrated load P at the tip, the shear stress reaches a maximum value of $\tau = \frac{3P}{2h}$. The maximum tensile stress $\sigma = \frac{6PL}{h^2}$ is at the clamped end where h is the beam height and L the span.

Under transient loading the shape of the pulse is strongly coupled with the structural dimensions such as the slenderness of the beam and the ratio of the length to width. It also strongly affects the wave motion and the wave interaction with the sandwich material discontinuities. A laminate composed of two layers of quasi-isotropic composite material is considered. Typical values of E_l, E_t, G and μ for a T300/5208 quasi-isotropic laminate are taken as [6]: E_1 = 53.3GPa, E_t = 14.3GPa, G = 7.1GPa and ν_{lt} = 0.28. The density of the composites is 1300.0 kg/m^3. The adhesive is an epoxy (ρ_a = 1000.0 kg/m^3) with Young's modulus of 3GPa and Poisson's ratio of 0.4. The two layers are assumed to be identical. The beam length is 0.1m and its overall thickness is 0.02m with the adhesive layer having a thickness of 0.1mm. The laminated beam is clamped at one end. The analysis is done by the finite element code ABAQUS. The verification of the code with this problem can be found in [7] for identical layers and [8] for different layers.

Figure 6 illustrates the ratio σ_T/τ for various impact profiles where σ_T is the maximum tensile bending stress in the layer and τ the maximum adhesive shear stress. Pulse shapes are rectangular, symmetric triangular, ramp triangle and abrupt leading edge triangle. Since the analysis is geometrically linear, the ratio is independent of the loading amplitude. Therefore, it is the shape of the pulse that has an important role in determining the failure modes. At the very beginning, the ratios are low because the shear stresses resulting from the initial wave fronts are high and the tensile stresses are rather small. Tensile stresses only build up near the clamped end. In the figure the static analysis result is also shown. It can be seen that adhesive shear failure is much more likely than tensile break of the bonded layer. With the same amount of impact energy, the ratio σ_T/τ is much higher in the dynamic than in the static case [9].

4.2 Effects of the adhesive material

The incorporation of thin, discrete layers of a tough ductile material at the interfaces provides greater resistance to impact by absorbing a considerable amount of energy. This is at the expense of large shear deformation of the adhesive layer or even adhesive failure.

As seen from the photoelastic tests [9] impact forces are greatly damped out when soft adhesives are used. Some intrinsic effects of the adhesive material are investigated by calculating the natural modes of the bonded beam. Table 1 lists the first eight modes (periods) for composite beams with different bonding strengths (G_a is the shear modulus of the adhesive material). The natural frequencies of either basic or high modes are reduced when soft adhesives are used. This, plus the test results clearly show the advantage of using a tough adhesive. In order to avoid excessively reducing the structure stiffness, the use of adhesive must be kept to a minimum. It seems by choosing an intermediate strength adhesive, an optimal design can be obtained.

4.3 Effects of the material of the bonded layer

The impedance and bending stiffness mismatches are the two main factors causing large interfacial shear failure. The former effect takes place at the initial wave propagation stage while the latter takes place at the stage when deflections start to occur.

	T (Natural periods) μs		
Modes	G_a=3 GPa	G_a=0.3 GPa	G_a=30 MPa
1	517.4	531.9	624.8
2	110.9	119.7	147.0
3	62.6	62.6	65.1
4	50.4	54.8	62.6
5	32.2	35.1	41.5
6	23.5	25.5	38.1
7	20.9	21.3	27.8
8	18.9	20.9	27.7

Table 1: Basic modes of different laminated beams

The sequence of layering is important. For an upper layer with high wave velocities, such as fiber reinforced composites and ceramics, waves travel a large distance before the load is transferred effectively across the layers through the bonding media. Thus, sharp shear concentrations arise and propagate very quickly along the interface region.

In the case of an upper layer of a relatively soft material, like aluminium which has low wave velocities compared to those in ceramics, adhesive shear concentrations do not travel as quickly as with a ceramic as the top layer. Therefore, localised shear failure would occur. This will be aggravated by bending mismatching which is caused by the immediate deflection of the aluminium layer in the impact region due to its flexibility.

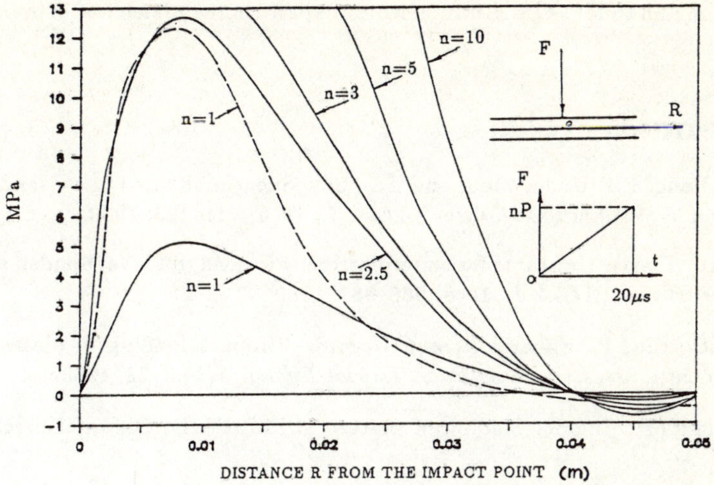

Figure 7: Adhesive shear stresses at 8μs for plates under impact. Solid lines represent the plate with ceramic backed by aluminium, while dashed line is for aluminium layer backed by ceramic. P=100KN

Figure 7 shows the shear stresses for plates consisting of two 10mm thick layers impacted at the origin. The choice of materials is quite arbitrary, only serving as a parameter for study. With the same impact force the interfacial shear stress is larger for the cases when the soft layer is at the top. This is due to the flexibility of this top layer. However, when impacted by a projectile, the resulting impact force and force rate depend on the top layer material and the structural stiffness. Therefore, it is much higher when there is ceramic at the top. This can be illustrated by considering the two extremes. Firstly, for a purely elastic, one dimensional collision the initial force is $v_0\sqrt{\rho E}A$ where v_0 is the impact velocity, A the projectile cross section area, and ρ and E are properties of the top layer. The elastic modulus and density for the ceramic are 310GPa and 2500kg/m^3, while for aluminium they are 69GPa and 2700kg/m^3, respectively. Therefore the impact force for the ceramic top layer is at least twice that for the aluminium top layer. Secondly, considering the ultimate compressive failure, the peak force can be written as σ_cA where σ_c is the ultimate compressive strength. For the ceramic, σ_c is about 5GPa whereas it is only 0.3GPa for the aluminium alloy. The impact force is even larger when the uniaxial strain state is taken into account. In the analysis, the adhesive has shear modulus of 1GPa.

Different force magnitudes are applied. From the figure, adhesive shear stress is seen to propagate much quicker for cases when a ceramic front layer is used. The bigger the impact force, the more severe the situation is. For neatness the rising parts of the stress for cases of $n = 3, 5, 10$ are not plotted. This failure mechanism and whether the adhesive layer could delay the ceramic shattering are now under extensive study.

Acknowledgement

The authors wish to express thanks to their supervisor, Dr. C.Ruiz for his help and direction and to Mr. G.K.Hillsdon whose experimental advice was of immense help.

References

[1] C.E.S. Ueng, E.E Underwood, and T.L. Liu. Shear modulus of new sandwich cores made of superplastic sheet. *AIAA J.*, 18(6):721–723, 1980.

[2] W.L. Ko. Elastic constants for superplastically formed diffusion bonded sandwich structures. *AIAA J.*, 18(8):986–987, 1980.

[3] R.S.J. Corran, P.J. Shadbolt, and C. Ruiz. Impact loading of plates: an experimental investigation. *Int. J. Impact Engng.*, 1(1):3–22, 1983.

[4] P.J. Shadbolt. Impact loading of plates. D.Phil thesis, Oxford University, 1981.

[5] F. Salvatorelli and M.A. Rollins. Some applications of ABAQUS to dynamic structural problems. In HKS, editor, *ABAQUS User's Conf.*, pages 377–396, HKS, Inc. Newport, Rhode Island, 1988.

[6] L. B. Greszczuk. *Impact dynamics, (ed) J.A. Zukas, et.al.*, chapter 3. John Wiley, New York, 1982.

[7] Y.R. Xia. Behaviour of two layered beams under impact loading. *Int. J. Impact Engng.*, 7(4):415–428, 1988.

[8] Y.R. Xia and C. Ruiz. Response of layered plates to projectile impact. In Harding. J., editor, *Fourth Int. Conf. on Mechanical Properties of Materials at High Rates of Strain*, page , Inst. of Physics, Oxford, 1989.

[9] Y.R. Xia and C. Ruiz. Shear effects between the layers of laminated beams under impact loading. Proceedings of the Inst. Mechanical Engineers, London, 1989, Pages 87–92.

Nonlinear Analysis of One-Way Stiffened Plates under Blast Loads

R.B. Schubak, M.D. Olson and D.L. Anderson
Department of Civil Engineering, University of British Columbia, Vancouver, B.C., Canada V6T 1W5

ABSTRACT

A simplified analytical procedure is developed to predict the dynamic response of one-way stiffened plates to high intensity load pulses by modelling them as rigid-plastic, singly symmetric beams with fully encastred ends. A linear approximation of the yield curve is used to uncouple the nonlinear problem into two separate linear ones. The new method is used to calculate the response of a five bay stiffened plate representative of those used in ship construction. The predicted response compares favourably with results from finite element analyses. The proposed method yields quick deflection estimates that are appropriately accurate for preliminary design of blast-loaded one-way stiffened plates.

INTRODUCTION

Rigid-plastic analysis has been used to derive the nonlinear dynamic response of rectangular beams [1,2], more general doubly symmetric beams [3,4], and plates [5]. A similar approximate solution for the nonlinear displacement response of one-way stiffened plates and singly symmetric beams is presented here.

The displacement response of one-way stiffened plates subjected to high intensity load pulses has been investigated experimentally and numerically [6,7,8]. For pulses of very high intensity (several times the static collapse pressure), it was observed that the displacement response of the plate and that of the stiffener were approximately the same. This result suggests that, away from the lateral edges, the stiffened plate behaves much like a singly symmetric beam with the plate acting as a large flange.

FORMULATION AND ANALYSIS

A one-way stiffened plate with uniform stiffener spacing a is considered (Fig. 1). The edges of the plate and stiffeners are fully clamped with constraints against in-plane displacements. The plate is subjected to a uniformly distributed blast-type, *i.e.* monotonically decaying, pressure pulse $p(t)$. An interior portion of the panel, which includes one stiffener and plating one-half of the distance to each neighbouring stiffener, will be assumed to behave as a beam of singly symmetric cross-section with mass per unit length m. The rigid-plastic beam assumptions used previously by the authors [4] are followed here as well.

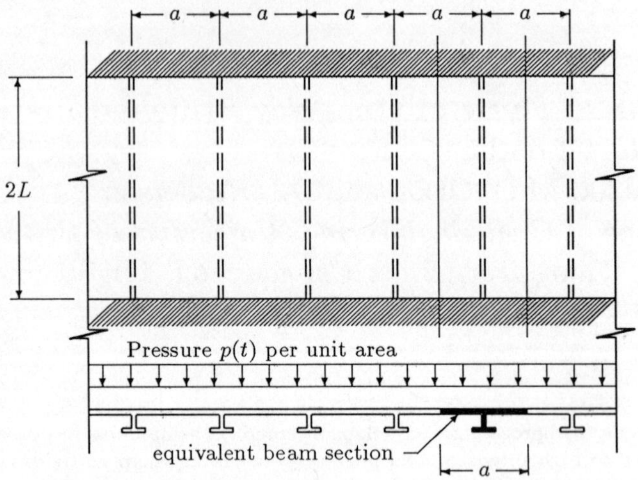

Figure 1: One-way stiffened plate subjected to uniformly distributed pressure pulse.

The beam will resist loading by some combination of axial force and bending moment stress resultants, N and M, respectively, located at the section centroid. Bending resistance will dominate in the early stages of deformation, and axial force or string resistance in the later stages. Dynamic response of the beam, then, will be comprised of some initial plastic hinge mechanism phase, with the stress resultants in the hinges determined by the interaction relation (or yield curve) for the section, and a final plastic string phase during which the beam has no bending resistance.

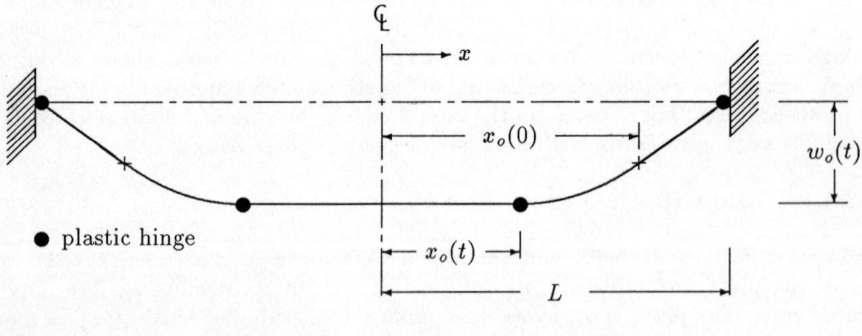

Figure 2: Initial deformation mechanism.

<u>Hinge Mechanism Response</u>

The beam will initially respond in the mechanism shown in Fig. 2. Hinges will form at the supports and at a time-dependent distance $x_o(t)$ to either side of midspan. The velocity field for this mechanism is

$$\dot{w}(x,t) = \begin{cases} \dot{w}_o(t), & \text{if } 0 \le x \le x_o(t); \\ \dot{w}_o(t) \cdot \left(\dfrac{L-x}{L-x_o(t)}\right), & \text{if } x_o(t) < x \le L, \end{cases} \tag{1}$$

where $w_o(t)$ is the midspan displacement. The extension of the beam at finite deflections is

$$e = 2\left[\frac{1}{2}\int_0^L \left(\frac{\partial w}{\partial x}\right)^2 dx - L\right]. \tag{2}$$

Differentiating with respect to time, the rate of extension is found to be

$$\dot{e} = 2\int_0^L \left(\frac{\partial w}{\partial x} \cdot \frac{\partial \dot{w}}{\partial x}\right) dx = \frac{2w_o\dot{w}_o}{L-x_o}. \tag{3}$$

This extension occurs in a discrete manner at the plastic hinges such that $\dot{e} = 2\dot{e}_m + 2\dot{e}_s$, where the subscript m denotes a field quantity at the travelling hinges and s, a quantity at the support hinges.

The beam will also be rotated through the finite angles θ_m at the travelling hinges and θ_s at the support hinges. If these angles are defined as positive in sagging, their time rates of change are easily found to be

$$\dot{\theta}_m = -\dot{\theta}_s = \frac{\dot{w}_o}{L-x_o}. \tag{4}$$

Combining Equations (3) and (4) yields the beam's kinematic relation:

$$(\dot{e}/\dot{\theta})_m - (\dot{e}/\dot{\theta})_s = w_o. \tag{5}$$

Assuming the hinges always move inward until the time they meet, the midspan segment will be flat. The bending moment is then constant and shear is zero through this segment so that the equation of motion for the segment is

$$m\ddot{w}_o(t) = a \cdot p(t) = q(t) \tag{6}$$

with initial conditions $w_o(0) = \dot{w}_o(0) = 0$, where $q(t)$ is the pressure resolved to act as a line load through the beam's axis of symmetry.

The free body diagram for the rigid end segment is shown in Fig. 3. Taking moments about the support point, the equation of motion is derived to be

$$\tfrac{1}{2}q(t)(L-x_o)^2 + M_s - M_m - Nw_o = \tfrac{1}{3}m(L-x_o)^2\left(\ddot{w}_o + \frac{\dot{x}_o\dot{w}_o}{L-x_o}\right) \tag{7}$$

where M_m and M_s are the bending moments—positive in sagging— in the travelling hinges and the supports, respectively. When reorganized and combined with Equation (6), Equation (7) yields

$$\frac{\partial}{\partial t}\left[(L-x_o)^2 \int_0^t q(\tau)\,d\tau\right] = 6\,(M_m - M_s + Nw_o) \tag{8}$$

which is then solved to give

$$x_o(t) = L - \sqrt{\frac{6}{\int_0^t q(\tau)d\tau}\left[(M_m - M_s)t + N\int_0^t w_o(\tau)\,d\tau\right]}. \tag{9}$$

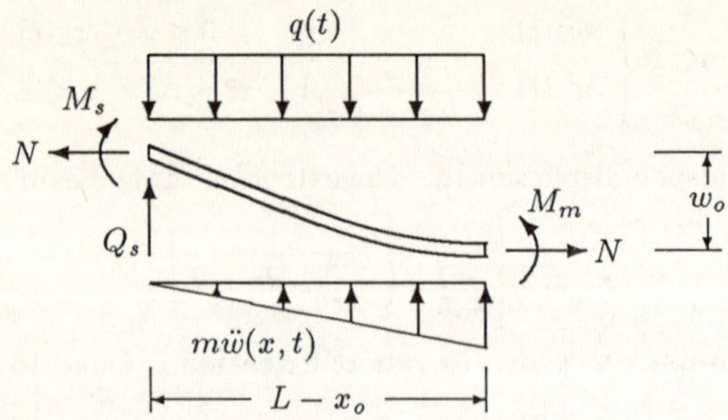

Figure 3: Free body diagram of the rigid end segment for the initial mechanism.

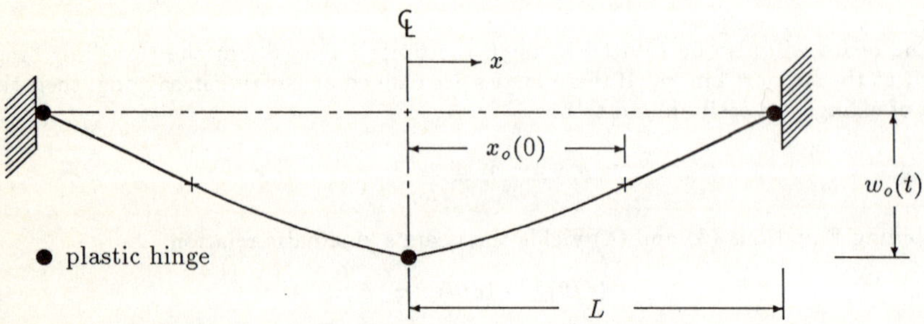

Figure 4: Midspan hinge mechanism.

Equation (9) shows that, for blast-type pulse loading and negative M_s, x_o must diminish with time and thus the earlier assumption to this effect is correct.

At the time t_2 the travelling hinges will meet at midspan and response will continue in the midspan hinge mode shown in Fig. 4. The rigid half-beams of this mode are similar to the rigid end segments of the travelling hinge mode with $x_o = \dot{x}_o = 0$. Putting these values into the previous results yields the same kinematic relation as before, Equation (5), and the equation of motion

$$\tfrac{1}{2}q(t)L^2 + M_s - M_m - Nw_o \;=\; \tfrac{1}{3}mL^2\,\ddot{w}_o, \tag{10}$$

where the subscript m now refers to the field state at the stationary midspan hinge.

The equations of motion for the initial response phase have heretofore been given in terms of the hinge stress resultants M_m, M_s and N. These resultants are determined by the beam section interaction relation and its associated flow rule. Fig. 5(a) shows a typical yield curve for a singly symmetric I-beam with a large top flange and a small bottom flange. N_o and M_o are the maximum values of N and M that may be developed within the section.

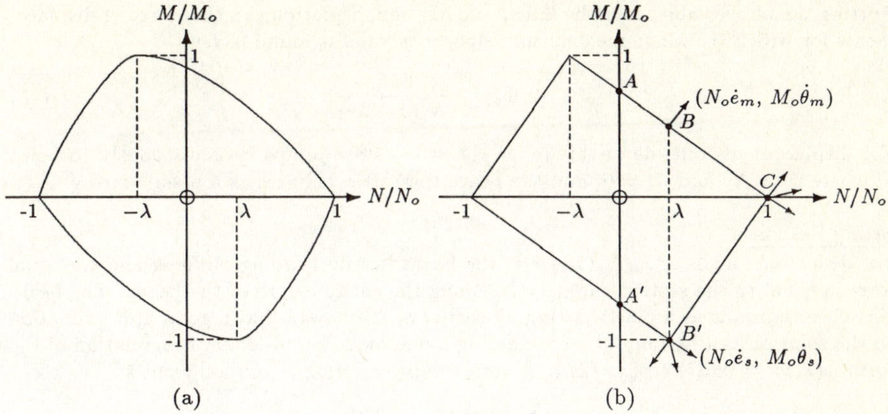

Figure 5: (a) Yield curve for a singly symmetric I-beam; (b) Linear approximation.

An approximation to the yield curve can be developed by inscribing the true yield curve with four linear segments, as shown in Fig. 5(b). Previous studies [3,4] have shown that such an approximation may be used to uncouple the nonlinear problem into two linear ones, thereby yielding a relatively simple and accurate solution.

Referring to Fig. 5(b), the initial stress states at zero deflection are points A at the midspan hinges and A' at the support hinges. Immediately at the onset of finite displacements, these stress states must jump instantaneously to points B at the midspan hinges, where $N/N_o = \lambda$ and $M/M_o = (1-\lambda)/(1+\lambda)$, and B' at the support hinges, where $N/N_o = \lambda$ and $M/M_o = -1$. Recalling Equations (6), (9) and (10), and defining q_o to be the linearized estimate of the static collapse load,

$$q_o = \frac{4M_o}{(1+\lambda)L^2}, \tag{11}$$

the equation of motion for the travelling hinge mode is, again, Equation (6) with

$$x_o(t) = L\left\{1 - \sqrt{\frac{3q_o}{\int_0^t q(\tau)d\tau}\left[t + \frac{\lambda(1+\lambda)N_o}{2M_o}\int_0^t w_o(\tau)\,d\tau\right]}\right\}, \tag{12}$$

and the equation of motion for the midspan hinge mode is

$$\ddot{w}_o + \frac{3\lambda N_o}{mL^2}\,w_o = \frac{3}{2m}\left[q(t) - q_o\right]. \tag{13}$$

As deformation proceeds, the stress states at the hinges stay as they are while the plastic deformation vector at the support hinges (point B') turns the corner. Hence, for the above equations to be valid we must have

$$(\dot{e}/\dot{\theta})_m = \frac{M_o}{N_o(1+\lambda)} \tag{14}$$

and

$$-\frac{M_o}{N_o(1-\lambda)} \leq (\dot{e}/\dot{\theta})_s \leq \frac{M_o}{N_o(1+\lambda)}. \tag{15}$$

Putting the above values into the kinematic relation, Equation (5), the range of displacements for which the hinge mechanism response is valid is found to be

$$0 \le w_o \le \frac{2M_o}{N_o(1-\lambda^2)}. \tag{16}$$

For displacements outside of this range the stress state jumps instantaneously to point C where $N = N_o$ and $M = 0$, and the beam thereafter behaves as a plastic string.

String Response

For deflections $w_o > 2M_o/N_o(1-\lambda)^2$, the beam has no bending stiffness and the axial force is equal to the section capacity N_o along the entire length of the beam. The beam therefore responds as a plastic string. Vaziri *et al* [2] showed that a good approximation to the solution can be found by considering a one mode sinusoidal approximation of the form $w(x,t) = w_o(t)\cos(\pi x/2L)$. A very simple equation of motion results:

$$\ddot{w}_o + \frac{N_o\pi^2}{4mL^2}\, w_o = \frac{4q(t)}{\pi m}. \tag{17}$$

As initial conditions at t_s, the time at which string response begins, the midspan displacement and kinetic energy of the beam are held constant through the sudden change in the deformation mode, as discussed in previous studies [2,4].

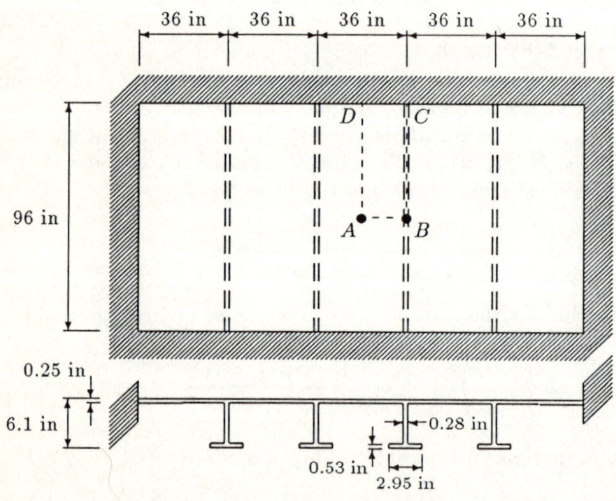

Figure 6: DRES panel geometry.

EXAMPLE APPLICATIONS

A five bay, T-beam stiffened steel plate, designed and constructed by Defence Research Establishment Suffield (DRES) for experimental purposes is considered. The geometry of the stiffened plate, hereafter referred to as the DRES panel, is shown in Fig. 6. The mass density of the panel is $\rho = 0.733 \times 10^{-3}\,\mathrm{lb\cdot s^2/in^4}$ and the yield stress is $\sigma_o = 54{,}400\,\mathrm{psi}$. The section properties of the equivalent beam representation are calculated to be $m = 8.84 \times 10^{-3}\,\mathrm{lb\cdot s^2/in^2}$, $M_o = 1.07 \times 10^6\,\mathrm{in\cdot lb}$, $N_o = 6.51 \times 10^5\,\mathrm{lb}$, and $\lambda = 0.538$. The static collapse load is $q_o = 1215\,\mathrm{lb/in}$.

HOB Test

The response of the DRES panel to blast loads has been investigated by Houlston and Slater [6] in a series of experiments at the DRES Height-of-Burst (HOB) facility. Trial 4 from this series is considered here. The measured pressure pulse incident on the panel is shown in Fig. 7 along with a trilinear approximation which is used in the rigid-plastic analysis.

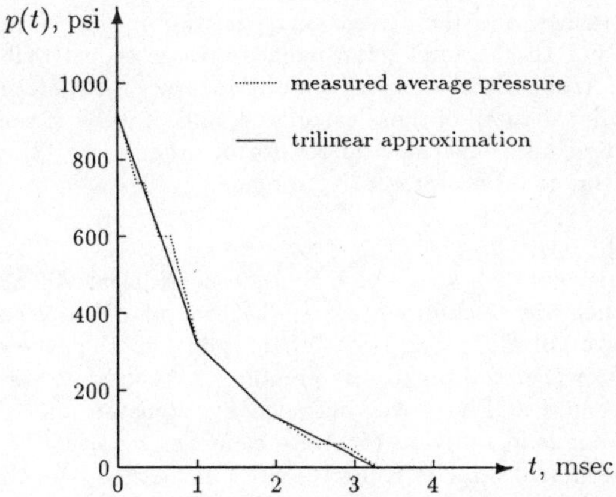

Figure 7: HOB test pressure pulse.

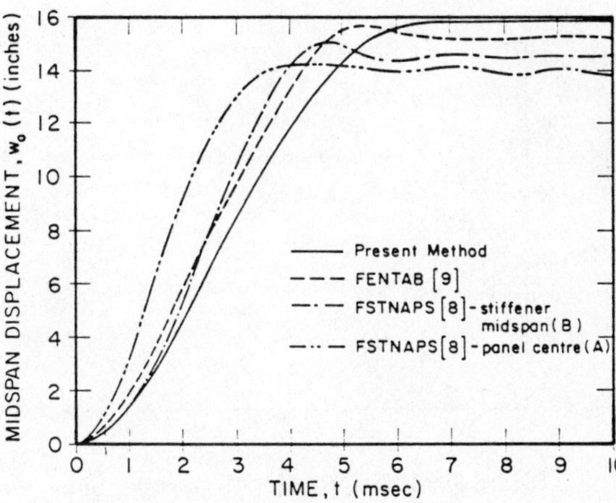

Figure 8: Predicted midspan response of the DRES panel for the HOB test.

The midspan displacement as calculated herein is plotted in Fig. 8. It is compared with the results from FENTAB, a beam finite element program developed by Folz *et al* [9], and FSTNAPS, a finite strip program developed by Khalil *et al* [8]. In these numerical analyses the material behaviour is considered to be elastic-plastic-strain hardening with elastic modulus $E = 30 \times 10^6$ psi and strain hardening modulus $E_T = 1.78 \times 10^5$ psi.

FENTAB models the DRES panel as a beam using ten elements on the half-span, so a comparison of its results with those of the present method is essentially a check on the error introduced by the rigid-plastic idealization, which is seen to be small for this case. FSTNAPS models one half of the panel using three strips for each stiffener and four strips for the plating between adjacent stiffeners, amounting to a total of 16 strips. The displacement response of the panel centre, point A in Fig. 6, and of the adjacent stiffener's midspan, point B, are considered. There is an initial disparity between the plating and stiffener response, but this disparity becomes small as the panel comes to rest. FSTNAPS predicts the permanent displacement of the stiffener midspan to be approximately 14.5 inches, and the present analysis, 15.8 inches — values that agree to within 10 percent. For this example, the panel response seems well approximated by beam response. Unfortunately the experimentally measured displacements for this case were very much smaller than all of those calculated, and, in view of the good comparison between the calculated and experimental results in other cases [8], it was felt that a meaningful comparison could not presently be made.

Event MISTY PICTURE

Another DRES panel was tested at the U.S. Defence Nuclear Agency's event MISTY PICTURE. The panel was flush-mounted on the ground. Heavy concrete walls were constructed above ground along two edges of the panel so that the shock front would be reflected from the walls. Averaging the pressures measured at various points on the panel, the pressure pulse in Fig. 9 was obtained by Houlston and Slater [7] and used as a spatially invariant load function for finite element analysis. The response to this pulse cannot be determined directly using the present method, since the sharp pressure rises during the first 10 msec, consequences of the reflected shock wave, would result in outwardly moving hinges. An alternative representation of the pressure pulse must be found.

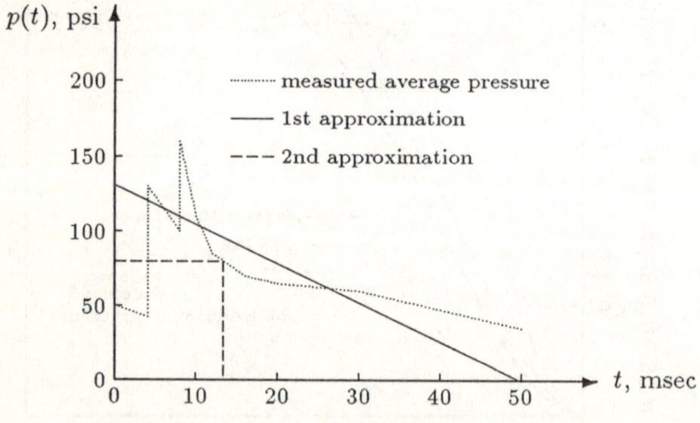

Figure 9: MISTY PICTURE pressure pulse.

The measured pulse is initially approximated by a triangular pulse with a peak intensity of 130 psi and a duration of 50 msec (Fig. 9), values chosen to provide an impulse equivalent to that of the measured pulse over the first 50 msec. Using this triangular pulse, the present method predicts a permanent midspan displacement of 13.75 inches after 11 msec. Examining the first 11 msec of the pressure history, it is apparent that the triangular pulse overestimates the impulse by almost 25 percent, and it follows that the response must also be significantly overestimated.

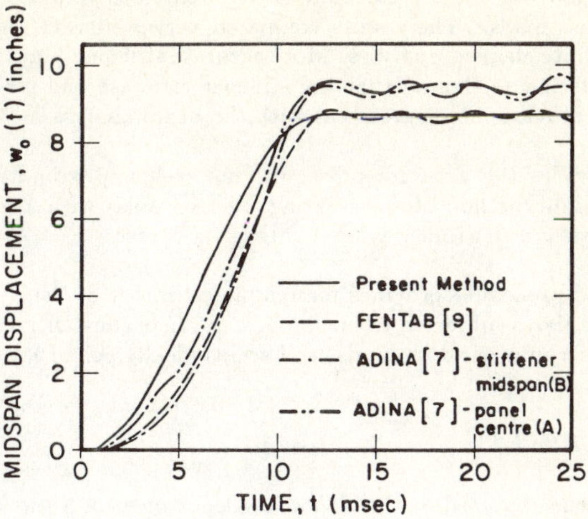

Figure 10: Predicted midspan response of the DRES panel for event MISTY PICTURE.

There is no blast-type pulse which outwardly resembles the measured pulse over the first 11 msec. However, using a procedure developed by Youngdahl [10], a good estimate of the permanent midspan displacement can be found by loading the panel with a rectangular pulse having the same impulse and centroid as the measured pulse. If it is assumed that the previously estimated response time of 11 msec is reasonable, the measured pressure pulse has an impulse of 1020 psi · msec and a centroid of 6.45 msec. The "equivalent" rectangular pulse is then determined to have an intensity of 79 psi over a duration of 12.9 msec, as shown in Fig. 9 by the dashed lines.

The displacement response is recalculated using this rectangular pulse. The panel is predicted to come to rest with a midspan displacement of 8.7 inches at 12.4 msec. This response is plotted in Fig. 10 along with a prediction using FENTAB and another by Houlston and Slater [7] using ADINA. In the ADINA analysis an interior portion of the panel, *ABCDA* in Fig. 6, was modelled by imposing conditions of symmetry at its edges. There is very good agreement between the analyses. The present method and FENTAB predict similar responses, validating the rigid-plastic material idealization and the rectangular pulse representation. ADINA predicts little difference in the permanent stiffener and panel displacements and compares well with the beam analyses, again validating the beam response assumption. The permanent displacement observed in the experiment was about 50 percent higher than predicted, but this was due to a failure of one of the boundary support beams which led to tearing of the panel's edge [7].

CONCLUSIONS

A simplified analytical procedure has been developed to predict the dynamic response of one-way stiffened plates to very high intensity load pulses. Interior portions of such panels were assumed to behave as rigid-plastic, singly symmetric beams with fully encastred ends. By inscribing the true beam yield curve with four linear segments, the nonlinear problem was uncoupled into two linear response phases, namely an initial plastic hinge mechanism phase and a final plastic string phase.

The new method was used to calculate the response of five bay stiffened plates subjected to two blast loads. The results compared very well with those obtained from nonlinear beam finite element analyses. More accurate stiffened plate models using finite strips and finite elements showed that the stiffener response and plating response were similar and were in reasonable agreement with the beam analyses.

In the case where the pressure pulse was not well approximated by a blast-type pulse, the Youngdahl method of representing the true pulse with a rectangular pulse of equivalent impulse and centroid was used with good success.

The developed procedure provides reasonable estimates of the dynamic response of one-way stiffened plates subjected to blast loads. It offers substantial savings in cost and time over nonlinear numerical methods, and seems ideally suited for preliminary design work.

ACKNOWLEDGEMENT

This work has been supported by the Canadian Department of National Defence through a contract from the Defence Research Establishment Suffield.

REFERENCES

1. Symonds, P.S. and Mentel, T.J. Impulsive Loading of Plastic Beams with Axial Constraints, J. Mech. Phys. Solids, Vol. 6, pp. 186–202, 1958.

2. Vaziri, R., Olson, M.D. and Anderson, D.L. Dynamic Response of Axially Constrained Plastic Beams to Blast Loads, Int. J. Solids Structures, Vol. 23, No. 1, pp. 153–174, 1987.

3. Schubak, R.B., Olson, M.D. and Anderson, D.L. Simplified Rigid-Plastic Beam Analysis, ASME J. Appl. Mech., Vol. 54, No. 3, pp. 720–722, 1987.

4. Schubak, R.B., Anderson, D.L. and Olson, M.D. Simplified Dynamic Analysis of Rigid-Plastic Beams, Int. J. Impact Engng., in press, 1989.

5. Jones, N. A Theoretical Study of the Dynamic Plastic Behavior of Beams and Plates with Finite Deflections, Int. J. Solids Structures, Vol. 7, pp. 1007–1029, 1971.

6. Houlston, R. and Slater, J.E. Structural Response of Panels Subjected to Shock Loading, 55th Shock and Vibration Bulletin, Part 2, pp. 149–163, 1985.

7. Houlston, R. and Slater, J.E. The Response and Damage of a Stiffened Panel Subjected to Free-Field and Confined Air-Blast Loading, Proceedings SUSI 89 Conf., Cambridge, Mass., 1989.

8. Khalil, M.R., Olson, M.D. and Anderson, D.L. Nonlinear Dynamic Analysis of Stiffened Plates, Computers & Structures, Vol. 29, No. 6, pp. 929–941, 1988.

9. Folz, B.R., Olson, M.D. and Anderson, D.L. Non-Linear Transient Beam Analysis on a Microcomputer, pp. 341–348, Proceedings of the Third Int. Conf. on Computing in Civil Engineering, Vancouver, Canada, 1988.

10. Youngdahl, C.K. Correlation Parameters for Eliminating the Effect of Pulse Shape on Dynamic Plastic Deformation, ASME J. Appl. Mech., Vol. 37, No. 3, pp. 744–752, 1970.

On the Elastic Response of Panels to Gas Explosions

B.R. Ellis* and D. Crowhurst**

**Structural Design Division, Building Research Station, Building Research Establishment, Garston, Herts., WD2 7JR, UK*

***Fire and Explosion Hazards Division, Fire Research Station, Building Research Establishment, Borehamwood, Herts., WD6 2BL, UK*

ABSTRACT

This paper describes the first part of an investigation concerned with the response of structural elements to gas explosions. The objective of the project is to gain a better understanding of the effects of gas explosions in buildings and provide information to justify or improve the design recommendations for buildings which may be subject to explosions. The work described is concerned with the elastic response of the panels, both how it can be predicted and how it can be measured during explosions.

The experimental work was conducted in a large scale test facility with a key factor in the study being the measurement of the structural characteristics of a panel. These characteristics along with the measured pressure time histories, provide the input information required to predict the panel displacements. For the explosions studied the predicted displacement compares well with the calculated displacement.

INTRODUCTION

Approximately twenty years ago a gas explosion occurred in Ronan Point which resulted in the progressive collapse of one corner of the building, killing four people and injuring seventeen others. The explosion and collapse are described in [1]. Although gas explosions in this type of building may be a rare occurrence, the resulting recommended changes to the design of tall buildings have become economically significant. After the failure at Ronan Point, it was recommended that "gas supplies should be disconnected from those existing tall buildings, the design of which renders them liable to progressive collapse, until they have been strengthened." The strengthening was later specified to provide for a standard static pressure of 5 psi (34kPa) where gas was used and when gas was not used the forces could be halved to 2.5 psi. As an alternative to strengthening the structure, the structure could be shown to be safe if there was an alternative load path in the event of a panel failing (ie the building is not liable to progressive collapse) or if there was a sufficient amount of tying steel connecting the elements. Although these alternative measures were used to demonstrate the safety of many structures the "5psi" rule is still employed.

After the explosion at Ronan Point a programme of work was started at the Fire

Research Station to measure the pressure time histories produced by a range of gas explosions in a large-scale explosion facility. The purpose of this work was to obtain a better understanding of the potential loads which a structure may have to withstand in an explosion. Further tests have been conducted on the pressures produced in explosions from portable gas canisters containing liquid petroleum gas (LPG) and aerosol cans. It has been assumed that the maximum pressure monitored in these tests is the factor which will relate to structural damage; but the time histories of the explosions are recorded, so considerable information is available regarding rates of loading. However, there is little information available regarding the response of structural elements to internal gas explosions, and the information regarding the parameters which relate to damage is, at best, confused.

In an attempt to gain a better understanding of structural response to gas explosions the authors have been monitoring a panel in the Building Research Establishment's explosion test facility at the Fire Research Station's Cardington laboratory. The first stage of the experiment was to monitor the elastic response of the panel to gas explosions and to verify that it could be predicted accurately. This paper provides a report of this work. The overall objective of the project is to justify or improve the design recommendations for buildings which may be subject to explosions.

EXPLOSION TEST FACILITY

Figure 1 shows a schematic plan view of the large-scale explosion test facility at Cardington, Bedfordshire. The explosion cell is a multi-compartment facility consisting of two 22 cubic metre rooms surrounded on three sides by a 1.2 m wide corridor. The modular construction of the cell allows building elements such as doors, windows and walls to be incorporated into the cell in a number of configurations, and the gas control and monitoring systems cater for a wide variety of flammable gas/air mixtures. This in-built flexibility allows explosions designed to simulate a wide variety of conditions to be undertaken.

The cell is extensively instrumented. Computer controlled transient data recorders linked to pressure, displacement and acceleration transducers are used to measure the corresponding time histories. Standard video and high speed film are used to record the development of an explosion.

TYPES OF EXPLOSION

For this investigation data from two types of vented explosion have been analysed.

1. Explosions generated by the ignition of LPG from erupting aerosol containers and small disposable LPG (butane) filled gas cartridges.

2. Explosions generated by the ignition of LPG from leaking refillable gas cylinders of the type found in portable domestic heating and cooking appliances.

To generate the type 1 explosion a full aerosol container or gas cartridge was loosely secured to a clamp and stand and positioned on a table at the centre or corner of one of the 22 cubic metre chambers. A small wooden crib was placed beneath the container and a second crib situated approximately 450mm away. Both cribs were ignited; the first to heat the container and cause it to rupture, and the second to rapidly ignite the discharged contents. This type of explosion is known to have caused damage to buildings in which similar events have occurred (Ames [2]). The products used were mainly household "dry" airfresheners, containing up to 180g of butane as propellant, or gas cartridges containing up to 480g of commercial butane.

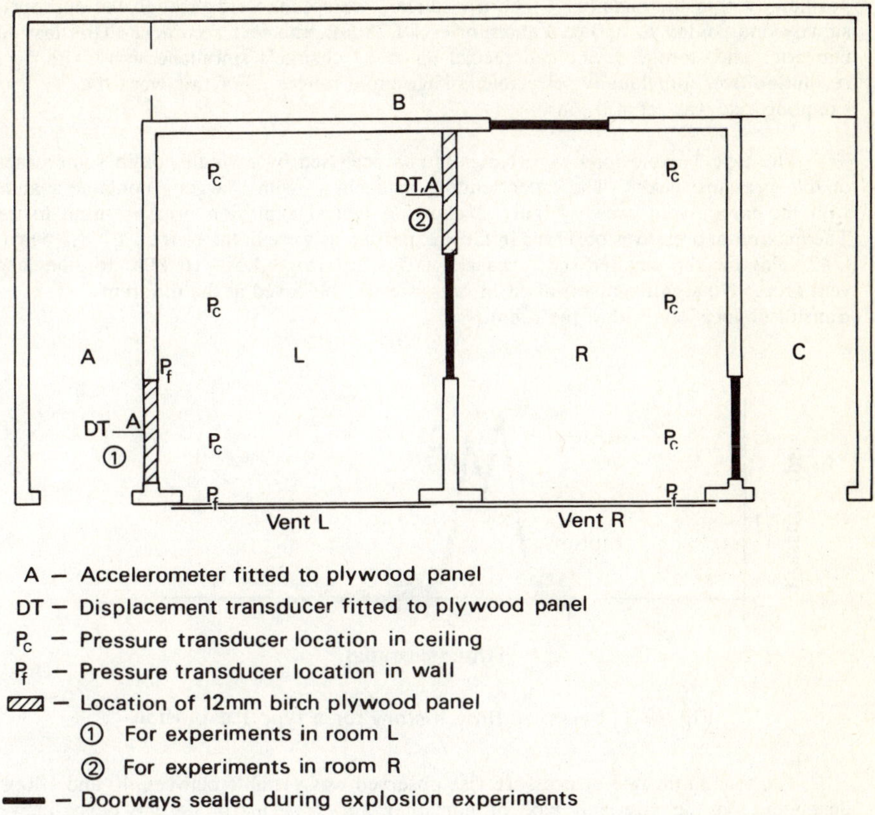

Figure 1 FRS explosion test cell, Cardington

The type 2 explosions were generated by the spark ignition of LPG allowed to leak into the test chamber through an open 4mm pipe. The leak rate was the maximum that could be achieved from a cylinder when leaking freely through the manufacturer's preset regulator. The concentration of LPG accumulating in the chamber was continuously monitored by infra-red analysis of samples withdrawn from different heights at the centre of the room. The concentration profile obtained was rarely reproducible, even when cylinders of the same size and initial weight were leaked under similar conditions. The cylinders were therefore allowed to leak for either a fixed time or until a predetermined weight of gas had been discharged.

For both types of explosion a relief vent was incorporated into the front wall of the test chamber. The vent coefficient, K, for the experiments was equal to either 4 or 8, where K is defined as the ratio of the area of the wall containing the vent divided by the vent area. The material used for the vent cover was 12mm medium density fibreboard. For the vent areas used this material has a failure pressure similar to that of a corresponding area of 4mm glass.

PRESSURE MEASUREMENTS

Measurement of the pressure developed inside the test chamber was made at up to 9 positions within the chamber. The transducers were connected through the appropriate signal conditioning to a Data Laboratories DL2808E transient recorder. This fast A/D converter and storage device can record up to 19 channels simultaneously, with 10 bit resolution over individually selectable voltage input ranges. For this work the sampling rate was set at 0.5ms.

The type 1 explosions were typically characterised by a single, or in some cases a double pressure peak. The latter tending to occur with larger containers and/or with the larger vent area. Figure 2 shows a typical explosion pressure-time history. The maximum pressures observed in these experiments were in the range 1.6 - 6.1 psig (11 - 43 kPa) for the smaller vent area and 0.5 - 2.3 psig (3.5 - 16 kPa) for the larger vent area. No significant variation in pressure was observed at the different transducer locations within the room.

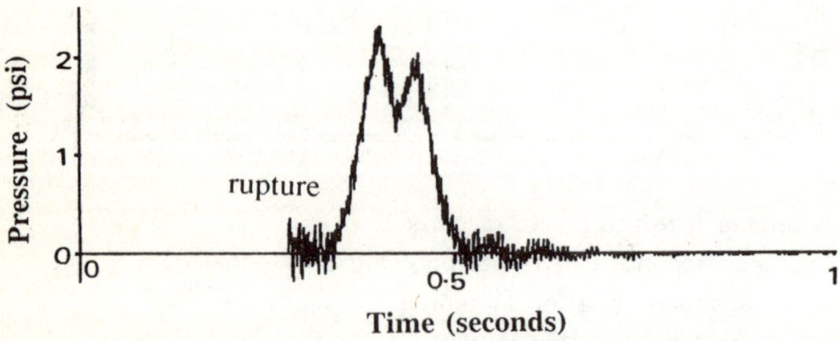

Figure 2 Pressure-time history for a type 1 explosion

The maximum rate of pressure rise observed was variable between 40 and 140psi/s dependent on the size and type of container and particularly on the delay between rupture of the canister and the ignition of the contents. The overall duration of the pressure pulse also varied considerably from 70 to 300ms.

Figure 3 shows a pressure-time history for one of the type 2 explosions. This clearly has a much more complex structure than observed for the type 1 event. Although the structure is complex it can be broken down into distinct phases, and each pressure peak related to a particular mechanism or process occurring during the explosion (see Cooper [3]). The first peak, Pv, is caused by the breaking of the vent cover. The second peak, Pe, arises from unburnt material expelled ahead of the advancing flame front being ignited as the flame passes through the vent. The third peak, Pa, sometimes termed acoustic peak, arises from the coupling of the combustion process with the acoustic properties of the test chamber. Although in the experimental facility this can be the dominant pressure peak, there is evidence to suggest that in most practical situations this sort of coupling is unlikely to occur [3]. The maximum pressures observed in these experiments were in the range 0.75 - 3 psig (3.5 - 20.7kPa) for experiments carried using a single room, (K=4), and 1 - 5 psig (6.9 - 34kPa) for experiments using two rooms separated by an interconnecting standard domestic door, with K=4 in the room where ignition and the leaking pipe were situated and K=8 in the second room.

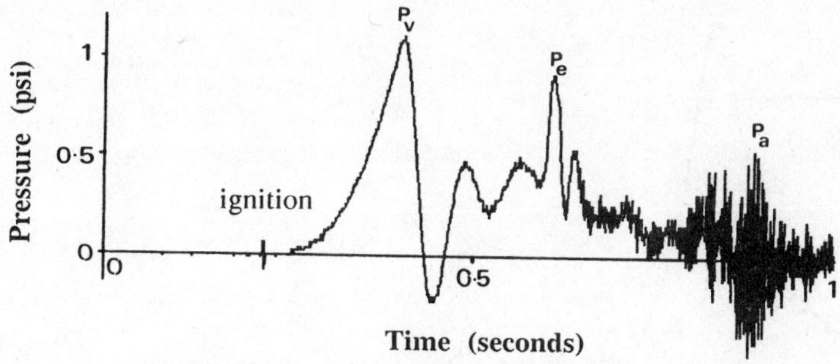

Figure 3 Pressure-time history for a type 2 explosion

RESPONSE MEASUREMENTS

In order to monitor the response of a panel to gas explosions one of the metal panels in the explosions rig at Cardington was removed and replaced by a birch plywood panel (2360 x 1015 x 12 mm). The position of the panel is shown in figure 1. To enable predictions of the panel response to be made, the structural characteristics of the panel had be be measured accurately, as well as the response of the panel to the explosions. The characteristics of the fundamental mode of vibration of the panel were measured using a forced vibration test.

a. Forced Vibration Testing

The testing method used was as follows:

1. A small vibration generator was fixed to the centre of the panel. The vibration produces a sinusoidal force normal to the plane of the panel.

2. The frequency of the force was incremented and the panel response monitored. Resonances were identified by peak values in the response.

3. At the resonance frequencies a second transducer was used to map the mode shape.

4. The frequency/response curve was analysed to determine the frequency, damping and stiffness of each mode.

The measured frequency/response curve is given in fig 4. The derived best-fit three degree-of-freedom curve is also included to illustrate the accuracy of the procedure.

The measured characteristics are actually those of the panel plus the vibration generator, and although the generator will not influence the stiffness of the panel, its mass will be added to the modal mass of the panel and will therefore decrease the resonance frequency. Although the mass of the generator has been kept to a minimum it still has a significant effect and in this case reduced the resonance frequency of the panel from 55Hz to 43Hz. The 55Hz figure has been checked using impact tests on the panel when the generator had been removed, and is the dominant frequency in the recorded response of the panel to explosions. However, the measured stiffness and mode shape can be expected to be reasonably accurate and these, together with the measured pressure time history, are the key parameters required for calculating the panel response.

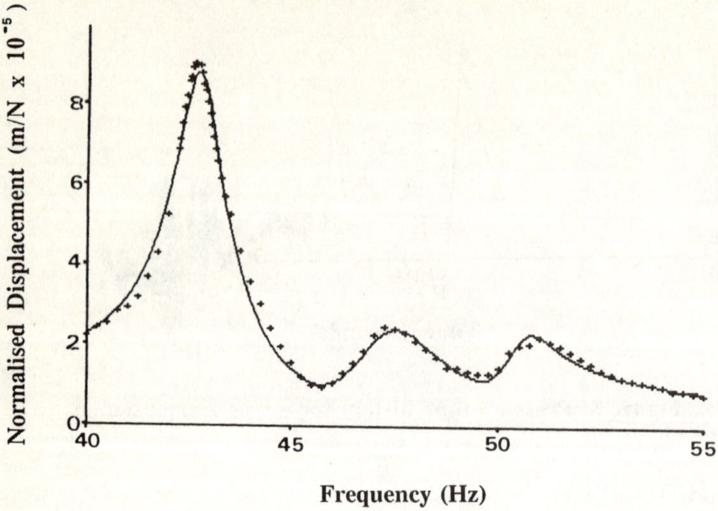

Figure 4 Measured frequency/response curve

b. Response of Panel to Explosions

The next stage in the experimental programme was to monitor the response of the panel to actual explosions. The parameters which can be recorded along with the pressure are acceleration and displacement of the panel. Theoretically it should be possible to integrate the acceleration to obtain the velocity and integrate the velocity to get the displacement. However, as will be shown later this is not a simple task. Initially, two accelerometers were fixed to the panel. One of the accelerometers was positioned in the centre of the panel and the other was positioned on the vertical centre line of the panel and one third of the height. After a number of tests it became apparent that the displacement of the panel should also be monitored, so a displacement transducer was fixed to the centre of the panel. The displacement transducer is not as convenient to use as an accelerometer because it has to be physically linked between the panel and a fixed reference point, which, for these tests, was provided by a tripod positioned outside the cell.

PREDICTIONS OF RESPONSE FROM MEASURED PRESSURES

A single degree-of-freedom elastic model was adopted to predict the response of the panel. The equation for a single degree-of-freedom system is

$$m\ddot{x} + c\dot{x} + kx = p(t)$$

As the mode shape has been measured and the shape normalised to the response of the centre of the panel, the measured acceleration and displacement at the centre of the panel are the same as the modal acceleration and displacement. The stiffness and damping of the fundamental mode, which were measured in the forced vibration tests, are used together with the measured pressure time history of an explosion. The modal mass was derived from the measured frequency of the panel excluding the vibration generator. The pressure, which varies with time, is assumed to be uniform over the panel. The modal force p(t) can be obtained by multiplying the pressure by the area of the panel and then by a factor which depends on the mode shape.

Given this information there are several methods of predicting the panel response, although for this paper only two will be examined. Firstly the Duhamel Integral method which is used to calculate the panel displacement, and secondly the predictor-corrector method which is used to calculate acceleration, velocity and displacement.

a. The Duhamel Integral Method

Perhaps the simplest method of obtaining the displacement of the system to the known loading is by using the method known as the Duhamel Integral. Basically the load history is considered to be a series of short impulses. For each impulse the response of the system is calculated. The overall response is obtained by summing the system response to each impulse. The calculated displacement and measured displacement for the type 2 explosion shown in figure 3 are given in figures 5 and 6. It can be seen that the calculated displacement compares favourably with the measured displacement. In fact this good correlation between measurement and prediction was obtained for each type of explosion with the exception of the tail of the pressure-time histories for some type 2 explosions when the acoustic peak, Pa, was significant. For the response of the panel to this acoustic peak the single degree-of-freedom model is probably inadequate.

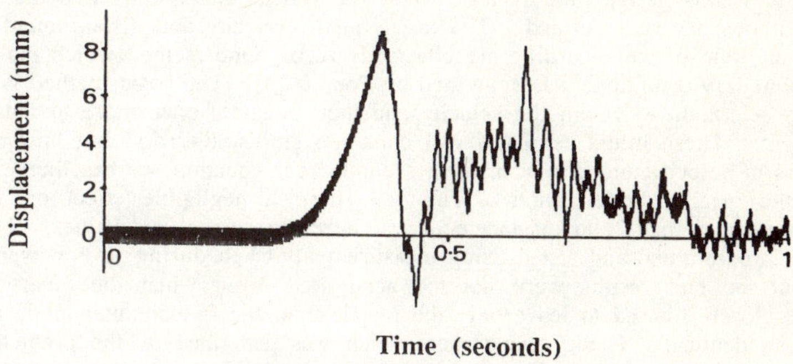

Figure 5 Measured displacement

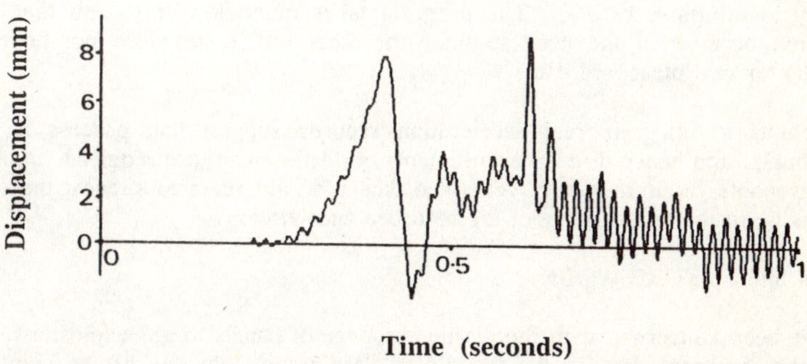

Figure 6 Calculated displacement

The Duhamel Integral method has the advantage of being numerically stable but as it uses the method of superposition it can only be employed with linear systems. Although this is adequate for the measurements reported herein, it cannot be extended to examine non-linear behaviour.

b. The Predictor-Corrector Method

There are many ways of solving the second order differential equation, but care has to be given to the numerical stability of the routines when dealing with a large number of increments. The method which has been used for this investigation is known as the predictor-corrector method (ie it makes use of one formula to predict the response and a second formula to check the accuracy and provide a correction.) The displacement is almost identical to that calculated using the Duhamel Integral. Unlike the Duhamel Integral this type of solution can be used to study non-linear behaviour.

NUMERICAL INTEGRATION OF ACCELERATION TIME HISTORIES

The easiest response parameter to measure is acceleration. If this can be integrated to determine velocity and displacement then it would only be necessary to measure acceleration to describe the structural response. However, numerical integration is known to be an unstable process, and over a large number of steps small errors multiply to produce false trends in the time series. With the system under consideration the boundary conditions are well defined. The acceleration, velocity and displacement at both the start and end of the recording are effectively zero. One method which makes use of these boundary conditions was suggested by Poppitz [4]. The basic method is to integrate the acceleration to obtain the velocity and then integrate once more to obtain the displacement. These initial estimates will contain significant errors, but knowing the final values of velocity and displacement a second order equation (or baseline) can be fitted to the original acceleration. This will have a negligible effect on the acceleration, but the velocity and displacement will have the correct end values. This procedure was programmed and tested using measurements taken during several type 1 explosions; but the end results were not too accurate. Even when the measured acceleration is closely filtered to leave only the response in the fundamental mode the result is almost identical. If the acceleration which was generated by the predictor-corrector program is used a reasonable result can be obtained.

A second method which was developed in California for integration of accelerations measured during earthquakes was evaluated. Again the problem is that of fitting a baseline to the original acceleration to eliminate low frequency trends. The program uses a series of least squares baseline fitting procedures coupled with high pass filtering to try to minimise errors. The program takes much longer to run that the previous version because of the need to filter the data, but it still does not provide accurate results for real measured data.

The attempts to integrate real acceleration records suggest that process is, at present, unreliable, and hence direct measurements of displacement are required. As the process is reasonable for numerically generated data it would seem to suggest that the main problems lie in the time history of the recorded acceleration.

DISCUSSION and FUTURE WORK

This paper has been concerned with the elastic response of panels to gas explosions. It has been shown, for one panel in the test rig at Cardington, that the displacement of the panel to a given explosion can be calculated accurately. This shows that the basic calculation methods work well and suggests that if the characteristics of any panel are measured accurately, say using a forced vibration test, then the elastic panel displacement to a given explosion can be determined accurately. A large number of forced vibration tests have been carried out on different building elements, so it is possible to calculate how they would respond to the explosions monitored in the rig.

The next stage of the experiment is to examine how panels actually fail under explosive loading and to see whether these failures can be predicted. It is expected that the correlation between measurements and predictions will be far worse for this stage than that achieved in stage 1. To obtain the experimental results a second explosion cell is being prepared at Cardington enabling panels of up to 2.4m square to be tested to destruction. The proposed initial testing scheme is to subject three concrete and three masonry panels to incrementally increasing peak loads until failure occurs. Measurements of pressure and response time histories will be made and after each explosion the characteristics of the panel will be measured to provide a detailed picture of the changes which occur as each panel is tested to failure. Numerical modelling of these experiments will also be undertaken.

The third and final stage of the project will be to monitor the response of some real buildings to actual explosions to verify that the conclusions reached in the controlled tests are equally applicable in the field.

ACKNOWLEDGEMENTS

The authors would like to thank their colleagues, Mr M Beak and Mr D Henderson for their help throughout the tests. The work reported in this paper is included in the research programme of the Building Research Establishment and it has been published by permission of the Director.

REFERENCES

1. Report into the Inquiry into the collapse of flats at Ronan Point, Canning Town, London. HMSO 1968.

2. Ames S and Crowhurst D. Domestic explosion hazards from small LPG containers. Jnl of Hazardous Materials, 19, 1988.

3. Cooper M G, Fairweather M and Tite J. On the mechanisms of pressure generation in vented explosions. Combustion and Flame Vol.65, p1-14, 1986.

4. Poppitz J V. Velocity and displacement of explosion-induced earth tremors derived from acceleration. Bull Seis. Soc. Amer. Vol 58, No 5, Oct 1968.

A Simplified Finite Element Model for the Impact of Thin Shells

A.S.L. Chan and G.A.O. Davies
Department of Aeronautics, Imperial College, London, SW7 2BY, UK

1 INTRODUCTION

The analysis of structural impact has been gaining momentum with the recent dramatic increase of computer power. As the knowledge of the structural behavior under impact conditions is becoming a primary concern in many fields of industrial and engineering activities, problems previously considered to be too complex and costly have been made possible by the availability of supercomputers. Modelling of the structure has also changed from the crudely idealistic spring/mass type of representation to the incorporation of many sophisticated and realistic details. Computer codes such as DYNA3D [1] and the like have the capability of dealing with very complicated situations and, when they are properly used, can give refined solutions formerly unimaginable. Nevertheless, for this kind of success to be possible, the first prerequisite is always the availability of very large computer resources to cope with the very demanding size and CPU time usually required.

However, there is one area of impact study where this may not be strictly necessary. It is when the primary objective is the assessment of damage and survivability, particularly when the structures involved are thin-walled shells. This type of structure includes aeroplanes, cars, railway carriages, ships, off-shore platform legs etc., and there are indeed special codes which deal exclusively with this type of impact problem , e.g [2]. Since the basic concern here is the deformation suffered on impact, the stresses are relevant only in so far as they give rise to the deformations. Their detailed distribution is relatively unimportant, as long as the strength of the structure is properly and adequately represented. For this reason, the problem can be solved by using even simpler finite element models than the conventional types.

The element proposed here is triangular in shape with three nodes at the corners. Its membrane behavior is conventional, but its bending stiffness is concentrated along the three edges. The number of freedoms per node is three, the smallest possible in a 3-D analysis. The number of unknowns in a model is thereby reduced from the usual thousands or tens of thousands to a matter of hundreds, with the consequent saving in the computing cost for the time consuming incremental solution of the dynamic equations. Moreover, monitoring of the impact-contact process also becomes less complicated because the distribution of the impact force on a triangle to its three nodes at the corners is statically determinate and therefore easy to be exact. A further advantage is that since no reduced integration is involved, the problem of spurious deformation ("hour-glassing"), usually present when using conventional elements in impact anlysis, does not arise here. All these combine to make the analysis a lot simpler, and enables it to be performed almost routinely with limited resources.

2. STRUCTURAL IDEALISATION

For this analysis, the shell is idealised into a mesh of flat triangular elements. Deformation of the element is described by three orthogonal displacements at every node at the junction of the triangles. In its own plane, each element behaves as a constant strain triangle deformed by the in-plane components of the displacements at its three vertices (Fig.1a). Bending of the shell is approximated by the rotations between the triangles about the edges which become hinge lines (Fig.1b), but the rotations are restrained by non-linear torsional springs representing bending stiffness of the actual shell. These rotations can be related to the normal (out of plane) displacement components at the nodes. Thus the membrane and the bending effects are assumed to be uncoupled, each obeying a non-linear stress/strain or moment/rotation law in accordance with the assumed elastic/plastic behavior of the material. However, coupling of these two effects is taken into account in the geometrical stiffness matrix for the large displacement calculation where local portions of the shell may well rotate through a right angle or more.

The exact details of the stiffness matrices will not be given here but can be found in Refs.[3,4]. Suffice to say that each triangular element has the same standard membrane stiffness as a constant strain 3-noded triangle. The bending stiffnesses of the hinge springs are calculated on the principle that the strain energy of the model, i.e. the hinge moment multiplied by the relative rotation between the two triangles, is the same as the strain strain energy in the two triangles when they are bent with a constant curvature to exactly the same rotation. Thus the degrees of approximation in the two modes of deformation are the same. They both produce constant strains in the triangle. The geometrical stiffness matrices are also obtained from first principle:- By considering how the equilibriums of the existing forces at the beginning of a time step are upset by the ensuing deformations during the time increment, and calculate the necessary additional forces required to restore equilibrium at the end of the step. The inclusion of these stiffnesses is important in the presence of gross deformation, in spite of the fact that the displacement increments an each step is small.

In the same way, reinforcements on the shell such as booms, stringers, frames, rings, etc. are idealised by an elasto-plastic 'beam' element compatible with the idealised shell. The elements are straight pin-ended bars lying along the edges of the triangles. Each bar has two end nodes and is considered to be hinged to the next bar, but rotations at the joints are again restrained by torsional springs. The three displacements at each end are resolved into three principal directions local to the bar. The displacement components in the axial direction activate stretching or compression of the bar, giving it the 'membrane' stiffness. The displacement components normal to the plane of the shell induce rotations about principal axes in the plane of the shell. They are constrained by hinges springs which contribute to the bending stiffness of the shell. Similarly, the third component of the nodal displacement that induces rotation at the far end of the bar about a normal to the shell is constrained by another torsional spring. Its stiffness which increases the membrane stiffness of the shell may sometimes be ignored if it is small. The stiffness of each spring is again derived from the energy principle: The strain energy in the spring, calculated from the hinge moment multiplied by the relative rotation between the two adjacent bars, is the same as the energy in the bars when they are bent with a constant curvature to the same angle between the two far ends. The representations of the bending stiffnesses for the shell and the beams are entirely consistent.

3.SOLUTION PROCEDURE

The mass of each element is assumed to be lumped at the nodal points and the assembled mass matrix is diagonal. The internal and the external (impact) force vectors are similarly assembled from the element contributions. The system of equations of motion that governs the impact behavior is solved in the time domain by the explicit central difference integration scheme, which involves very simple vector operations since all equations are uncoupled.

For impact between two shells, one of them is arbitrarily denoted as the impactor and the other the target. Each node on the contact surface of the impactor may contact the target on a node, or an edge between two nodes, or the surface of an element. In all cases, the forces acting on the impactor node may be distributed to the node or nodes of the target body by a simple static consideration. Friction force in the direction of the relative tangential motion is allowed. The displacements of the impactor node in relation to the target body can be calculated analogously. No penetration through the target surface is allowed, but sliding in the tangential direction is constrained only by the friction force. The calculations become a great deal easier if the target is not a thin walled shell but an object which can be assumed to be rigid.

Contact of a potentially possible contact node with the target is initiated by an examination of its geometrical position relative to the candidate target elements. It is released from contact either when the impact force becomes positive (tensile), or when the geometry indicates that the two bodies have separated at that point.

4. EXAMPLES

A. Impact of a cylinder with a rigid rectangular rail.

This example is similar to the one given in the DYNA3D Manual [1]. Unfortunately the end cross sections have not been made rigid in this calculation although the geometry and the mass are the same. The analysis was performed with 564 elements and 844 degrees of freedom, taking 45 mins. on a CYBER 710-855 for a total impact duration of 6.5 msec, involing 13000 time steps. The deformed shape at 4 msec is shown in Fig.2.

B. Impact between two identical pipes.

The pipes are 50" long, 0.432" thick, with a diameter of 6.625". They are both horizontal and at right angles to each other. The one on top drops on the other symmetrically at an initial speed of 2500 in/sec while the pipe at the bottom is fixed at both ends. The material is steel and assumed to be elastic/ perfectly plastic with a yield stress of 100,000 psi. Because of the double symmetry, a quarter of each pipe is idealised by 300 elements with a total of 948 degrees of freedom for both. The time taken for the 7 msec impact duration is about 70 mins on the CYBER for 3500 steps. The deformed configurations are shown in the series of figures in Fig.3.

C. Oblique impact of a reinforced cylinder.

A typical steel cylinder 10" in radius and 60" long, with a flat bulkhead at one end and open at the other, is impacted onto a smooth rigid edge over half of the bulkead as shown in Fig.4. As this is only a demonstration to show how the gross deformation can be studied easily, the half cylinder is divided very crudely into 30 degree segments around the circumference and 6 segments along the length. The thickness of the shell is 0.01" everywhere except for a 10" band at the top where it is considered to be 2" thick. This is a simple way of representing a heavy mass on top of a 50" long cylinder and it provided enough energy to ensure massive distortion at the impact end. The initial velocity is selected as 17 m/sec, equivalent to a drop from 15 m.

Comparison is made with an identical cylinder reinforced by beam stiffeners and rings placed on the othogonal set of grid lines. The sectional properties of all these are the same, having areas of 0.15 in^2 and a second moment of 0.025 in^4 in the two orthogonal directions, with an elastic limit moment of 2×10^3 lb.ins. The material is also assumed to be elastic/ perfectly plastic.

The progress of the impact simulation is shown in the Figures of 5 and 6. It is not meant to be a realistic case, but enough to illustrate the effect of the reinforcements. The calculation for the reinforced cylinder took 11 mins to obtain the deformation in 18 msec of real time and involving 1800 time steps.

5. REFERENCES

1. Hallquist J.O. & Benson D.J.
DYNA3D User´s Manual. UCID-19592 Rev.2. Lawrence Livermore National Laboratory. March,1986.

2. Winter R. & Pifko A.B.
Finite Element Crash Analysis of Automobiles and Helicopters. Structural Impact an Crash Worthiness Vol.II. pp. 278-309 Conference Papers (Ed. J.Morton). Elsevier.

3. Chan A.S.L. & Li F-N
Impact of thin-walled shells using a simplified finite element model. To be published.

4. Chan A.S.L.
Reinforcements in the Impact Analysis of Thin-Walled Shells Using a Simplified Finite Element Model. To be published.

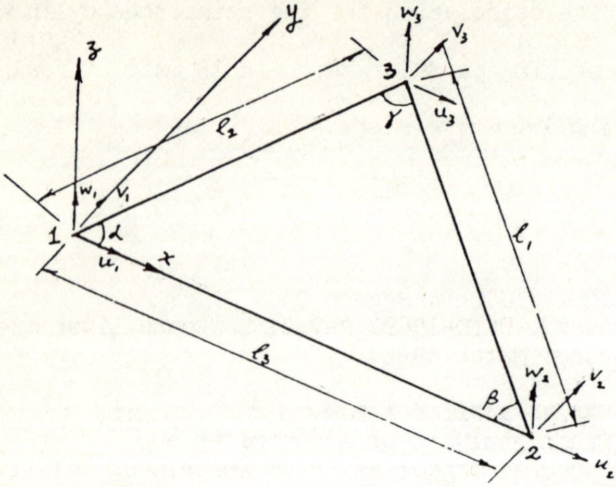

Fig. 1a. Triangular Element

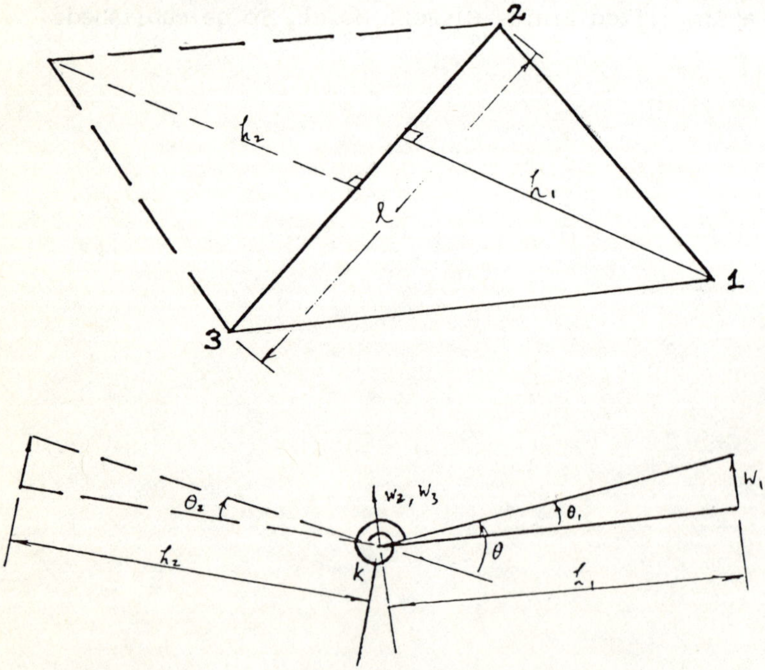

Fig. 1b. Bending between Two Triangles

Fig. 1b. Bending between Two Triangles

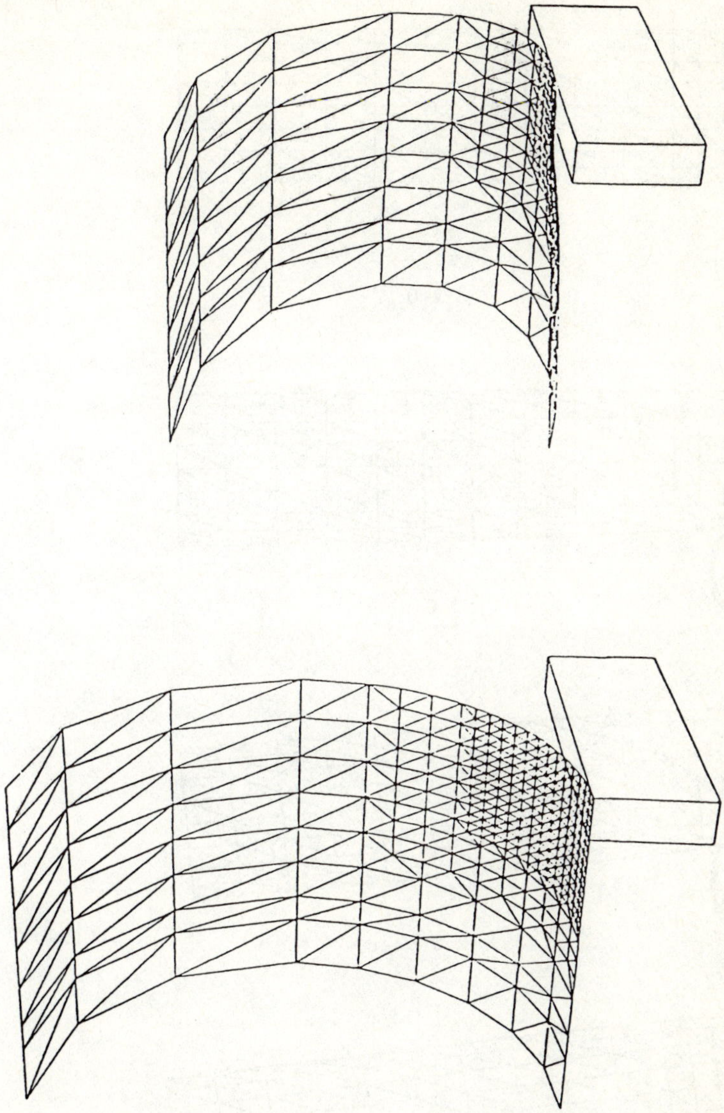

t = 4.0 ms

t = 0.0 ms

Fig. 2. Impact of Cylinder on Rigid Rail

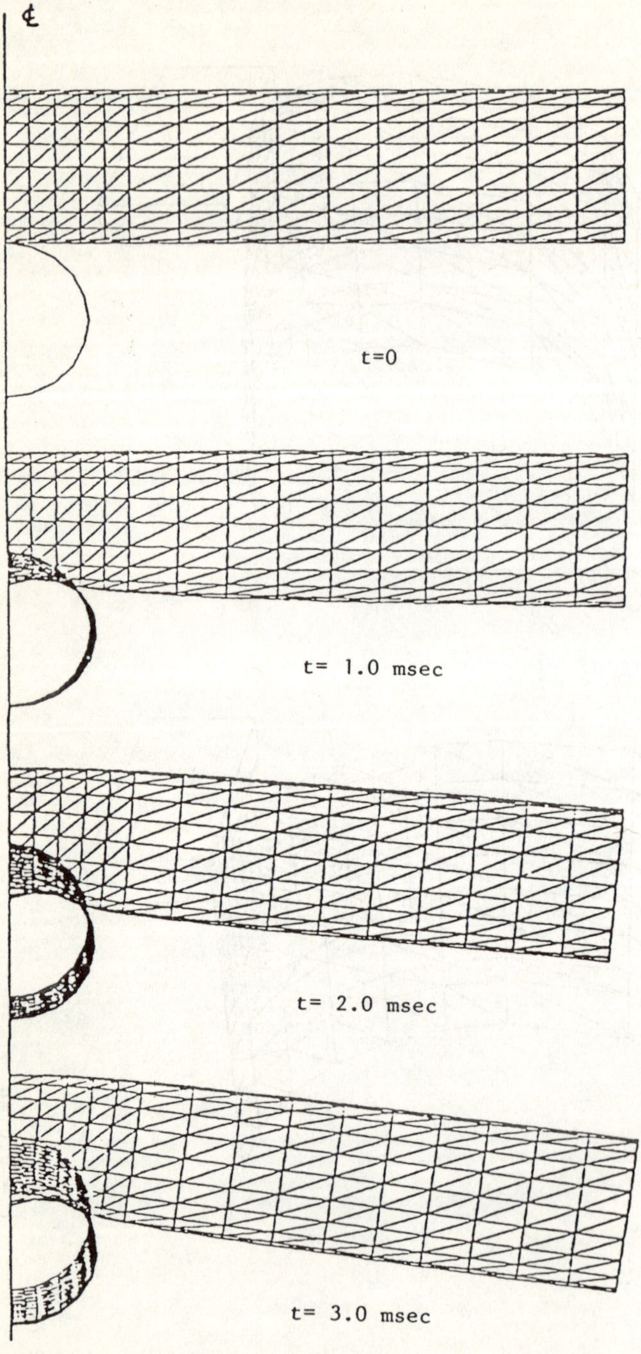

Fig. 3a. Impact between Two Pipes

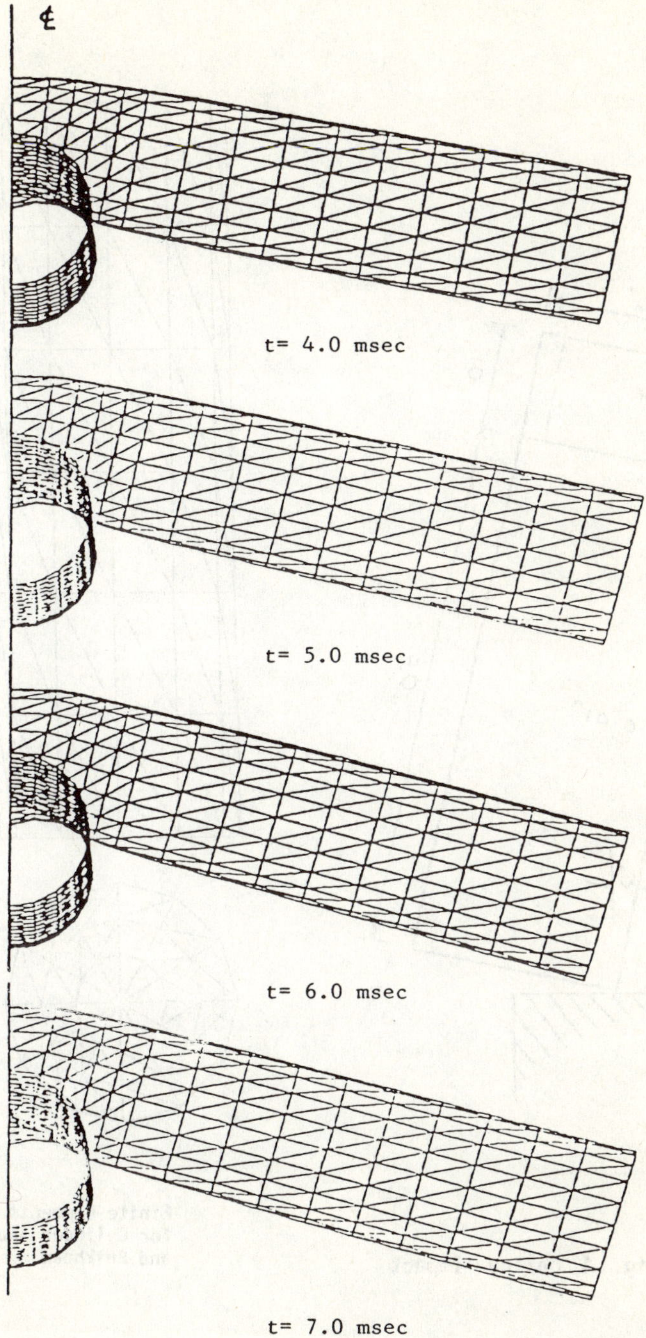

Fig. 3b. Impact between Two Pipes (contd.)

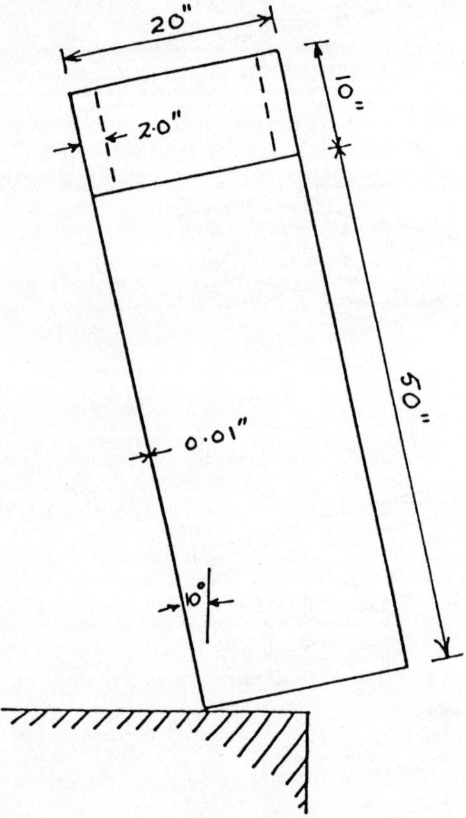

Fig. 4 Oblique Impact

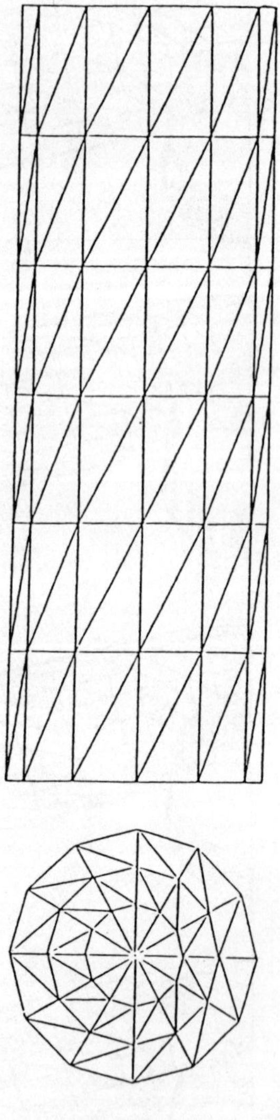

Finite Elements
for Cylinder and
and Bulkhead

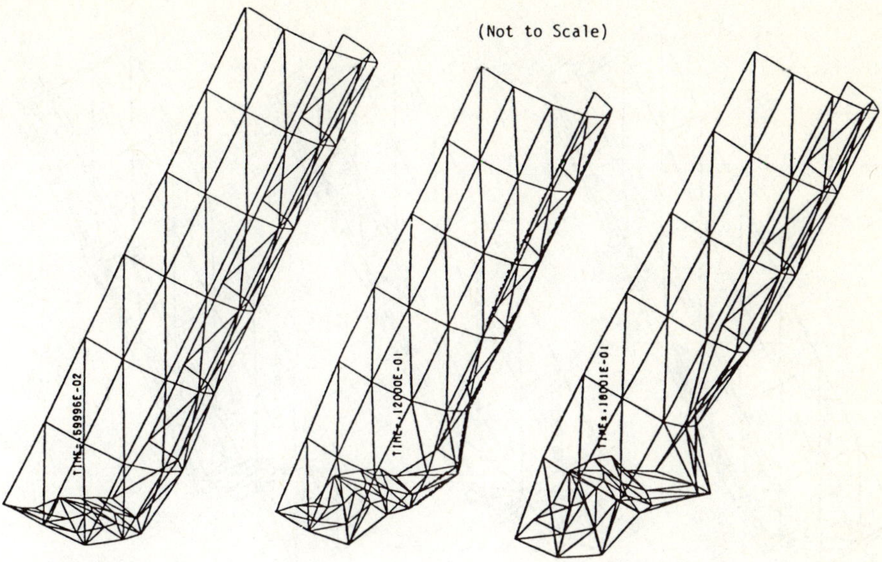

Fig. 5. Deformation of Cylinder

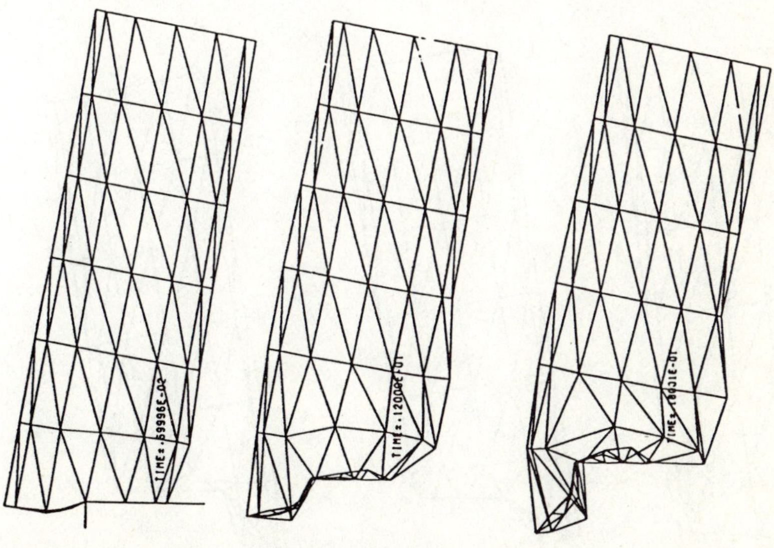

No Reinforcement.

t = 6 m.sec.

No Reinforcement.

t = 12 m.sec.

t = 18 m.sec.

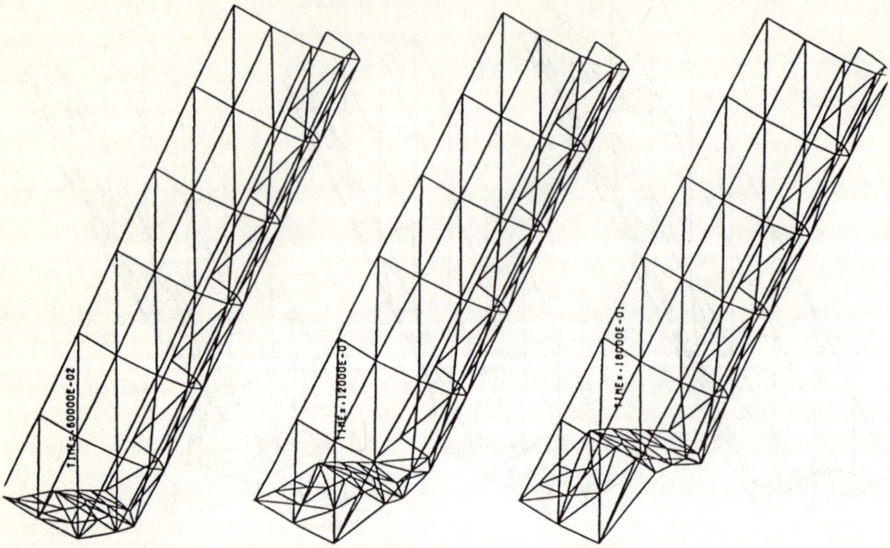

t = 6 m.sec. t = 12 m.sec. t = 18 m.sec.

Fig. 6. Deformation of Reinforced Cylinder

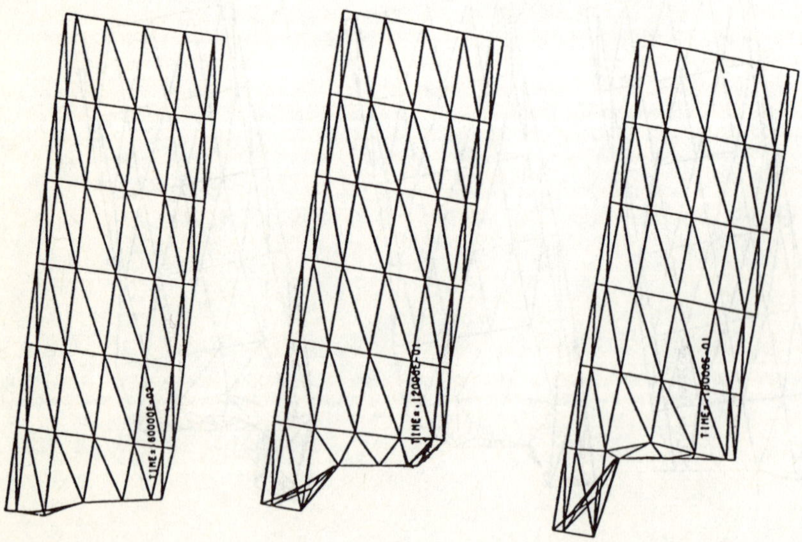

t = 6 m.sec. t = 12 m.sec. t = 18 m.sec.

Dynamic Response and Failure of Single and Sandwich Cylindrical Shells under Lateral Blast Loading

F. Salvatorelli-D'Angelo

University of Oxford Department of Engineering Science, Oxford, OX1 3PJ, England, UK

Abstract

A series of experiments were performed in order to identify the failure modes of thin single and sandwich shells under lateral blast loading. High speed photography has been used to view the events. The failure modes consisted of dynamic pulse buckling followed by large inward deflection of the front face for the single shell. An outward motion due to the increase of the internal pressure was observed. For the double shells the situation is more complex due to the effect of the sandwiched material on the response.

1 Introduction

The problem of dynamic buckling of thin cylindrical shells arises in situations such as cylindrical tanks, aircraft or submarines subjected to the lateral blast loading caused by an external explosion or, when a tank is entirely within a hydrocarbon gas cloud, to a rapid increase in the external pressure due to a deflagration. The response of an *LNG* tank to blast loading caused by an unconfined vapour explosion, is of great interest in Britain, where many such tanks are close to large population centres.

The initial response of the shell is elastic, followed by plastic deformation that may eventually lead to its rupture depending on the duration of the pressure pulse and its amplitude. Many authors have studied this problem, their treatments differing in scope and range of applicability depending on the simplifying assumptions that were chosen.

Anderson and Lindberg [1] suggest that on the basis of observed behaviour on axisymmetric blast loading, the results can be applied to cases of non-uniform sideways pressure provided that the pressure does not vary greatly over one buckled wavelength. This is, in fact, contrary to the experimental results of the present investigation. The buckling was defined in terms of a large amplification of assumed initial imperfections. Short duration, high amplitude pulses tended to cause plastic deformation whereas long duration low pressure buckling was elastic. In the case of very thin shells, with $\frac{a}{h} \approx 2000$, the elastic branch dominates and pre-buckling plasticity does not occur except for very short duration high pressure pulses.

In this paper, experimental results are given for the dynamic buckling of thin single and sandwich cylindrical shells under lateral blast loading. The loading arrangement and the shell modelling is described. The effect of the insulating material in the dynamic response and failure is explained and compared with the results for a single shell under the same loading conditions.

2 Blast wave loading

2.1 General aspects

As soon as the blast wave reaches the cylinder there is a reflection, which increases the effective load. This is followed by the diffraction of the blast along the circumference of the cylinder, during which a complex pattern of shock reflections has been observed [2]. It is generally accepted that this diffraction phase lasts about three times the time taken for the shock to travel across the shell diameter [3]. A theoretical determination of the pressure variation with time and position in the general case is very difficult but the analysis is relatively simple when the shock can be considered to be travelling on idealised acoustic media [4]. The most common approach is to model the load distribution as a time-modulated cosine distribution($n \geq 1$), thus assuming an instantaneous engulfment of the shell by the blast load:

$$p(\theta, t) = \begin{cases} p_0 \cos^n \theta f(t), & \text{for } -\frac{\pi}{2} < \theta < \frac{\pi}{2}; \\ 0, & \text{for } -\frac{\pi}{2} > \theta > \frac{\pi}{2}. \end{cases} \tag{1}$$

From a series of tests, in which models of cylindrical tanks were exposed to the blast wave of an open shock wave compressed air gun it was found that this approximation was inaccurate. An approach was suggested in which a cosine distribution was applied step by step over a gradually increasing arc as the shock travelled along the circumference [5].

At any given time, the portion of the shell bearing the load, will tend to move inwards and in doing so will meet with the resistance of the rest of the shell. A simple model of what happens is provided by a single degree-of-freedom system [6].

2.2 Description of the apparatus

The "explosion" is modelled by the sudden release of compressed air from a pressure vessel, sealed by a bursting diaphragm. The energy density is clearly considerably less than in a TNT explosion and the release time is longer than the corresponding reaction time.

The pressure pulses originated after the unconfined explosion of an LNG vapour cloud were modelled for this study with the device. Reflected pressure on a rigid cylindrical model was measured as a function of the angular position and height on the model. Measurements were made over a range of static pressure in the chamber between 689 $mbar$ (10 psi) to 2758 $mbar$ (40 psi). Additional pressure measurements were taken to obtain the incident pressure against the model. Details of the pressure instrumentation system and the methods for supporting the instrumented rigid cylinder will be described in the following sections. Similar fixtures were to be used again in the testing of the flexible models.

The apparatus consists of an aluminium pressure chamber sealed by a 210 mm diameter Melinex diaphragm with a thickness of up to 100 μm, clamped between a pair of flanges with "O" rings, see Figure 1. The chamber is pressurised to any required pressure within the range 0.639 bar to 2.758 bar and the shock wave is generated by bursting the diaphragm. This is achieved by releasing a spring-loaded knife with a cross-shaped end, see Figure 2.

The diaphragm opens up along two mutually perpendicular diameters releasing an air jet over 100 μsec to 200 μsec. This is analogous to the operation of a closed shock tube in which gas in a high pressure section(Driver) of the tube is suddenly released into a lower pressure section(Expansion) of the tube driving a shock wave in the front.

A series of tests were conducted to investigate the characteristics of the shock wave. Piezoelectric pressure transducers were fixed at the positions shown in Figure 3. The signal from the pressure transducers was stored in transient recorders, triggered by an accelerometer fixed to the pressure chamber flange. The purpose of these tests was to find the relationship between the pressure in the chamber and the peak value of the incident pressure and the velocity of the shock wave. The results are summarised in Figure 4. A diaphragm thickness of 25 μm was used for up to 0.9 bar, 50 μm for the range $0.9 - 1.2\,bar$, 75 μm for $1.2 - 2.7\,bar$ and 100 μm above 2.7 bar.

The relationship between $\overline{P_0}$ pressure in the chamber and P_i, the peak incident pressure is,

$$P_i = 0.0866\overline{P_0}. \tag{2}$$

at 210mm from the diaphragm and,

$$P_i = 0.1078\overline{P_0} - 46. \tag{3}$$

at 280 mm(All pressures in mbars). The shock wave velocity was estimated from the response of the lower transducers and of the top single transducer and found to be $340 \pm 20\,m/sec$.

After this a number of tests were done to determine the variation with time of the pressure distribution along the generator oriented between $\theta = 0°$ and 180° with respect to the centre line of the shock tube, at 22.5° intervals. Typical plots of pressure-time are shown in Figure 5.

The peak diffracted pressure when the shock touches the cylinder ($\theta = 0°$) is the same as the reflected pressure from a flat wall. It drops off as θ increases until it becomes approximately equal to the incident pressure at $\theta = 90°$. Over the backside of the cylinder it remains below the incident pressure. Oscillations caused by mechanical ringing are detected in the proximity of the pressure chamber. The distribution of pressure around the rigid cylinder has been approximated by the following empirical expression:

$$P(\theta) = \begin{cases} (P_r - P_i)\cos^2\theta + P_i, & \text{if } -\frac{\pi}{2} < \theta < +\frac{\pi}{2}; \\ P_i, & \text{if } +\frac{\pi}{2} < \theta < +\frac{3\pi}{2}. \end{cases} \tag{4}$$

where P_i and P_r are the peak incident and reflected pressures at a given position [1]. In these expressions, it is assumed that the peak pressure is reached simultaneously at all points of the structure. The tests show that the front face $(-\pi/2 < \theta < \pi/2)$ sees the shock front almost instantaneously, with a delay of less than 0.12 $msec$. The back face, which is subjected to a lower pressure, reaches peak values below the incident pressure and between 0.12 $msec$ and 0.5 $msec$ after the first peak. A time delay of 0.5 $msec$ has to be seen in the context of the actual response of the structure itself. For example, a thin walled cylindrical shell model with a wall thickness of 0.05 mm has a first mode vibration period of around 2.5 $msec$ and it could be accepted that the pressure rise is instantaneous. On the other hand, as the rigidity of the model increases(By increasing the wall thickness for example) the vibration period could become equal to or smaller than the time delay so that equations 4 would not be strictly applicable. In spite of this obvious limitation it is customary to express the distribution(Pressure, time) in the form,

$$P(\theta, t) = P(\theta)P_0(t). \tag{5}$$

where $P(\theta)$ is given by equation 4 and $P_0(t)$ is the measured variation of pressure with time at $\theta = 0°$(Including reflection). The state of stress of the structure may be estimated fairly simply by assuming that the pressure distribution is as described by these equations. The exact treatment of a plane pressure pulse moving through an acoustic fluid, striking a cylindrical shell and subsequently engulfing it, is however possible [4].

The pressure distribution also varies along the axial direction, from the base of the cylinder to the roof. In particular there is a large difference between an open cylinder, where a region of negative pressure is generated and cylinders with a domed roof, see Figure 5. The analysis for the cylinder may be based on the pressures measured in the mid section.

It has been reported [3] that both the incident and the reflected pressure decay exponentially with time, thus it may be assumed that:

$$P_0(t) = P_r e^{-\left(\frac{t}{t_0}\right)}. \tag{6}$$

this relation does not include the transit delay of the blast wave while travelling along the obstacle. The variation of the pressure with time is not, in fact, the same on the back surface, where it rises relatively slowly, as on the front where the rise is very sharp. More detailed relations are explained by Baker [7].

Ross et al. [8] studied the effect of the blast waves on cylinders with both ends clamped and concluded that the pressure distribution could be modelled by a modified form of the preceding equations:

$$P(\theta) = \begin{cases} P_r \left[\left(1 - \frac{P_i}{P_r}\right) \cos^{1.8}\theta + \frac{P_i}{P_r} \right] \left(1 - \frac{t-t_0}{\tau}\right) e^{-\alpha\left(\frac{t-t_0}{\tau}\right)}, & \text{if } -\frac{\pi}{2} < \theta < -\frac{\pi}{2}; \\ P_i \left(1 - \frac{t-t_0}{\tau}\right) e^{-\alpha\left(\frac{t-t_0}{\tau}\right)}, & \text{if } +\frac{\pi}{2} < \theta < +\frac{3\pi}{2}. \end{cases} \tag{7}$$

where t_0 is the engulfment time based on the shock wave velocity, α is the decay rate of the plane wave and τ is the duration of the positive pressure phase of the plane wave. This corresponds to the Friedlander decay for the positive phase. A comparison between the results obtained experimentally and those of Anderson and Lindberg [1] (equations 4, 5 and 6) and of Ross et al. [8] (equation 7) is shown in Figure 6. The agreement is substantially good up to $\theta = 90°$ although both sets of equations fail to identify the effect of a variation in the test chamber pressure. On the back surface, the experimental results show that the assumed distribution, i.e., $P = P_i$(Constant), is incorrect.

A computer code named *FLARE* [9], has been used to simulate the experimental set-up. The pressure distribution obtained from the numerical results is also plotted in Figure 6. The agreement between experimental and numerical results is good when the obstacle is at a distance from the bursting disk equal to two diameters. It is only fair when the obstacle is very near to the disk, 1.5 diameters. No detailed investigation of the reasons for this disparity has been undertaken since to do so would require instrumentation that is not available. The poor time response of the pressure transducers, the inadequacy of the $2D$ numerical model used to simulate a $3D$ wave whose shape and behaviour will be increasingly more complex as the bursting disk is approached, are likely contributing factors. A corollary is that experimental measurements should only be taken with obstacles sufficiently far from the bursting disk.

3 Experimental results

The model geometrical reduction was $1 \div 300$ resulting in a cylinder of 140 mm diameter and for convenience the height was the same as the diameter. This resulted in wall thicknesses of 0.07 mm to 0.1 mm. Electroformed copper shells sealed at the top with a domed roof and fixed at the bottom to a stiff ring were subjected to blast loading. The tank models were almost imperfection and residual stress free and can only be tested once. The models were attached to the experimental rig via a base plate made in aluminium that holds tightly the bottom ring and placed frontal to the generated pressure pulse. The domed roof was soldered to the top ring. The pressure transducers that record the incident pressure were positioned on the sides of the model in such a way as to record only the incident pressure.

The model and the relative position of the transducers with the open end of the tube are shown in Figure 7. One of the main interests of the test was to observe the displacement of the shell wall during the response. In order to increase the contrast and appreciate the deformation a grating with a 5 mm pitch was painted over the shell surface.

In this case painting a sufficiently fine grating on the surface was the main difficulty. A grating of the order of 1 line per mm as a minimum is needed. It is almost impossible to paint this grid by hand and the risk of damaging the model during it might increase. It is also required to use a very intense flash in order to provide enough light to obtain reasonable quality pictures.

Model	*Magnitude*	*Dimension*
Inner shell	Diameter	126.0 mm
	Thickness	0.75 mm
	Ring Diameter	123.0 mm
	Ring Dimensions(square)	3.0 mm
Outer shell	Diameter	140.0 mm
	Thickness	0.75 mm
	Ring Diameter	137.0 mm
	Ring Dimensions(square)	3.0 mm
Double-shell without insulation	Diameter	126.0 mm
	Separation between shells	7.0 mm
Double-shell with insulation	Separation between shells	7.0 mm
	Mass of the insulation	25.0 gr

Table 1: Model geometrical properties

The fast cine camera located approximately $3\,m$ away from the model. The angle between the centre line of the field of view of the camera and the axis of the shock tube was about 20 degrees and the filming speed was set at 10000 $frames/sec$. A timing device consisting of a micro light was connected to the camera and set on during testing to check the speed at which the event was filmed and to time the frame sequence. In some cases due to early trigger the event was recorded at a lower speed. The filming background was blue and the film selected after several trials with different trademarks and sensitivities was Ilford $FP4$.

A programme of three sets of main tests each one consisting of four types of models was chosen so that the dynamic effects and insulation contribution on the response and strength of the shells could be introduced in progressive states. The three sets of tests consist of a weak, intermediate and a strong blast. This is in order to test in the range from quasistatic regime to the highly impulsive load, progressing upwards in gradual steps until a definite dynamic buckling condition was detected. Usually dynamic buckling was accompanied by noticeable permanent plastic set, as seen in the test state of the models after the test. The models for each one of the main test can be subdivided in:

- Single wall test (inner wall).
- Single wall test (outer wall).
- Double wall test (without insulation).
- Double wall test (with insulation).

The models were assembled in anticipation of each series of tests. An evenly distributed amount of perlite was placed to ensure that the density of the material matched the density of the real scale model. The geometrical characteristics are resumed in Table 1. The peak reflected pressures between rigid and flexible models at very early times are very similar but subsequent deformation will presumably affect to some extent these values. This effect is neglected here, as no experimental technique for obtaining such data directly exists and it is assumed that the same pressure-time history and decay times observed for the rigid replica can be applied to the deformable shells.

The shells in all the experiments are empty, which is the most dangerous case for lateral load. Considering them full of liquid involves the study of the fluid/structure interaction, which complicates the analysis considerably [10]. Another factor to consider is the possibility that the trapped air will escape freely once the process of deformation is started. Presumably this will reduce the deformation of the shell wall.

The model does not display complete dynamic similitude but the results from previous model studies indicate that this small copper model can be useful in studying qualitatively the effect of static and dynamic loads [11]. The objective of each series of tests

Test number	*Chamber pressure (psi)*	*Shell type*	*Incident pressure mbar*	*Applied impulse (mbar msec)*	*Number of lobes n*	*Buckled circum. %*
9	40	Inner	-	-	8	33.0
10	40	Outer	251.3	119.3	8	42.0
11	40	Double, no ins.	274.6	120.5	8	45.0
12	40	Double, with ins.	268.6	108.2	8	33.

Table 2: Summary of tests results.

within a shell group was to determine the maximum blast load that could produce no damage, and the minimum blast load that could produce significant damage.

This significant damage was arbitrarily selected to be a maximum permanent inward deflection equal to a percentage of the shell radius of the shell or the shell separation in the case of the double shells. This central deformation can be related to the possibility of fracture of the internal wall here defined as a failure of the structure. Damage to the models is given by a qualitative description plus the maximum radial deformation of the shell in percentage of the radius [5].

4 Discussion of the experimental results

On the completion of the test series, it was found that the blast pressure pulse had an approximate rise time of 50 μsec and an approximate linear decay of 1.0 $msec$ to 2.0 $msec$ and the pulse duration varied between 1.0 $msec$ and 2.5 $msec$. These values depend on the initial pressure in the driver section of the open shock tube. At a chamber pressure of 20 psi no permanent deformation was found. However for pressures in excess of 20 psi large deformations, of the order of several times the wall thickness, were detected.

The transducer signals were recorded for the majority of the tests. However on a few occasions due to problems with the trigger of the transient recorder the pressure-time history could not be stored. The pulse profile from 1.0 $msec$ to 10.0 $msec$ shows a typical blast wave profile consisting of a positive phase of approximately 2.0 $msec$ and a negative phase of about 5.0 $msec$. Many further reflected shocks were subsequently received. For a pressure of 40 psi the experimental results are shown in Figure 8 for the outer shell, Figure 9 for the double shell without insulation and Figure 10 for the double shell with insulation.

A summary of the test results appears in Table 2. The circumferential mode number, n, represents the number of full buckled waves around the shell circumference. This number, however, is not truly representative of the deformation process, in that only a portion of the shell buckles circumferentially. The percentage of the shell deformed was relatively constant. About 30 % to 40 % of the cylinder circumference was plastically deformed.

The deformation at the front face of the single shells clearly indicates a sudden rise in displacement that corresponds to a snapping phenomenon over the loading range considered. The dynamic buckling pattern has resemblance with the static buckling situation only at very early times. This corresponds to the diffraction phase of the shock and a considerable number of lobes on the shell wall could be seen. At first, the film shows a symmetric pattern of deformation in all four cases. Later, it is clearly shown that a snapping phenomenon arises. Here, the shell reaches a very high strain that is compressive in the circumferential direction and tensile in the axial direction. It remains deformed, retained by the drag force for approximately 2.0 $msec$, before springing back after several post-buckling oscillations. The movement completely stops due to the combined effect of damping and increased stiffness as the shell wall hardens. The event, as it can be observed, is highly nonlinear.

A similar behaviour was observed for the double shells. Without insulation the initial response of the outer shell is in essence exactly the same as that for the single shell, but the large deformations will eventually cause contact with the inner shell that is still at rest. This contact completely changes the deformation pattern. What is now observed is a complex response where waving of the outer shell is the predominant feature.

With insulation the deformation observed is significantly less due to the increased stiffness now that the shells are coupled. However, an interesting phenomenon was detected. In all cases, the inner shell suffered comparable damage to the outer shell and this damage was considerable higher than that observed without insulation. This must be attributed to the momentum gained by part of the insulating material when suddenly loaded by the pressure wave. In all the experiments, buckling was followed by plastic deformation that remained as a plastic set after the action of the blast.

The main interest in this set of experiments is to obtain some qualitative information to validate a numerical technique. This in turn will be used to predict the response of structures of much more complex geometry. The differences due to dynamic response scaling between model and prototype are of little significance at this stage.

The experimental results show in all cases that there is very little effect of the blast on the roof. Additionally, it has been observed that the very stiff top ring does not allow any coupling between cylindrical wall and hemispherical cap. This could simplify the numerical analysis because each structure could be analyzed independently.

The tank failure always started at the leading generator of the cylinder facing the blast load. The formed hinges spread circumferentially in two symmetric directions with movement also in the axial direction. The extent of this failure region spreads about 50 degrees each side of the leading edge. The maximum deflection occurred along the mid-length.

Often at low pressure in the driver chamber, circumferential dynamic buckling did not occur, oscillations were only observed with large initial amplitudes and there was no subsequent damage to the shell wall. This suggests that for a given type and size of shell there exists a critical load that determines if vibration, buckling or a fundamental collapse pattern is to be formed. Experimental determination of dynamic critical loads would require extensive testing.

The failure as defined in this study, was considered to correspond with material rupture. Fracture on the low pressure tests was not observed, however in the high pressure tests excessive wrinkling of the shell wall indicates that fracture might have ocurred. It is believed that failure could be related to the centerpoint deflection, and for a given shell there exists a critical centerpoint deflection for failure.

5 Conclusions

The open shock tube described here offers a simple means of producing a blast loading on a structural model. It is possible to control the characteristics of the pressure profile by varying the chamber pressure, the chamber volume and the thickness and material properties of the membrane. It has also been shown that the variation of pressure with time and position obtained in the experiment can be modelled quite accurately by numerical analysis. This offers the opportunity of interpreting experimental results on the one hand and reducing the number of physical experiments replacing them by numerical simulations on the other.

A direct scaling from a model to the full size structure presents serious practical difficulties and is best done by developing a mathematical model which is first checked against the experiment and then used to analyse the full size structure. The experiment is then required to reproduce only the most significant features of the real event. In this case these are the ratio between peak pressure and dynamic buckling pressure of the structure, the pressure distribution and the ratio between the pressure pulse duration and the fundamental vibration period of the structure. Provided that the values of these parameters are similar in the model and in the full-size structure, the experiment will be representative of the real event and may be used to validate the mathematical model. The ability to vary the pressure profile with the experimental technique that has been described is therefore of great advantage over the more difficult methods involving high explosives or hydrocarbons in small scale or large scale tests. It is, of course, clear that this technique is not meant to model the explosion itself, but only its effects on a structure.

Experimental results show that, the buckling of a shell can occur at a certain load amplitude or a specific impulse. Collapse of the shell can be induced by either applying a large load with short duration of a smaller load with longer duration resulting in the same impulse. Such results suggest that for a given tank, some critical load, and impulse produces a transition between the buckled and the collapsed patterns.

Based on the tests performed only 30 % to 45 % of the front face of the shell is deformed. On the other hand failure might ocur at the ends of the shell due to excessive tensile stress generated during the inward motion. It is important to notice that the insulation in some cases can have a detrimental effect on the safety of the internal shell. ■

References

[1] D. L. Anderson and H. E. Lindberg. Dynamic pulse buckling of cylindrical shells under transient lateral pressures. *AIAA Journal*, 6(4):589–598, April 1968.

[2] V. J. Bishop and R. D. Rowe. *The interaction of a long Friedlander shaped blast wave with an infinitely long circular cylinder.* AWRE Report 0-38/67, Atomic Weapons Research Establishment, Aldermaston, Berkshire, England, 1967. Incident blast wave 20.7 psi. Positive duration 50 μsec and a 16 cm diameter cylinder.

[3] S. Glasstone and P. Dolan. *The effects of nuclear weapons.* U. S. Department of Defense and U. S. Department of Energy, third edition, 1977.

[4] R. G. Payton. Transient interaction of an acoustic wave with a circular cylindrical elastic shell. *Journal of the Acoustical Society of America*, 18(6):722–729, June 1960.

[5] F. Salvatorelli-D'Angelo. *Structural stability under dynamic loading of LNG tanks.* PhD thesis, Engineering Science, Oxford University, Exeter College, Trinity Term 1988.

[6] C. Ruiz, F. Salvatorelli-D'Angelo, and V. K. Thompson. Elastic response of thin-wall cylindrical vessels to blast loading. *Computers and Structures*, 1988. To be published.

[7] W. E. Baker. *Explosions in air.* University of Texas Press, Austin, second edition, 1973.

[8] C. A. Ross, W. S. Strickland, and R. L. Sierakowski. Response and failure of simple structural elements subjected to blast loadings. *The Shock and Vibration Digest*, 9(12):15–26, December 1977.

[9] S. F. Hall, D. Martin, and J. Mackenzie. *Gas cloud explosions and their effects on nuclear power plants, FLARE User Manual.* EUR, first edition, 1984. EUR 8955 EN.

[10] W. K. Liu and D. C. Ma. Coupling effect between liquid sloshing and flexible fluid-filled systems. *Nuclear Engineering and Design*, 72(3):345–357, 1982.

[11] V. K. Thompson. *Structural integrity of liquid natural gas storage tanks.* PhD thesis, Engineering Science, Oxford University, Balliol College, Trinity Term 1985.

Figure 1: General arrangement of test rig.

Figure 2: Detail of the open shock tube.

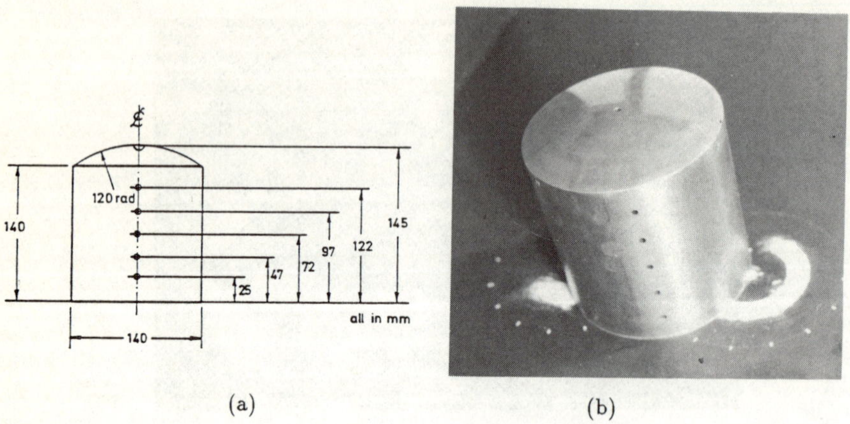

Figure 3: Cylindrical model, general dimensions.

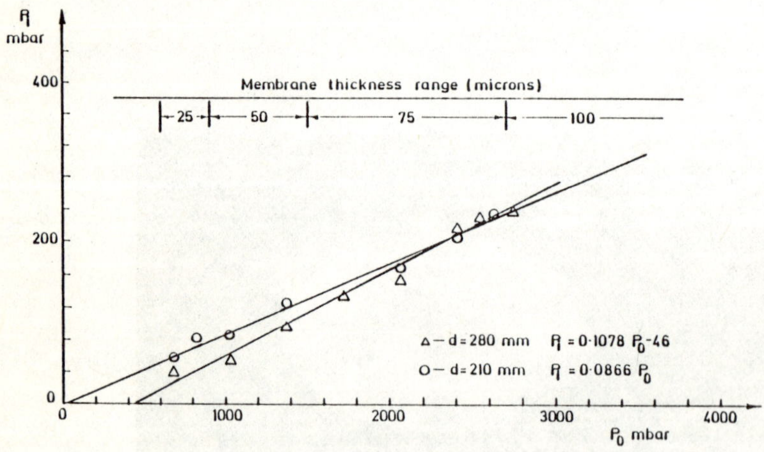

Figure 4: Calibration curves for the open shock tube.

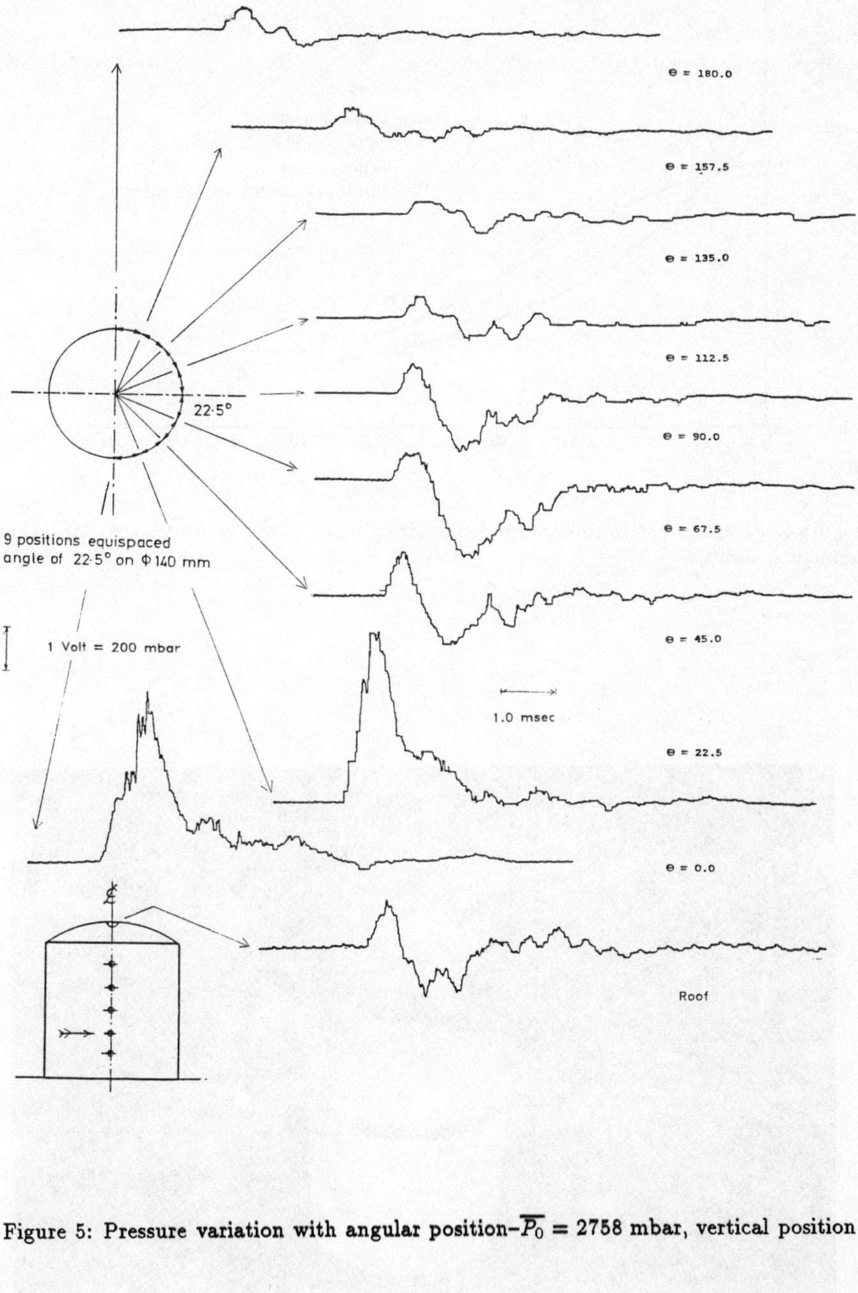

Figure 5: Pressure variation with angular position-$\overline{P_0}$ = 2758 mbar, vertical position 2.

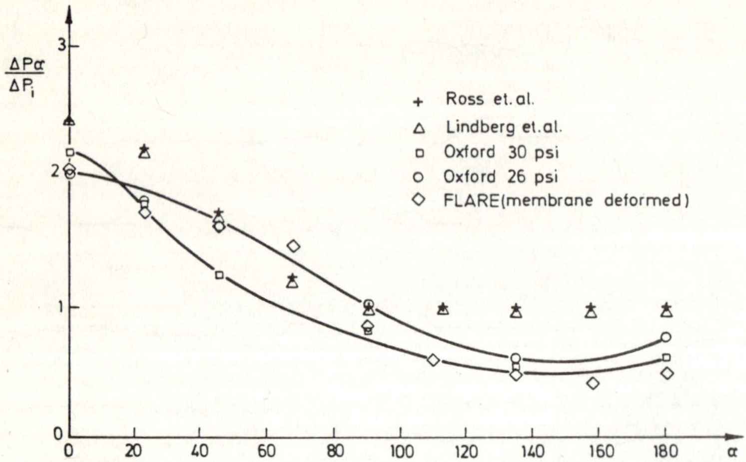

Figure 6: Pressure distribution around circular cylinder ($d = 280\,mm$), experimental and numerical results.

Figure 7: Model and pressure instrumentation on the rig.

Figure 8: Deformed shape for outer shell, $\overline{P_0} = 40\,psi$.

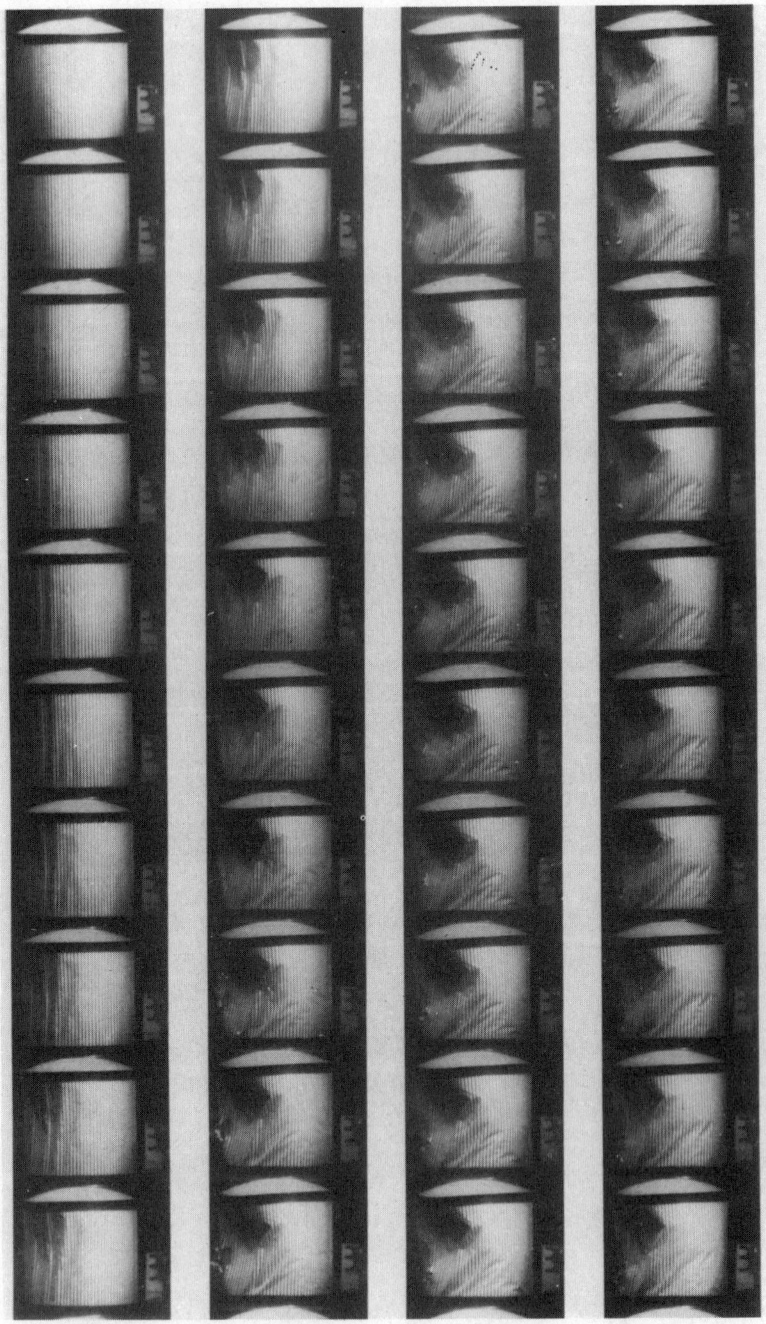

Figure 9: Deformed shape for double shell without insulation, $\overline{P_0} = 40\,psi$.

Figure 10: Deformed shape for double shell with insulation, $\overline{P_0} = 40\,psi$.

SECTION 6 - BUILDINGS AND DEMOLITION

Structural Assessment of Blast Damaged Buildings

M.S. Williams*, T.S. Lok** and G.C. Mays**

* *W.S. Atkins Engineering Sciences, Epsom, Surrey, UK*

** *Civil Engineering Group, Royal Military College of Science, Shrivenham, UK*

ABSTRACT

Methods for assessing the integrity of blast-damaged buildings are presented and discussed. A strong emphasis is placed on the use of empirical expressions and methods which allow a rapid evaluation of structural elements to be carried out by non-specialist engineering personnel. The paper is primarily concerned with assessment techniques for steel and reinforced concrete, as these materials are widely used in the majority of modern construction. However, brief guidelines on the assessment of timber and masonry structures are also presented.

INTRODUCTION

In most instances of explosive damage to buildings, one of the most immediate requirements is for a rapid assessment of the integrity of the damaged structure, in order to establish whether measures such as evacuation, temporary propping or demolition are necessary. The aftermath of an explosion is frequently a chaotic event, and it is unlikely that there will be an opportunity for a detailed, accurate assessment of building integrity. It is therefore advantageous to have available simple assessment procedures, so that rapid numerical checks can be carried out.

Prior to carrying out a numerical assessment of the stability of the damaged structure, it is necessary to physically assess the degree of damage and its effect on structural behaviour. Damage may be due to the blast itself, or the fire which frequently accompanies an explosion. Excellent guidance on visual assessment is given by Rhodes [1], while advice on non-destructive testing techniques can be found in [2]. Having gained an estimate of the effectiveness of damaged joints, the degree of structural continuity and any possible reduction in material strength, it is possible to proceed to the numerical assessment phase.

The requirements for integrity of a damaged building are broadly in line with those for the original structure. However, since the existence of the structure in its damaged state is necessarily a short-term event, it is reasonable to adopt much lower factors of safety than would be used in design. Additionally, it is unlikely that design live and wind loads will be achieved in the short duration being considered; reduced values may therefore be assumed.

REINFORCED CONCRETE STRUCTURES

Nature of damage

Reinforced concrete structures are likely to display substantial resistance to blast loading. The material has high strength, ductility and the ability to rapidly redistribute loads. Additionally, its massive nature makes buckling and local instability problems unlikely. For explosions other than those occurring in contact with the structure, it is probable that the damage will be concentrated at the joints with adjacent elements. The level of damage due to an explosion is therefore likely to be dependent on the quality of reinforcement detailing at a joint. In this respect, structures assembled from precast elements are likely to be particularly vulnerable.

If the explosive event leads to a fire in the building, this can also cause damage to concrete elements. The high temperatures can cause reductions in material strength for both the steel and concrete, though there will be some recovery on cooling.

Numerical assessment of blast damage

Since we are primarily concerned with the likelihood of collapse, the most appropriate assessment technique is yield line analysis. The method, developed by Johansen [3], is well-known to practising civil engineers. Basically, the engineer postulates a collapse mechanism, based on rotation at a number of hinges and yield lines, at which the plastic moment of the member has been reached. By equating the work done by the external loads to the internal energy dissipated in the yield lines, it is possible to calculate the load required to cause collapse.

In the absence of safety factors, the plastic moment of a singly reinforced concrete element may be taken as

$$M_p = f_y A_s d (1 - 1.1 f_y A_s/f_{cu}bd) \tag{1}$$

where A_s = area of tension reinforcement,
b = breadth of section,
d = depth to tension reinforcement,
f_y = yield strength of tension reinforcement,
f_{cu} = characteristic concrete cube strength.

In assessing a damaged structure, one starts by calculating the plastic moment of the damaged element. For an in-situ assessment it may be difficult to ascertain the amount of reinforcement present in a member. However, it should be possible to assess bar spacings using a covermeter. It is then necessary to make a conservative assumption of bar diameter and so deduce a reasonable estimate of the reinforcement area. Numerous techniques are available for assessing concrete strength [2]. Having assessed the plastic moment, the energy dissipated in the yield lines or hinges can be expressed in terms of an assumed hinge rotation θ. In doing this, due attention should be paid to the possible effect of any damage on the likely yield line pattern, and on the magnitude of the plastic moment. For example, it may be reasonable to assume that a damaged connection is free to rotate. The plastic moment at this point should therefore be taken as zero, thus reducing the overall capacity of the member.

It is then necessary to calculate the work done by the applied loads. The work done can be expressed in terms of an unknown loading, w, and the hinge rotation, θ. This expression can be equated to the yield line energy, θ eliminated and hence the value of w required to cause failure can be found.

Finally it is necessary to assess the actual loads on the structure, for comparison with the calculated collapse load. For short-term stability of damaged structures, loads can be assumed to be as follows [4].

- full dead load;
- 1/3 of the design live load, except for cases where that load is largely permanent (eg. warehouses), where the full live load should be used;
- 1/3 of the design wind load.

The factor of safety against collapse is given by the ratio of the collapse load to the actual load. A value of greater than 1.2 would normally be sufficient for short-term safety.

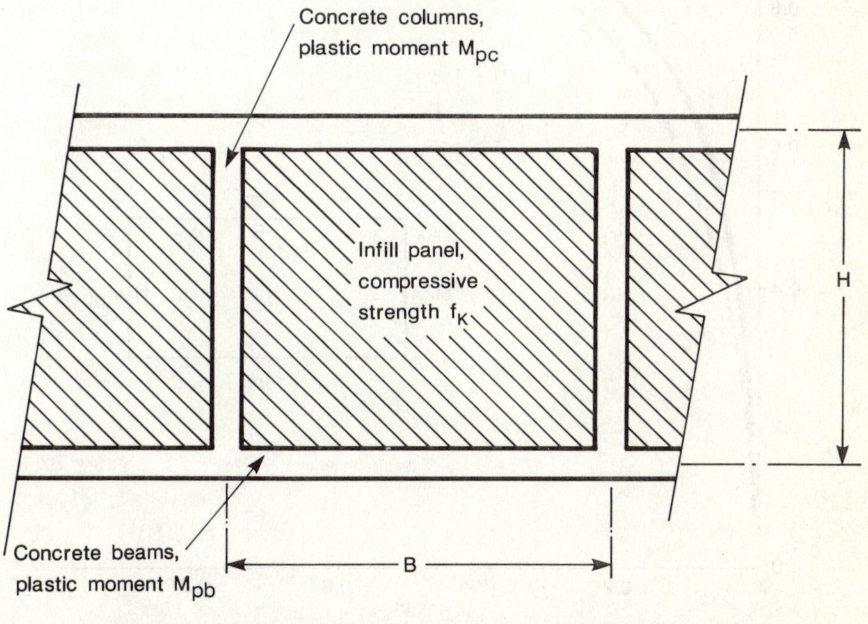

Figure 1 Typical reinforced concrete frame with infill panels.

In most instances, reinforced concrete frames are infilled with panels of lower strength materials such as masonry. If these are intact after the blast, then they can considerably enhance the strength of the frame, a phenomenon which it would be excessively conservative to ignore. The effect can be incorporated by calculating an enhanced plastic moment for the adjacent columns. A simplified method for doing this has been developed from the work of Wood [5].

Consider the simple beam–column frame shown in Figure 1. We can calculate the plastic moments for the beams (M_{pb}) and columns (M_{pc}) as before, assuming no infill is present. Let the smaller of these be M. Additionally, we must assess the compressive strength of the infill, f_k. For masonry infills, in the absence of more detailed guidance, it is suggested that new, good–quality brickwork is assumed to have a strength of 6.4 N/mm^2, while older brickwork, with deteriorating mortar or evidence of poor workmanship, is taken as having a strength of 5.2 N/mm^2. The strength ratio for the infilled frame of infill thickness t can then be calculated as

$$m = \frac{12M}{f_k B^2 t} \tag{2}$$

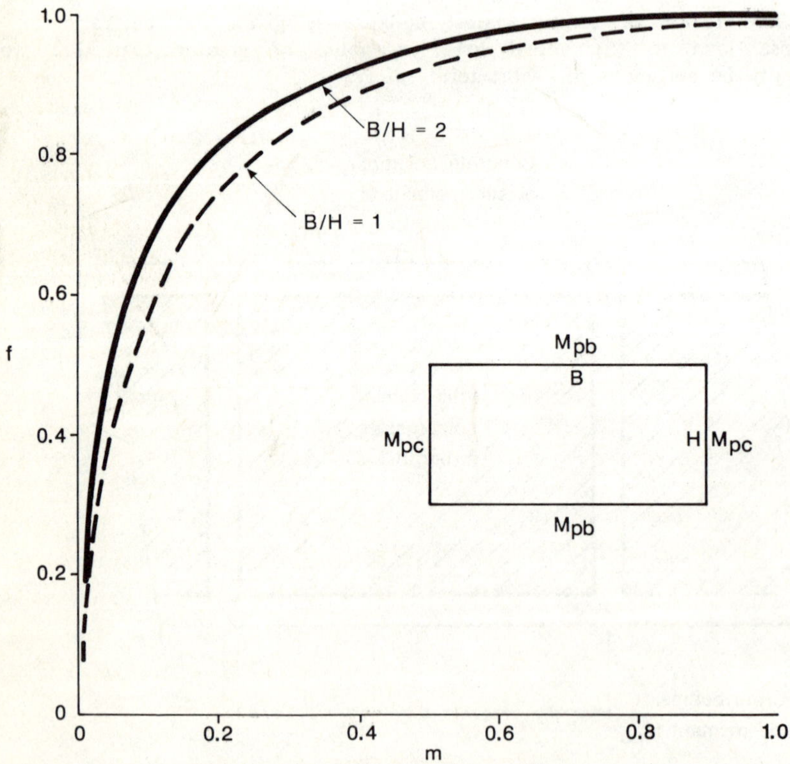

Figure 2 Loading coefficient (f) as a function of strength ratio (m) for infilled panels.

A load coefficient, f, can then be read from the appropriate graph in Figure 2, depending on the geometry of the panel. Finally, the enhancement factor for the column plastic moment may be calculated using the formula

$$k = f\ (B/H + 1/m) \tag{3}$$

STEEL STRUCTURES

Nature of damage

Structural steelwork's inherent tensile ductility enables the material to extend beyond the linear regime. For example, the post plastic tensile strain can reach 10 to 15 times of the yield strain without appreciable increase in stress. Beyond this range, strain hardening occurs and failure occurs soon after at the ultimate load. It is the ability of structural steel to undergo such substantial permanent deformation that makes it an appropriate blast-resistant material. However, some higher steel grades are less suitable because of low ductility and the possibility of premature fracture. The purpose here is to discuss the criteria and procedures for rapid assessment of the stability of elements of a damaged steel structure and overall frame stability. The discussion is therefore confined to statical considerations.

Assessment procedures

Damaged steel structures should be assessed according to the practice and standard of the appropriate country. In the UK and Europe, the modern standards are BS5950 [6] and the ECCS Code [7] respectively. A summary of various national steelwork design codes, steel specifications and dimension and properties of sections may be found in reference [8]. The philosophy of each nation's design Codes does not vary significantly from those contained in BS5950.

Assessment of beam elements To proceed with assessment, the fundamental properties of the cross-section have to be ascertained. These are the steel strength grade (from tests) and the cross-section dimensions. Both properties are necessary in controlling local buckling of outstands and web plates since sections are essentially made up of plates.

For a short simply-supported I-beam, with effectively restrained compression flanges, the following criteria must be satisfied:-

a. The maximum shear force must not exceed the shear capacity of the section (in the case of thin webs, the shear buckling resistance).

   It can be shown by elastic analysis of symmetrical I-sections that the maximum shear stress occurs at mid-depth of the web and is approximately given by

$$\tau_{max} = 1.6\ \tau_j \tag{4}$$

   where τ_j is the shear stress at the flange/web junction. τ_{max} is rarely calculated and in practice the average shear stress is used instead. Providing the web slenderness is low, the limiting shear capacity is reached when the web has attained a fully yielded condition.

The stress at which a material yields in shear is related to the tensile (design) strength. By Von Mises' distortional strain energy theorem, the shear strength

$$p_v = 0.6\ p_y \tag{5}$$

where p_y is the design strength. The shear capacity is

$$P_v = p_v \text{ x depth of web x web thickness} \tag{6}$$

The section is adequate in shear when the shear force

$$F_v < P_v \tag{7}$$

Shear buckling often governs the capacity of plated girders and guidance is given in BS5950:Pt. 1. At the present time, such criterion will not govern the design of any hot-rolled Universal Beam sections in grade 43 mild steel [9].

b. The maximum bending moment must not exceed the moment capacity of the section.

The maximum moment, M due to the applied loads is compared with the plastic moment of resistance, M_p of the section. The plastic moment is attained when the full depth of the cross section has yielded and for satisfactory assessment,

$$M < M_p \tag{8}$$

It is worth noting that although the plastic moment is being taken as the limiting capacity of a section, this does not imply that plastic analysis has to be used to determine the moments due to the applied loading. The moments can generally be determined by elastic analysis, with plastic analysis based on a mechanism of hinges being an alternative for continuous beams and rigid jointed frameworks.

Under the combined action of flexure and shear, part of the web's capacity is employed in resisting the shear force. The moment capacity is therefore less than the full plastic moment in the absence of shear.

The reduction in the total moment capacity due to shear is relatively small since the flanges are the main contributors to the flexural capacity of the cross section.

Similarly, there will be a reduction in the moment capacity in the presence of an axial force. However, axial forces in beam elements are small and this effect may be neglected for assessment purposes.

c. Bearing stress and other local stress concentrations at supports and load points must not be exceeded. This criteria rarely controls the load carrying capacity of a damaged structure and will not be discussed further.

d. Deflections should be limited so as not to impair the serviceability or strength of the structure. The tolerant level may be taken as between 0.01 to 0.003 times the member length for a damaged structure.

e. Local buckling prevents the cross section from developing its plastic moment of resistance and in some cases from attaining its yield moment. Buckling loads may be calculated by treating the compression flange outstand of the web as an individual plate. The elastic analysis of such plates may be found in textbooks on instability [10]. In practice, however, the limiting criterion for outstands and web plates are more severe because imperfections such as initial curvature and residual stresses are present in steel sections. Thick welded plate girders are affected rather more severely than hot-rolled sections.

The lack of effective compression flange restraint in a damaged beam reduces further the load carrying capacity of that element. Assessment of such an element is governed by lateral torsional stability.

The most appropriate treatment for lateral torsional stability is to consider a simply supported beam subjected to equal end moments with its ends prevented from twisting and displacing laterally. However, both ends are free to warp so that the flanges are permitted to rotate in a horizontal plane as shown in Figure 3.

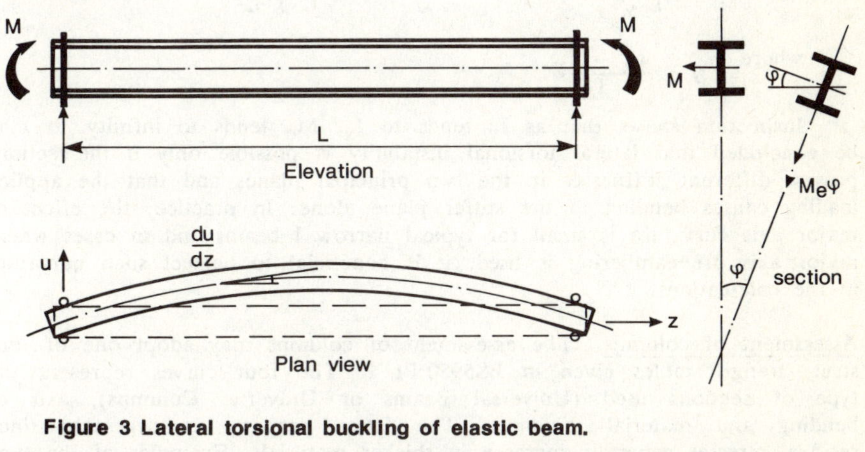

Figure 3 Lateral torsional buckling of elastic beam.

Bifurcation of stability occurs when the limiting moment (elastic critical moment) is reached. At this value, the beam buckles by twisting and deflecting laterally.

By considering the equilibrium of the buckled beam, the governing differential equations may be written as

$$EI_y \frac{d^2u}{dx^2} = -M_e \cdot \varphi \tag{9}$$

$$GJ \frac{d\varphi}{dz} - EI_w \frac{d^3\varphi}{dz^3} = M_e \frac{du}{dx} \tag{10}$$

where G is the modulus of rigidity, J is the torsional constant, E is Young's modulus of elasticity, I_w is the warping constant, I_y is the lower second moment of area and M_e is the elastic critical moment. φ and u are the angle of twist and lateral displacement respectively.

Equations (9) and (10) are solved simultaneously by assuming the following displacement functions

$$u = \delta \sin \frac{\pi z}{L} \tag{11}$$

$$\varphi = \alpha \sin \frac{\pi z}{L} \tag{12}$$

It can be shown that the elastic critical moment is given by

$$M_e = \frac{\pi}{L} \sqrt{E I_y G J} \cdot \sqrt{1 + \frac{\pi^2 E I_w}{G J L^2}} \tag{13}$$

It was assumed that deflection in the plane of the major axis had no effect on lateral torsional behaviour. If this effect had been considered, then Equation (13) is modified to

$$M_e = \frac{\pi}{L} \sqrt{\frac{E I_y G J}{\gamma}} \cdot \sqrt{1 + \frac{\pi^2 E I_w}{G J L^2}} \tag{14}$$

where $\gamma = \frac{I_x - I_y}{I_x}$

Inspection shows that as I_y tends to I_x, M_e tends to infinity. It may be concluded that lateral torsional instability is possible only if the sections possess different stiffnesses in the two principal planes and that the applied loading causes bending in the stiffer plane alone. In practice, the effect of major axis curvature is small for typical narrow I–beams and in cases where major axis pre–cambering is used, it is beneficial to neglect such curvature in the calculations.

Assessment of columns The assessment of columns may adopt one of four strut strength tables given in BS5950:Pt 1. The four curves represent the type of sections used (Universal Beams or Universal Columns), axis of bending and material thickness. The latter property is necessary since residual stresses generally increase in thicker materials. Examples of the type of strut strength table to adopt are shown in Table 1.

Section	Thickness	Axis of Bending	a
Rolled UB	All	x–x	2.0
Rolled UB	All	y–y	3.5
Rolled UC	< 40 mm	x–x	3.5
Rolled UC	< 40 mm	y–y	5.5
Rolled UC	> 40 mm	x–x	5.5
Rollec UC	> 40 mm	y–y	8.0

Table 1. Strut curves for different applications

The parameter 'a' is known as the Robertson (imperfection) constant and is varied to give the desired shape of the curve.

The other most apparent feature in column design is that squashing of the member is permitted provided the slenderness is low. The limiting slenderness depends only on the design strength of the section.

The limits on web slenderness for avoiding local buckling are more restrictive than for beam elements because uniform stress is experienced in compression members rather than the linear variation in beams. Guidance for such criteria is available in BS5950.

Multi-storey steel frameworks
Most multi-storey frames are braced by an internal core such as a lift shaft or stairwell, and occasionally by bracing struts or shear walls. However, modern structures tend to be lightly clad with open interior construction. In an event where effective restraints are removed, it is appropriate to assess the damaged structure as an unbraced framework. Although this form of assessment is conservative, the authors believed that such action is justified. In such cases, stability of the whole structure is the most important criterion. Storey stiffness calculations may be performed speedily using the design charts provided by Wood and Roberts [11]. To ensure adequate strength of an unbraced framework, the Merchant-Rankine formula may be used provided the requirements of reference [12] are met.

The calculation of the ultimate load carrying capacity of an unbraced framework using the Merchant-Rankine approach requires the prior determination of the lowest elastic critical load. If the minimum failure load is taken as unity under appropriately factored load conditions, then all that is required is proof that the plastic collapse load factor is satisfactory. A lower bound plastic analysis approach is most appropriate in this case.

BRICK AND MASONRY STRUCTURES

Unreinforced masonry buildings generally consist of substantial load-bearing walls, which support their own weight, together with transmitted dead and live loads from the floors and roof. It is likely that the high lateral pressures developed by blast loading will cause failure of the mortar joints between the bricks or blocks. It is thus quite possible for damage to be sustained locally within a single panel.

It is common for masonry elements to be breached by proximate explosions. Provided that sufficient structural weight is available, masonry has the capacity to transmit loads around these damaged areas by arch action [1]. Stability can be quickly assessed using simple statics. In order for the structure to remain stable, it is necessary for the resultant of the member dead weight and the arch thrust to reach the ground without passing outside the structure, as illustrated in Figure 4.

For explosions occurring at longer ranges, damage is more likely to be in the form of cracking, lean and bulge. Masonry has the capacity to sustain quite severe cracking without major stability problems. Typically,

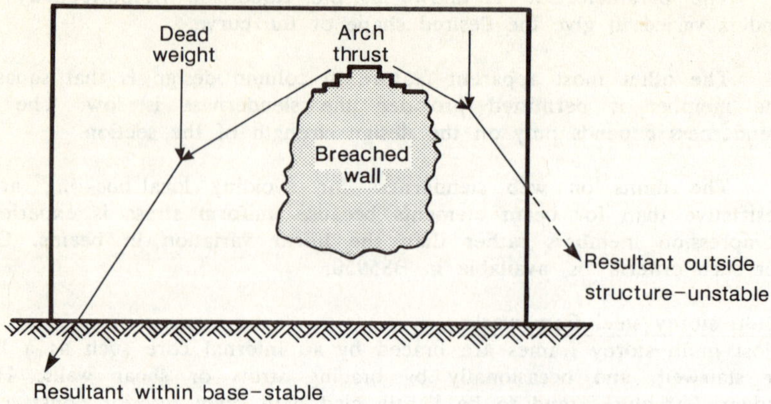

Figure 4 Arch action in blast-damaged masonry

cracks up to 15mm in width will cause severe serviceability problems and will require some rebuilding in the long term, but are unlikely to indicate imminent collapse [13].

Masonry can also withstand a certain amount of lean and bulge without collapse [14]. The amount is dependent on the quality of mortar between the individual blocks. Generally, older structures have mortars with very low cement content, and therefore low adhesion. In these cases, the maximum eccentricity of a free-standing wall should be limited to 0.85 times its width under self-weight loading only, and 0.5 times its width in the presence of additional vertical loads. For newer structures, where the mortar has better adhesion, the figure of 0.85 may be used for a wall with or without additional vertical loading.

TIMBER FRAMED STRUCTURES

Timber is used almost exclusively for small-scale housing, which tends to be of rather lightweight construction. Such timber structures therefore have a very limited resistance to blast loading.

Rapid assessment of timber structures is made difficult by the wide range of timber strengths used in construction. No simple procedure is available for assessing timber strength in situ. If a numerical assessment is to be carried out, it is therefore necessary to make assumptions which may prove excessively conservative.

CONCLUSIONS

Damage to buildings due to explosions can be caused by the initial blast loading or by a possible ensuing fire. The nature of the damage may vary with the structural form and with the construction material used. The engineer therefore needs to have a range of assessment tools at his disposal, many of which have been outlined in this paper.

Given the necessity of making certain simplifying assumptions, it is usually possible to carry out a reasonable and rapid assessment of a damaged structure. Established analysis techniques are often appropriate, though they may need some simplification and modification before they are suitable for short-term stability assessment.

REFERENCES

1. Rhodes, P.S. The Structural Assessment of Buildings Subjected to Bomb Damage, Structural Engineer, Vol. 52, pp. 329-339, 1974.

2. Institution of Structural Engineers. Appraisal of Existing Structures, I. Struct. E., London, 1980.

3. Johansen, K.W. Yield Line Theory, Cement and Concrete Association, London, 1962.

4. British Standards Institution. BS8110: The Structural Use of Concrete, Part 1: Code of Practice for Design and Construction, BSI, London, 1985.

5. Wood, R.H. Plasticity, Composite Action and Collapse Design of Unreinforced Shear Wall Panels in Frames. Proc. Instn. Civ. Engrs., Part 2, Vol. 65, pp. 381-411, 1978.

6. BS5950:Pt. 1 Structural Use of Steelwork in Buildings. Code of Practice for Design in Simple and Continuous Construction: Hot Rolled Sections. BSI London 1985.

7. European Convention for Constructional Steelwork: European Recommendation for Steel Construction, London, The Construction Press, 1981.

8. International Structural Steelwork Handbook. BCSA London, Chameleon Press Ltd, 1983.

9. Steelwork Design Guides to BS5950:Pt. 1:1985, Vol. I CONSTRADO, London 1985. Vol. II The Steel Construction Institute, London 1986.

10. Allen, H.G. and Bulson, P.S., Background to Buckling, McGraw-Hill Book Co (UK) Ltd. 1980.

11. Wood, R.H. and Roberts, E.H., A Graphical Method of Predicting Sidesway in the Design of Multistorey Buildings, Proc. Instn. Civ. Engrs., Part 2, Vol. 59, pp. 353-372, June 1974

12. ECCS-TC8, Structural Stability TWG 8.2 - Systems, Ultimate Limit State Calculation of Sway Frames with Rigid Joints, Publication 33, ECCS 1984.

13. Building Research Establishment. Assessment of Damage in Low-Rise Buildings, BRE Digest No. 251, London, 1981.

14. Construction Industry Research and Information Association. Structural Renovation of Traditional Buildings, CIRIA Report 111, London, 1986.

Demolition by Controlled Explosion as a Dynamical Process

E. Yarimer

Department of Civil and Municipal Engineering, University College London, Gower Street London WC1E 6BT, UK

ABSTRACT

The equation of motion for a building in free fall disintegrating by contact with the ground is set up by two different methods. Solution of the equation shows that there is a critical initial mass - initial velocity boundary separating the case of incomplete destruction from the case of complete disintegration.

INTRODUCTION

Starting from 1981 five residential tower blocks have been demolished in the London area by the controlled explosion method. In one of these five cases, the 21-storey building was incompletely destroyed, with the 10 upper storeys landing intact on top of the broken up remains of the first 11 floors. This building had to be finished by the ball and chain method.[1] The other four projects were entirely successful.

In a demolition by controlled explosion, the disintegration of the building is not due to the explosions themselves, but rather to the contact and collision with the ground at the lower end of the collapsing body. The explosions initiate and sustain the fall and also impose a desired pattern on the collapse mechanism. The breaking up of the components into rubble will require a large input of energy. The main source of this energy is the gravitational potential. However the presence of an adequate reservoir of energy is not enough for the process to be self-sustaining: as in the phenomemon of crack arrest during brittle fracture, the motion may slow down to a halt if the rate of release of energy into the fracture zone becomes insufficient.

In the current application of this demolition method, structures are pre-weakened to a degree where the safety factor against collapse by crushing under self-weight is barely above unity. This is commonly achieved by taking down some of the internal loadbearing walls. The external cladding is left intact in order to contain the flying fragments and to muffle the noise of the explosions. One of the requirements for a successful demolition is that the floor slabs should become fragmented into small pieces. Therefore the explosions are sequenced to occur from the central core outwards, thus causing a destructive shearing distortion of the storeys.[2]

DERIVATION OF THE GOVERNING EQUATION

Continuous model

The elasticity of the material is ignored. Therefore the enhancement of local stress due to wave propagation effects in the building is neglected. The shearing mechanism induced by the sequencing of the explosions is also disregarded: the motion is uniform over the extent of each floor, and indeed over the entire falling body. The bottom layer of the body is assumed to break up at a constant level of the reaction force P from the ground. This is taken to be a factor times the weight above. The factor may be less than 1.0, representing the case where the strength is reduced to below the freestanding value by blasting just before the collapse process starts. In the example presented below the factor was taken equal to 1. A density α is defined such that a height of 1 metre becomes α metres when collapsed.

At the simplest level the process may be described in terms of the motion of a rigid body with changing mass, as in the rocket flight problem of classical mechanics.[3] The equations developed here are analogous to those derived by Dehn[4] in the theory of the penetration of projectiles impacting fixed targets. However due to the presence of an extra term, the solution method used by Dehn, namely separation of variables, is not applicable here, and a different solution technique is adopted.

Let the instantaneous velocity be V (downwards positive) and let M be the current mass. In Figure 1 the drawing on the left shows the situation just before the crushing of mass element dM. At this time the element dM is still part of the body moving at velocity V, and consequently the momentum is

$$(M + dM)\,V$$

The middle drawing shows the situation dt seconds later, when mass dM has come to rest. The velocity of the main body has changed to V+dV. The momentum is:

$$M\,(V + dV)$$

The impulse delivered during this time is:

$$-P\,dt + M\,g\,dt$$

where P is the force acting from the ground onto the crushing layer, positive if in the direction shown. Equating the change of momentum to the impulse and dividing by dt, one obtains

$$M\dot{V} + V\dot{M} = -P + M\,g \tag{1}$$

Furthermore,

$$\dot{M} = -\rho\,\alpha\,V \tag{2}$$

In the above, ρ is the mass per unit height of solid material (no voids) of the building, and α is the density (volume of solids / total volume). Equations (1) and (2) form a pair of nonlinear ordinary differential equations in the unknown functions M(t) and V(t). It is possible to eliminate V, say, and obtain a single equation in M. This is not done here, because there is no closed form solution for the resulting equation in any case. The equations may be written in the form

$$\dot{M} = -\rho\alpha V$$

$$\dot{V} = \frac{1}{M}(-\dot{M}V - P)$$

$$= \frac{1}{M}(\rho\alpha V^2 - P) + g \quad (3)$$

This system of first order nonlinear differential equations is to be solved by numerical methods or by approximate algebraic methods. When the solution has been obtained, it may be used to obtain the force acting between the crushing layer and the still intact falling mass. From the right hand side drawing in Figure 1, writing the momentum change equation for the layer itself, one gets:

$$(-\dot{M})V = P - F \qquad \text{or} \qquad F = P + \dot{M}V \quad (4)$$

It may be seen that F is also the quantity appearing in brackets in Eq. 3. Normally, the term $\dot{M}V$ is negative, and so F is less than P.

<u>Discrete model</u>

In this model the collapse of each storey is treated as one stage of a discretised process. Let N be the number of storeys still upstanding, and let the mass of one storey be m, and its height h. The potential energy change on reducing the height from N h to (N-1) h is found to be $\left(N - \frac{1}{2}\right)(1-\alpha)\, m\, g\, h$ where $(1-\alpha)$ arises from the fact that the crushed storey occupies a height of αh. Also, if the velocity changes from V to V+dV during this stage, the kinetic energy increase may be shown to be

$$(N-1)\, m\, V\, \Delta V + (N-1)\frac{1}{2}m(\Delta V)^2 - \frac{1}{2}m\, V^2 \quad (5)$$

Finally, the work done in collapsing one storey is taken to be $\gamma(1-\alpha)\, N\, m\, g\, h$, that is a factor γ times the weight above, multiplied by displacement. Writing the energy balance one obtains a quadratic equation for ΔV:

$$A\,\Delta V^2 + B\,\Delta V + C = 0$$

with:

$$A = \frac{N-1}{2}m$$

$$B = (N-1)mV$$

$$C = \left[\frac{1}{2} - (1-\gamma)N\right](1-\alpha)mgh - \frac{1}{2}mV^2$$

Putting $h = V\,\Delta t$, $\Delta V = \dot{V}\,\Delta t$ it is possible to reconcile this equation with Eq. (3) up to a factor of 1/2 in the V^2 term. This discrepancy is thought to be due to the discretisation error. Also, the term in ΔV^2 is only present in the discrete formulation, since higher order infinitesimals will tend to zero in the continuous formulation. The qualitative behaviour of the two formulations is found to be the similar, although a quantitative comparison has not been made so far. In the remainder, only the solution of the continuous equation will be discussed.

RESULTS AND DISCUSSION

Equation (3) may be integrated by the Runge Kutta method. However an analytical solution would be preferable since it would allow general predictions to be made more easily. An exact analytical solution has not been obtained, but an approximate solution based on a series expansion of the 1/M nonlinearity has been achieved. The details are outside the scope of this presentation. However the salient features of the solution can be demonstrated in the phase plane diagram in ($\dot{M}$, M) plane. Because of (2), this is also the (V , M) plane. For a particular building, Figure 2 shows the vector field of local gradients of the phase trajectories of the system. It is clear that there is a separatrix above which the velocity grows steadily, and below which the system decelerates and comes to a halt with a finite mass remaining. A particular system must be launched from a point above the separatrix in order to get a complete disintegration. The curve in Figure 2 is a plot of initial velocities versus initial masses available on the assumption that the bottom n storeys are suddenly removed (by blasting) and that the rest falls as a rigid body until it reaches ground level, when the process starts. The intersection of the separatrix with this curve would be a critical point inasmuch as it would set a lower limit for the number n for the launch to be successful.

Finally, it should be noted that the rate of velocity increase in parts of Figure 2 acquires very high values. Evidently, this is impossible from the physical point of view; Equation (3) shows that the calculated rate of change of velocity may exceed g if the quantity F turns negative. It is concluded that the equations are only valid up to this point, and that beyond this point the solution must be continued with a form of unimpeded free fall.

REFERENCES

1. "Ball brings the house down in Hackney", Construction News, November 14, 1985
2. Williams, G.T. Demolition of Tall Buildings by the Controlled use of Explosives, in Decommissioning (Ed. Whyte, I.L.) pp.180-185, Proceedings of the First International Conf. on Decommissioning Offshore, Onshore and Nuclear Works, U. of Manchester March 22-24, 1988. Thomas Telford Ltd., London, 1988.
3. Burghes, D.N. and Downs, A.M. Classical Mechanics and Control, Ellis Horwood Ltd., Chichester, 1975.
4. Dehn, J., A Unified Theory of Penetration. Int. J. Impact Engng.,Vol. 5, pp. 239-248, 1987

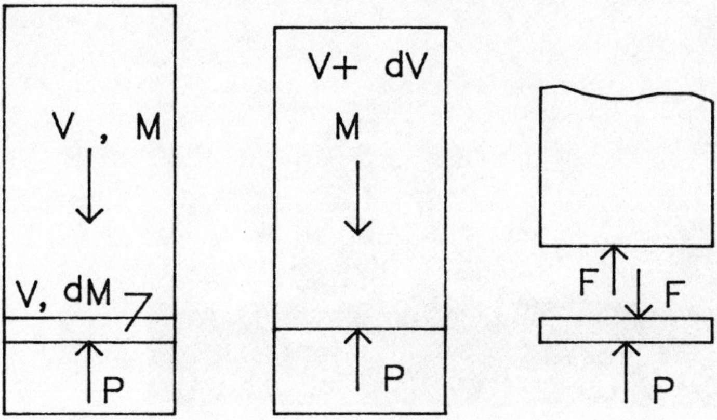

Figure 1. Definition sketch

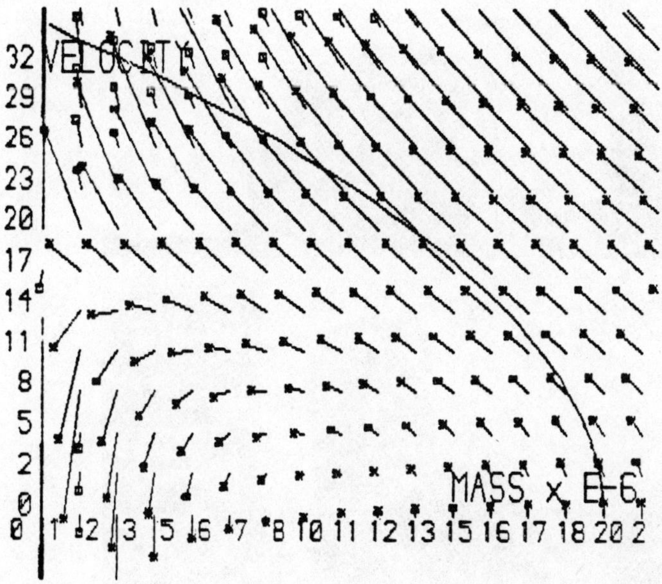

Figure 2. Phase portrait

Figure 3. Demolition of Highworth Point
Hackney, London 1986

Analyses of Shothole Detonation for the Dismantling of Concrete Structures

M. Freiman, N.J. Krutzik, C. Schmitz, and R. Tropp
Siemens, KWU Group, Federal Republic of Germany

ABSTRACT

The suitability of nonlinear analyses for examining the effects of dismantling using explosive charges was investigated. The procedure consists of two steps: a) investigation of the detonation process inside the shothole and calculation of the pressure time histories at the shothole boundaries; b) nonlinear analysis of the reinforced concrete structure using the pressures obtained in the first step. Having evaluated the crack patterns and the corresponding maximum strains the most probable contour of the break was determined. The results of the analyses were found to be in good agreement with the experimental results.

INTRODUCTION

Within the framework of the HDR dismantling test program [1], a number of concrete specimens were tested using different cutting techniques initiated by detonation in shotholes. Most of the specimens were reinforced-concrete slabs measuring 2.0 x 2.0 m with a thickness of 70 cm. The concrete properties and the reinforcement ratio are the same as those used for the biological shields of the first nuclear plants scheduled to be decommissioned. In view of the time and cost required to find an optimum explosive charge for each spatial configuration, initial considerations were made regarding the feasibility of employing computational methods. Consequently an analysis of cutting with explosives was included in the test program.

The test slab shown in Figure 1 was to be dismantled by a simultaneous explosion in 16 shotholes which were drilled from the front surface at equidistant positions. 45 g of PETN explosive were used for each shothole. The arrangement of the explosive and of the high-strength mortar filling material is shown in the same figure. Determination of the pressure in the shotholes and its stripping effects on the concrete were of major interest.

ANALYSIS OF THE DETONATION PROCESS

Considering an isolated shothole, there are basically three material zones to be addressed: the explosive, the air gap and the boundaries consisting of concrete or filling material at the top of the charge. Assuming that the boundaries are rigid until the shock front has been established, the peak pressure on the concrete can be evaluated using the hydrodynamic equation of state for the explosive and the ideal gas equation of state for air.

Because of the axisymmetric arrangement of the explosive within the cylindrical hole, the solution can be reduced to a one-dimensional analysis. In each horizontal layer the overdriven detonation starts at the centerline and proceeds to its boundary.

Equations of State

The Jones-Wilkins-Lee (JWL) equation of state was used to describe the pressure-volume-energy behavior of the explosive. The equation for pressure P is:

$$P = A\left(1 - \frac{\omega}{R_1 V}\right)e^{-R_1 V} + B\left(1 - \frac{\omega}{R_2 V}\right)e^{-R_2 V} + \frac{\omega E}{V} \qquad (1)$$

where

A, B and C	= equation-of-state parameters in Mbar
R_1, R_2, and ω	= equation-of-state parameters
$V = v/v_o$	= volume of detonation product/volume of undetonated explosive
P	= pressure in Mbar
E	= detonation energy per unit volume in $(\mathrm{Mbar\text{-}cm^3})/\mathrm{cm^3}$

The characteristic equation of air can be written as:

$$P \,.\, V = C \,.\, E_o \qquad (2)$$

where

$C = \Gamma - 1 \qquad \left(\Gamma = \frac{\partial \ln P}{\partial \ln v} = 1.402\right)$

$E_o = 2.4876 \,.\, 10^{-6}\ (\mathrm{Mbar - cm^3})/\mathrm{cm^3}$

Method of Analysis

The analysis was performed using the computer code HEMP [2], a two-dimensional finite difference program developed by Wilkins for elastoplastic and hydrodynamic analysis of structures under transient and impulse loading. The chemical energy to be released according to Eq. (1) is stored in each explosive zone as an initial energy E_o. The time for the detonation front to reach a specific zone is calculated in advance from the known detonation velocity D and the distance from the point of detonation to the center of the zone. A burn fraction F is calculated so as to spread the burn front over several zones in the same way as an artificial viscosity smears a shock. For one-dimensional analyses the burn fraction is defined as

$$F = (1 - V)/(1 - V_{CJ}) \qquad (3)$$

where

V_{CJ} = relative volume at Chapman-Jouguet point

For the model shown in Figure 2, the following values for the explosive were used [3]:

ρ	=	1.54 g/cm^3 (specific mass)	A	=	6.253 Mbar
P_o	=	0.18 Mbar	B	=	0.2329 Mbar
D	=	0.713 cm/µs	R_1	=	5.25
E_o	=	0.081 $(Mbar\text{-}cm^3)/cm^3$	R_2	=	1.6
Γ	=	2.788	ω	=	0.28

In order to check the sensitivity of the burn calculation, the input related to the size of the zones and the detonation parameters was varied. Only one reliable presssure measurement was available to verify the established shock front. The measurement was obtained from a test in which 22.5 g of PETN were used. Hence a corresponding computation was carried out for the same size of charge. In Figure 3 the computed and measured pressures are compared. Both curves have a steep peak (108 kbar calculated, 90 kbar measured) followed by a second maximum pressure that might indicate the beginning of the rarefaction wave. As the measurement broke down about 10 µs after the detonation and the computation time was limited by the assumption of rigid boundaries, the actual duration of the second maximum pressure was uncertain.

For the analysis of the test slab, a pressure time history with two different descending branches I and II, as shown in Figure 4, was used. The peak pressure of 208 kbar was calculated considering the actual quantity of the explosive in the shotholes.

ANALYSIS OF THE BLASTING PROCESS

With regard to the equidistant spacing of the shotholes in the slab, the 3D-model for calculating dismantling effects can be reduced to the space comprising half the distance to the neighboring hole in both horizontal directions. Because of the axisymmetric nature of loading, only a quarter of the model was actually computed. The model consists of 390 six- and eight-node quadrilateral solid elements to represent concrete including the reinforcement and the filling material.

Method of Analysis

The Lagrangian code NONDYN [4] was used to calculate the local material response. For the idealization of structures it employs a Popov-Galerkin finite element formulation using four-, six- or eight-node quadrilateral solid elements. A diagonal mass matrix is used. Structural and material damping is neglected. An explicit central-difference time-integration algorithm is used. The code uses an artificial viscosity to smear shock fronts over several mesh widths in the calculation.

Materials at an interface can be designated as slave and master and procedures are built into programs to permit contact, separation and sliding between master and slave surfaces. A special procedure was employed to take into account the wave interference on the model boundaries caused by the propagating detonation front from the neighboring shotholes.

Material Model for Reinforced Concrete

The material model includes the basic mechanism of reinforced concrete by blending the individual properties of plain concrete, cracks and reinforcement into a single composite. Considering the two general types of failure, tensile and compressive, and the classical orthotropic plasticity framework, the plastic model of concrete is based on an off-centered yield criterion, a plastic flow rule and a hardening rule. In general, the failure surface is represented by three stress invariants.

The maximum tensile stress criterion is used for the adjunction of the crack model. A linear kinematic softening law is used to model the post-yielding behavior. The finite tension stiffening slope aims at modeling progressive crack propagation instead of a brittle rupture. The crack model may account for up to three cracks at a point. The model has the capacity to close the crack if the tension ceases.

Elastoplastic behavior in the state of uniaxial stress is assumed for the reinforcement. The steel directions are assumed to be orthogonal and to coincide with the coordinate system obtained by orthonormalization of the natural coordinates of each element. In this way the steel directions rotate with the element when it undergoes large displacements.

The material input data are shown in Figure 5. They are based on the experience obtained in high-velocity impact calculations. On account of the considerably higher strain rates caused by an explosive, the influence of increased concrete compressive strength was also investigated [5].

Results of Analysis

Due to the shock of 208 kbar (20,800 N/mm^2), which exceeds several times the compressive strength of concrete at high strain rates, the elements near the detonation front are violently loaded and the space with the explosive charge extends rapidly. Because of the top filling which can withstand the pressure wave for only a few microseconds, vertical upward forces and subsequently shear stresses are induced. With regard to the triaxial state of stress, the concrete is also subjected to considerable tensile forces.

The damage to the concrete near the explosion is visible after about 100 μs (Figure 6). As its distance from the explosion point increases, the power of the shock wave decreases. On the boundaries of the model the waves propagating from the neighboring detonation points cause interference effects. Thus the wave is reflected several times until its full attenuation is reached.

The probable break contour can be determined after about 600 μs. While damage of the concrete near the explosive charge is obvious, the shape of the total break can be conveniently predicted by evaluating the element strains. The crack strains exceeding 5 ‰ are shown in Figure 7. The circumferential and vertical strains of 15 ‰ and 24 ‰ respectively were found as minimum values in the 8th element layer. This implies a complete break within this layer.

In Figure 8 the measured mean pressure in concrete and the computed stress intensity are compared. Both curves are in good agreement. The break obtained in the experiment and the calculated break contour are shown in Figure 9. Both shapes and the resulting quantities of material removed conform to each other.

It should be noted that the calculated crack strains were considerably lower when loading curve II was used. In this case the dismantling effects were reduced to the local region of the explosive charge only. In a third calculation loading curve I was used again but the compressive strength of the concrete element adjacent to the explosive charge was increased by a factor of 10. Slight differences were revealed in strains. The predicted break had the same shape.

CONCLUSION

The computations of the shothole detonation for dismantling purposes were found suitable for tracing the explosion process and its effects on reinforced-concrete structures. The calculations were performed using standard software for high-velocity simulations and the rather small number of experimental results available at the moment. Increasing activities in the field of dismantling with explosives will broaden the existing experimental database and the scope of the corresponding analyses as well.

REFERENCES

1. HDR-Sicherheitsprogramm
Beton-Zerlegungsversuche in einem geschlossenen Containment
HDR-Versuchsgruppe ZER
PHDR-Arbeitsbericht Nr. 62-85

2. HEMP/KWU
General Description and Program Organisation,
Internal Report, June 1977

3. B. M. Dobratz, LLNL Explosives Handbook
Properties of Chemical Explosives and Explosive Simulants,
LLNL Report UCRL-52997, March 16, 1981

4. NONDYN, Rechenprogramm zur nichtlinearen Berechnung von Stahlbeton und elastoplastischen Werkstoffen
KWU Arbeitsbericht R 621/84/0039

5. LASL Shock Hugoniot Data, Stanley P. Marsh, Ed.
University of California Press, 1980

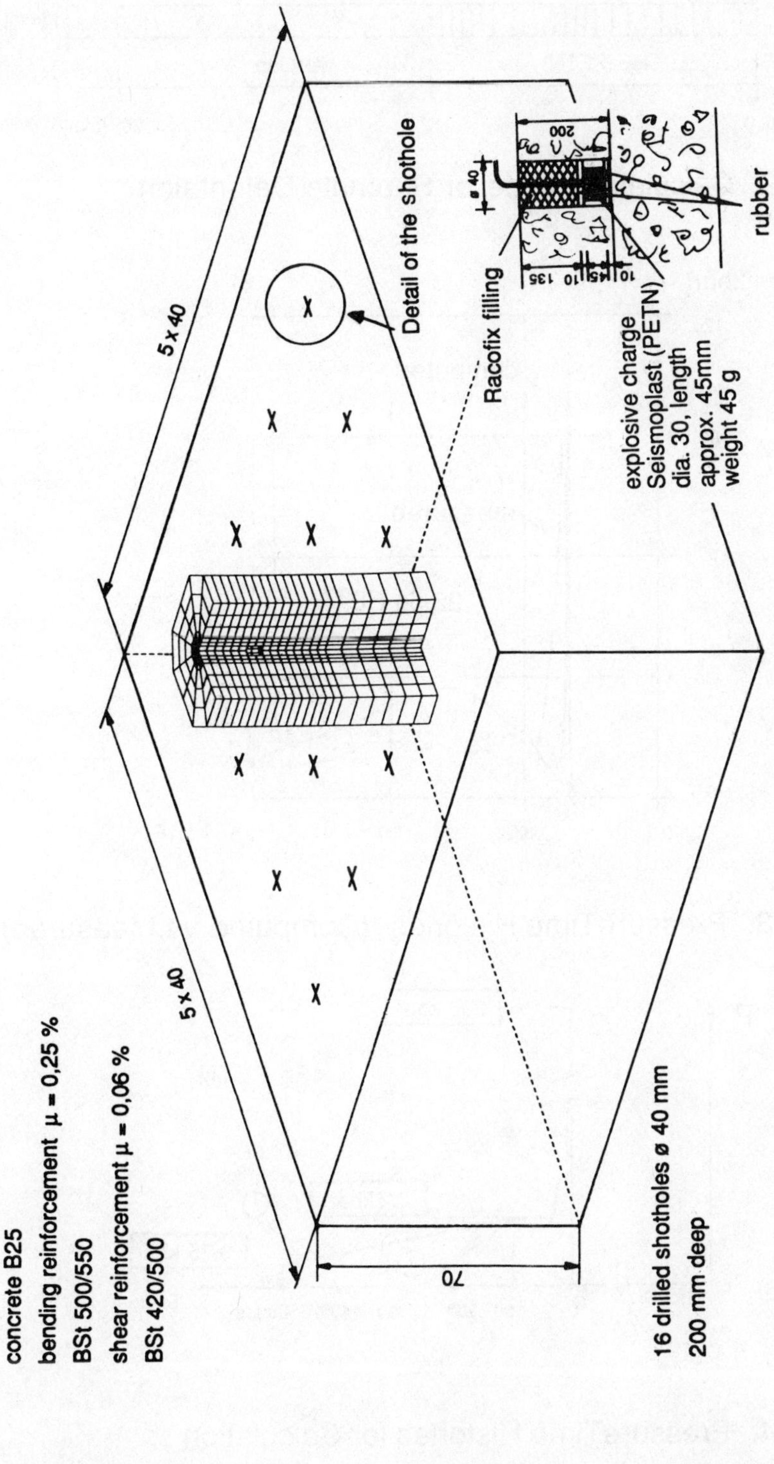

Figure 1. Test Slab

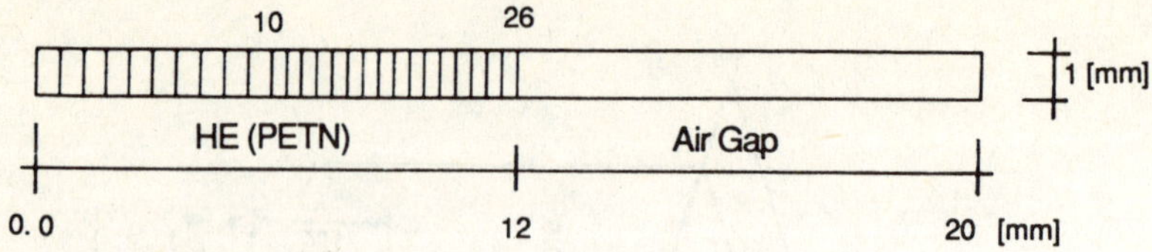

Figure 2. Calculation Model of Shothole Detonation

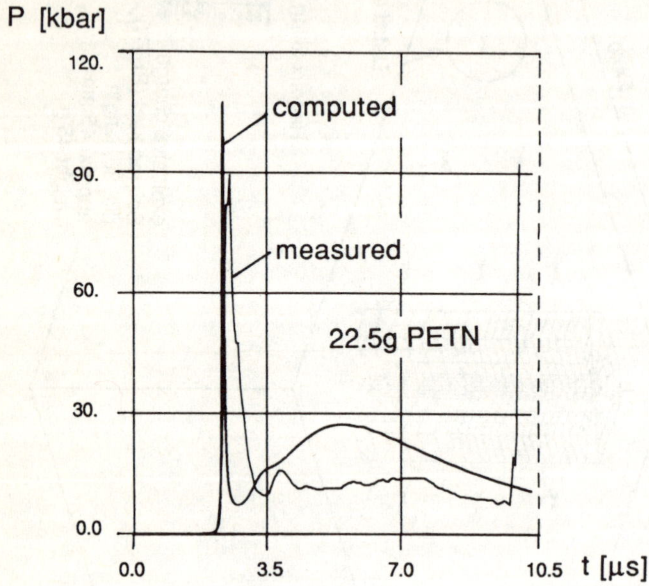

Figure 3. PressureTime Histories (Computed vs.Measured)

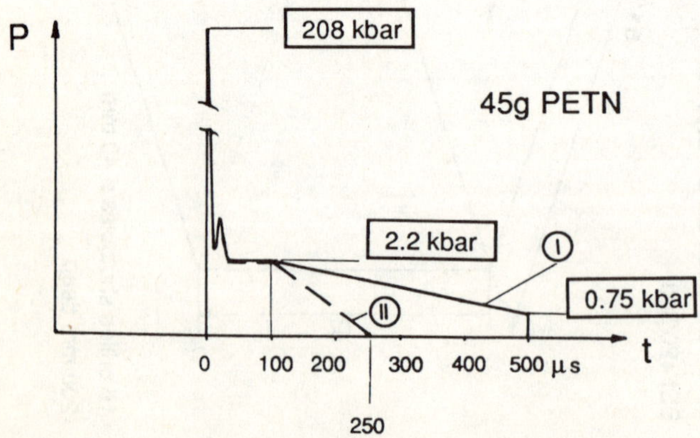

Figure 4. PressureTime Histories for Calculation

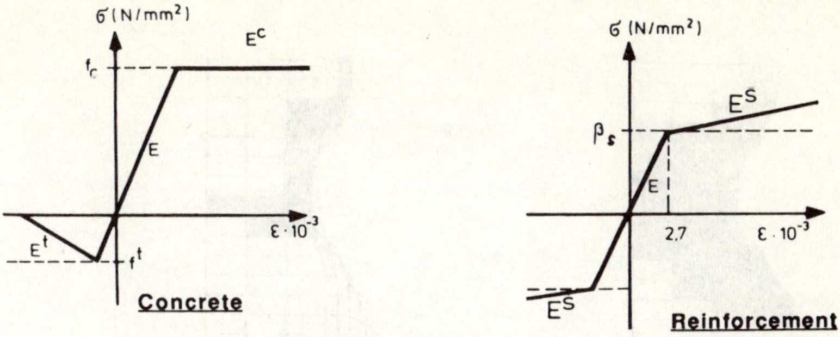

Concrete

$f^c = 50\ N/mm^2$

$f^c = 500\ N/mm^2$ at the shothole

$f^t = 3.5\ N/mm^2$

$E = 36000\ N/mm^2$

$E^c = 0$

$E^t = 6500\ N/mm^2$

Reinforcement

Bst 420 / 500, 500 / 550

$\beta_s = 567\ N/mm^2$

$E = 210000\ N/mm^2$

$E^s = 1050\ N/mm^2$

Figure 5. Material Model for Reinforced Concrete

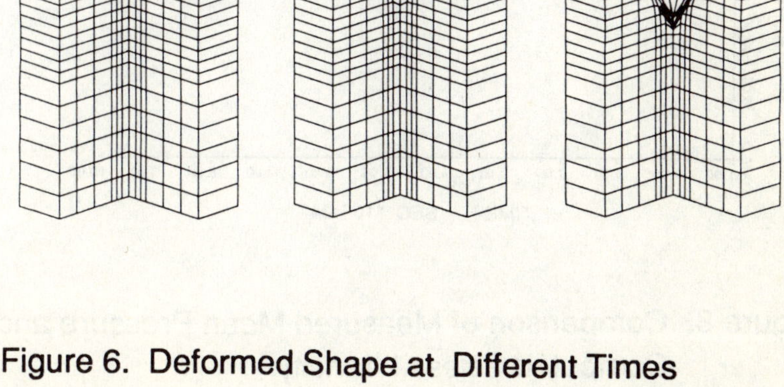

Figure 6. Deformed Shape at Different Times

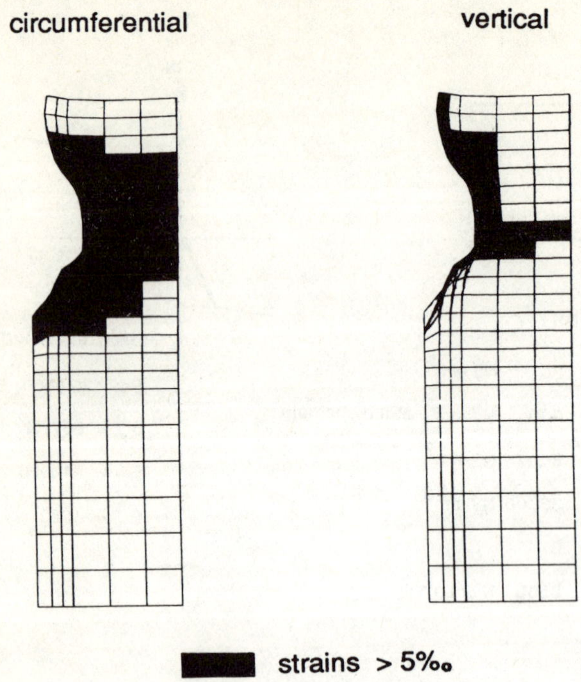

Figure 7. Crack Strains at t = 590 μs

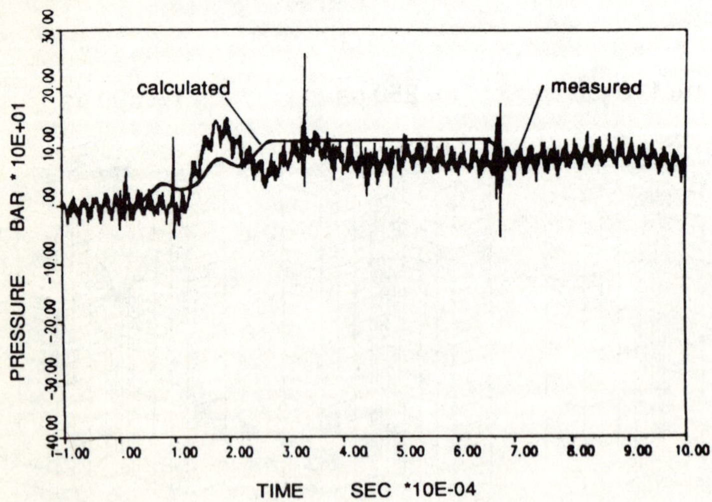

Figure 8. Comparison of Measured Mean Pressure and Computed Stress Intensity

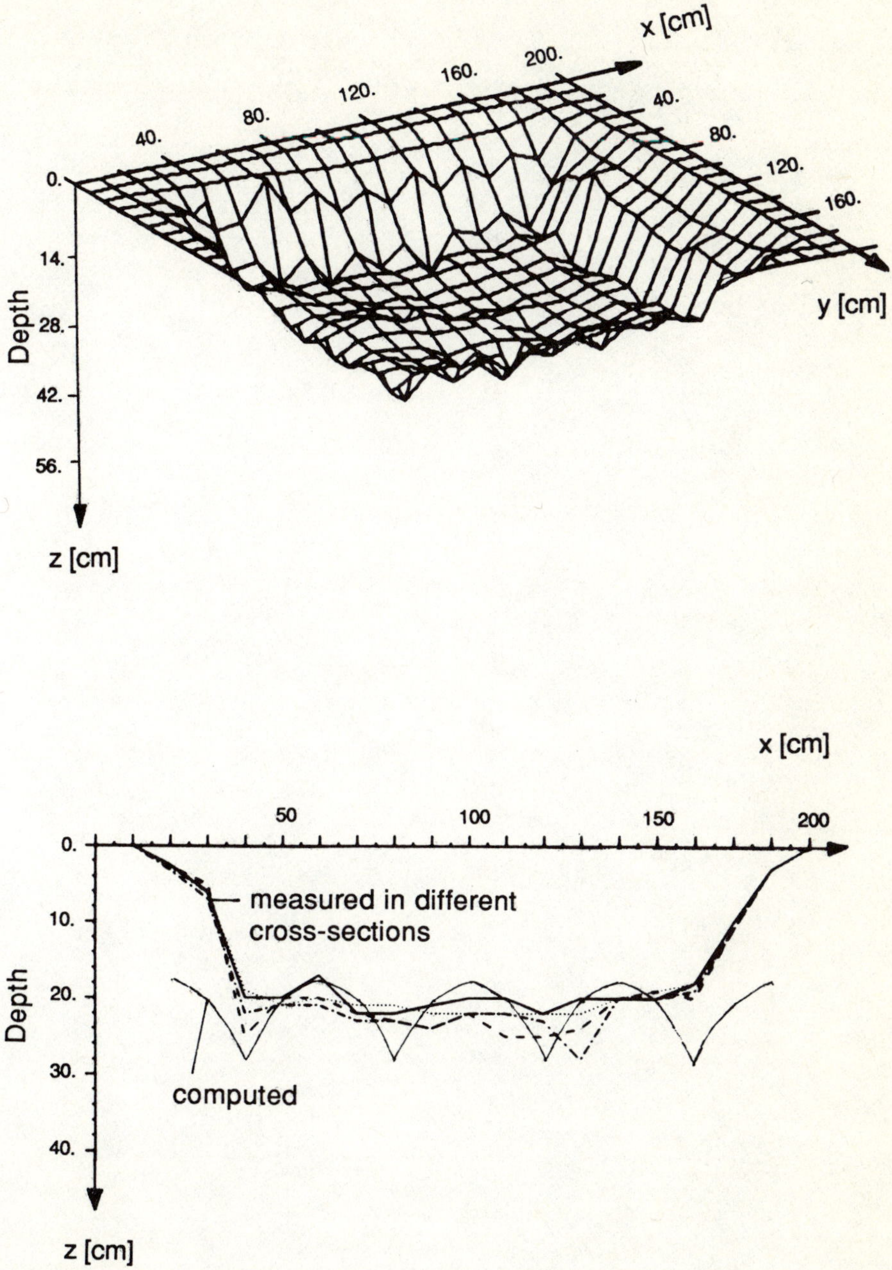

Figure 9. Computed vs. Measured Break Contour

SECTION 7 - UNDERGROUND STRUCTURES AND SOILS

Physical Modelling of Impact Loading on Buried Structures

M.C.R. Davies
School of Engineering, University of Wales College of Cardiff, Wales, UK

ABSTRACT

The results of an experimental investigation into the effect of impact loading on buried structures are presented. Instrumented model cylinders were buried at varying depths and subjected to an impact load at the surface. Measurement of the stresses in the cylinder and the soil adjacent to it provided a valuable insight into the dynamic soil structure interaction. A quantitative indication of the attenuation of the stress wave in the soil was obtained from comparison of stress measurements in tests conducted at different cover depths. A SDOF method for predicting the dynamic magnification factor is presented. Predicted values were in agreement with those measured and values presented in the literature.

INTRODUCTION

Buried structures, such as sewers, pipes, tunnels and protective structures have to be designed to support not only the dead loading of the soil providing cover but also a live loading at the ground surface. This can be in the form of a static load, moving loads or dynamic loading. Surface impact loading of buried structures may be caused by a number of situations e.g. an axle load impacting on the ground surface above a buried structure (as the result of going over a bump in the road) or the blast wave which, may be accompanied by a kinetic component from debris, in an explosion. An impact at the ground surface causes a stress wave to travel through the soil which results in impulsive loading on the buried structure.

Investigations into the impulse loading of buried structures have tended to concentrate on the effects of dynamic overpressures. Comparisons of the static and dynamic overpressures necessary to cause failure of buried cylindrical shells have been reported by a number of workers; e.g. Bulson[1], Marino and Riley[2],

Albritton et al[3]. An analytical approach using a dynamic version of the finite element method has also been developed, Dunns and Butterfield[4]. The results of this analysis indicated that for buried cylindrical shells of varying stiffness, with depths of cover greater than 0.5 of the radius, a value of 1.2 was obtained for the dynamic magnification factor - a ratio of the maximum dynamic response to the static response of the system. This value compares well with the values of dynamic magnification factor obtained experimentally by Marino and Riley[2] and Albritton et al[3]. It was concluded by Dunns and Butterfield[4] that a dynamic magnification factor of 1.2 could be generally applied in design as a means of avoiding the computational effort involved in a full finite element analysis for each application. However, although there was variation in the stiffness and depth of burial of the structure analysed, the comparisons of the experimental and finite element results were for a single loading boundary condition only, i.e. an overburden pressure.

An experimental study was conducted to obtain a greater understanding of the response of buried flexible structures to localized impact surface loading. Special attention was paid to the capacity of soil in damping and absorbing impact energy in order to assess the ability of soil to prevent impact loads being transmitted to buried structures.

EXPERIMENTAL STUDY

Apparatus and experimental procedure

The structure used in the laboratory tests was a brass cylinder of diameter D = 100mm and wall thickness t = 0.127mm; resulting in a very flexible structure. The models were placed in a 1.5m cubic tank, Figure 1, and buried at varying depths of cover. The cylinders were fabricated from five lengths - joined together with adhesive tape - to span the tank. Segmental construction prevented the instrumented central section from experiencing any

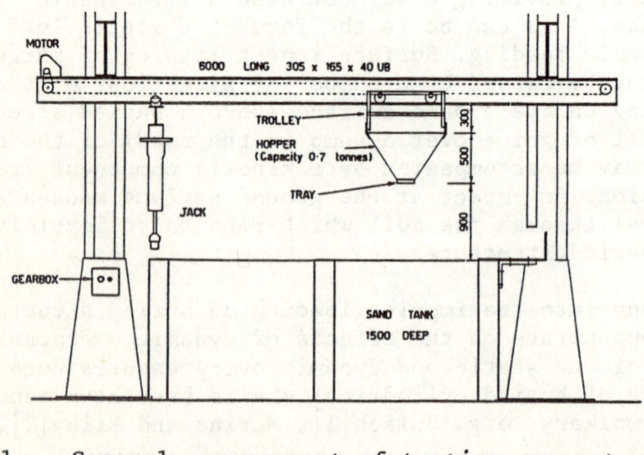

Figure 1. General arrangement of testing apparatus

end effects (as a result of confinement in the tank); thus ensuring plane strain conditions for the instrumented section. Circumferential strains in the cylinder wall were measured by 9 pairs of strain gauges (one on the inside and the other outside) positioned symmetrically around the middle segment. Stresses in the soil around the cylinder were measured using miniature pressure cells located in the soil.

A dry sand foundation with a repeatable uniform density of 16.8 ± 0.1kN/m^3 was achieved using a hopper. The 700kg capacity hopper was mounted on rails above the tank, Figure 1, and driven at a constant rate by an electric motor using a chain drive. A baffle tray fixed below the base of the hopper ensured a uniform rain of sand, which fell through at least 1.0m into the tank.

Two series of buried cylinder tests were performed. The geometry of the models was similar in both cases, as shown diagrammatically in Figure 2. Tests were conducted on cylinders buried at different depths of cover, z. In the first series, the ground surface

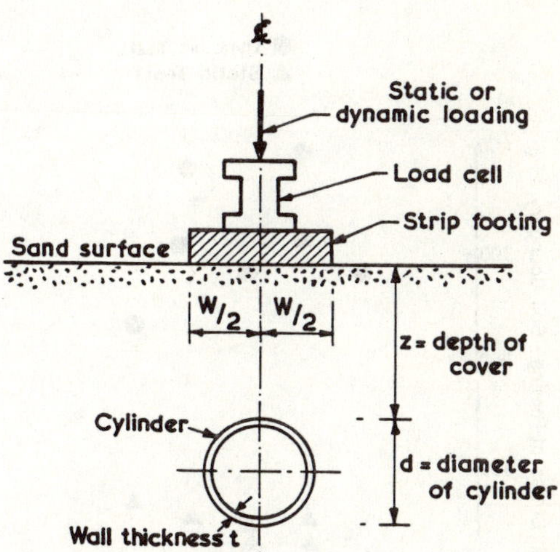

Figure 2 Geometry of model

above the buried cylinder was loaded with an increasing static pressure until the cylinder failed. Surface loading was applied using an hydraulic jack acting on a footing of width, W = 100mm; the force acting on the footing being measured using a load cell and the stress in the soil by pressure cells located in the soil immediately below the footing.

The second series of tests repeated the model configurations of the static tests. Dynamic loading of the soil/structure system was an impulsive force provided by dropping a weight (a beam to maintain plane strain conditions) onto the footing. Cylinders buried at different depths were subjected to the same surface impact allowing an assessment of the effects of stress wave attenuation on the structural response.

Results of Loading Tests

Dynamic Magnification Factor The results of the two series of tests were used to quantify the dynamic response of the soil/structure system in terms of a dynamic magnification factor, D, which for an elastic system this can be shown to be :

$$D = P_S/P_D \tag{1}$$

where P_S and P_D are the static and dynamic loads to produce the same structural response, respectively.

Values of peak stress recorded at the soil/footing interface in the dynamic loading tests are plotted v. depth of cover in Figure 3, together with static stresses at the equivalent cylinder responses. The results indicate that to produce the same structural response the peak dynamic stresses were much larger than

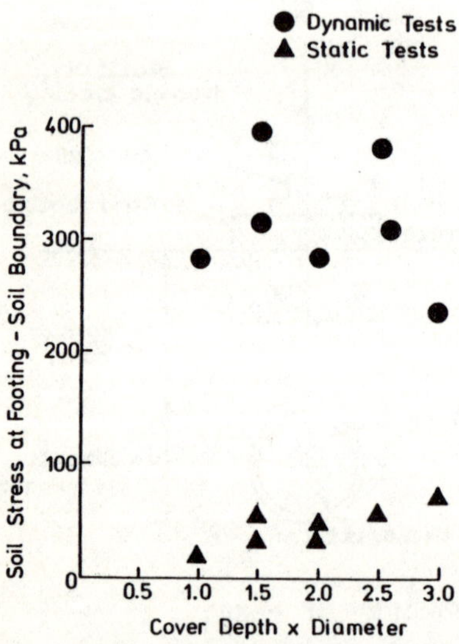

Figure 3. Dynamic and static soil/footing interface stresses at equivalent cylinder responses

static values. The dynamic magnification factor for the soil/structure system was D = 0.15. If this value is compared with the dynamic magnification factor for the case of an overburden pressure, i.e. D = 1.2, (e.g. Bulson[1]) it is clearly evident that a large variation in the value of dynamic magnification factor results from a change in loading boundary conditions.

Attenuation of Stress Wave A "free field" test was conducted to investigate the effect of the structure on the stress propagation through the soil. The soil was instrumented with pressure cells located below the centreline of the footing at varying depth. The peak stress recorded in each pressure cell is plotted v. depth in Figure 4. When elastic body waves are propagating

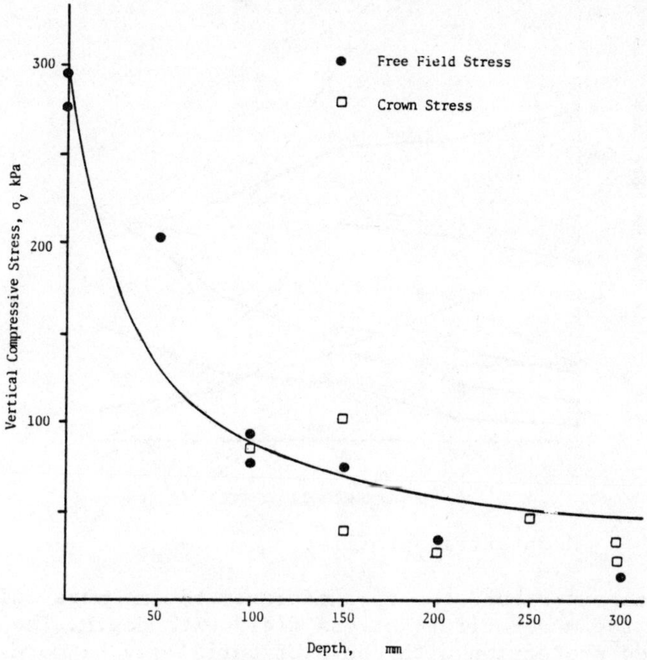

Figure 4. Attenuation of vertical stress

along a cylindrical wave front (as is the case for the plane strain surface loading) attenuation due to geometric damping is proportional to $x^{-0.5}$, where x is the distance from the source of the impact. A curve of this form is plotted together with the results in Figure 4. The agreement is very good although at greater depths the measured attenuation was slightly greater than that predicted - probably due to the effect of viscous damping.

The results indicate that the stress immediately above the crown of the cylinder was very similar to the free field value, Figure 4. Pressure cells located adjacent to the springing of the cylinder indicated that at depths greater than 1.5d the lateral stresses were of similar value to the crown stresses. Thus for depths greater than 1.5d, radial stresses acting on the cylinder may be reasonably represented by the value of free field vertical stress at the crown.

Structural Response The peak normal (i.e. hoop) strains in the cylinder, plotted against half circumference, are shown in Figure 5. As the cover depth was increased so the distribution

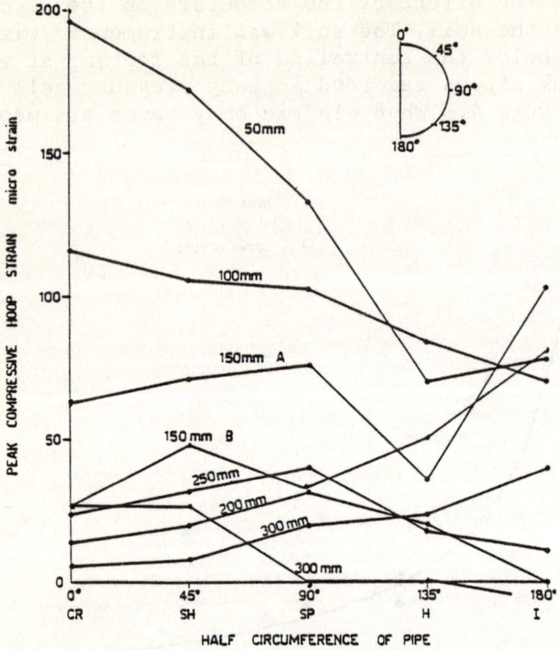

Figure 5. Hoop strain profiles

of normal strain around the circumference became more uniform, reflecting the more uniform stress field with depth. The effect of increased protection with depth of burial may be more clearly observed by comparison of the cylinder crown strains, Figure 6.

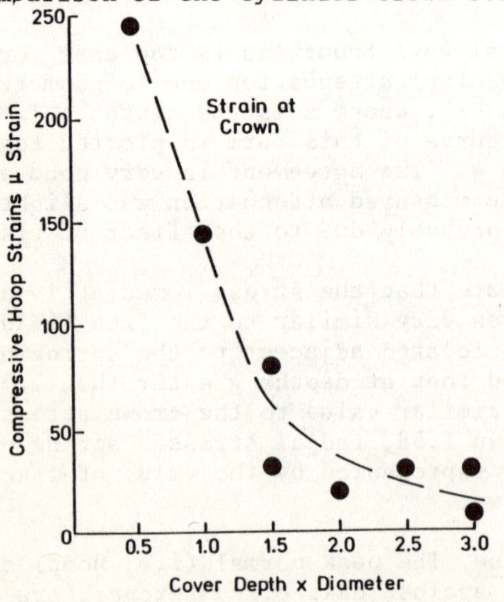

Figure 6. Decay of hoop strain with depth

With increasing depth of burial the value of peak strain initially decreased very rapidly. At depths greater than 1.5 diameters increasing depth of burial only resulted in a small reduction in measured strain. Thus for the range of cover depths in the experimental programme 1.5d cover appears to be optimum.

ANALYTICAL STUDY

Method of analysis

The experimental evidence indicates that if the response of a buried structure to static loading is known it is possible to predict the dynamic response by applying a dynamic magnification factor. The static response of a buried structure may be found either through measurement or predicted using analytical techniques, e.g. Gumbel et al [5]. If the dynamic magnification factor for the system can be obtained then the dynamic response may be predicted.

Prediction of dynamic magnification factor, D

Dynamic loading in the experimental study was an impulsive force (in the form of a half-sine-wave) of duration, t_1 = 1.22ms. When considering the response of a system to a dynamic impulse, the maximum response will be reached in a very short time. During this time there will be very little energy absorbed in the system by damping forces. Consequently damping forces will not be considered in this analysis.

A solution has been presented by Clough and Penzien |6| for the dynamic magnification factor, D, of an undamped single degree of freedom system subjected to impulsive loading. For the type of impulse observed in the experiments the dynamic magnification factor is given by :

$$D = \frac{v_{max}}{P_D/k} = \frac{2\beta}{1-\beta^2} \cos \frac{\pi}{2\beta} \qquad (2)$$

where v_{max} is the maximum dynamic response, P_D is the peak pressure and β the ratio of the applied load frequency, to the natural free-vibration frequency viz :

$$\beta = \frac{\bar{\omega}}{\omega} \qquad (3)$$

In order to calculate the dynamic magnification factor, D, it is necessary, therefore, to obtain the natural free-vibration frequency of the system under impact loading.

Calculation of natural free-vibration frequency, ω

The natural frequency, ω, of a free-vibrating system may be obtained using Rayleigh's method. The method assumes that the system deforms at the fundamental frequency - which corresponds to the frequency in the first mode of vibration. The results of a rigorous dynamic finite element analysis (Dunns and Butterfield |4|) indicated that the first mode of vibration was by far the most dominant, supporting this assumption. It is necessary to assume the shape that the system takes up in the fundamental frequency. For an elastic system this may be considered to be the deflected static shape under the same loading bound conditions as the dynamically loaded system. By assuming the deflected shape the multi degree of freedom system is reduced to a single degree of freedom. This also results in the solution for dynamic magnification factor, D, Equation 2, to be used.

In a free-vibrating system the total energy in the system does not vary, i.e. the sum of potential energy, V, and the kinetic energy, T, is a constant, hence,

$$\frac{d}{dT}(T + V) = 0 \tag{4}$$

and, therefore, the maximum kinetic energy equals the maximum potential energy. In a free-vibrating system the kinetic energy is a function of the natural frequency, hence, if the mode shape is known this equality may be used to calculate the natural free-vibrating frequency.

If an elastic system is freely vibrating with a frequency, ω, the maximum variation of potential energy in the system is given by,

$$V = \frac{1}{2} \int \delta\sigma_i \cdot \delta\varepsilon_i \, d \, vol \tag{5}$$

where σ_i is the vector of stresses in the system and ε_i the vector of maximum strains. The maximum value of kinetic energy of the system is given by,

$$T = \frac{1}{2} \int m \cdot \delta v_i^2 \cdot \omega^2 \, d \, vol \tag{6}$$

in which m is the mass/unit volume and v_i the maximum value of the assumed displacement function. By equating Equations 5 and 6 the free-vibration frequency may be found from,

$$\omega^2 = \frac{\int \delta\sigma_i \cdot \delta\varepsilon_i \, d \, vol}{\int m \cdot \delta v_i^2 \, d \, vol} \tag{7}$$

Analysis of experimental results

A program was written to calculate the natural free-vibration frequency of the soil foundation in the model tests. The foundation was divided into elements to form a mesh and, for the appropriate loading boundary conditions, the normal and shear stresses and strains - and hence the displacements - for each element in the mesh were calculated using an elastic stress distribution for the free field condition. The value of ω was obtained by summing the contribution of each element to the potential energy and kinetic energy for the whole foundation.

The forced frequency due to the impact was 2575rad/s and the calculated circular natural frequency of the soil 297rad/s. This resulted in a dynamic magnification factor, D = 0.23. The calculated dynamic magnification factor predicted the trends in the experiments, i.e. $P_D >> P_S$, and was of similar magnitude to the experimentally obtained value of 0.15.

An analysis of experiments with different loading boundary conditions was conducted to further assess the accuracy of this method for calculating the dynamic collapse pressure of buried cylinders. Bulson[1] published results of static and dynamic overburden pressure tests on flexible cylinders buried in sand. Using the method described above the free-vibration frequency of the model foundations subjected to overpressure loading was found to be ω = 224rad/s. The dynamic loading in these tests was provided by an explosion, the blast surface pressure v. time curve being of the typical Friedlander form, as shown in Figure 7, with a duration, t_1 = 20ms. This approximated to a triangular loading impulse, for which there is a solution for the dynamic magnification factor, of a similar form as that in Equation 2 for sinusoidal loading (Clough and Penzien[6]).

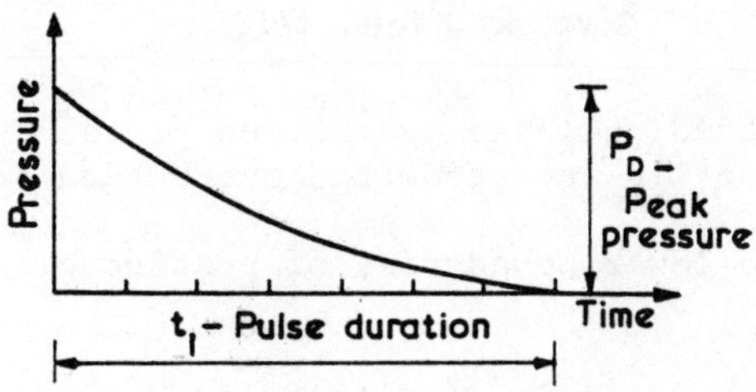

Figure 7. Pressure duration curve resulting from an explosion

A dynamic magnification factor of D = 1.31 was obtained from this analysis. This result is consistent with the experimentally obtained value of 1.2 obtained in Bulson's experiments[1] and the complex numerical calculations of Dunns and Butterfield[4]. Results of static and dynamic collapse surface pressures from Bulson's tests are plotted together with the predicted values in Figure 8. The experimental results for the static tests showed some scatter, with the data falling between the upper and lower scatter boundaries shown in the figure. Consequently, two prediction lines for the dynamic collapse pressure, $P_D = P_S/D$, are displayed in Figure 8; one for each of the boundaries of scatter from the static test data. The dynamic test results shown in Figure 8 constitute the limits of a dynamic "collapse zone". The analysis correctly predicted that the dynamic collapse pressures were lower than the static values at all depths of cover. The majority of the experimentally obtained dynamic collapse pressure data lies between the plots

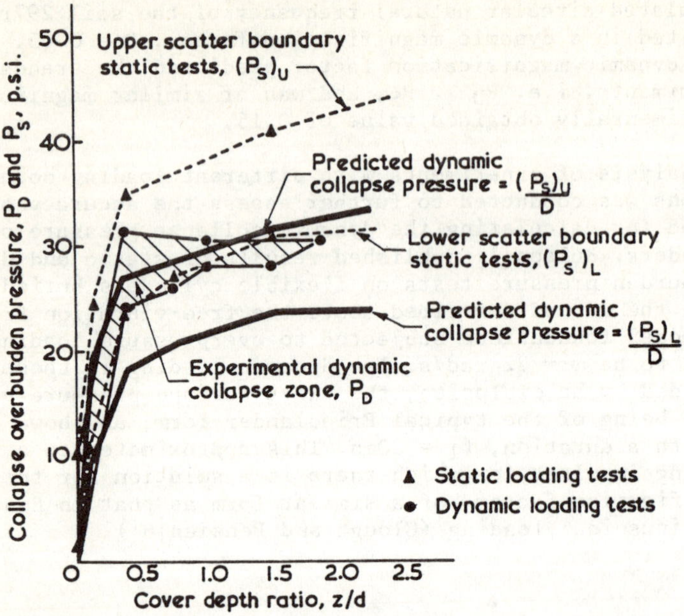

Figure 8. Measured collapse overburden pressures, after Bulson [1], and predicted overburden pressure

of the upper and lower boundaries of predicted collapse pressure in Figure 8.

Predictions of the dynamic magnification factor of the soil/ structure system in both the cases presented above are most encouraging, the trends in the two tests series being correctly modelled. In the impact loading tests, described herein, the sinusoidal loading was applied over a duration much shorter than the natural period of the soil mass. The impulsive overpressure, caused by the shock pressure wave resulting from an explosion, was a triangular wave form, the duration of which, 20ms, was only slightly less than the natural period of the soil mass (which was calculated to be 28ms). The different loading boundary conditions, i.e. the nature and duration of the impulse, resulted in the varying response.

CONCLUSIONS

A series of experiments was conducted to investigate the response of buried flexible structures to surface impact loading. The following conclusions can be made :

1. The experimental observations indicated that a dynamic magnification factor, D, could be used to predict the dynamic collapse pressure from the static value.

2. Geometric damping dominated the attentuation of the stresses in the foundation above the buried structure. Stresses immediately above the crown were similar in magnitude to the free field values.

3. The experiments indicated that 1.5 diameters was the optimum depth of cover.

4. Assuming that the foundation layer vibrates as a single degree of freedom system and the stress distribution is given by the free field response, Rayleigh's method may be used to calculate the free-vibration natural frequency of the foundation. Hence the dynamic magnification factor may be obtained.

5. Dynamic collapse loads predicted from the static values closely followed the trends observed in the experiments, giving reasonably accurate quantitative results.

6. In addition to providing an understanding of soil structure interaction during impact loading the data obtained in the experiments also forms a valuable data set for the validation of more complex analytical and numerical procedures.

ACKNOWLEDGEMENT

This work has been carried out with the support of the Procurement Executive, Ministry of Defence.

REFERENCES

1. Bulson, P.S. Stability of Buried Tubes Under Static and Dynamic Overpressure. Part I: Circular Tubes in Compacted Sand, Military Engineering Experimental Establishment, Research Report RES 47.5/7, 1966.

2. Marino, R.L. and Riley, W.F. Response of Buried Structures to Static and Dynamic Overpressures, Proc. Symp. on soil-structure interaction, University of Arizona, Tucson, 1964.

3. Albritton, G.E., Kirkland, J.L., Kennedy, T.E. and Dorris, A.F. The Elastic Response of Buried Cylinders, U.S. Army Waterways Experiment Station, Technical Report 1-750, 1966.

4. Dunns, C.S. and Butterfield, R., Flexible buried cylinders - Part II: Dynamic response, Int. Journal Rock Mech. Min. Sci. Vol. 8, pp 601-612, 1971.

5. Gumbel, J.E., O'Reilly, MP., Lake, L.M. and Carder, D.R., The development of a new design method for buried pipes, pp 87-98, Proc. Europipe Conference, Basle, 1982.

6. Clough, R.W. and Penzien, J., Dynamics of Structures, McGraw-Hill, 1975.

Experimental Investigation of the Dynamic Loading of Soil and the use of Finite Element Simulation for Analysis

A.D. Pullen*, L. Ianucci*, J.B. Newman* and S.H. Perry**
*Department of Civil Engineering, Imperial College, London, UK
**Department of Civil Engineering, Trinity College, Dublin, Ireland

ABSTRACT

A programme of work to determine the near-field response of soil to conventional high explosive blast involves dynamically loading soil specimens using both explosive and impact drivers. An experiment is described which involves impact loading a small confined specimen of soil and determining the stress/strain history within it. The finite element code *DYNA2D*, together with its pre- and post-processors *MAZE* and *ORION*, respectively, are used to model the experiment and the finite element and experimental results are compared.

INTRODUCTION

A programme of work was initiated in 1983 to determine the near-field response of soil to conventional high explosive blast. The purpose of the project is to provide the means to predict blast wave propagation in soils, including modification experienced at a soil/structure (concrete) interface. One of the principal objectives is to produce and verify complex material models for soil which will allow the use of hydrocodes to predict the response of more complex real structures.

Overview of experimental programme.
In order to describe fully the material behaviour under dynamic loading it is necessary to record displacement and pressure histories at a sufficiently large number of locations within a specimen to allow the displacement and pressure histories at any other location to be interpolated or extrapolated with confidence. A simple specimen geometry and load history reduces the number of locations required. The primary transducers being used are commercially available miniature accelerometers and in-house produced diaphragm pressure gauges. Typically, an accelerometer is mounted on the back of a pressure gauge and both are embedded in the soil, orientated such that they measure accelerations and

pressures propagating either radially from the HE source or parallel to the direction of impact. Three principal experiments are employed to produce the experimental data required.

The Conical Shock Tube (CST). This apparatus consists of rigid-walled conical container filled with soil. A pseudo-spherical blast wave is initiated in the soil by a small explosive charge detonated at the apex of the cone. Theoretically, a 10g charge in the CST produces a blast wave equivalent to over 5kg in the free-field. The principal advantages of this experiment are that it is rapid and cheap to repeat. (Pullen and Perry [1].)

Field trials. Primarily, these are intended to verify the CST results. Typically, a spherical charge of between 0.5 and 3.6kg is detonated at depths of between 0.5 and 2 metres below the ground surface, either in-situ on an outdoor range or, when a suitable range containing the soil of interest cannot be found, in an artificial cuboid of soil confined by concrete walls.

The Impact rig, which will form the basis of this paper. The apparatus consists of two parallel 50mm diameter, 6m long steel guide rails mounted vertically 160mm apart and about 300mm from a 250mm square "H-column", which acts as rigid backbone. A suitable projectile can be dropped from up to 6m on to a variety of specimens or targets, the velocity at impact approaching 11 m/s. Higher velocities are achieved by use of a catapult. Targets up to about 1m cube in size can be accommodated. Uses of this rig relevant to this project include investigation of the response of soil to planar shock loading and determination of the dynamic characteristics of various transducers embedded in soil. It will also be used to investigate the dynamic pore pressure response of saturated soils under shock loading.

Overview of computational programme
The computational analysis part of this project makes use of the explicit hydrodynamic codes DYNA2D and DYNA3D with their corresponding pre- and post-processors. Also, the implicit codes NIKE2D and NIKE3D are used for inertia type problems and numerical calibrations.

The above codes have been ported to a SUN UNIX workstation, which is used for all 2D analysis and some 3D analysis. The CRAY X-MP at the University of London Computer Centre is used for all large 3D problems.

Rather than simulate each of the above experiments with existing material models within the above hydrocodes, which can produce non-unique results, a technique commonly known as the Lagrangian Method is employed. The Lagrangian method solves for continuity, for Newton's second Law and for energy within a one dimensional flow field. Hence, if three or more strain-measuring transducers and three or more stress-measuring transducers are placed within the flow field, that is, along the axis of the CST or in a specimen on the impact rig, a complete dynamic constitutive relationship can be derived for use within the hydrocodes. The hydrocodes can then be used to model both the original experiment and more complex flow fields. Static material constitutive constants are derived from static tri-axial testing of specific soils, using a novel steel jacket to provide passive confinement.

Also, the FE technique was used to design and numerically calibrate the diaphragm pressure gauge. Figure 1A shows the mesh and figure 1B the first natural frequency. Other natural frequencies were calculated and used to determine the frequency response of this type of transducer.

Figure 1A

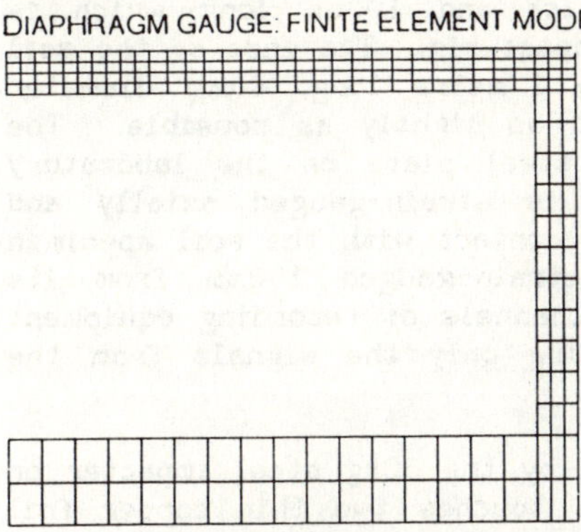

Figure 1B

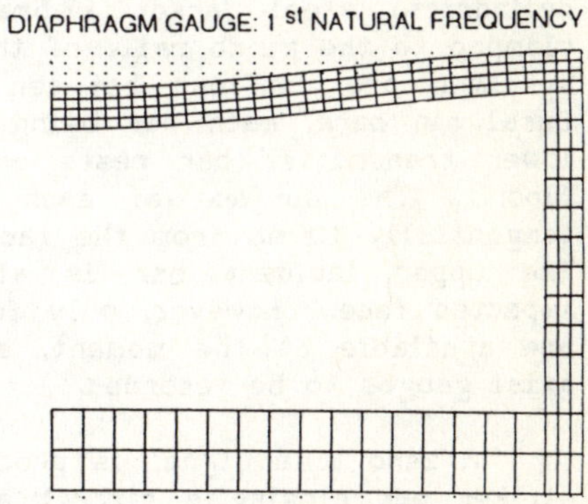

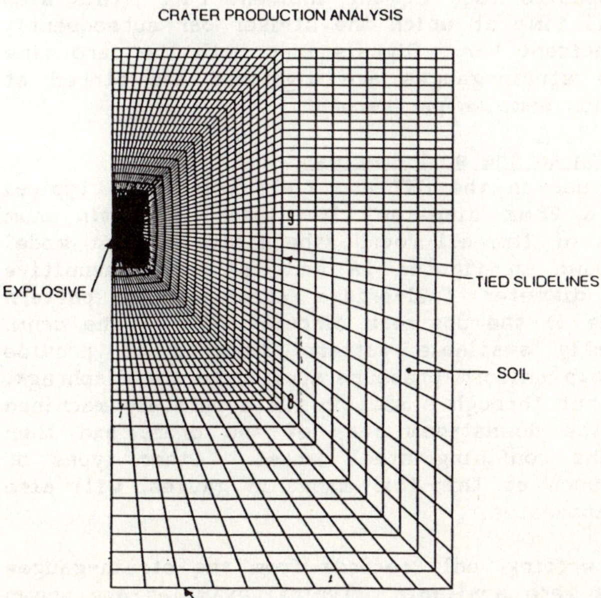

Figure 1C
A typical free field mesh used to validate the CST and material constitutive model.

IMPACT EXPERIMENTS

The purpose of the experiment described here is to determine the stress/strain history in a dynamically loaded cylindrical soil specimen. Subsequently, the response of pressure/stress gauges placed in this soil specimen will be compared with the known stress history. The experiment must be repeatable and this has been found to be far easier to achieve with the impact apparatus than by blast loading in the CST.

The soil specimen and its containment.
In the prototype apparatus, shown in figure 2, the soil specimen is a cylinder 50mm long by 40mm in diameter and has a mass of 113 grams. Circumferential confinement is provided by a cylindrical steel jacket 17.5mm thick and 100mm long which is clamped to the guide rails of the impact rig. The ends of the soil specimen are confined between two 1 metre long, 40mm diameter Duralumin bars, each bar being held as lightly as possible. The lower transmitter bar rests on a steel plate on the laboratory floor. The surface of each bar is strain-gauged axially and tangentially 100mm from the face in contact with the soil specimen The upper incident bar is also strain-gauged 100mm from its impacted face. However, only four channels of recording equipment are available at the moment, allowing only the signals from the axial gauges to be recorded.

A zero time signal is produced by the 30kg steel impacter or striker bar closing a circuit as it touches two thin copper foil conductors on the impacted face of the incident bar. This also serves to indicate the time at which the striker bar subsequently separates from the incident bar. Signals from both the zero time circuit and from the strain-gauges are digitized and stored at rates of up to 1 million samples per second.

The transducer embedded in the soil specimen
This is intended for use in the CST and field trials. A typical example consists of a 28mm diameter 11mm thick Duralumin drum with a wall thickness of 1mm all round (the finite element model of this gauge is shown in figures 1A and 1B). The sensitive element is a 24mm diameter full-bridge strain gauge pattern attached to the inside of the upstream circular face of the drum. This is a commercially available pattern designed to provide maximum sensitivity to a uniform pressure applied to the diaphragm. Lead wires are taken out through a slot in the separately machined lid, which provides the downstream face of the drum, and then through a hole in the confining steel jacket. Other types of pressure transducer, such as thin-foil manganin gauges, will also be evaluated in this apparatus.

At the time of writing, only records from the strain-gauges on the duralumin bars were available. Typical examples are shown in figures 4A to 4F and are discussed later.

Figure 2. Experimental apparatus for the impact loading of soil.

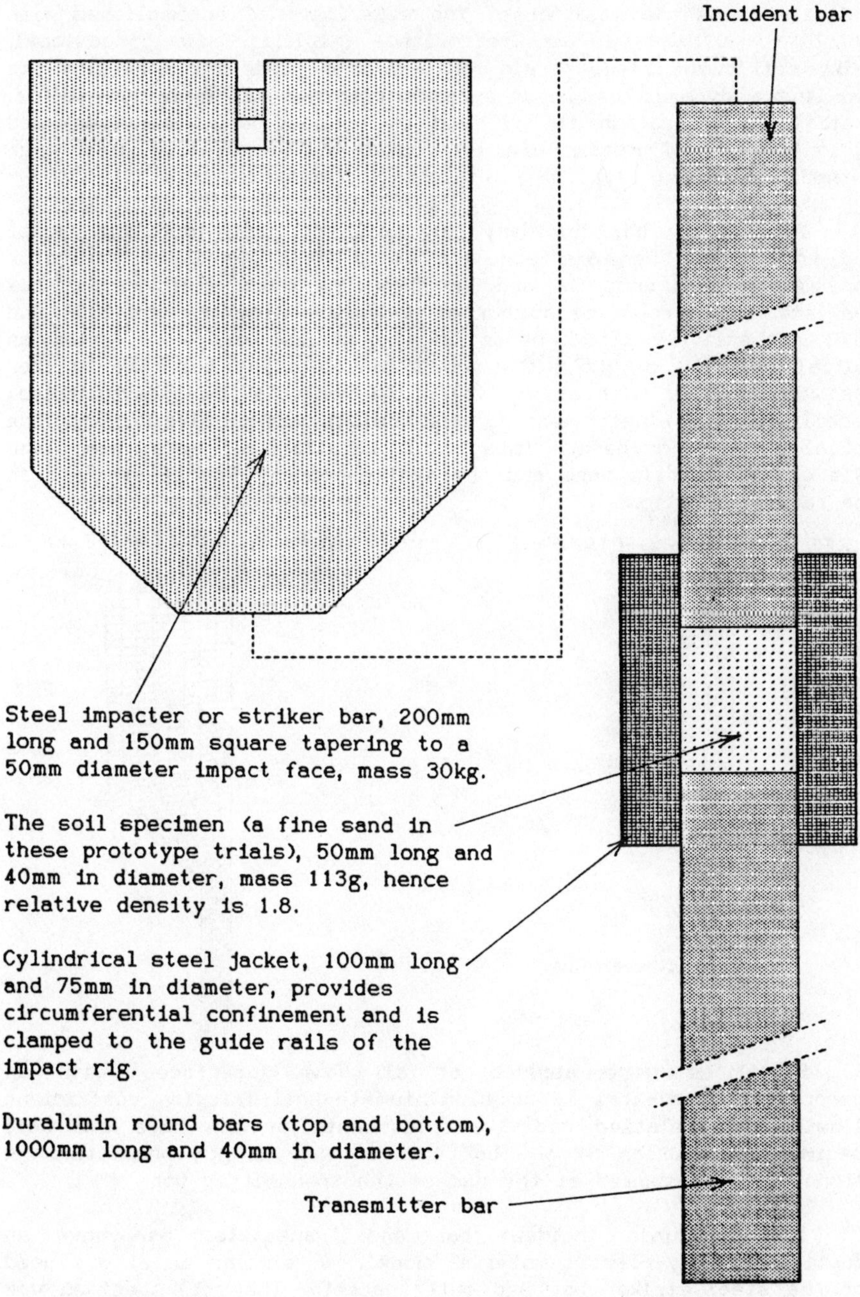

COMPUTATIONAL WORK

A finite element simulation of the experiment is accomplished with DYNA2D. DYNA2D is a vectorized, explicit, two-dimensional, axisymmetric and plane strain finite-element code for analysing the non-linear dynamic and hydrodynamic response of inelastic solids. Spatial discretization is achieved by 4-noded solid elements, and the equations of motion are integrated by the central difference method. (Hallquist [2])

The striker bar, incident bar, soil specimen, confining jacket and transmitter bar are modelled using MAZE, an input generator for DYNA2D. Figures 3A and 3B show the mesh employed in this analysis. The resulting number of nodes and elements are 2950 and 3372 respectively. All nodes in the striker bar were given an initial velocity of 8.5 m/s towards the incident bar. The striker bar was modelled with an equivalent radius to allow axisymmetrical modelling. The kinetic energy and height are equivalent to the actual striker bar before impact. In addition, all nodes along the axis of the elastic bars and the soil specimen were constrained in the radial direction.

FIGURE 3A IMPACT RIG FE MODEL

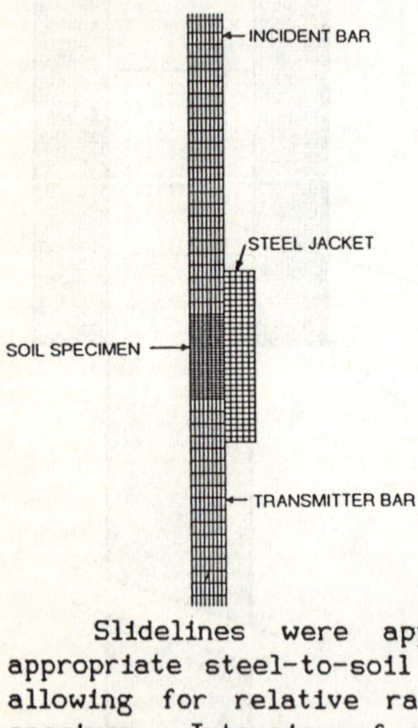

FIGURE 3B IMPACT RIG FE MODEL

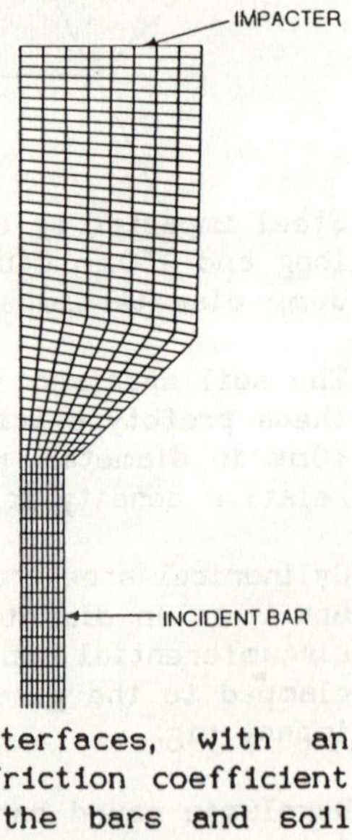

Slidelines were applied at all five interfaces, with an appropriate steel-to-soil or aluminium-to-soil friction coefficient allowing for relative radial displacements of the bars and soil specimen. Intrusion of one bar into another is not permitted. A "stonewall" is assumed at the end of the transmitter bar

The duralumin incident bar and transmitter bar have an elastic/perfectly-plastic material model. A similar model was used for the steel striker bar and outer jacket. The soil specimen was modelled using the soil and crushable foam material model within DYNA2D. The material constants were derived from similar static triaxial tests on a soil specimen.

COMPARISON OF THE RESULTS OF EXPERIMENT AND COMPUTATION

Experimental and hydrocode-calculated strain histories are shown in figures 4A to 4F, the line tagged with an "E" indicating the experimental data. Figures 4C and 4D show experimental strains records from the same gauge location but from different impact tests, as do figures 4E and 4F. For the first 200 micro-seconds after impact the behaviour of the incident bar is not dependent on the unknown soil behaviour and is, therefore, well predicted by the hydrocode. The pulse amplitudes differed slightly on the upper gauge on the incident bar (figure 4A). This was considered to be due to the modelling of the striker bar and a possible error in the axial strain measurement caused by gradual debonding of this gauge before failure. During the following 200 micro-seconds the wave reflected from the lower end of the bar can be seen to reach the gauges at the appropriate times. Pulse amplitudes and pulse lengths are remarkably similar for this period. After 400 micro-seconds, contact between the striker bar and incident bar no longer exists. At 980 micro-seconds contact is re-established, compared with the DYNA2D calculated time of 920 micro-seconds. Initial contact times agree well with experimental measured times. On the second contact a similar pulse is transmitted along the incident bar. Poor correlation between experimental and hydrocode predicted arrival times is observed beyond 800 micro-seconds. Figures 4C to 4F indicate unreliable and unrepeatable experimental data beyond this time, that is, after the second contact. However, pulse shapes and amplitudes are similar.

CONCLUSIONS

It has been shown that the use of a simple constitutive model for soil within DYNA2D can simulate complex one-dimensional wave propagation behaviour. Good correlation for the initial compressive strain pulse with experimental results is observed. However, using simply derived static soil data for a dynamic analysis, where a complex interaction exists between different materials, was found to be inadequate. Calculations of increased accuracy should be obtained by deriving the material constitutive constants from dynamic experiments by Lagrangian analysis. Simple modifications to existing material models, such as strain-rate and strain-rate history dependencies with a movable deviatoric yield surface on loading and unloading and a corresponding movable cohesion constant, will increase the accuracy for complex dynamic analysis. Further experiments will be carried out incorporating improvements to the existing apparatus, in particular to the data aquisition system.

ACKNOWLEDGEMENTS

The work described in this paper was supported by the Procurement Executive, Ministry of Defence.

Figure 4A. Axial gauge 100mm below impacter.

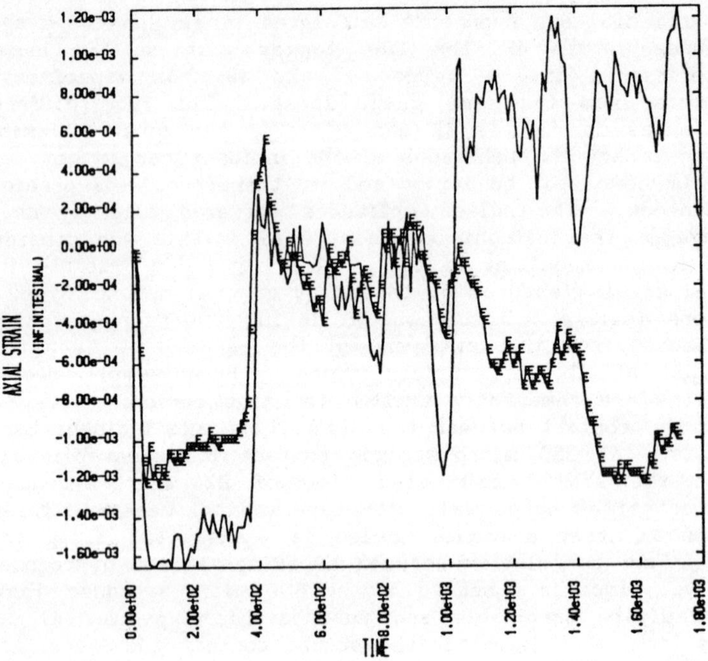

Figure 4B. Tangential gauge 100mm above soil.

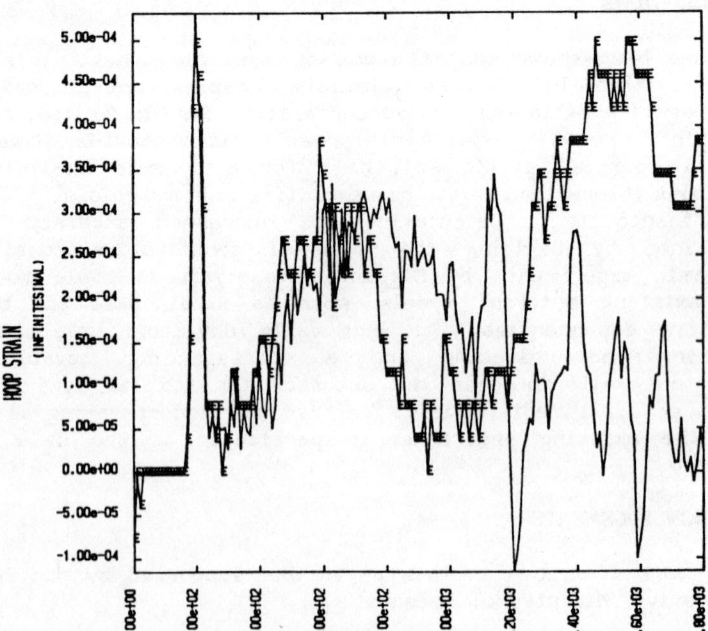

Figures 4C and 4D. Axial gauge 100mm above soil, experimental records from identically set up experiments.

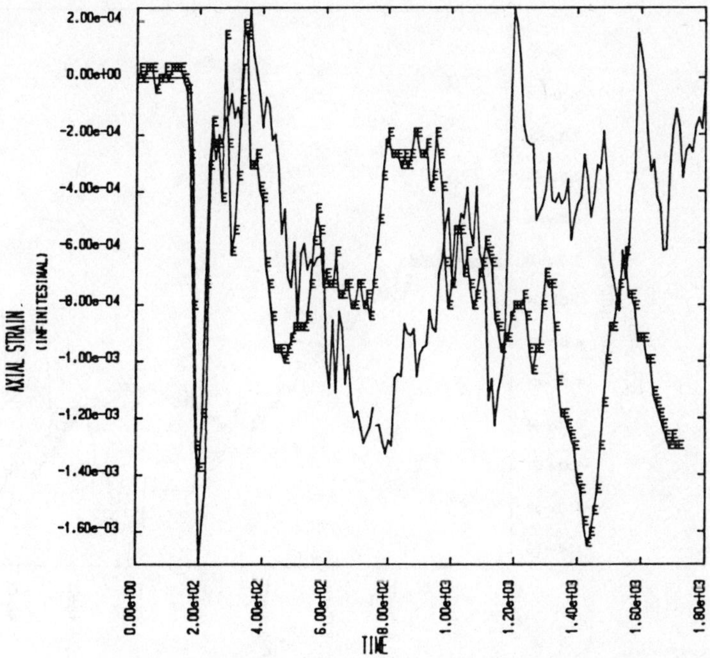

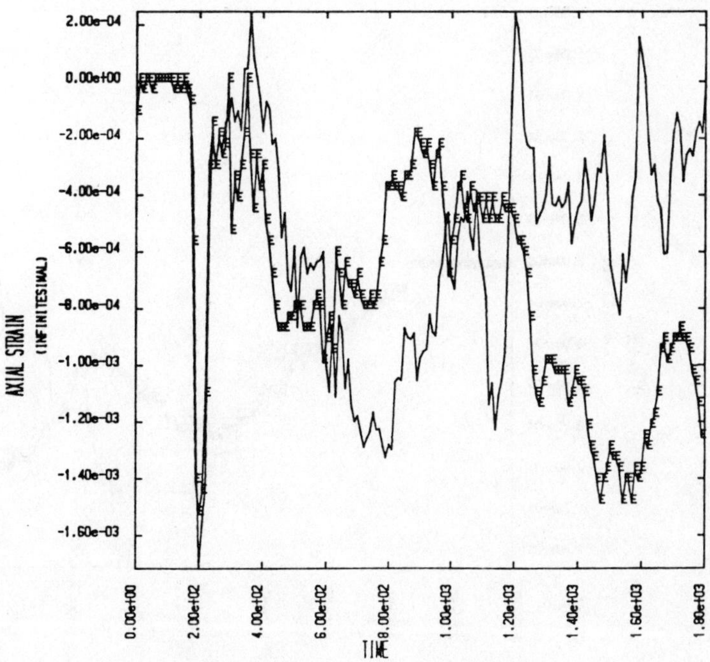

Figures 4E and 4F. Axial gauge 100mm below soil, experimental records from identically set up experiments.

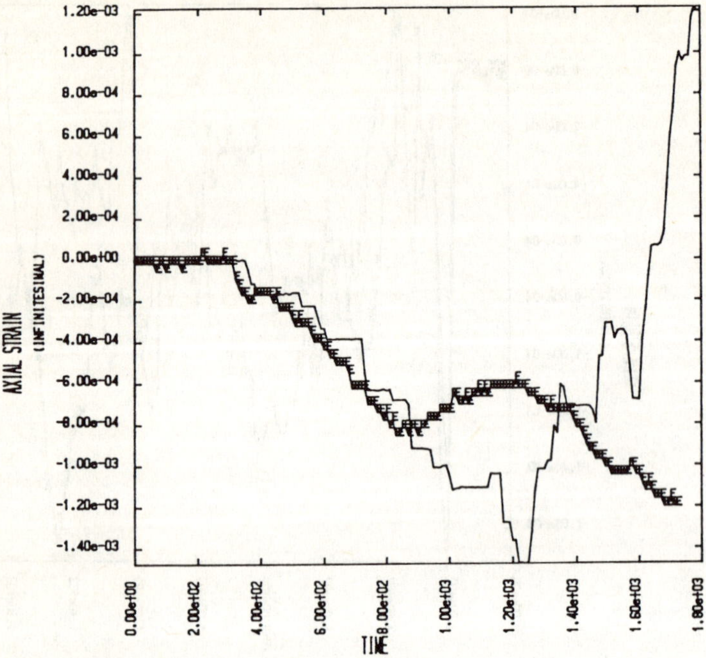

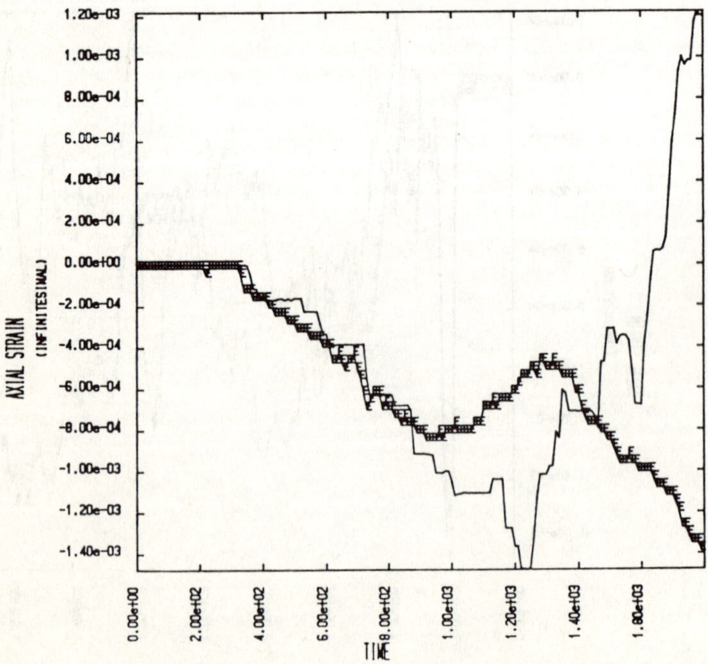

BIBLIOGRAPHY

1. Pullen, A. D. and Perry, S. H. A Laboratory Method for the Investigation of Near-Field Soil Response to Conventional Explosive Blast. Proceedings of the International Symposium on the Interaction of Non-Nuclear Munitions with Structures, Mannheim, 1987.

2. Hallquist, J. O. User's Manual for DYNA2D - An Explicit Two-Dimensional Hydrodynamic Finite Element Code with Interactive Rezoning. University of California, Lawrence Livermore Laboratory, Report UCID-18756 Revision 2, 1984.

Effect of Deformation on the Shock Response of Buried Structures Subject to Explosions

E.E. Hinman

Weidlinger Associates, 333 Seventh Avenue, New York, NY 10001, USA

ABSTRACT

An approximate method is presented for determining the rigid body response of buried structures subject to explosions; the effects of exterior wall deformation and interface shear friction are included. The structure is modeled by a three-degree-of-freedom system. The two exterior walls perpendicular to the blast direction are modeled by lumped masses and elasto-plastic resistance functions. These are attached to either side of a third rigid mass, representing the remainder of the structure. A decoupling procedure is used to model structure-medium interaction effects.

INTRODUCTION

Extensive research in the area of groundshock due to nuclear explosions has resulted in analysis methods which have been successful in predicting the response of buried structures in simulation tests. These problems are solved by using the "soil island" method (Nelson [8]) in which the soil, the structure, and the interface are modeled using finite elements (FEM) and velocities are applied at the boundaries. The models have many degrees of freedom and require constitutive equations for all the materials used.

These calculations are time consuming and have to be performed on mainframe computers. One method for simplifying the computations is to decouple the response of the soil and the structure (Wong and Weidlinger [10]). This eliminates the need for modeling the soil surrounding the structure, significantly simplifying the finite element model. In this method a decoupled exciting force is applied directly to the structure, consisting of the interface free-field stress as modified by structure-medium-interaction (SMI) effects. Results using this method have been shown to compare well with the those of the "soil island method."

These methods have been used with success for the case of nuclear explosions. Unfortunately, the free-field stress is not well defined for non-nuclear groundshock problems and there are only limited test data available. Ref.(2) provides a semi-empirical expression for the free-field stress based on test data from the early 1970's. Reservations have been expressed regarding the uncertainty of some of the parameters in this free-field definition, particularly use of seismic velocity to characterize the soil. Considering the degree of uncertainty in the definition of the free-field source function, elaborate FEM computations are not justified at this time. However, lumped mass models may be used effectively to gain insight into the physical behavior of these systems.

Lumped mass models have been used in recent years to provide an approximate solution for both the deformation response and the shock response of buried structures subject to non-nuclear explosions. In Refs. (3), (7) and (9) the decoupling procedure of Ref. (10) is used to determine the deformation response of single-degree-of-freedom (SDOF) systems. Ref. (9) also provides an approximate closed form solution for the shock response of a buried structure modeled as a rigid mass, and in Ref. (6) the effect of adding above ground stories to the structure is examined.

In these studies the two principal response modes, the deformation response and the rigid body response, are treated separately. However, in certain circumstances the interaction of these two modes is significant. In this paper, the effect of deformation response on the rigid body motion of the structure is explored. The rigid body response model of Ref. (9) is improved by including the deformation of the exterior walls perpendicular to the blast, and also the effect of interface friction. Deformation of the exterior walls is modeled using the SDOF method of Refs. (7) and (9).

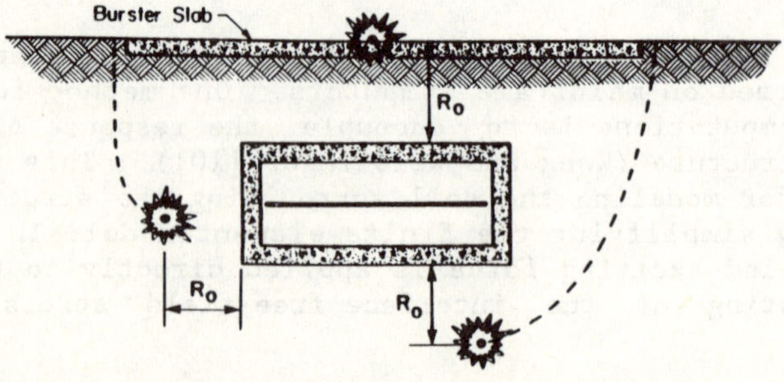

Figure 1 - Buried Protective Structure

An example problem is solved using numerical methods, and the effect of varying the stiffness of the walls is investigated. The effect of soil seismic velocity on the response is also explored. Results of parametric variations are compared with the solution for rigid exterior walls and the solution method of Ref. (2). Peak response comparisons, time histories, and response spectra are presented.

PROBLEM FORMULATION

A buried, box-shaped structure is subjected to an explosion located at a range, R_o, from the nearest surface. Three generic configurations are shown in Fig. (1).

The explosion produces a radially expanding shock wave that propagates through the soil. As the stress wave envelops the structure, it moves as a rigid body. This motion is used to define the internal shock environment. It is affected by the threat, the deformation response of the structural components, rigid body rotation, interface shear friction, structure mass, and the soil characteristics.

The primary exciting force is the pressure acting on the surface nearest to the explosion. It is assumed that the walls and slabs are able to survive this loading. The peak pressure varies along this surface [see Fig. (2)]. Only a portion of the wall or slab directly across from the explosion responds through plastic deformation. The remainder responds elastically.

Depending on the threat, the soil properties, and the structure properties gap may form between the structure and the soil, causing the structure to become unloaded. Relative motion between the soil and the structure also causes interface friction to occur on the sides parallel to the blast direction.

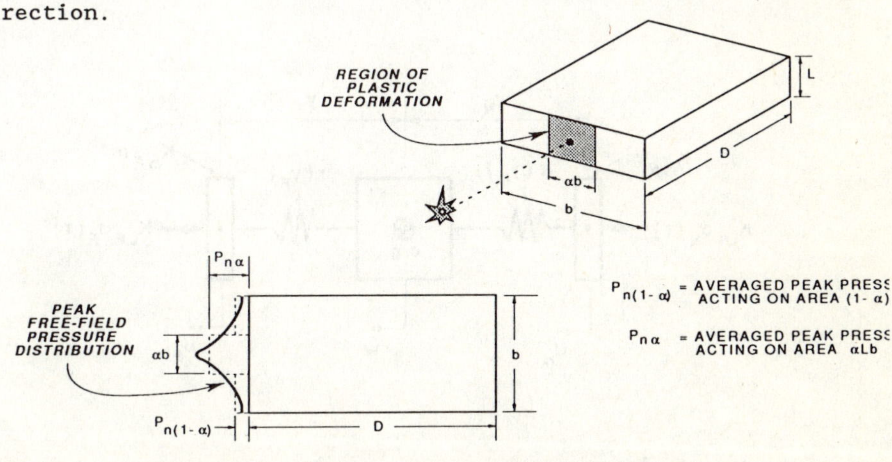

Figure 2 - Effective Width of Near Wall

The rigid body response is used to predict the response of the contents in the building by using the structure velocity as the exciting function on a linear elastic oscillator of variable frequency. This is the in-structure shock. Fragility curves of equipment are used to determine the frequency ranges in which shock isolation is needed.

LUMPED MASS MODEL

The lumped mass model shown in Fig. (3) is used. It models the deformation response of the exterior walls perpendicular to the blast [7,9] and the interface friction. The two walls are referred to as the "near wall" and the "far wall" to indicate their locations with respect to the charge, and it is assumed that the center of gravity (c.g.) of the explosive source is located directly across from the mid-point of the near wall.

The lateral motion of the exterior walls, as a function of time, t, consists of two parts: a) the deformation with respect to the box structure, y(t); b) the rigid body displacement of the c.g. of the structure, w(t).

The total displacements of the near (n) and far walls (f) are therefore

$$u_n(t) = y_n(t) + w(t) \quad (1)$$

$$u_f(t) = y_f(t) + w(t) \quad (2)$$

The displacement $y_n(t)$ in Eq. (1) is the displacement of the near wall along the effective width αb [see Fig. (2)]. The remaining width of the near wall is assumed to remain rigid.

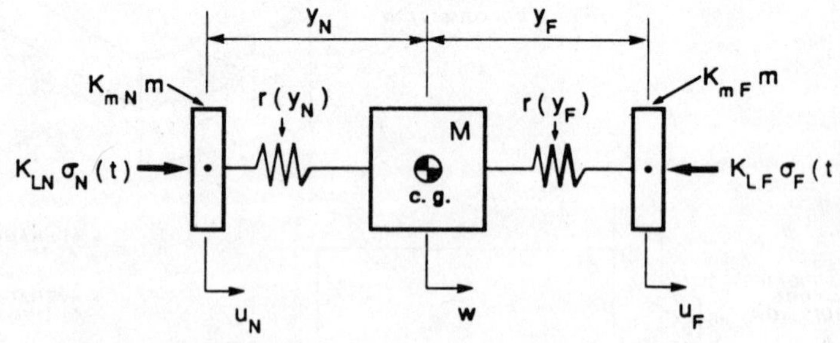

Figure 3 - Lumped Mass Model

The interface pressures [7,9] acting on these walls, are

$$\sigma_n(t) = 2P_{n\alpha}(t) - \rho c\dot{u}_n \tag{3}$$

$$\sigma_f(t) = \qquad\qquad - \rho c\dot{u}_f \tag{4}$$

where (.) indicates differentiation with respect to time, ρc is the average acoustic impedance of the soil and, $P_{n\alpha}$, is the average free-field stress acting on the effective near wall area αbL. Ref. (2) provides formulas and tables for evaluating $P_{n\alpha}$ and ρc.

For walls of mass, m, elasto-plastic flexural resistance, r, subjected to the interface pressures of Eqs. (3) and (4), the equations of motion of the near and far walls [7,9] are respectively,

$$K_{LMn}\, m\ddot{y}_n + \rho c\dot{y}_n + r(y_n) = 2P_{n\alpha}(t) - m\ddot{w} - \rho c\dot{w} \tag{5}$$

$$K_{LMf}\, m\ddot{y}_f + \rho c\dot{y}_f + r(y_f) = \qquad\qquad - m\ddot{w} - \rho c\dot{w} \tag{6}$$

where K_{LM} is the ratio of the mass factor, K_M, to the load factor, K_L, [1] and all terms are expressed in units of force per unit area.

Eqs. (5) and (6) must satisfy the initial conditions

$$u_n(0) = u_f(0) = w(0) = 0 \tag{7}$$

$$\dot{u}_n(0) = \dot{u}_f(0) = \dot{w}(0) = 0 \tag{8}$$

and the load cut-off conditions:

$$\text{when } \sigma_n < 0, \text{ we set } \sigma_n = 0 \tag{9}$$

$$\text{and when } \sigma_f > 0, \text{ we set } \sigma_f = 0 \tag{10}$$

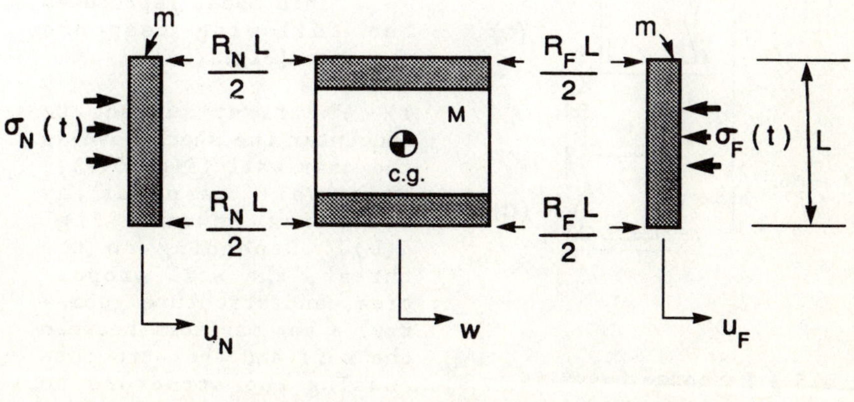

Figure 4 - Exterior Wall Reactions

The loading acting on the exterior walls is transmitted to the box structure by: the elasto-plastic reactions, R_n and R_f, acting over the effective width, αb; the average pressure, $P_{n,1-\alpha}$ acting along the remainder of the width, $(1-\alpha)b$, and the interface friction, F [see Fig. (2)].

The equation of motion for the box structure is:

$$M\ddot{w} + F = \alpha(R_n + R_f) + (1-\alpha)P_{n,1-\alpha} \tag{11}$$

where

$$R_n = 2P_{n\alpha} - m\ddot{w} - \rho c\dot{w} - K_{Ln}[m\ddot{y}_n + \rho c\dot{y}_n] \tag{12}$$

$$R_f = - m\ddot{w} - \rho c\dot{w} - K_{Lf}[m\ddot{y}_f + \rho c\dot{y}_f] \tag{13}$$

and the total mass of the structure is

$$M_t = M + 2m \tag{14}$$

The interface friction expressed as an equivalent pressure is taken to be

$$F = \Phi\, g\, M_t\, D/L \tag{15}$$

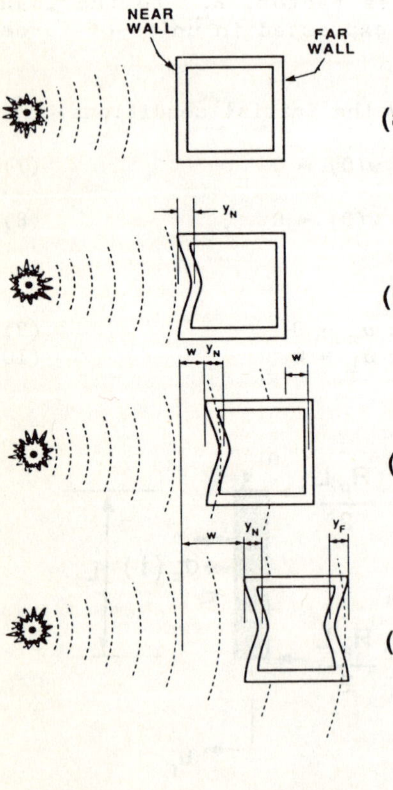

Figure 5 - Response Sequence

where Φ is the friction coefficient, g is the acceleration of gravity, D is the structure length, and L is the structure height.

Limitations of the model are discussed in Refs. (4) and (5) and the effect of shear friction on the response of this system is explored in Ref. (4).

This model reproduces the following response sequence [Fig. (5)]:

1) The first surface to encounter the shock wave is the near wall [Fig. 5(a)]. This wall responds by deforming inward [Fig. 5(b)]. Depending on the threat, the soil properties, and structure geometry, a gap may form between the soil and the structure causing the structure to become unloaded.

2) The loading is transmitted to the box through the near wall reactions, Rn, and $P_{n,1-\alpha}$ [Fig. 5(c)] causing the box to move.

3) The box motion causes the far surface to be loaded with pressure $\sigma_f(t)$, resulting in far wall deformation [Fig. 5(d)]. Load cut-off may occur at this face.

4) The box velocity is reduced by the reactions, R_f, of the far wall and the interface friction, F.

NUMERICAL EXAMPLE

The rigid body response of a buried structure subjected to a side-on burst is considered, for the special case when α=1 and F=0. Input parameters are shown in Fig. (6).

Eqs.(5)-(15) are solved using numerical methods. The peak displacement, velocity, and acceleration are determined to be

w_{max} = 1.4 in $\dot{w}_{max}$ = .01 in/msec $\ddot{w}_{max}$ = 3.4 g's

Detailed discussion of this problem is given in Ref. (5).

EFFECT OF WALL STIFFNESS AND RESISTANCE

The effect of stiffness and ultimate resistance of the walls is considered by varying the wall thicknesses from h=20" to h=90" while keeping the total mass, M_t, constant. The peak displacement, velocity, acceleration, and deceleration of the box structure are plotted against the wall thickness, h, in Fig. (7).

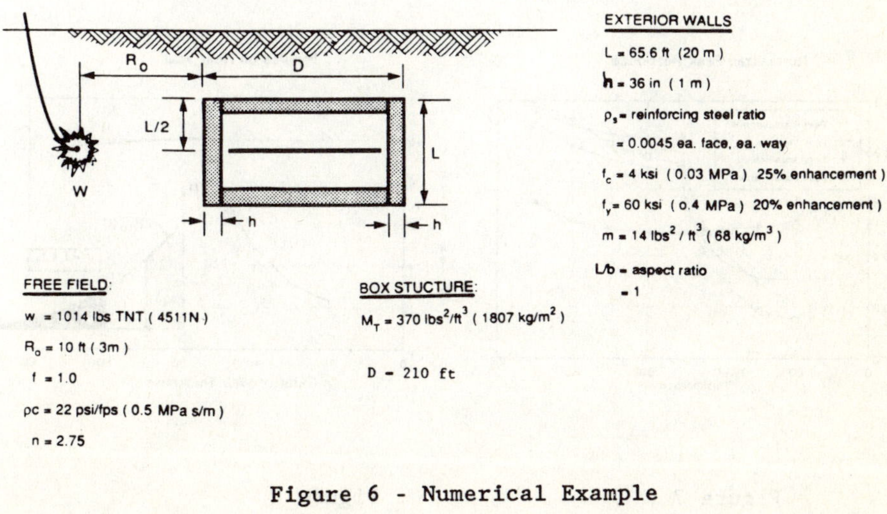

Figure 6 - Numerical Example

Response values in Fig. (7) are normalized with respect to the following rigid (i.e., non-deforming) wall solution:

$$w_{max} = 4.1 \text{ in} \qquad \dot{w}_{max} = .029 \text{ in/msec}$$

$$\ddot{w}_{max} = .0026 \text{ in}^2/\text{ms} \qquad \ddot{w}_{min} = -.00031 \text{ in}^2/\text{ms}$$

Fig. (7) shows that there are two distinct response regimes. For more flexible slabs (h < 51" or h/L < 1/15.4) the peak displacement, velocity, and deceleration are markedly affected by wall deformation -- with peak values that are 20-80% of the rigid wall solution. For stiffer slabs (h > 51" or h/L > 1/15.4) these quantities are less affected by wall deformation -- with values that are 80-100% of the rigid wall response.

The sensitivity to wall stiffness is due to the load cut-off condition of Eq.(9). The load cut-off behavior affects the impulse acting on the system, as shown in Fig. (8). Note that the response curves of Fig. (7) are similar to the pressure impulse of Fig. (8). The peak acceleration shown in Fig. (7) is insensitive to stiffness because it occurs before load cut-off (at $t \approx 0$).

Load cut-off occurs at early times for flexible slabs because these walls experience large plastic deformations which cause a gap to form between the soil and structure, and consequently cause the unloading of the system. Stiffer slabs, experiencing smaller deformations maintain contact with the soil at these times. Load cut-off occurs at later times for stiffer slabs due to the increased rigid body motion of the box structure.

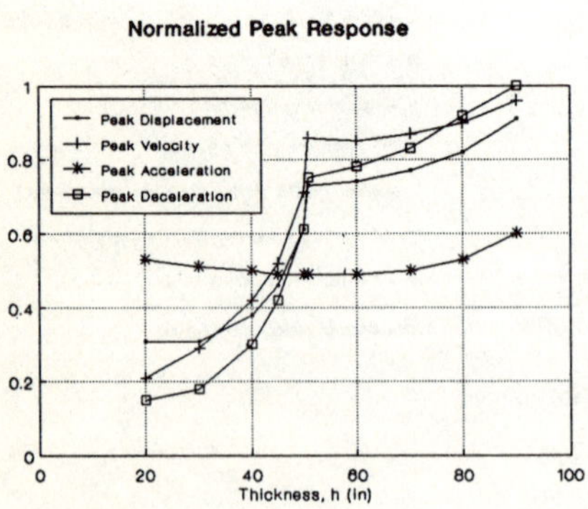

Figure 7

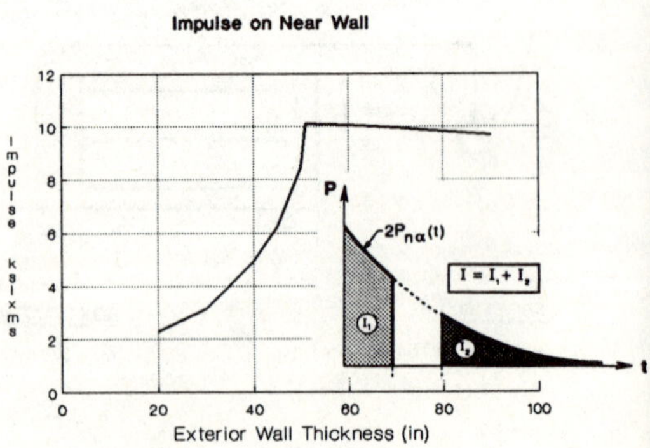

Figure 8

The velocity time histories for the box structure are used as input to predict the response of a linearly elastic oscillator. Damping is taken to be 5% of critical in this study. Results are plotted in the form of response spectra in Fig. (9), for a structure having flexible walls, h=36", stiff walls, h=72", and the rigid wall case. Results are compared with the semi-empirical solution method given by the WES manual [2]. (This method is independent of wall thickness.)

In Fig. (9) the in-structure values show trends that are similar to the peak response values of Fig. (7). The TM5-855-1 manual solution shows greater in-structure displacements and accelerations and a smaller pseudo-velocity for this particular numerical example.

SENSITIVITY TO SEISMIC VELOCITY

In Refs. (7) and (9), it was found that the deformation response for exterior walls was relatively insensitive to the soil characteristics. To determine whether this is also true for rigid body response, seismic velocities from c=.5 ft/msec to 5 ft/msec (using the corresponding attenuation coefficients from Ref. (2)) are considered for a range of depth-to-span ratios.

In Fig. (10) peak in-structure displacement, velocity, and accelerations are plotted against seismic velocity, c, for the four cases: h = 36", h= 72", rigid walls, and the WES manual solution.

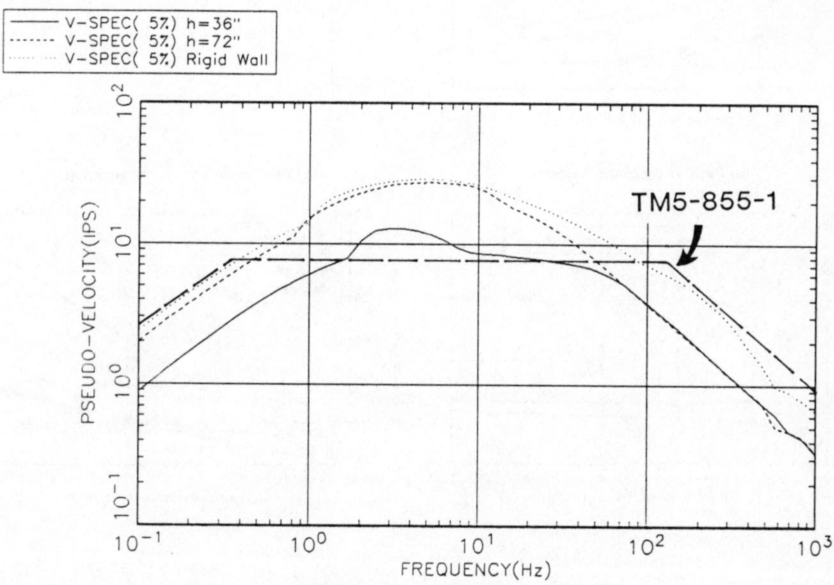

Figure 9 - Response Spectra

The in-structure displacement [Fig. 10(a)] and acceleration [Fig. 10(c)] for the lumped mass system is lower than the WES manual solution, and the in-structure velocity [Fig. 10(b)] is higher. This is consistent with the findings of Fig. (9). Deviation between the two solution methods is apparent for the in-structure accelerations and velocities in the seismic velocity range, c>1000fps [see Fig. 10(b) and 10(c)].

CONCLUSION

The results of this study indicate a sensitivity to the depth-to-span ratio of the exterior walls/slabs. Load cut-off effects have a mitigating effect on the rigid body response of structures having flexible walls. In-structure shock is similarly affected. In-structure velocities and accelerations are found to be less sensitive to the seismic velocity of the soil for c >1000 fps than the TM5-855-1 solution. Validation by detailed finite element analysis and explosive testing is urgently needed.

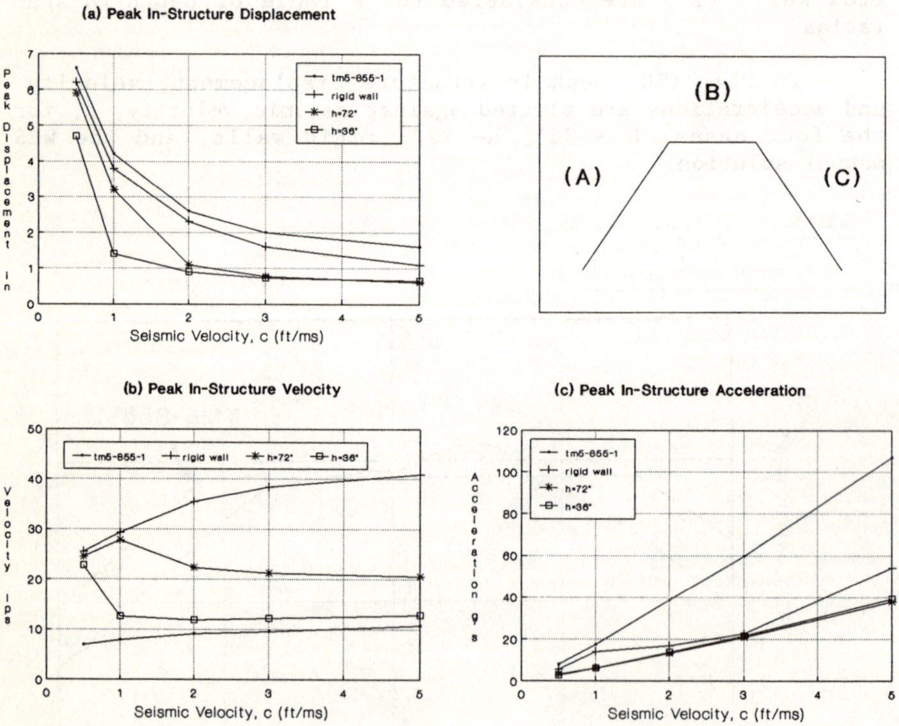

Figure 10 - Effect of Seismic Velocity on In-Structure Shock

REFERENCES

1. Biggs, John, M., Introduction to Structural Dynamics, McGraw-Hill Book Company, New York, 1964.
2. Fundamentals of Protective Design for Conventional Weapons, TM5-855-1, Headquarters, Dept. of the Army, Wash. DC, Nov. 1986.
3. Drake, J.L., R.A. Frank and M.A. Rochefort, "A Simplified Method for the Prediction of the Ground Shock Loads on Buried Structures," Proc. International Symposium on the Interaction of Conventional Munitions with Protective Structures, Mannheim, West Germany, March 09-13, 1987.
4. Hinman, E., "Interaction of Deformation and Shock Response For Buried Structures Subject to Explosions," Proc. International Symposium on the Interaction of Non-Nuclear Munitions With Structures, Panama City, FL, April 17-21, 1989.
5. Hinman, E., "Shock Response of Buried Structures Subject to Explosions," Proc. ASCE Speciality Conference on Structures for Enhanced Safety and Physical Security, March 8-10, 1989, Arlington, VA.
6. Hinman, E., "Shock Response of Underground Structures to Explosions," presented at the Eighth International Colloquium on Vibration and Shock, Jun 14-16, 1988, Ecole Centrale de Lyon, Ecully, France.
7. Hinman, E., and Weidlinger, P.,"Single Degree of Freedom Solution of Structure-Medium Interactions," Proc. International Symposium on the Interaction of Non-Nuclear Weapons With Structures, Mannheim, West Germany, March 9-13, 1987.
8. Nelson, I, "Numerical Solution of Problems Involving Explosive Loading," Proc. Dynamic Methods in Soil and Rock Mechanics, Karlsruhe, 5-16 Sept., 1977, Vol. 2, A.A. Balkema Rotterdam, Holland, pp. 239-297.
9. Weidlinger, P., and Hinman, E., "Analysis of Underground Protective Structures," Journal of Structural Engineering, ASCE, Vol. 114, No. 7, July 1988, pp. 1658-1673.
10. Wong, F.S., and Weidlinger, P., "Design of Underground Protective Structures," Journal of Structural Engineering, ASCE, Vol. 109, No. 8, August, pp. 1972-1979.

Analysis of Shock-loaded Buried Reinforced Concrete Arches using a Hybrid Approach and Advanced Constitutive Models

David J. Stevens* and Theodor Krauthammer**

**Department of Civil and Environmental Engineering, Clarkson University, Potsdam, New York 13676, USA*

***Department of Civil and Mineral Engineering, University of Minnesota, Minneapolis, Minnesota 55455, USA*

INTRODUCTION

The physical processes that govern the dynamic interaction between a soil continuum and an abutting or embedded structure are very complex and, often, highly nonlinear. Some of the phenomena typically observed are: wave attenuation, reflection and refraction; material and geometric nonlinearities; rate effects; and, interfacial effects, such as slippage, debond, and rebond. For highly idealized situations, such as elastic material, simple geometry, and small deformations, closed form solutions for the response of the soil and structure are possible, but, in the majority of cases, the severity of the nonlinearity requires a numerical treatment. The success of any such numerical approach is determined by its ability to accurately reproduce/predict the physical nonlinearities, and, also, by its computational efficiency. Thus, with every improvement made in material models and in computational algorithms, the goal of developing practical, reliable numerical approaches, capable of calculating *a priori* the significant responses observed in soil-structure interaction problems, is brought closer.

The numerical approach presented in this paper represents one attempt at expanding the capabilities of dynamic soil-structure interaction modeling. In this approach, a Hybrid numerical method that merges the Finite Difference Technique (FDT) with the Finite Element Method (FEM) is combined with a rate-dependent cap model for dry sand, a nonlocal continuum damage/plasticity model for concrete, and an elastic/strain hardening plasticity model for steel. The resulting numerical approach is then applied to the analysis of a field test program, in which buried, Reinforced Concrete (RC) arches were subjected to surface generated blast loads (Smith, Betz, and Baird [1]).

HYBRID APPROACH

The Hybrid approach combines the FDT for modeling the soil with the FEM for modeling the structure to yield an efficient computational method. As is well recognized, the FDT is well suited for analyzing wave propagation in continuous nonlinear media with large deformations; the FEM is the obvious choice for structural analysis. The Hybrid approach is briefly presented here; for more complete descriptions, see Stevens and Krauthammer [2].

The combination of the FEM and FDT is motivated by an examination of the discretized form of the equation of motion. In two dimensions, the equation of motion can be written as:

$$\rho \, \underline{\ddot{u}} = \mathrm{Div}\, \underline{\underline{\sigma}} + \rho \, \underline{b} \tag{1}$$

where $\underline{\ddot{u}}$ is the acceleration vector, ρ is the mass density (per unit area), Div $\underline{\underline{\sigma}}$ is the divergence of the two dimensional Cauchy stress tensor $\underline{\underline{\sigma}}$, and $\underline{b}$ is the external acceleration vector. The spatial discretization of this equation is achieved by replacing the divergence operator with an integral definition of the partial derivative (Wilkins [3]) and the acceleration vector with a central difference approximation, as presented in Stevens and Krauthammer [2]. By defining a grid point as the centroid of a small discretized area of the soil domain, and, by applying the Wilkins' technique and the central difference approximation to Eq. 1, it can be shown that, at the grid point,

$$\underline{\dot{u}}^{(t+\Delta t/2)} = \underline{\dot{u}}^{(t-\Delta t/2)} + \underline{F}^{(t)}\Delta t/m + \underline{g}\Delta t \tag{2}$$

where

$$\underline{F}^{(t)} = \underline{P}_E^{(t)} + \underline{F}_S^{(t)}, \qquad \underline{F}_S^{(t)} = \sum_{n=1}^{N} {}^n\sigma_{ij}^{(t)} \, \epsilon_{jk} \, {}^n\Delta x_k \, \hat{1}_i \, ,$$

$\underline{\dot{u}}^{(t+\Delta t/2)}$, $\underline{\dot{u}}^{(t-\Delta t/2)}$ are the velocity vectors at time t - $\Delta t/2$ and t + $\Delta t/2$, respectively, Δt is the time step, $\underline{F}^{(t)}$ is the total out-of-balance force vector on the grid point at time t, $\underline{P}_E^{(t)}$ is the externally applied load on the grid point at time t, $\underline{F}_S^{(t)}$ is the soil force on the grid point at time t (the summation is taken over a discretized closed curve around the grid point), ${}^n\sigma_{ij}^{(t)}$ are the average soil stresses on side n of the discretized curve, ϵ_{jk} is the permutation tensor, ${}^n\Delta x_k$ are the components of the vector representing side n, m is the mass of the grid point, and $\underline{g}$ is the gravity vector.

Furthermore, to include the effects of an adjacent or abutting structure, the force vector generated by the structure at time t, $\underline{F}_{ST}^{(t)}$, can be added to the out-of-balance force vector, $\underline{F}^{(t)}$ and, also, the equivalent lumped structural mass can be added to the mass of the grid point. The calculation of the structure force vector, $\underline{F}_{ST}^{(t)}$, presents a natural opportunity to introduce the FEM. As Eq. 2 shows, the displacements of the grid points at the next time step can be calculated from quantities known at the current time step; if a subset of the grid points are "attached" to a structure, then the structure's displacements and, hence, strains and stresses at time t are also known and the structural resistance force vector $\underline{F}_{ST}^{(t)}$ can be calculated. On the FEM level, the forces associated with the degrees of freedom are equivalent, in a virtual work sense, to the internal element stresses (Stevens and Krauthammer [2]). No limitations are placed on the type of structure, which may be materially and geometrically nonlinear.

The Hybrid method was originally implemented and evaluated by combining a general purpose FEM code with an available FDT code (NESSI, Cundall [4]) in which the Wilkins' differencing technique is implemented. The efficacy of the Hybrid method was validated by Stevens and Krauthammer [2] through comparisons between predictions for simple interaction problems, with and without material nonlinearity, using the Hybrid approach, the FEM, and the FDT. For these simple problems, execution time savings of 50 percent were found for the

Hybrid method relative to a pure FEM Analysis; for more realistic and complex problems, the time savings should be greater.

NONLOCAL CONTINUUM DAMAGE/PLASTICITY MODEL FOR PLAIN CONCRETE

Given the complex nonlinear behavior of concrete, it is not surprising that many distinctly different constitutive theories for the dynamic and static response of concrete have been proposed over the years, as discussed in Stevens and Krauthammer [5, 6]. In the approach used here, the theories of continuum damage mechanics and plasticity are merged to form a rate-independent Continuum Damage/Plasticity (CDP) model.

It must be emphasized that the CDP concrete model discussed here is rate-independent. While it is accepted that the rate of straining can affect the response of concrete, the magnitude, significance and underlying mechanics of this effect are not understood, due to the scarcity of adequate experimental data. As pointed out by Hegemier and Read [7] in their review article on the subject, the apparent rate dependency has been observed most often in uniaxial stress and strain tests wherein strain rates at levels associated with seismic loading were applied; unfortunately, the amount of experimental work that has been performed for blast loading strain rates is sparse. Hegemier and Read [7] question the validity of the uniaxial stress test results by arguing that wave propagation effects in relatively long specimens can greatly alter the specimen's response. They conclude that the existing data for high loading rates ".... probably does not reflect the true rate-dependence of this material." Recently, however, two test programs (Gran et al. [8]; Malvern et al. [9]) have been undertaken in which the radial inertia effects can be minimized or controlled. At the time of this research, the preliminary results from these apparati were still under evaluation, and, as such, do not constitute a reasonable base from which to develop a suitable rate-dependent model. Therefore, the model developed herein is taken as rate-independent, and, as the successful results presented later show, this assumption did not prove to be unreasonable. However, this issue is being addressed in the continuing development of the model.

Under static loading, the nonlinear response of concrete is created through the combination of microcrack growth and frictional slip. The microcrack-induced strength and stiffness degradation can be modeled with the theory of continuum damage mechanics, while the plastic flow and prepeak nonlinearity of concrete, created by frictional slip, is accommodated with the theory of plasticity. The continuum damage mechanics portion of the CDP model based loosely on the scalar (or isotropic) damage approach proposed by Frantziskonis and Desai [10], in which the key assumption made is that the material behavior can be decomposed into two parts, an undamaged portion (or topical portion, in the words of Frantziskonis and Desai [10]) and a damaged portion. Assuming that microcracking reduces the area of contact over which both shear <u>and</u> normal stresses act, the effective stress $\underline{\underline{\sigma}}$ can be defined as

$$\underline{\underline{\sigma}} = (1 - d)\underline{\underline{\sigma}}^t \tag{3}$$

where $d \equiv A_D/A$ is the local scalar damage parameter, A_D and A are the damaged and total areas, respectively, and $\underline{\underline{\sigma}}^t$ is the topical stress. It is also assumed that the topical and damaged stress tensors are symmetric, and, that the strains in the damaged and topical components are equal to the average strains. The evolution of the damage parameter d is governed by a strain based damage surface which incorporates an equivalent tensile strain, defined as the norm of the positive projection of the strain tensor $\underline{\underline{\epsilon}}^+$,

$$\underline{\underline{\epsilon}}^+ = \sum_{a=1}^{3} <\epsilon^{(a)}>^+ \underline{e}^{(a)} \otimes \underline{e}^{(a)} \tag{4}$$

where $\epsilon^{(a)}$ are the eigenvalues of $\underline{\underline{\epsilon}}$, $\underline{e}^{(a)}$ are the normalized eigenvectors of $\underline{\underline{\epsilon}}$, and $<x>^+ = (x + |x|)/2$.

Eq. 3 determines the reduction in stiffness and the development of strain softening in the damaged concrete. As is well known, the application of softening, rate-independent constitutive models leads to mesh dependency, localization of strains, and erroneous predictions of energy dissipation in numerical calculations. Since strain softening in concrete is the result of the formation of discrete internal structure due to microcracking, concrete can no longer be represented as a continuum and some method for enforcing finite energy dissipation must be incorporated. One method is the "nonlocal" approach, in which an appropriately chosen dependent variable is defined as a function of the average or weighted values of the independent variables in the neighborhood of a point. Such an approach was formulated for concrete in a continuum damage mechanics model by Pijaudier-Cabot and Bazant [11], who used a nonlocal definition of the damage variable. The CDP model also incorporates a nonlocal definition of the damage variable, calculated as a weighted average over a representative volume (Stevens and Krauthammer [5]).

The topical stress $\underline{\underline{\sigma}}^t$ is determined through a plasticity approach, developed by Schreyer and Bean [12], in which a strain hardening modified Drucker-Prager failure surface is combined with a curved cap to form a smooth and continuous yield surface. The modified Prager-Drucker yield surface, f^{PD}, is plotted in the $J_2^{1/2}$, P plane in Fig. 1 (J_2 is the second invariant of the topical deviatoric stress tensor and P is the hydrostatic pressure). Over the majority of the yield surface, the plastic strains evolve through the use of an associated flow rule. However, when combined with a pressure dependent yield surface, an associated flow rule predicts dilatational strains, which, in a plane strain state, creates an unrealistic confinement stress in the out-of-plane direction, leading to erroneous predictions of shear and tensile stresses. Therefore, the CDP model employs a nonassociated flow rule in the region of low compressive and tensile hydrostatic stresses; also, as discussed in Stevens and Krauthammer [5, 6], the nonassociated flow rules are modified to remove the possibility of spurious energy generation, which exists when nonassociated flow rules are used. The plastic flow potential surface g is taken as nearly horizontal, as shown in Fig. 1.

ELASTIC/STRAIN HARDENING PLASTICITY MODEL FOR STEEL

The choice of a material model for mild steel is made by considering the specific goal of analyzing RC arches in a blast environment. In RC beams (and arches), a portion of the transverse shear is resisted by the dowel effects of the reinforcement; it is assumed here that the contribution of the steel reinforcement to the shear resistance is small and that the steel response can be taken as uniaxial. With this assumption, a rate-independent, uniaxial, elastic/strain hardening plasticity model is chosen for the steel. Rate effects are not included; a compilation and review of experimental data from strain rate tests on steel (Soroushian and Choi [13]) shows that strain rates up to 10 s^{-1} do not appear to significantly affect material properties for steels with yield strengths over 50 ksi. The steel model used here was proposed by Park and Paulay [14] and it is presented in Fig. 2.

VISCOUS CAP MODEL FOR SOIL

A number of distinctly different approaches have been proposed for modeling the static and dynamic response of soils, as discussed by Chen and Baladi [15]. In the approach used here, a rate-dependent model is formed by combining a soil cap model with a linear viscosity treatment.

Again, as for concrete, it is well accepted that the rate of straining strongly affects the response of most soils; unfortunately, again, the quantity and quality of experimental results for soils under different loading paths and high strain rates are small and, thus, the appropriateness of any rate-dependent model is also open to question. While some of the viscoplastic approaches can approximate the available experimental data, the computational costs of such models are very high, and, due to the limited available data, there is no clear reason to believe that soils behave viscoplastically. Therefore, in order to approximately reproduce the rate effects exhibited by soils under high rates of loading, while retaining computational simplicity, a two-invariant associated cap model is combined with a linear viscosity approach, as suggested by Baladi [16]. While there is no physical reason to believe sand is a viscous material in the classical sense, the use of viscosity does result in a reasonable approximation to the energy dissipation seen in field tests of blast-loaded soils.

In this model, the decomposition of the soil stress $\underline{\underline{\sigma}}^S$ is assumed as:

$$\underline{\underline{\sigma}}^S = \underline{\underline{\sigma}}^{EP} + \underline{\underline{\sigma}}^V \tag{5}$$

where $\underline{\underline{\sigma}}^V$ is the viscous stress tensor and $\underline{\underline{\sigma}}^{EP}$ is the elastic/plastic stress tensor computed with an associated Drucker-Prager cap model, presented in detail by Chen and Baladi [15]. As Fig. 3 shows, the cap failure surface consists of three parts: 1) a modified perfectly plastic Drucker-Prager yield surface f^{MDP}, 2) an elliptic strain hardening cap f^C, and 3) a tension cutoff surface f^T. The viscous stress tensor may be written as:

$$\underline{\underline{\sigma}}^V = \lambda \, (\mathrm{Div}\, \dot{\underline{u}}) \, \underline{\underline{\delta}} + \mu \, [\nabla \dot{\underline{u}} + (\nabla \dot{\underline{u}})^T] \tag{6}$$

where λ and μ are viscosity coefficients, and $\nabla \dot{\underline{u}}$ is the gradient of the velocity vector $\dot{\underline{u}}$. Assuming that the bulk velocity is zero (Baladi [16]) yields $\lambda = 2/3 \, \mu$, and Eq. 6 involves only one material parameter. Since laboratory test results for high rates of loading in dry soils are meager and contradictory, the viscosity parameter is chosen based on the measurements recorded during field testing at a location distant from any structure. The definition of $\underline{\underline{\sigma}}^V$ given in Eq. 6 admits the possibility of both compressive and tensile viscous stresses; in order to obviate this, a spectral decomposition of the soil stress $\underline{\underline{\sigma}}^S$, similar to Eq. (4), is used to remove the tensile stresses. As shown in Stevens and Krauthammer [6], the viscous Drucker-Prager cap model for the soil provides a reasonable and computationally efficient method of including strain rate effects in soil.

INTERFACE MODEL

In general, the accurate representation of the interface between two materials is as crucial to a successful analysis as is the accurate modeling of the materials themselves, as discussed in numerous published papers dealing with contact problems, such as construction joints in concrete, faults in rocks, or metal to metal contact between machine parts. These examples of interface problems contain well defined surfaces over which discontinuities in the tangential and normal displacements are possible. However, under static and dynamic loading, the response of the interface between soil and concrete, two physically similar geomaterials, does not usually involve discrete jumps in the tangential displacements across the interface but, instead, the deformation localizes into a narrow shear band a small distance away from the interface (Desai, Drumm, and Zaman [17]; Mueller [18]; Schreyer and Chen [19]). Recently, Mueller [18], using a unique high pressure, high rate torsional device, found that the response of a sand/rough grout interface was essentially elastic/plastic with a slight amount of hardening. An examination of static and dynamic test results also showed that the yield points of the interface were clustered around the yield curve of the sand and that the post-test surfaces of the specimens were covered with soil; this suggests that failure occurs within the sand, not at the interface, and that the strength properties of the interface are close to the sand's strength properties. Therefore, in this approach, no special interface element is implemented; instead, regular FDT zones and the soil cap model are used. The cap model predicts zero tensile strength so that a state of debond is created when volumetric dilational strains are present, and, compression stresses will likewise be predicted under compaction strains; also, the perfectly plastic modified Drucker-Prager yield surface will predict the elastic/near perfectly plastic shear response seen by Mueller [18]. As shown in the successful results of the analysis, this approach appears reasonable.

BURIED ARCH WITH INTEGRAL FLOOR (KDAT PROGRAM)

In the Kachina Dynamic Arch Test Program (KDAT), three 1/4 scale buried RC arches were tested by the Air Force Weapons Laboratory and the New Mexico Engineering Research Institute at the McCormick Ranch test site, Kirtland Air Force Base, New Mexico in 1983 (Smith et al. [1]). The arch chosen for analysis was 5.58 feet long with an integral floor and was buried 24 inches deep without footings. The geometry of this arch is presented in Fig. 4a, which shows that flexural reinforcement, but no shear reinforcement, is used in the arch and the floor. As shown later, the lack of shear reinforcement in the arch directly affects its response.

A High Explosive Simulation Technique (HEST), using conventional explosives, was employed to simulate a 1.56 KT yield nuclear explosion at a peak overpressure of 1740 psi, traveling at 10880 feet per second across the soil surface; see Smith et al. [1] for a discussion of the loading.

During the test, the KDAT arch sustained a large amount of damage as Fig. 5 shows; the view in Fig. 5(b) is equivalent to slicing the arch through the center of the floor in the direction of the long axis and then folding it out and laying it flat so that the interior surface is displayed; 0° represents the crown of the arch. Large tensile cracks can be seen at the crown and at the floor center (180°). The concrete spalled on the interior surface at 45°

to 60° and the interior reinforcement in this area buckled. In Fig. 5(b), shear cracks can be seen between the crown and 45°. The joint connecting the arch and floor (at 90°) remained intact, but some spalling was observed.

Hybrid Model

Since the arch was located several meters from the perimeter of the test bed, the unloading wave generated at the test bed edge does not reach the arch before the initial vertical pulse of the shock wave passes through the arch. Therefore, a state of plane strain is assumed, and a two dimensional model is used for analysis. The FDT mesh of the Hybrid model used to model the soil is shown in Fig. 4(b). Although the surface blast pressure travels with a finite velocity across the test site, it is assumed that the speed is sufficiently fast such that the pressure history occurs simultaneously at all surface points; thus, only half of the domain is modeled. The FDT portion of the Hybrid approach uses a "mirror" approach to subtract off the wave as it reaches the boundary, minimizing the effect of wave reflections (Cundall [4]); thus, the left and bottom edges of the grid are modeled as transmitting boundaries. On the right hand edge, the vertical displacement is free and the horizontal displacement is fixed. The FEM mesh is shown in Fig. 4(a); eighteen large displacement, layered Timoshenko beam elements (described in Stevens and Krauthammer [5]) are used for the arch and eight beam elements are used for the floor. At the upper and lower right hand nodes, the horizontal displacement, bending rotation, and shear rotation are fixed at zero.

Loading

As mentioned above, chemical explosives were used to simulate a 1.56 KT yield nuclear explosion at a peak overpressure of 1740 psi. The average measured peak overpressure was found to be 2030 psi. For this analysis, the blast load is simulated with the Brode triple exponential function (Brode [20]). Fig. 6 presents the surface blast pressure history used in the analysis; the Brode coefficients are listed in Stevens and Krauthammer [6].

Material Properties

Only the uniaxial strength of the concrete and the yield and ultimate strength of the steel were reported for the KDAT test program (Smith et al. [1]); thus, "typical" values for the concrete and for the steel are assumed. Also, since only the angle of friction for the soil at the test site was available, typical values for the cap model are assumed. The modulus of elasticity, Poisson's ratio, and the viscosity for the soil are chosen to match the arrival times and the stress and velocity histories measured in an earlier test program at the same site with the same type of blast load. Using the measured surface pressures, a one dimensional analysis of a soil column was performed to obtain these values. The material constants used for the steel, concrete and soil are listed in Stevens and Krauthammer [6].

Comparison of Analytical and Experimental Results

The results of the Hybrid analysis are compared with the strain gage and accelerometer data recorded during the testing of the KDAT arch. In the figures that are discussed, compression is taken as positive, 0° refers to the crown of the arch and 90° refers to the joint between the floor and the arch.

The outer and inner reinforcement bar strain histories at 0°, 45°, and 74° are shown in Figs. 7, 8, and 9, respectively. At 0°, there is excellent agreement between predicted and computed strains for the outer bar; the predicted and measured values for the inner rebar also agree very well in arrival time and peak, but the measured value drops off to a residual value of 4000 microstrain. It appears that the local buckling observed in the real arch at approximately 60° may relieve the strain at this point. However, overall, the agreement at 0° is excellent.

At 45°, the agreement in peak strain and time of arrival are excellent for the outer and inner reinforcement bars, but, after the peak strains, the two measured strains display different behavior. Again, this may be due to the buckling of the inner bar that occurred at 60°, midway between the gages at 45° and 74°. However, the agreement is very good.

At 74°, the peak strains and the overall shape of the curves match well for both bars, but the predicted time of peak strain in the inner bar occurs approximately 2 msec too soon, and, in the outer bar, the difference is about 6 msec. This time shift is almost certainly due to the buckling of the inner reinforcement at 60°; the buckling delays the transfer of the compressive thrust down to the joint, as the strain gage data implies. Unfortunately, while the Timoshenko beam element can model the flexural and shear motions, it cannot reproduce the localized buckling of the reinforcement after spalling of the concrete. In any event, the predicted values do supply a good estimate of the peak strains that occur at 74°.

The next comparisons are between the computed and measured vertical velocities at 0°, 45°, 90°, and at the floor center, as shown in Fig. 10. At 0°, the peak value and the time of arrival agree well but the calculated value drops off, unlike the measured velocity. At 45°, the time to peak velocity and the overall shape of the curves agree well but the magnitudes do not, with the measured value showing more motion. At 90°, the peak velocity and the general shape of the curves agree somewhat but the times of peak velocity disagree. Lastly, at the floor center, the shape and values agree well but the curves diverge after 17 msec. Again, all of the discrepancies between the measured and predicted values could be due to the local buckling at 60°, which reduces the shear and thrust that can be transferred to the joint.

CONCLUSIONS

A Hybrid numerical approach, combining the FEM and FDT, is summarized for the modeling of dynamic soil-structure interaction problems, and, for the specific problem of analyzing shock-loaded, buried RC arches, a nonlocal continuum damage/plasticity model for concrete, a rate-dependent cap model for soil, and an elastic/strain hardening plasticity model for steel are presented.

As the comparisons between analytical and experimental values show, the predictions made <u>a priori</u> by the Hybrid approach with the Timoshenko beam element and advanced constitutive models agree well with the measured responses. It must be emphasized that this is an <u>a priori</u> analysis and that the predictions shown are the result of analyses in which the material values, loads, and masses are chosen based on the available pretest data or, in the case of the concrete and soil properties, on available "typical" values.

The one shortcoming of this numerical approach lies in its inability to predict the compression buckling observed at 60°. Due to this, the responses predicted at 74° and at 90° diverge from the measured values but, for the other locations, the agreement overall is good. In future developments of the analytical approach, a method of incorporating buckling into the Timoshenko beam element will be developed and implemented.

For the real arch, it is obvious that shear reinforcement in the areas of high compressive strain (at 45° through 90°) would prevent buckling of the reinforcement bars, thus improving the structural integrity of the arch and reducing the velocities and displacements. Since buckling is a sudden and catastrophic event, it cannot be considered an acceptable failure mechanism even if it allows the arch to deform away from the pressure without a complete loss of structural integrity. In order to control the damage and displacements, energy dissipation should be accommodated through ductile modes of failure, not brittle. In future testing, arches with shear reinforcement should be evaluated.

ACKNOWLEDGEMENTS

This work was sponsored by the Defense Nuclear Agency under Contract No. DNA001-85C-0410 with the University of Minnesota.

APPENDIX I. REFERENCES

1. Smith, J.L., Betz, J.F., and Baird, G.T. Kachina Test Series: Dynamic Arch Test-3 Analysis Report, AFWL-TR-85-63, AFWL, Kirtland AFB, NM, 1986.
2. Stevens, D. J., and Krauthammer, T. A Finite Difference/Finite Element Approach to Dynamic Soil-Structure Interaction Modeling, Comp. and Str., 29(2), 199-205, 1988.
3. Wilkins, M.L. Calculations of Elastic-Plastic Flow, UCRL-7322, Rev. I, U. of Calif., Lawrence Livermore Radiation Lab., 1963.
4. Cundall, P.A. Users Manual and Report on NESSI, Norwegian Geo. Inst. Explicit Soil-Structure Interaction, Shell Research KSEPL, Rijswijk, 1983.
5. Stevens, D. J., and Krauthammer, T. Nonlocal Continuum Damage/Plasticity Model for Impulse Loaded RC Beams, J. Str. Eng., ASCE, Ac. for Pub., 1989.
6. Stevens, D. J., and Krauthammer, T. Development of an Advanced Computational Approach for the Analysis of Buried Reinforced Concrete Structures Subjected to Severe Stress Transients, CME Rep. ST-88-05, U. Minn., Minneapolis, MN, 1988.
7. Hegemier, G.A., and Read, H.E. On Deformation and Failure of Brittle Solids: Some Outstanding Issues, Mech. of Mat., No.4, pp. 215-259, 1985.
8. Gran, J.K., Florence, A.L., and Colton, J.D. Dynamic Triaxial Compression Experiments on High-Strength Concrete, in Proceedings of the ASCE 6th Annual Structures Congress, Orlando, FL 1987.
9. Malvern, L.E., Tang, D., Jenkins, D.A., and Gong, J.C. Dynamic Compressive Strength of Cementitious Materials, Mat. Res. Soc. Symp., Vol. 64, pp. 119-138, 1986.
10. Frantziskonis G., and Desai, C.S. Elastoplastic Model with Damage for Strain Softening Geomaterials, Acta Mechanica 68, 151-170, 1987.
11. Pijaudier-Cabot, G., and Bazant, Z.P. Nonlocal Damage Theory, Rep. No. 86-8/428n, rev. January 1987, Center for Concrete and Geomaterials, Northwestern U., IL, 1987.
12. Schreyer, H.L., and Bean, J.E. Plasticity Models for Soils, Rock and Concrete, Rep. to the New Mexico Engrg. Res. Inst., Albuquerque, NM, 1987.
13. Soroushian, P., and Choi, K.B. Steel Mechanical Properties at Different Strain Rates, J. Str. Eng., ASCE, Vol. 113, No. 4, pp. 663-672, 1987.

14. Park R., and Paulay, T. Reinforced Concrete Structures, Wiley-Interscience, John Wiley and Sons, NY, 1975.
15. Chen, W.F., and Baladi, G.Y. Soil Plasticity, Theory and Implementation, Elsevier Science Publishing, NY, NY, 1985.
16. Baladi, G.Y. Personal Communication, October, 1987.
17. Desai, C.S., Drumm, E.C., and Zaman, M.M. Cyclic Testing and Modeling of Interfaces, J. Geo. Eng., ASCE, Vol. 111, No. 6, pp. 793-815, 1985.
18. Mueller, C.M. Shear Friction Test Support Program; Laboratory Friction Test Results for WES Flume Sand Against Steel and Grout: Report 3, USAE WES, Tech. Rep. SL-86-20, Vicksburg, MS, 1986.
19. Schreyer, H.L., and Chen, Z. One-Dimensional Softening with Localization, J. Appl. Mech., ASME, Vol. 53, pp. 791-797, 1986.
20. Brode, H.L. Analytic Expression for Pressure Time History, PSR Note, 1982.

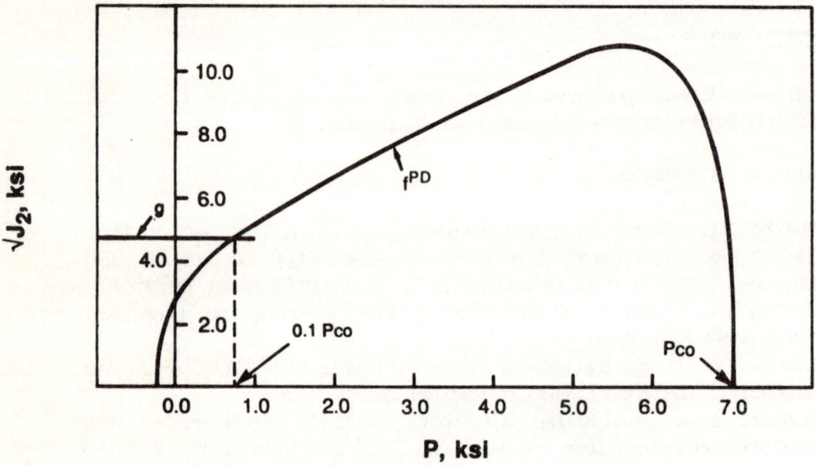

FIG. 1. Prager-Drucker Yield Surface, f^{PD}, with Plastic Potential, g.

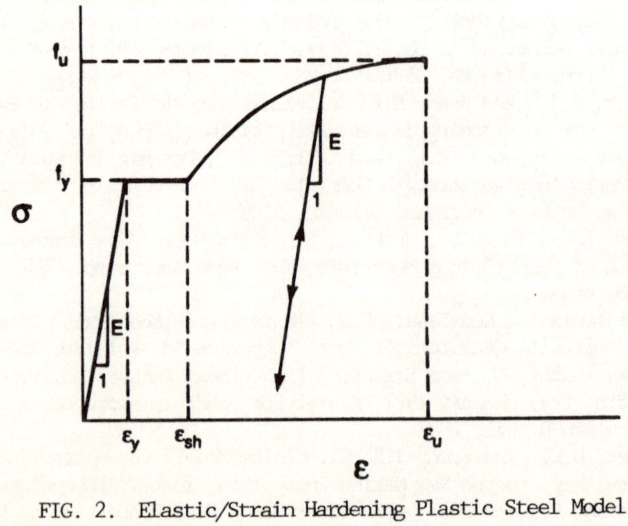

FIG. 2. Elastic/Strain Hardening Plastic Steel Model.

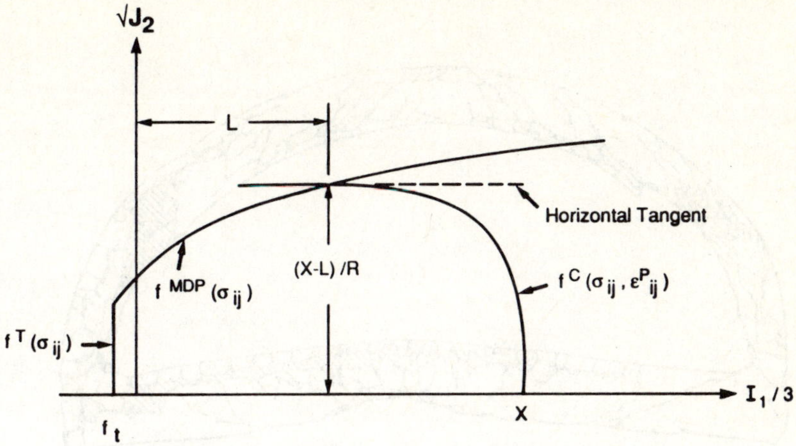

FIG. 3. Drucker-Prager Cap Surface for Sand.

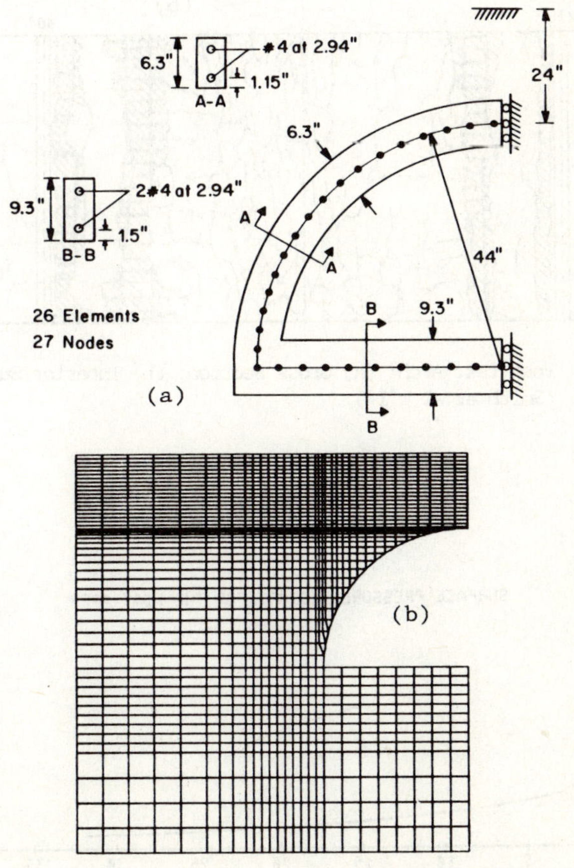

FIG. 4. KDAT Hybrid Model: (a) FEM Mesh; (b) FDT Mesh.

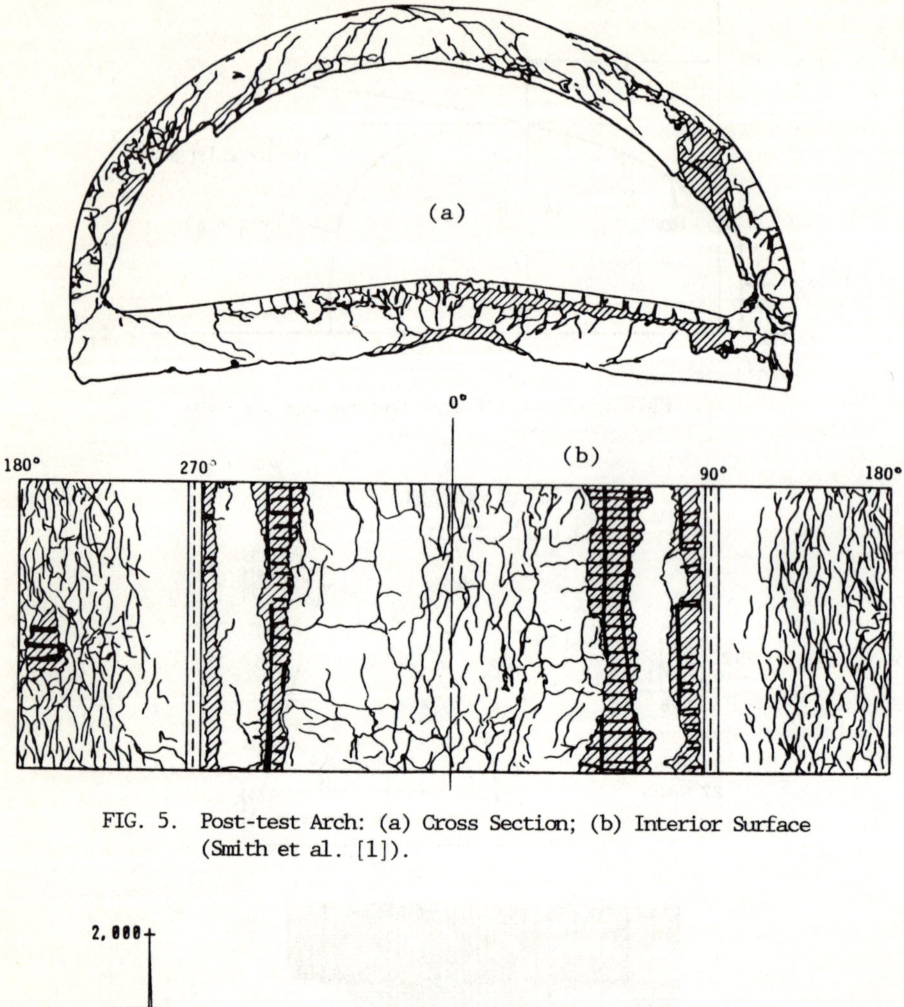

FIG. 5. Post-test Arch: (a) Cross Section; (b) Interior Surface (Smith et al. [1]).

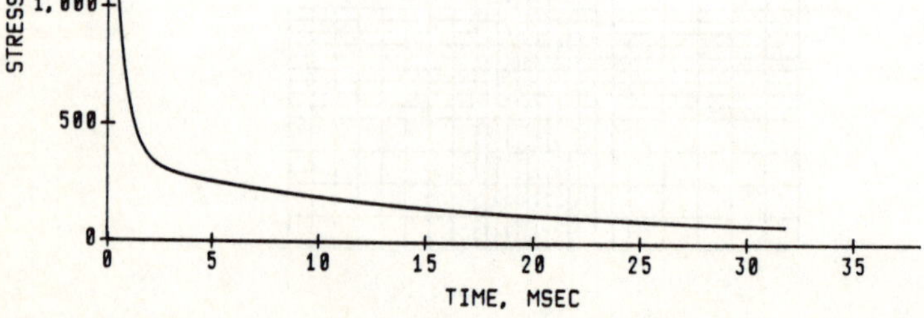

FIG. 6. Applied Surface Pressure History.

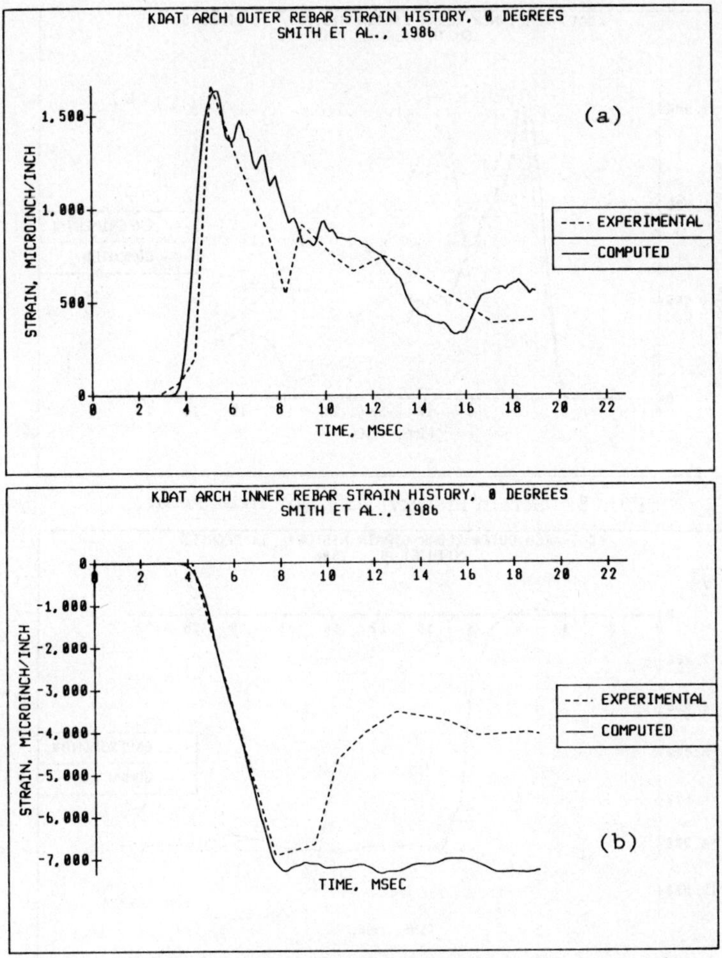

FIG. 7. Strain History, 0°: (a) Outer Rebar; (b) Inner Rebar.

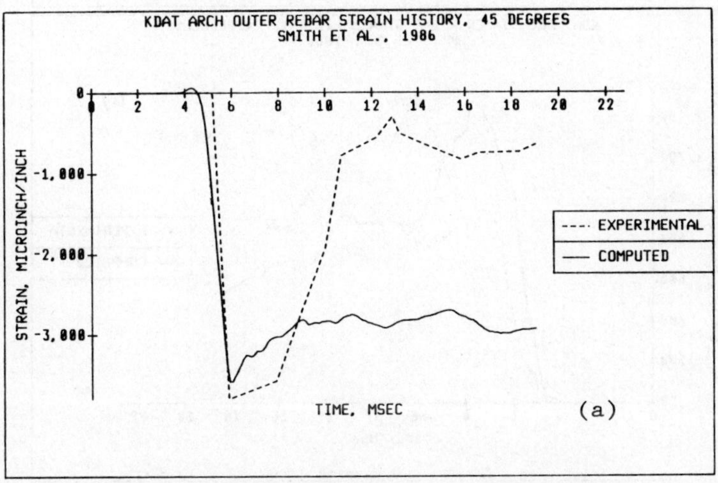

FIG. 8. Strain History, 45°: (a) Outer Rebar.

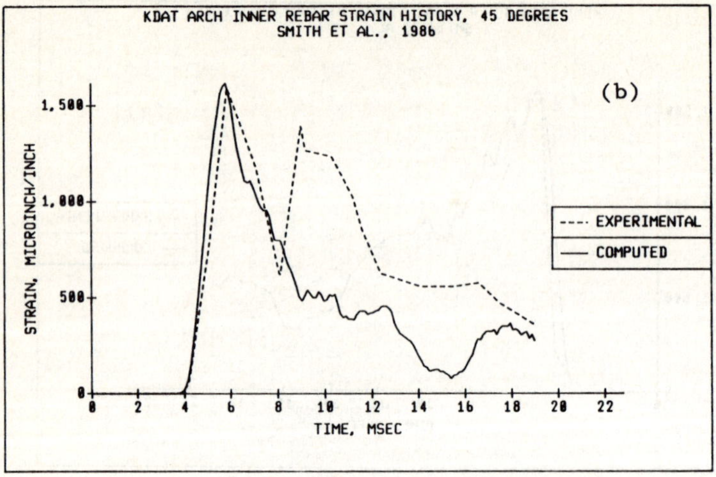

FIG. 8. Strain History, 45°: (b) Inner Rebar.

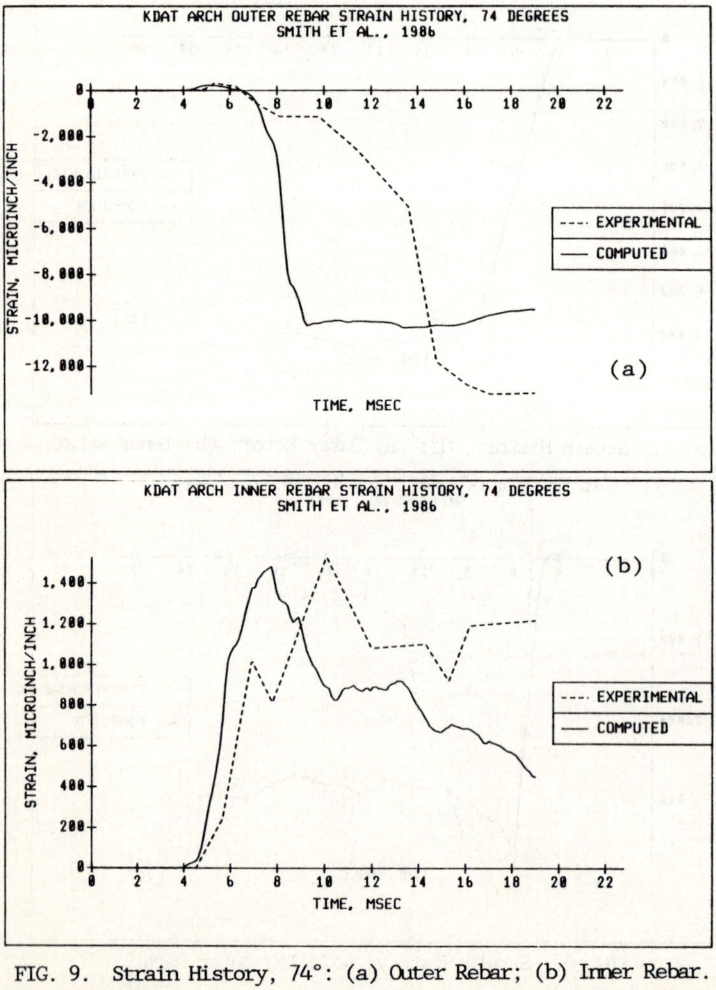

FIG. 9. Strain History, 74°: (a) Outer Rebar; (b) Inner Rebar.

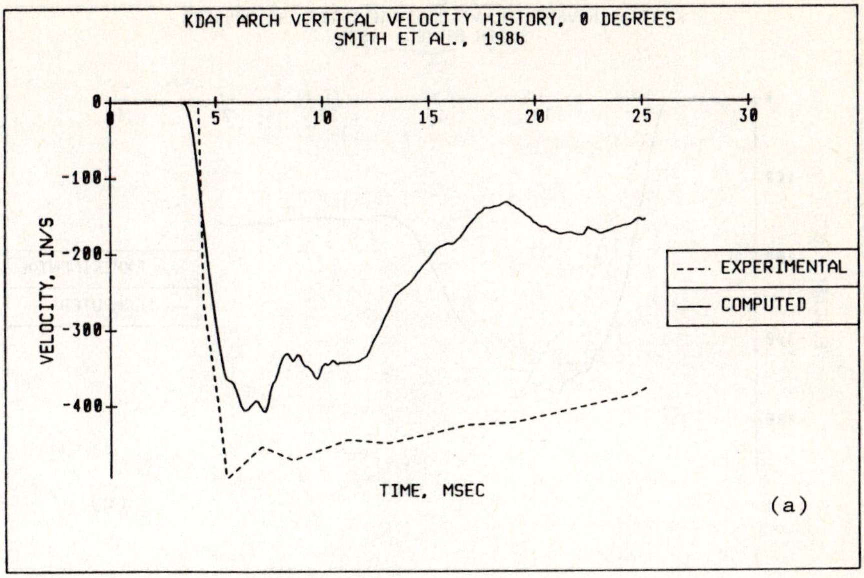

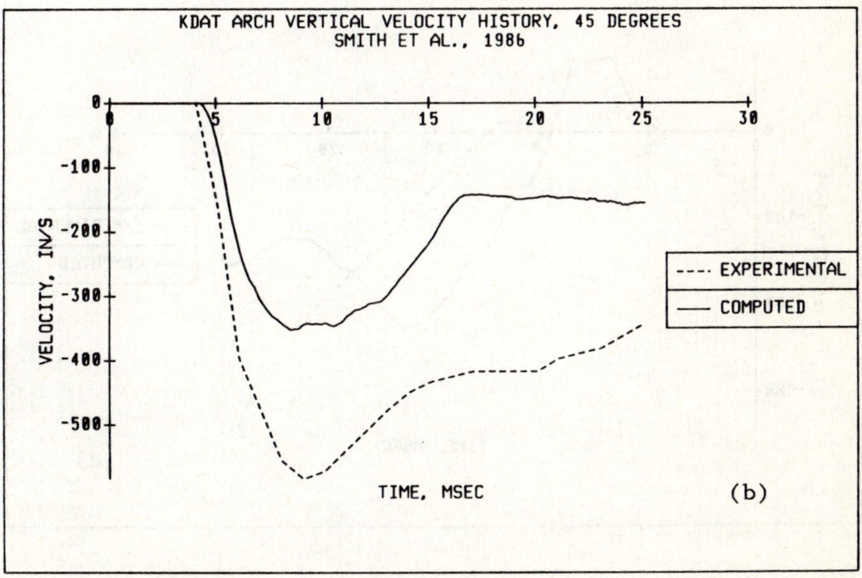

FIG. 10. Vertical Velocity Histories: (a) 0°; (b) 45°.

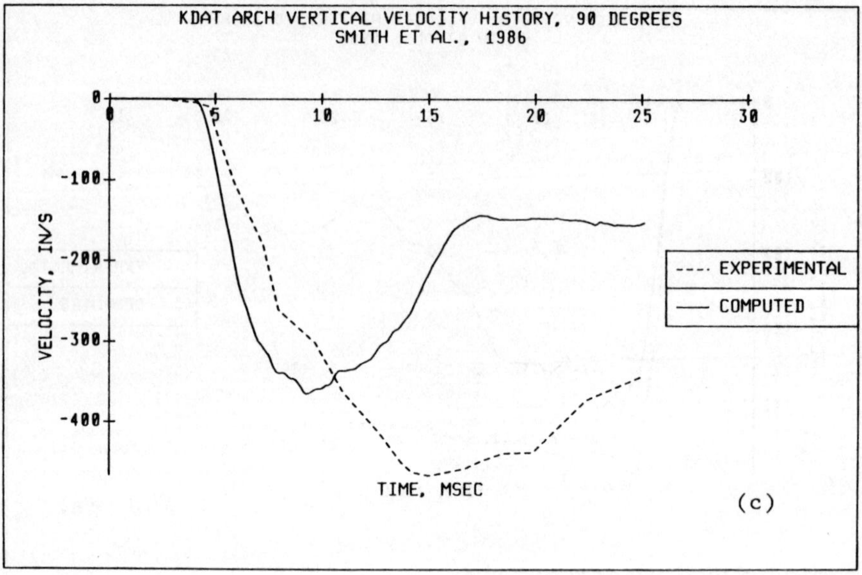

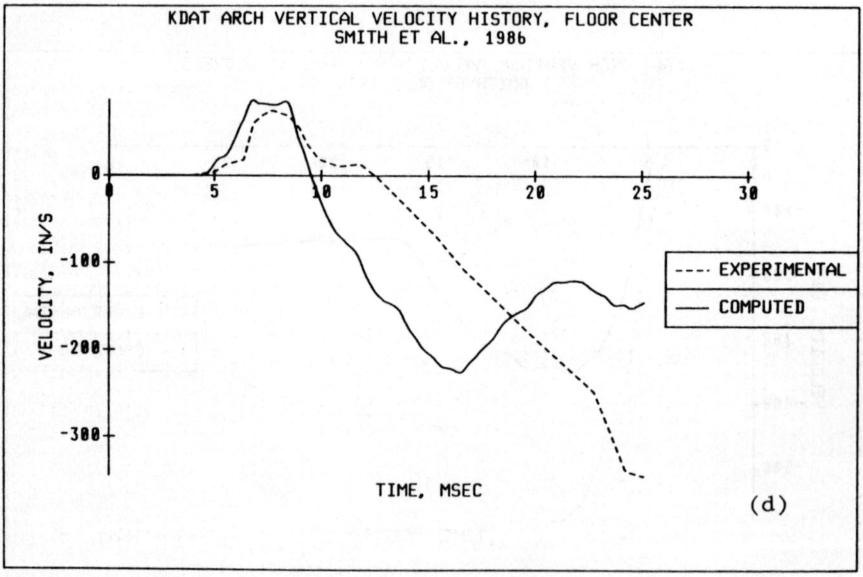

FIG. 10. Vertical Velocity Histories: (c) 90°; (d) Floor Center.

SECTION 8 - GENERAL STUDIES

Simulation of the Dynamics of Clearing Stationary Bodies from Roads with Scraper Blades

P.W. Spence* and J.G. Hambly**

*Hebron & Medlock, Structural Analysis Division, Bristol, UK

**Countermine Section Leader, RARDE, Christchurch, UK

ABSTRACT

A study has been made of the behaviour of scatterable mines when impacted by a scraper blade on a vehicle. Equations of motion have been developed for the mine and blade to determine the influence of many parameters on the behaviour of the mine after impact. However, the analysis has been limited to showing only what type of motion occurs after impact, predicting the complete motion of the mine has not been attempted. Results are presented to show how the mine will have either a stable sliding motion, with some tipping, or a random tumbling motion depending on such factors as impact speed, friction between the mine and the ground or blade, mine geometry and mass properties and orientation of the blade. From these studies it was concluded that it will be possible to devise a clearance strategy for particular mines.

1 INTRODUCTION

Modern warfare has seen the advent of scatterable mines on the battlefield and more specialised area denial sub-munitions on airfields post attack. These are surface laid mines and are in no fixed pattern on the ground. The mines can be remotely delivered by a variety of means over considerable ranges. The result is that all classes of vehicle in all areas of the battlefield need a capability to break-through scatterable minefields.

The simplest technique for breaking through such minefields is to displace the mines with a vehicle mounted scraper blade leaving a cleared path for following vehicles.

The aim of this work was to provide a better physical understanding of the interaction of mines with a scraper blade. The method adopted was to model the mine blade interaction in a general manner incorporating as variables all identifiable parameters. The nature of the mines' motion post impact was considered to be most important as it is this that determines whether or not a mine is cleared from the vehicle path.

The model has been used to identify the relative importance of the blade and mine parameters and the sensitivity of performance to variations in these parameters.

The model is presented for two generic mine types and some typical results described. Further work with the model is planned to optimise mechanical clearance devices.

2. GENERAL APPROACH

The studies were broken down into two main areas of work: 2-D and 3-D studies of the problem. In each case a computer program was written to calculate the motion of the mine during and after impact with the blade. In the 2-D studies the blade was treated as being normal to the path of the vehicle, hence the mine would also move in the same direction. This greatly simplified the representation of the blade and enabled a better understanding of the dynamics of the impact.

A further area of complexity lay in devising routines to check on the geometry of the blade/mine interaction. With complex geometries for the mine, and particularly for the 3-D analysis, there were obvious benefits to be gained from simplifying the initial stages of the analysis.

The motion of the mine and forces acting on it were the principal items of interest in the response. The computer program was written so that values of these parameters were displayed as the analysis progressed. In this way a check could be made for either unexpected trends or enabling the user to stop the analysis at a convenient point.

The performance criteria that could be applied were to determine whether a particular case resulted in the mine sliding away from the blade, tumbling, or slowing down for additional impacts. The aim was to generate results showing the transition between the different types of behaviour as functions of the mine and blade parameters. The analysis was limited to predicting the initial motion of the mine and did not cover the full tumbling motion.

Test data were available in the form of video recordings from a number of trials and a report analyzing the mine motions from the recordings [1]. The nature of the experiments and the difficulty in controlling many parameters meant that there was considerable scatter in the results. However, some clear trends were evident in the recordings and attempts were made to match these with the analyses.

The models were used to determine the values of some parameters that gave the best correlation with the experimental results. These values were then used to explore the effects of other parameters and results are presented as indicated in Section 7.

As these studies progressed it became apparent that the 3-D studies did not greatly extend the knowledge gained from the 2-D studies. Consequently this paper presents only the results from the 2-D studies.

3 DEFINITION OF MINE CHARACTERISTICS

Two mine shapes were analysed in these studies. A cylindrical mine (Figure 1) had a mass of 1.7 Kg and was 120 mm diameter and 70 mm high. The C.G. was assumed to be in the centre of the mine and the moment of inertia about a horizontal axis through the C.G. was 0.0022 Kg.m^2.

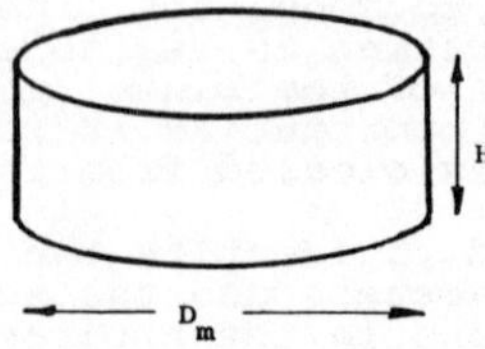

Figure 1 Cylindrical Mine

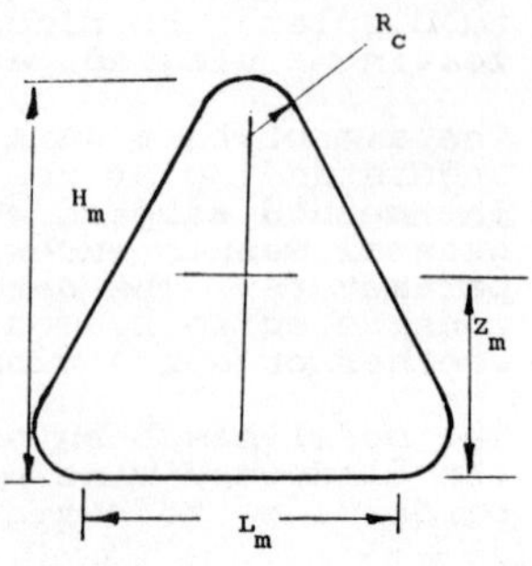

L_m = 40 H_m = 70
R_c = 8 Z_m = 25

Figure 2 Cross-Section of Triangular Mine

The second mine was a triangular prism (Figure 2) with flat sides and radiused corners. The mass was 2 Kg and the moment of inertia about an axis normal to the triangle and passing through the C.G. was 0.0005 $Kg.m^2$.

4 DEFINITION OF BLADE CHARACTERISTICS

The blade was treated as a flat surface attached at a point on its upper edge to a vehicle (assumed to have infinite mass). The overall dimensions, mass properties and stiffnesses are shown in Figure 3.

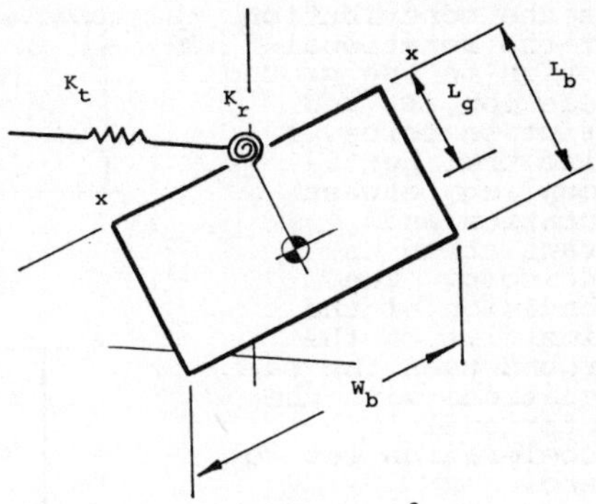

Mass = 25 Kg I_{xx} = 1.44 Kg. m^2
L_b = 600 mm L_g = 350 mm W_b = 900 mm
K_t = 139000 N/m K_r = 16700 Nm/rad

Figure 3 Blade Geometry

5 EXPERIMENTAL RESULTS

The experimental results [1] show some trends that can be compared with simulations results. Because of the nature of the experiments it was very difficult to accurately control some of the parameters - particularly the height of the blade above the ground. Consequently there was considerable scatter in the results. Four main classes of behaviour were identified with transition between the classes related to increasing velocity of impact:

1. The mine 'sticks' to the blade and slides along the ground

2. The mine jolted forward then impacted again with more jolts but less severe than the first and eventually 'stuck' to the blade

3. The mine jolted severely and tumbling, with subsequent impact with the blade. There appeared to be a transition between class 2 and 3 behaviour at impact velocities of 10 - 15 Kph.

4. The mine knocked clear of the blade at the first impact and tumbling.

Additional tests were carried out to determine the coefficient of friction between a mine and the surface of the test track. These tests showed that the coefficient of sliding friction was 0.5 and about 33% higher for static friction.

6 EQUATIONS OF MOTION FOR THE MINES

The equations of motion for the mines have been developed for three conditions:

mine flat on the ground
mine tipping nose-down
mine tipping nose-up

The geometry and equations for the cylindrical mine are shown in Figure 4. The translation and rotation of the mine are x and a with a being -ve for nose-up tipping. The interface forces between the blade and mine are P_i and F_i.

In each case the possibility of the mine being stationary in terms of the forward motion but still rotating has to be considered as well as the contribution of the rotational motion to the ground reaction, R, and friction force, F_r. Also the inertia coupling between rotation and translation is included. The condition of the mine flat on the ground uses the same equations with the rotational acceleration set to zero.

Figure 5 shows the geometry and equations for the triangular mine. The equations of motion for this mine are virtually the same as for the cylindrical mine. The only significant difference is that the moment of inertia for the mine and the distance between the C.G. of the mine and the point of contact on the ground have to be re-calculated at each time step.

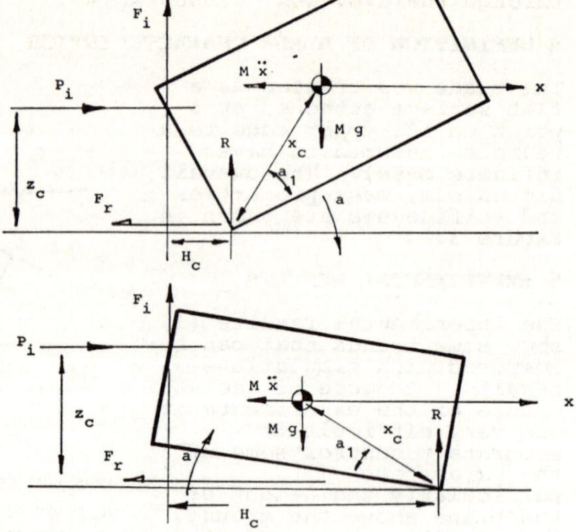

$$a_2 = |a| + a_1$$
$$R = M\,g - F_i - M\,a\,x_c \cos(a_2)\,\mathrm{sign}(a)$$
$$\ddot{x} = [P_i - F_r - M\,\ddot{a}\,x_c \sin(a_2)] \,/\, M$$
$$\ddot{a} = [P_i\,z_c - M\,g\,x_c \cos(a_2)\,\mathrm{sign}(a) - M\,\ddot{x}\,x_c \sin(a_2) + F_i\,H_c\,] \,/\, I_e$$
$$I_e = I_o + M\,x_c^2$$
$$\dot{x} \neq 0, \quad F_r = \mu\,R\,\mathrm{sign}(\dot{x})$$
$$\dot{x} = 0, \quad F_r = \text{lesser of } [P_i - M\,\ddot{x} - M\,\ddot{a}\,x_c \sin(a_2)] \text{ or } \mu\,R\,\mathrm{sign}[P_i - M\,\ddot{x} - M\,\ddot{a}\,x_c \sin(a_2)]$$

Figure 4 Equations of Motion for Cylindrical Mine

7 EQUATIONS OF MOTION FOR THE BLADE

The geometry and equations for the blade are shown in Figure 6. The translational motion is an absolute motion but the rotational motion is defined as relative to an initial position. The blade has translational and rotational stiffnesses and in order to counteract gravity forces there is an initial torque in the rotational spring.

8 BLADE-MINE INTERACTION GEOMETRY

Motion of the mine begins when the blade makes contact with the mine. A check is made at each time-step within the program to determine if contact is occurring. Different geometries have to be considered according to whether the mine is flat or tipped and these are shown in Figure 7 for the cylindrical mine and Figure 8 for the triangular mine.

The general procedure was to use co-ordinate geometry to define a line and a point and to calculate the shortest distance between the two. The line would be either the surface of the blade or the mine (or a tangent to the curved surface of the triangular mine) and the point would be either the edge of the

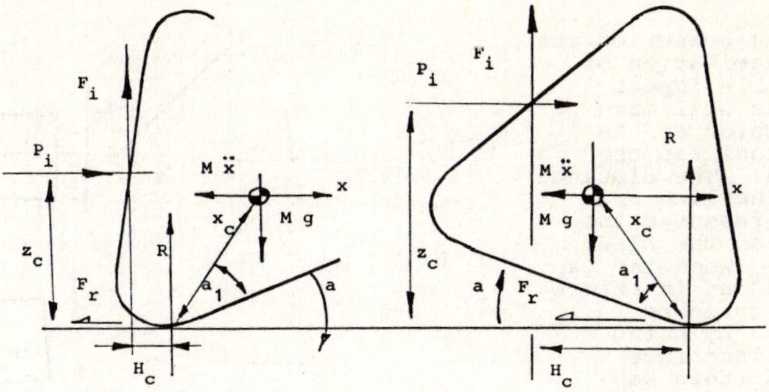

$$x_c = [(z_m - R_c[1 - \cos(a)])^2 + (L_m/2 + R_c \sin(a))^2]^{1/2}$$
$$I_e = I_o + x_c^2$$
$$a_1 = \tan^{-1}[(z_m - R_c[1 - \cos(a)])/(L_m/2 + R_c \sin(a))]$$
$$a_2 = |a| + a_1$$
$$R = M g - F_i - M \ddot{a} x_c \cos(a_2) \operatorname{sign}(a)$$
$$\ddot{x} = [P_i - F_r - M \ddot{a} x_c \sin(a_2)] / M$$
$$\ddot{a} = [P_i - M g x_c \cos(a_2) \operatorname{sign}(a) - M \ddot{x} x_c \sin(a_2) + F_i H_c] / I_e$$
$$\dot{x} \neq 0, \quad F_r = \mu R \operatorname{sign}(\dot{x})$$
$$\dot{x} = 0, \quad F_r = \text{lesser of } [P_i - M \ddot{x} - M \ddot{a} x_c \sin(a_2)] \text{ or } u R \operatorname{sign}[P_i - M \ddot{x} - M \ddot{a} x_c \sin(a_2)]$$

Figure 5 Equations of Motion for Triangular Mine

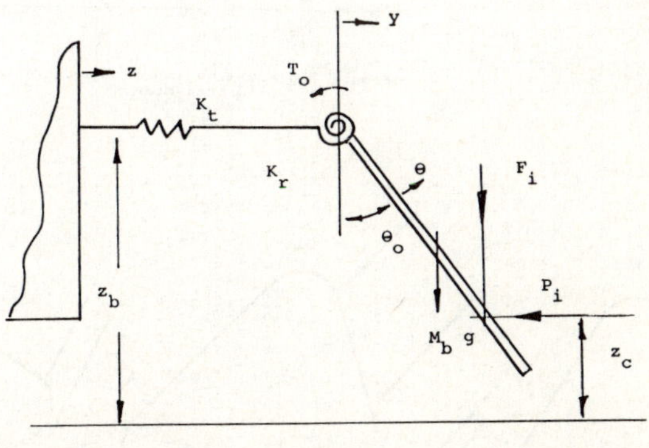

M_b = Mass of blade

T_o = Torque required to establish initial angle O_o

$$y = [K_t(z - y) - P_i] / M_b$$
$$O = [T_o - M_b g L_g \sin(O + O_o) - P_i(z_b - z_c) - F_i(z_b - z_c) \tan(O + O_o) - K_r O] / I_{xx}$$

Figure 6 Equations of Motion for the Blade

blade or a corner of the mine. Examination of the possible impact geometries will make it evident which of the combinations was used in each case. The distance between the line and point represented the gap between the blade and mine. When the gap was closed an interface force P_c could be calculated by using a notional interface stiffness (this was required in order for the mathematical model to operate - it represents a local stiffness for deformation of the edge of the blade and the surface of the mine).

There will be a force F_c normal to P_c due to friction between the blade and mine. In order to determine in which direction it was acting the relative velocity between the blade and mine had to be calculated.

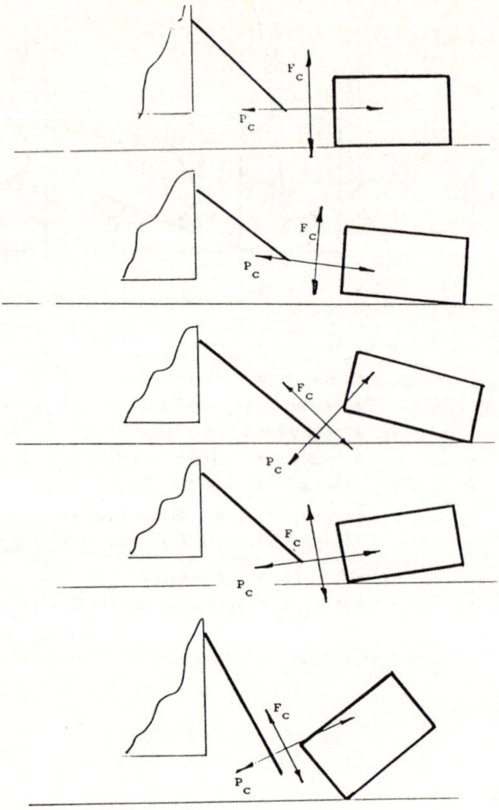

Figure 7 Blade-Mine Impact Configurations for The Cylindrical Mine

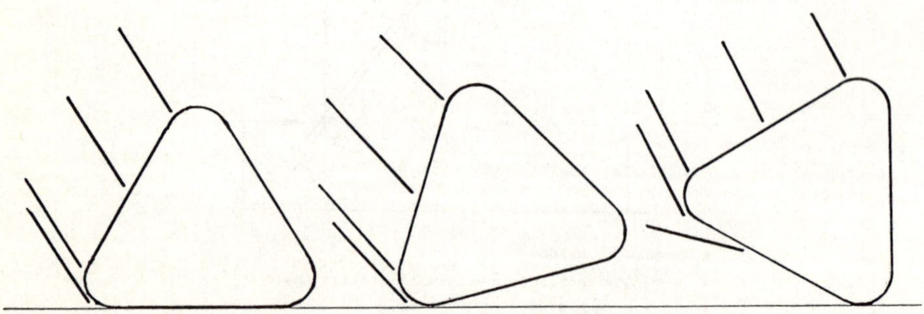

Figure 8 Blade-Mine Impact Configurations for The Triangular Mine

9 SOLUTION OF EQUATIONS OF MOTION

The non-linear characteristics of the problem tend to preclude the use of sophisticated time-step solution routines since the step change in, for example, friction forces could occur in the middle of a step. We have decided to use the simplest form of integration procedure, viz

$$\dot{x}_{i+1} = \dot{x}_i + \ddot{x}_i . \Delta t \qquad x_{i+1} = x_i + \dot{x}_i . \Delta t + \ddot{x}_i . \Delta t^2/2$$

where $\dot{x}_i$ and x_i are the values of velocity and displacement at the beginning of a time step, Δt is the time interval and $\ddot{x}_i$ is the acceleration during the time-step. Using very small values of Δt, typically 10^{-5} sec, satisfactory solutions can be obtained. A facility is included in the program to increase the time-step when certain conditions are satisfied. If the gap between the blade and mine is greater than 10 mm and the mine is flat then the time-step is increased by a factor of 10. If the mine is tipping, then if the angle is greater than about 2 then the time-step is increased by a factor of 10. If the mine is flat and the gap is greater than 50 mm then the time-step is increased by a factor of 25.

10 SIMULATION RESULTS - CYLINDRICAL MINE

There were an enormous number of possible combinations of parameters that could be studied and some arbitrary decisions had to be made to restrict them. The parameters that were of greatest interest were as follows:

- Interface stiffness between blade and mine
- Coefficient of friction between mine and ground
- Coefficient of friction between mine and blade
- Height of blade edge above the ground
- Initial pitch angle of the blade

The height of the blade above the ground was not a parameter that could be controlled in the trials in any way because pitching of the vehicle and local surface irregularities had a major influence on its value at impact. However, from witness marks on the mine it appeared to lie in a range of +/- 10 mm about the C.G. of the mine. This range was used to set maximum and minimum values in many of the simulations.

The principal results that were noted during each run were:

- Maximum velocity of the mine
- Maximum angle of the mine (+ve for nose down, -ve for nose up)
- Maximum acceleration of the mine
- Maximum interface force

Typical results are shown in Fig 9 in which time histories of some of the parameters are shown. This example shows a succesion of impacts with the mine in which the average speed of the mine increased towards that of the blade - eventually the mine will be constantly pushed by the blade.

Maximum Velocity - The maximum velocity of an initially stationary mine after impact was between 1.7 and 2 x the blade velocity. This result was almost independant of the interface stiffness and mostly influenced by the height of the edge of the blade relative to the C.G. of the mine and friction between the mine and the ground. It should be noted that the theoretical maximum velocity of the mine is 2x the velocity of the blade.

As friction increases the maximum velocity will decrease, however the onset of tipping of the mine means further investigation of this parameter is of little use, particularly

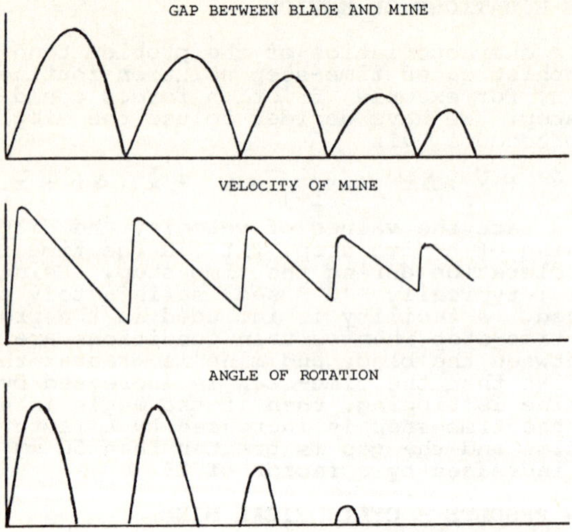

Figure 9 Typical Time-Histories of Motion of Mine

at high values of friction and/or impact velocity when tipping occurs almost immediately. The analysis of the video recordings appeared to show that the velocity of the mine after impact was about 95% of the velocity of the vehicle. This result was influenced by the fact that the measurement of velocity had to be made over several frames of the recording and so was reduced by the deceleration of the mine.

Maximum Angle of Tipping - If the mine is struck above the C.G. then it will tend to tip nose-down and vice-versa. The maximum angle of tipping for this mine was between $30^\circ - 50^\circ$ in either case after which it would overturn and tumble.

These results were strongly dependent on friction and an experimental investigation yielded a value of 0.5 with an increase of 33% for static friction. Figure 10 shows the effect of friction on the maximum angle of rotation with the blade edge set at 25 mm and 45 mm above the ground, i.e 10 mm below or above the C.G. of the mine.

Introducing friction between the blade and the mine also had significant effects. Figure 11 shows the simulation results with the ground coefficient of friction set at 0.5. It is clear from considering the initial movement of the blade against the mine at impact that the vertical interface force will be downwards thereby increasing the ground reaction force and friction force. Also the location of the ground reaction force will move towards the rear edge of the mine which will introduce a stabilising effect. This is seen in the results which show that for impact above the C.G. of the mine a small amount of friction between the blade and mine gives a significant improvement in the stability of the mine. For impact below the C.G. there is a slight improvement in what was already shown to be a generally stable response.

A further investigation was carried out on the effect of the interface stiffness. Some results are shown in Figure 12 and show that while it is not very significant it would be useful to have a representative value for future work. However, knowing that the coefficient of friction between the mine and the ground is about 0.5 the best match of parameter values to give agreement with the experimental results were blade coefficient of friction 0.05 and interface stiffness 10^7 N/m.

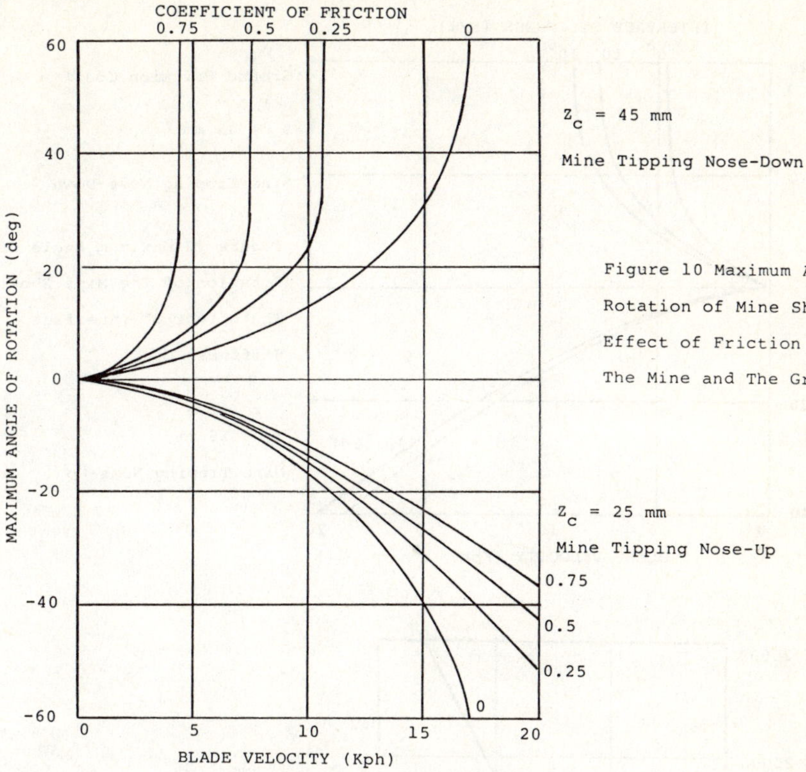

Figure 10 Maximum Angle of Rotation of Mine Showing Effect of Friction Between The Mine and The Ground

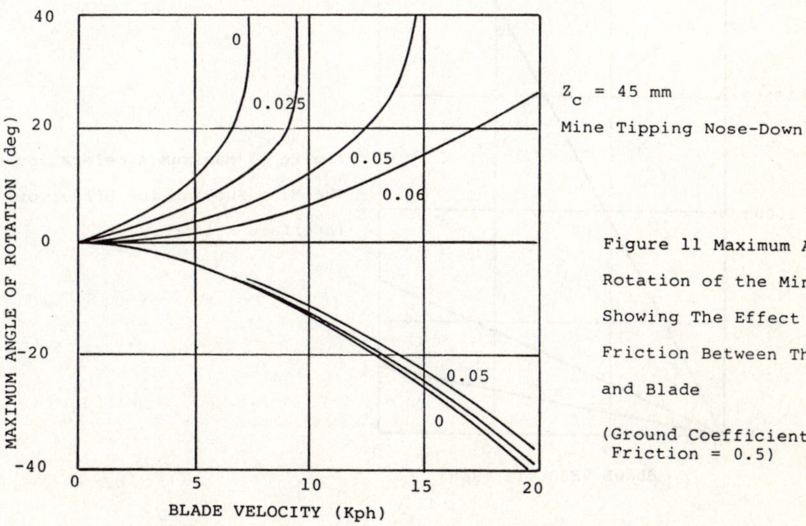

Figure 11 Maximum Angle of Rotation of the Mine Showing The Effect of Friction Between The Mine and Blade

(Ground Coefficient of Friction = 0.5)

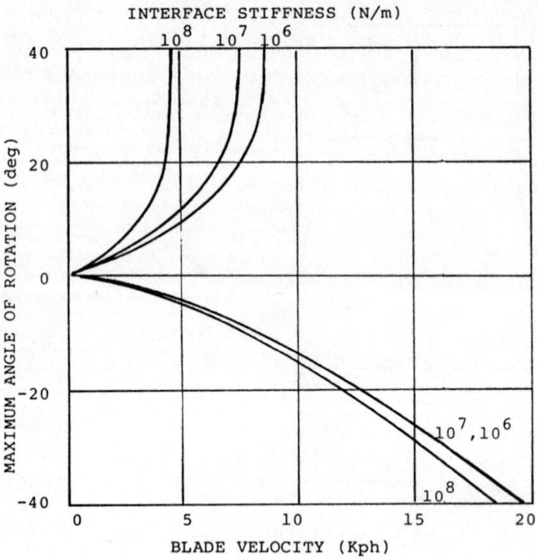

Figure 12 Maximum Angle of Rotation of The Mine Showing The Effect of Interface Stiffness

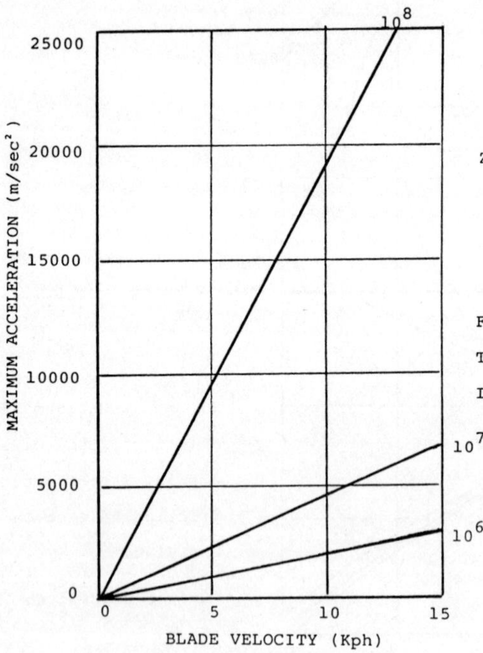

Figure 13 Maximum Acceleration of The Mine Showing The Effect of Interface Stiffness

Maximum Acceleration of The Mine - the maximum acceleration of the mine was very sensitive to the interface stiffness. Figure 13 shows results for three values of interface stiffness with zero friction between blade and mine and nominal friction between blade and ground. The blade height above the ground had a small influence on the results which show that there will be very large acceleration levels but for very short durations.

These results show that it might be possible to determine a representative value for the interface stiffness. At the time that this work was being done a programme of trials was being planned to measure the peak acceleration during impacts.

11 SIMULATION RESULTS - TRIANGULAR MINE

This shape of mine introduces a more varied pattern of behaviour after impact which is very dependent on friction and the height of the blade.

The height of the C.G. of the mine was assumed to be 25 mm above the ground and the radius of curvature of the corners was 8 mm. Hence if the edge of the blade was less than 8 mm above the ground impact other than at very low velocity resulted in nose-down tipping.

With the blade edge higher than 8 mm then there could be nose-up tipping and higher than 40 mm there was nose-down tipping again. This is shown in Fig 14 which presents the maximum angle of rotation for two impact heights as a function of impact velocity and ground friction coefficient. Very varied behaviour occurs and a very interesting phenomenon was noted - for impacts above the C.G. there was a form of 'limit-cycle' behaviour in which the mine oscillated between an angle of 0 and about 40 of rotation. This motion is shown schematically in Figure 15 - an impact above the C.G. is followed by another impact as the lower half of the mine rises underneath the edge of the blade.

Figure 16 shows stability envelopes for the mine as functions of impact velocity and ground friction coefficient. As might be expected, there is a range of blade heights relative to the C.G. of the mine that result in stable sliding motion (with some rotation).

12 CORRELATION WITH EXPERIMENTAL RESULTS - TRIANGULAR MINE

This can only be done on a qualitative basis as there has been no analysis of the video recordings. In general the behaviour predicted by the model matches the experimental behaviour of the mine.

13 3-D SIMULATION STUDIES

A model was written for the cylindrical mine to include motion normal to the motion of the vehicle that would result from the blade being set at an angle to this direction.

In general the results of these studies did not add to the understanding of the motion of the mine that had been obtained from the 2-D model. Also, there was insufficient experimental data available to make other than very superficial comparisons with the simulations. More detailed tests are planned for the future.

14 CONCLUSIONS

It has been shown that 2-D simulations of the impact between a vehicle mounted scraper blade and a mine on a hard, level surface can match experimental data very satisfactorily and can be used to predict the type of motion of the mine. This leads to the very useful possibility of using this type of model to assist in defining a clearance strategy to suit different types of mine.

15 REFERENCES

1. Fryer, S. Unpublished M.O.D.(P.E.) Report, August 1988

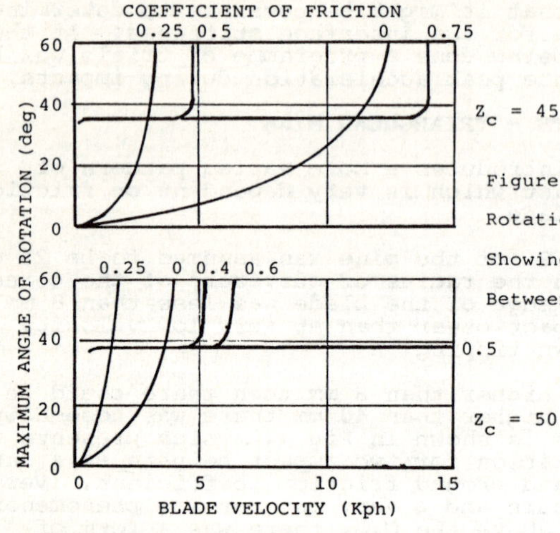

Figure 14 Maximum Angle of Rotation of Triangular Mine Showing Effect of Friction Between Mine and Ground

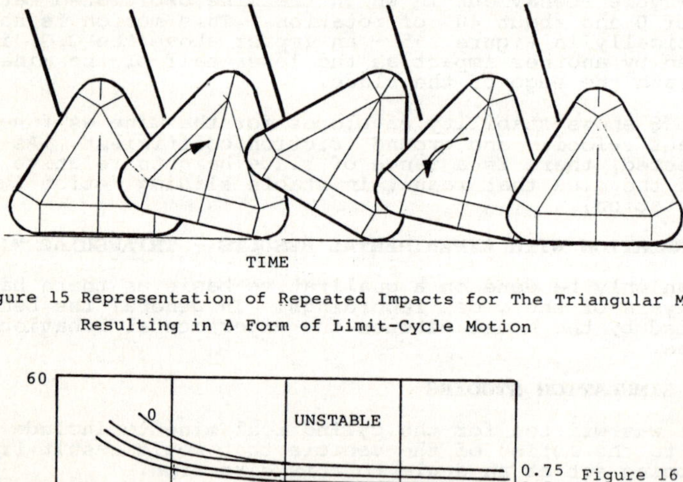

Figure 15 Representation of Repeated Impacts for The Triangular Mine Resulting in A Form of Limit-Cycle Motion

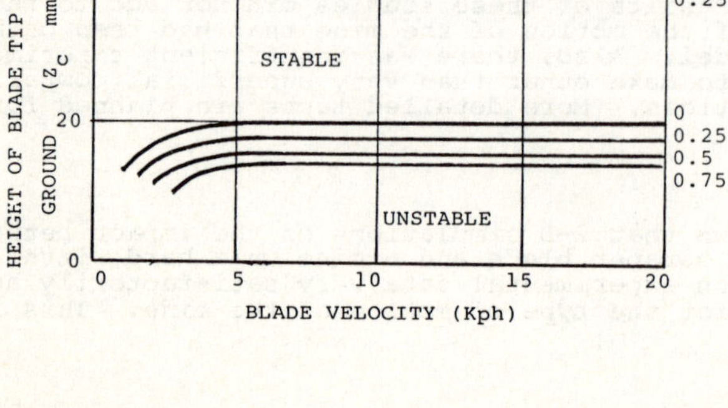

Figure 16 Effect of Friction Between Mine Ground on Stability Boundaries

Development of High Explosive Resistant Safes

U. Hornemann* and J. Wienke**

**Ernst-Mach Institut, D-7858 Weil am Rhein, West Germany*

***Bode-Panzer AG, D-3000 Hannover, West Germany*

ABSTRACT

New safe walls were developed to withstand attacks using mechanical tools and/or explosives. Model experiments with concrete slabs were carried out to obtain an insight into the influence of armor elements on the resistance capability of reinforced concrete walls. For both the damage criteria spalling and perforation the critical charge masses were determined. The results of the slab tests were transferred to 1 m thick prototype safe walls. These were loaded with an in-wall 5 kg charge. Various parameters were tested and the protection reserves demonstrated by detonating a second 5 kg charge after the first loading. The results of the prototype experiments led to a new standard construction for safe walls of the highest safety category.

INTRODUCTION

In the last few years both a changed conduct of criminals and new kinds of burglary tools have led to an enormous increase of the vulnerability of strong boxes and safes. In consequence, materials and constructions had to be adapted to the increased threat. More and more an invulnerability against attacks using high explosives is demanded. The following paper will give a report about the development of explosive resistant safe walls made of concrete material.

Today one of the most effective modern burglary tools is the diamond fitted core drill. In conventional concrete walls with a thickness of 800 mm, boreholes with a diameter of 400 mm can be drilled in less than one hour. A conventional reinforcement with normal steel is only a small obstacle to the drilling process. Therefore, armor elements that can be integrated into newly built safe walls have been developed in the last few years by several industries for the repulsion of attacks by core drill. These elements are essential components of protection. Provided that a safe wall contains an armor element, the core drill can only penetrate into the wall up to the surface of the armor element. In order to obtain in spite of this wall perforation quickly, the borehole in front of the armor element can be filled with an explosive charge. The result of this situation was the demand that under the attack of an explosive mass up to 5 kg a perforation must be prevented. However, the question arises whether an attack with 5 kg of explosives is realistic at all because safes are usually placed in buildings. An explosive mass of 5 kg is generally accepted as an upper limit. The confinement of a charge in a concrete borehole reduces the noise as well as the effects of the detonation gases enormously.

TEST PROCEDURE

In order to develop explosive resistant safe walls at low cost the first experiments were carried out with scaled down models. The scaling laws of explosions demand similarity in geometry, time and mass. If λ is taken as the geometric scaling factor, the result of the dimensional analysis will be that areas are reduced by λ^2, volumes by λ^3, and forces by λ^2, whereas pressures and densities remain constant. For an explosive mass of 5 kg, assumed as the maximum threat, and a safe wall with a size of 3 by 3 meters and a thickness of 1 meter, Table 1 indicates the corresponding masses of explosives and wall thicknesses for different scaling factors. With a scaling factor of only λ = 1/2, the detonation experiments can be carried out in a laboratory.

Table 1. Explosive masses and dimensions of concrete walls for various scaling factors

Scaling factor [λ]	Explosive mass [kg]	Concrete wall Area [m]	Concrete wall Thickness [m]
1	5.000	3.0 x 3.0	1.00
1/1.4	1.822	2.1 x 2.1	0.71
1/2	0.625	1.5 x 1.5	0.50
1/3	0.185	1.0 x 1.0	0.33

One of the main questions concerning experiments with reduced models is whether the physical and mechanical properties of the construction elements simulate those of the prototype structure. This becomes valid especially for a high dynamic load which is hardly accessible by measurements and at which the force time histories and accelerations occurring on each volume element of the model cannot be given easily. Another problem is that the construction elements must be adapted to the technical facts. The armor elements, for example, which are integrated in the concrete walls cannot be reduced to a correct scale size because their components cannot be provided within a reasonable time at reasonable costs. Metal plates and reinforcing bars have to be accepted in those sizes which are commonly available. Using specially manufactured sizes would mean that an experiment would no longer be in a reasonable cost relation to the prototype experiment. In order to account for these problems as well as for the technical facts, the following engineering procedure was applied. The experiments were not carried out with a constant scaling factor. Only some structural components were scaled down. Then, in a series of experiments the functional correlation between the interesting parameters was determined. This correlation was verified experimentally within a certain range of variables. The demanded values for the prototype model are finally determined by extrapolation.

MODEL EXPERIMENTS

Different series of experiments with concrete elements were carried out in order to obtain an insight into the influence of armor elements on the resistance capability of safe walls and also to enable the extrapolation to an explosive mass of 5 kg. The size of the slabs was 1000 x 1000 mm^2, the thicknesses were D = 120, 180, 260, and 360 mm. The quality of the concrete was B 55. The slabs were surrounded by a U-steel frame. The reinforcement consisted of three reinforcing mats built on 14 mm thick steel bars with a mesh size of 100 x 100 mm^2. In relation to the two outer ones, the middle mat was shifted in both horizontal directions by 50 mm. The two outer mats lay 10 mm (D = 120 mm) and 25 mm (D > 120 mm) below the surface of the slabs, respectively. The middle layer was always positioned in the center of the slabs. The reinforcement corresponded to that of original safe walls. As the mass of the reinforcement steel remained constant, the proportion of reinforcement per volume unit changed as follows with increasing thickness of the slabs:

$$D = 120 \text{ mm} \; - \; \mu = 604 \text{ kg/m}^3; \quad D = 180 \text{ mm} \; - \; \mu = 403 \text{ kg/m}^3;$$

$$D = 260 \text{ mm} \; - \; \mu = 279 \text{ kg/m}^3; \quad D = 360 \text{ mm} \; - \; \mu = 201 \text{ kg/m}^3.$$

Plasticized PETN explosive with a density of $\rho = 1600 \text{ kg/m}^3$ was applied. The charge had a cylindrical geometry with diameter/height ratio of unity. It was set upon the slab and covered with moist sand ($\rho = 1000 \text{ kg/m}^3$). The mass of the sand was proportional to the mass of explosive and covered an area of 6 times 6 and a height of 3 charge diameters (Figure 1).

Series B had an armor element included, while series A was done without. In both cases, the charges were covered with sand. Series C was carried out without any surrounding mass and without armor elements to obtain a comparison with values given by literature [1]. The armor element was simulated by a steel construction consisting of two steel plates with a thickness of 2 mm and rectangular steel pipe pieces of size 20 x 30 x 1.5 mm^3 narrowly placed side by side (Figure 2). It was placed between the slab and the charge. The main effect of the element is the attenuation of shock waves in the slab. Two damage criteria were determined as a function of the explosive mass: The first one is the beginning of spalling at the rear of the slab, the second is the beginning of perforation. Perforation was determined as the existance of an optical contact throughout the slab.

During the course of the experiment, the explosive mass was increased or decreased, respectively, until the damage criterion was sufficiently approached. Afterwards, by means of a qualitative evaluation, a decision was taken whether or not the threshold was closer to the upper or the lower experimental value. The mass of explosive in the experiment went up to 2 kg. Figures 3 and 4 show examples of a slab and an armor element, both loaded. The thresholds of experiments without attenuation layer can be approximated in a double logarithmic representation by bilinear functions (Figure 5). The bilinearity is due to the decrease in specific reinforcement with increasing slab thickness. Thick slabs show a threshold shifted in direction to lower explosive masses. The threshold curves for spalling and perforation are not exactly parallel. This is obviously in contradiction to literature [1] for unconfined charges. In experiments with attenuation layers higher masses of explosives were used than in those without. In the first approximation, the threshold curves are running linear in a double logarithmic plot (Figure 5). Due to influence of the attenuation layer the concrete wall thicknesses, at a given mass of explosive, can be reduced to 36% for spalling and between 21 and 28% for the perforation criterion. The limit curves can be described by the equation $D = mW^n$, where D is the thickness of the concrete wall [m], W is the mass of explosive [kg] and both m [m/kg] and n are constants which are given in Table 2.

An extrapolation to 5 kg explosive mass indicates that behind the armor element a concrete wall of at least 450 mm thickness is necessary to prevent any perforation. Another observation demonstrated that it is impossible to prevent a spalling of the inner side of a safe wall with a thickness of 1 m. Consequently, the armor element would have to be located at a depth of only 250 mm. However, such a depth is completely unacceptable for reasons of drilling protection because the amount of work which is necessary to uncover a larger area of the armor element is by far too small. The threshold curve for perforation with uncovered charges in the absence of attenuation layers can be described by the equation

$$D = 0.21\ W^{1/3}.$$

Table 2. Constants of equation $D = mW^n$

Criteria	Series	Slab thickness [m]	m [m/kg^n]	n [-]
Spalling	A_S	$0.12 \leq D \leq 0.18$	0.35	0.26
Spalling	A_S	$0.18 < D \leq 0.36$	0.51	0.41
Spalling	B_S	$0.12 \leq D \leq 0.36$	0.33	0.43
Perforation	A_P	$0.12 \leq D \leq 0.18$	0.22	0.34
Perforation	A_P	$0.18 < D \leq 0.36$	0.26	0.54
Perforation	B_P	$0.12 \leq D \leq 0.18$	0.19	0.42
Perforation	B_P	$0.18 < D \leq 0.36$	0.19	0.47
Perforation	C_P	$0.12 \leq D \leq 0.36$	0.21	0.33

It is parallel to the Hader curve [1]. However, it is shifted in direction to smaller explosive masses (Figure 6). The reason for the shift lies in the definition of the damage criterion.

DIFFERENT TYPES OF ARMOR ELEMENTS

The extrapolation of the results of the model experiments to 5 kg explosive indicates that the attenuation effect of the armor element has to be increased in order to get sufficient reserves of protection. Therefore, different types of armor elements were tested and evaluated qualitatively by means of the damage degree. For the tests a 2 kg charge was applied (Figure 7). For the prototype walls an armor element was chosen which consists of two chambers and provides a very high protection against the diamond fitted core drill (Figure 8). The front part consists of a metal plate with a thickness of 4 mm. In the following free space with a depth of 52 mm vertically standing, freely movable twisted steel bands (size 40 x 8 mm^2) were installed. The free space is closed by a metal plate of 3 mm thickness. The following second chamber contains an aluminum-corundum cast plate and is closed also by a metal plate of 4 mm thickness.

A load of a 2 kg charge did not cause a perforation, although in the rear metal plate a 300 mm long and 12 mm wide crack was generated. The front chamber was deformed and the front steel plate was heavily damaged. At the initial location of the charge the aluminum-corundum cast plate was broken into several big fragments.

EXPERIMENTS WITH PROTOTYPE WALLS

From the results of the model experiments and of the tested armor elements the design of prototypes was derived which represented parts of safe walls in full size with a thickness of 1 m. The size of the prototype walls was 2300 x 3100 x 1000 mm^3 (Figure 9). At the top and the bottom part of the wall ceiling and floor attachments had been integrated. The armor element with a thickness of 100 mm was placed at a depth of 350 mm from the front. The front part of the wall contained two and the rear part three reinforcement mats with a mesh width of 100 mm and 150 mm, respectively. The mats were made from steel bars with a diameter of 14 mm. In front of and behind the armor element two layers of horizontally lying and twisted steel bands (size 40 x 8 mm^2) were fixed at a distance of 150 mm. These steel bands provide an additional protection against mechanical tools. Within a circle of 400 mm diameter three boreholes with a diameter of 90 mm were drilled down to the armor element by means of a core drill. An explosive mass of 5 kg was divided into three equal parts and filled into the boreholes. The rest of the boreholes was stemmed with wet clay. The charges were initiated at the same time.

The detonation caused a crater with a diameter of about 800 mm at the front (Figure 10). The crater extended to the armor element which was deformed but a hole was not generated. The rear side of the wall showed a number of radial cracks and some pieces of concrete had been spalled from the surface. The result demonstrated that the limit for spalling had just been reached. Comparing this with the extrapolated result of the model series, an increase of the resistance for spalling of 55% is achieved if the criterion for this comparison is based on the explosive masses. It can be assumed that the limit for perforation will be increased on the same scale. Under the same conditions a perforation would be reached if a mass of explosive higher than 14.5 kg would be used. The reason for the increased thresholds in comparison to the model experiments is mainly due to the thicker and stronger armor element. The fact that the wall provides considerable protection reserves was clearly confirmed by a second test in which a 5 kg charge was placed in the crater. After the detonation no perforation was observed. A slightly higher damage of the wall was obtained when the 5 kg charge was placed in a single borehole (152 mm in diameter). At the rear of the wall the spalling area was increased from 0.05 m^2 to 1.5 m^2 and the bending from 20 mm to 100 mm, respectively.

In the same way a safe wall with a thickness of 600 mm containing identical inner construction was loaded with 5 kg explosive. The armor element was placed at a depth of 200 mm. Due to the relatively weak confinement of the charge a smaller crater than the one created in the 1 m wall was generated at the front. The same damage to the armor element was observed.

The spalling area at the rear covered an area of 1.4 m^2 and had a thickness up to 80 mm. Consequently, the spalled layer extended at some spots behind the reinforcing mat which lay at a depth of 50 mm. With a second 5 kg load a perforation of 320 mm in diameter resulted.

Normally large safe constructions are equipped with a ventilation system. In consequence, ventilation tubes have to pass the walls of 1 m thickness and can be used as available boreholes for explosives. Within the ventilation tubes the charge can be placed in front as well as behind the armor element. Therefore, the question arose what kind of dimension and arrangement for the ventilation tubes has to be chosen in order to achieve a sufficient safety for the wall. Therefore, three tubes placed parallel with an inner diameter of 50 mm and a center distance of 200 mm were filled with 5 kg explosive. The charge was placed symmetrically in the wall. Two craters were generated by the detonation, one in front and one in the rear of the wall. The armor element was still covered with concrete. The ventilation tubes were partly broken and the diameter was enlarged. An optical contact through the wall was produced. The wall still had sufficient protection reserves after the detonation.

CONCLUSIONS

The experiments and investigations have clearly demonstrated that the method of tests with reduced models which in part contain full-size elements lead to acceptable results. The prototype tests resulted in a standard construction for safe walls of the highest safety category which provided a high protection resistance against attacks with mechanical and thermal tools as well as high explosives.

REFERENCE

[1] Hader, H.J.: Effects of Bore and Cased Explosives Charges on Reinforced Walls, pp. 221-226, Proc. of the Symposium on the Interaction of Non-Nuclear Munitions with Structures, U.S. Air Force Academy, Colorado, USA, 1983.

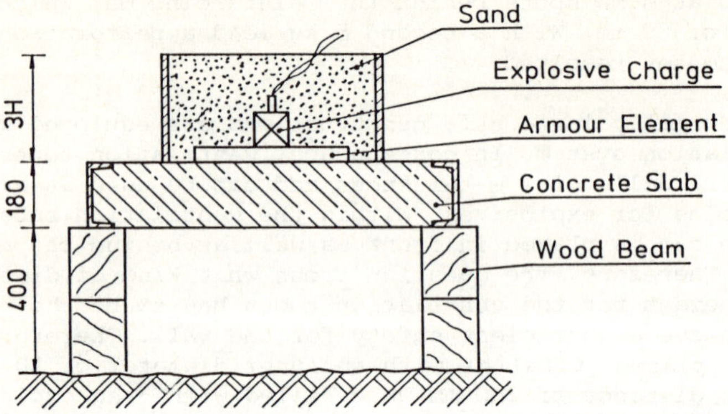

Figure 1. Experimental set-up for loading concrete slabs by an explosive charge covered with sand.

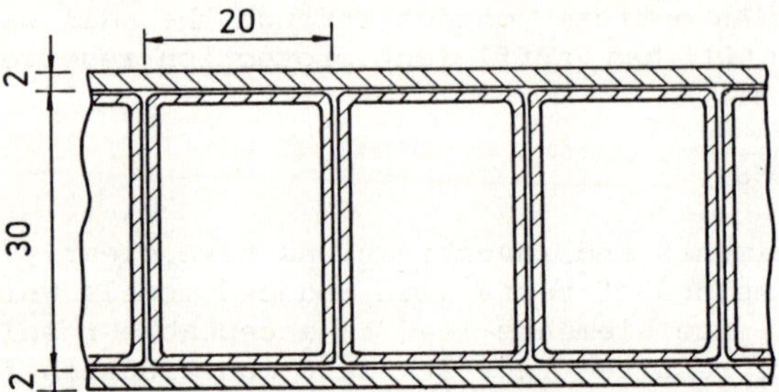

Figure 2. Design of the attenuation element placed between the explosive charge and the concrete slab.

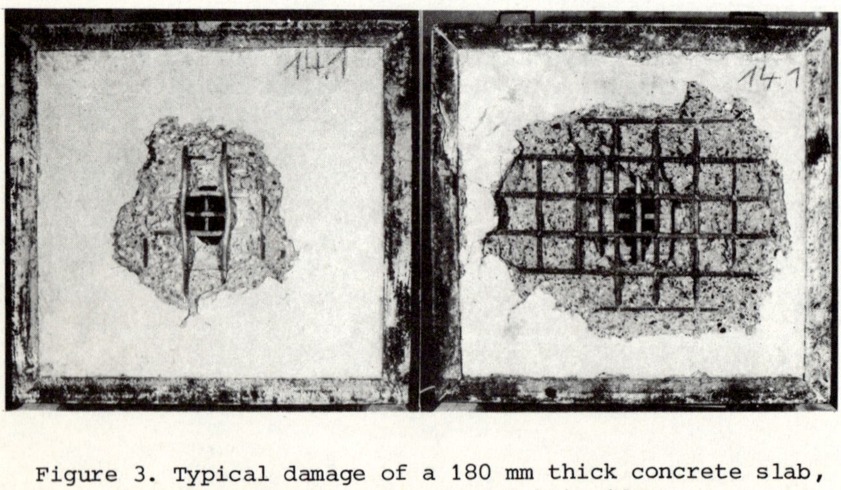

Figure 3. Typical damage of a 180 mm thick concrete slab, front (left) and rear (right) side.

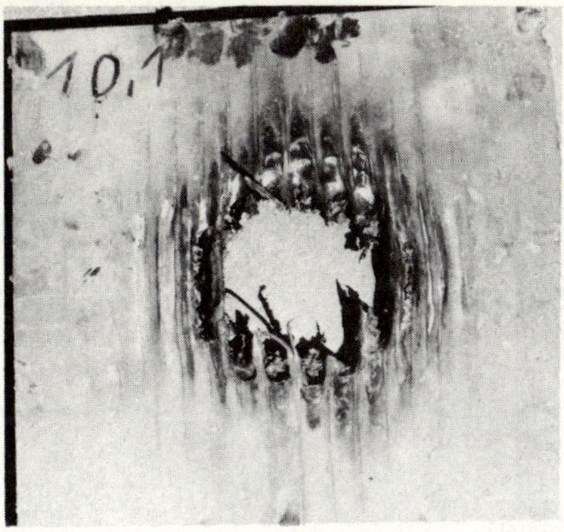

Figure 4.
Typical damage of an attenuation plate, front side.

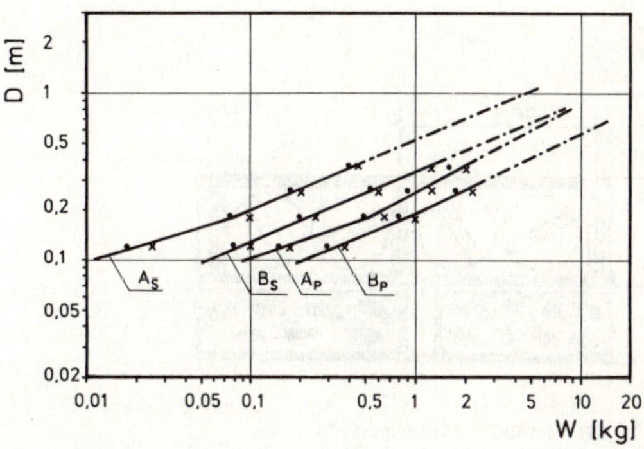

Figure 5. Spalling (index S) and perforation (index P) threshold curves for both series, without (A) and with (B) attenuation plates and sand covered charge. D - slab thickness and W - explosive charge mass.

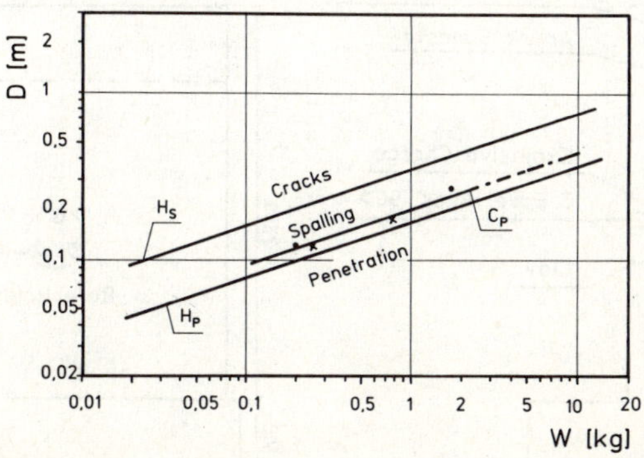

Figure 6. Spalling (H_S) and perforation (H_P) threshold curves from Hader [1] and from this work (C_P). The explosive charges were uncovered and in contact with the concrete surface.

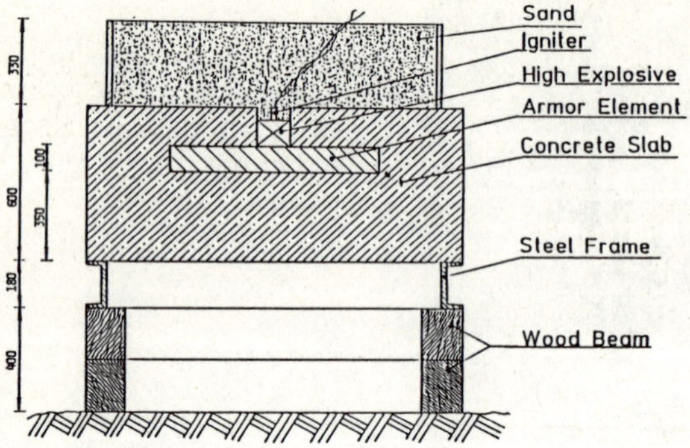

Figure 7. Experimental set-up for testing various armor elements.

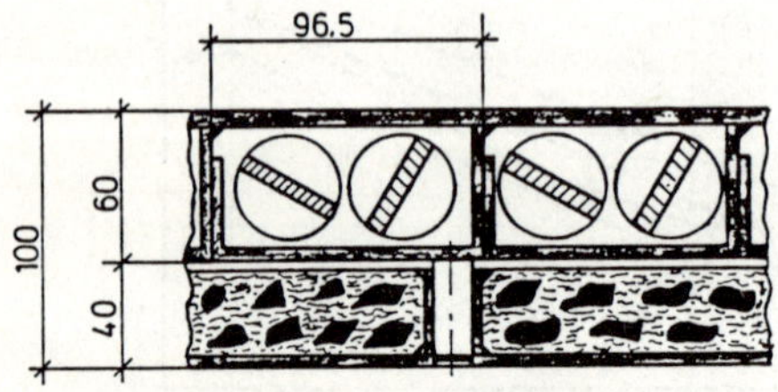

Figure 8. Design of an armor element.

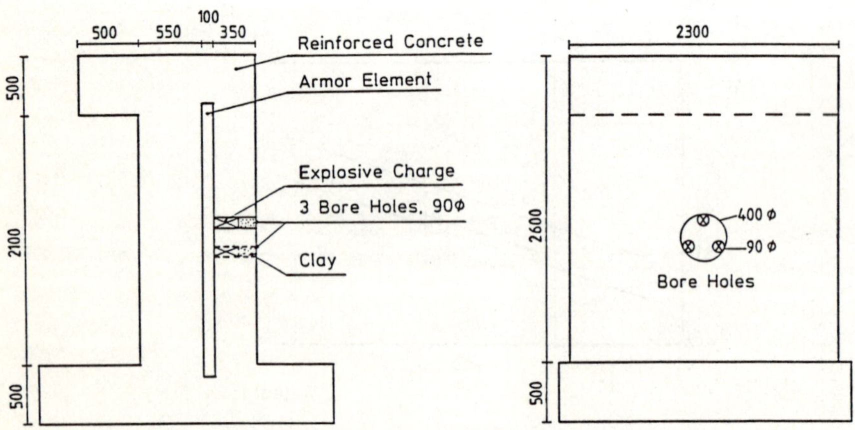

Figure 9. Dimensions of the prototype safe wall and the placing of the armor element and the explosive charges.

Figure 10. Damage of the prototype safe wall after a 5 kg detonation.
Left: front side; right: rear side

Design and Verification of Blast Resistant Structures

G. Mazzà*, P. Molinaro* and L. Spoletini**

**ENEL - CRIS, Milan, Italy*

***SNIA - BPD, Rome, Italy*

ABSTRACT

The design and verification of structures to resist the effects of blast waves is a matter of some interest both in industrial and military fields. In the present paper a case history relevant to a steel blast-resistant door is examined in detail both from a numerical and experimental point of view. Moreover, in order to evaluate the possibility of using materials lighter than steel, a method for the dynamic characterization of fiber composite elements is shown.

INTRODUCTION

Since 1983 a study for the design and verification of armoured doors was started at the Hydraulic and Structural Research Centre of ENEL (Italian National Power Board). The main aim of such barriers was to prevent the entry of saboteurs or terrorists in buildings that, because of their importance, had to be protected. At any rate, in considering the safety for personnel and equipment relevant to hazardous industrial activities, the installation of blast-resistant doors could also be taken into account.

In previous papers [1] and [2], the behaviour of some steel doors subjected to the explosion of TNT charges of different weight was examined. Numerical simulations were carried out both using FEM and 1-DOF (one degree of freedom) models. In those cases only partial residual plastic displacements measured on the back of the doors were available. Moreover the stress-strain law for the characterization of the steel was known only approximately. Hence the comparison between numerical and experimental results, even if fairly good, was based on an incomplete set of data. The case history presented in this paper is more complete because the residual plastic displacements are known in great detail, the steel is characterized performing impulsive dynamic tests and, as the structure was provided with a set of instruments during the test, strains time histories are also available.

THE EXPLO-1 SYSTEM

The availability of experimental data relevant to a set of tests carried out on actual structures has allowed us to validate a software package developed for the numerical analysis of structures subjected to blast waves.

In Fig.1 the simplified flow-chart of the computer program is shown. Blast wave parameters can be computed according to experimental correlations based on extensive test data carried out by several authors [3], [4], [5] and [6]. For a preliminary design the use of 1-DOF models could be advisable because of its simplicity and capability to give the order of magnitude of displacements and boundary reactions. The use of FEM models is more suitable for a final design or for a verification of an actual structure. The results of the analyses can be obtained both in numerical and graphic form.

THE CASE HISTORY

Only numerical results relevant to FEM analyses are presented in this paper. The structure was a box-shaped steel door subjected to a contact explosion due to a TNT charge of 20 kg (44.5 lb). Because of the negligible strength of the so-called stand-off chamber (see Fig.2) its contribution to the overall stiffness of the structure is ignored, while its mass is taken into account by increasing coherently the mass of the steel slab closer to the chamber. In Fig.2 the finite element meshes of the door are shown. In the first mesh only one row of elements links the two main slabs; in the second mesh three rows of elements, with a slight eccentricity, were considered in order to make possible their bending. The back slab of the door was provided with a set of accelerometers and strain-gauges installed to record the behaviour of the structure during the test (see Fig.3). Residual plastic displacements were measured on the door in great detail. The qualitative deformed shape of the door after the explosion is also shown in Fig. 3. In order to evaluate the strain rate influence on the stress-strain law relevant to the steel, a set of tests were carried out. Fig.4 shows the characteristics of the specimen used for tests and typical stress-strain curves for different strain rates.

Many numerical analyses were carried out to evaluate the influence of different parameters on final results. Figs.5 and 6 show the superposition of residual plastic displacements with those computed for different times by the FEM analysis. The only difference between the two cases is relevant to the mesh used for computation according to the scheme shown in Fig.2. A comparison between computed and recorded strains is shown in Fig.7 only for three strain gauges. In some cases, as the correspondence between the position of the gauge and the node of the mesh is not perfect, the strain time history curves relevant to closer nodes are shown. The agreement between experimental data and numerical results is satisfactory. Hence the authors consider EXPLO-1 as a system that could be used with some confidence for design and verification purposes at least for this kind of structure and for a range of TNT charges up to 20-30 kg.

In spite of the good behaviour of such a kind of structure, problems of manoeuvrability can arise, because of their weight if they have to be operated very often . A possible way to build lighter structures is to take into account the use of composite materials. Metal (alumina), ceramic, aramid fibers seems to be the more suitable materials for the construction of energy absorber sandwich elements [7]. Experimental data on composite materials behaviour to blast loadings are not available at the moment. The dynamic characterization of such materials could be carried out using the technique shown in the following paragraphs.

EXPERIMENTAL TEST DESCRIPTION

A quantitative evaluation of the proposed materials capability to absorb the shock waves produced by the detonation of high explosives can be experimentally carried out using the so called "AQUARIUM TECHNIQUE".Test arrangement consists in a perspex cylinder, filled with water, see Fig.8, in direct contact, at one of its ends, with a specimen of the material to be studied.The shock absorbing material is, hence, located between the water filled cylinder and an H.E. charge. The explosive charge is initiated by a plain wave generator. The shock wave, produced by the explosive detonation, is transmitted to the water filled cylinder after having passed through a given thickness of the absorbing material. The shock wave proceeding in the liquid medium is recorded on a streak camera film.

SHOCK WAVE PARAMETERS DERIVATION METHOD

The streak camera film, read with an optical comparator, gives couples of coordinates space/time representing the distance covered by the shock wave in water versus the time.
An analytical translation of such points, using an exponential regression method, is then performed, in order to determine the coefficients of the following relation:

$$s = A_1 e^{A_2 t} + A_3 t + A_4 \quad (1)$$

where t is the time and s the space.
All the magnitudes, associated with the shock wave transmission in the water can, now, be calculated provided that the water equation of state is known.
In fact, assuming for the water equation of state the following relationship:

$$Up = 0.607 + 0.372\ Us + 0.0283\ Us^2$$

where Up is the water particle velocity and Us is the shock velocity, Us = ds/dt can be calculated from the equation (1), while the shock wave pressure (P1) can be calculated with the known relation:

$$P_1 = \varrho_0\ Us\ Up$$

where ϱ_0 is the water undisturbed density.
The curve pressure versus distance could be drawn.

SHOCK WAVE ABSORBING MATERIALS ASSESSMENT METHOD

As known, in a medium passed through from a shock wave, a difference in the internal energy can be expressed by the relation:

$$\Delta E = \tfrac{1}{2}\,(P_1+Po)\,\left(\frac{1}{\varrho_0} - \frac{1}{\varrho_1}\right) \simeq \tfrac{1}{2}\,Up^2$$

where: ΔE is the water internal energy increase, Po is the water undisturbed pressure and ϱ_1 is the water density after the shock wave. In particular both pressure transmitted to the water and water internal energy increase can be calculated in correspondence with the specimen/water interface (s=0).

These two magnitudes could be taken as indicators of the transmission of the energy produced by high explosive detonation. Also if experimental results are not yet available, some preliminary firing tests have been performed.

In fig.9 an example of a streak record is reported where it is possible to recognize the space/time curve described by the shock wave transmitted to the water after having passed through the specimen.

ACKNOWLEDGEMENTS

The authors express their gratitude to the LIPS-VAGO Company (RACAL -CHUBB Group) that have kindly allowedus to use experimentalresults. The authors also acknowledge the Commission of the European Communities JRC of ISPRA (ITALY)that allowedus to use the Computer Code EURDYN for finite element analyses [9] and for impulsive tests carried out to characterize dynamically the steel of the door.

REFERENCES

Paper in Conference Proceedings

1. G. Mazzà, R. Minasola, P. Molinaro, A. Papa; Analysis of the behaviour of armoured doors sujected to contact explosions: experimental measurements compared with results computed by FEM and ONE-DOF numerical models, Proceed. of the Int. Conf. on Computational Methods and Experimental Measurements, Porto Carras, Greece September 2-5, 1986.

2. G. Mazzà, R. Minasola, P. Molinaro, A. Papa; Dynamic Structural Response of Steel Slab Doors to Blast Loadings, Proceed. of the Int. Conf. on Computational Plasticity, Barcelona, Spain, April 6-10, 1987.

Books

3. J. Henrych (1979) The Dynamics of Explosions and its Use, Elsevier Scientific Publishing Company, New York.

4. W.E. Baker et Al. (1973) A short Course on Explosion Hazards Evaluation, SouthWest research Institute, San Antonio and Houston.

5. W.E. Baker (1973) Explosions in Air,University of Texas Press, Austin, Tx.

6. Department of the Army, the Navy and the Air Force (1969) Structures to Resist the Effects of Accidental Explosions.

7. J.W. Weeton, D.M. Peters, K.L. Thomas (1986) Engineers' Guide to Composite Materials, American Society for Metals, Metals Park, Ohio 44073.

8. M.H. Rice, R.G. Mc Queen, J.M. Walsh Solid state physics, VOL. 6

Report

9. J. Donea, S. Giuliani, J.P. Halleux (1986) Theoretical Aspects of the EURDYN Computer Programs for Non - linear Transient Dynamic Analysis of Structural Components, EUR 5473 e, Commission of the European Communities - JRC - Ispra, ITALY.

10. J.K. Rigdon, I.B. Akst; Sylas Mason co USA An analysis of the "AQUARIUM TECHNIQUE" as a precision detonation pressure measurement gauge.

11. M.A. Cook, R.I. Keyes, W.O. Ursenbach; UNIVERSITY OF UTAH USA Measurements of detonation pressure.

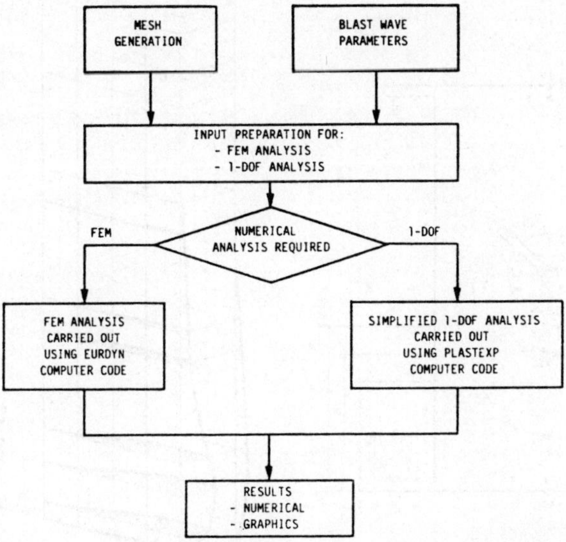

Fig. 1 - Explo-1 system: simplified flow-chart

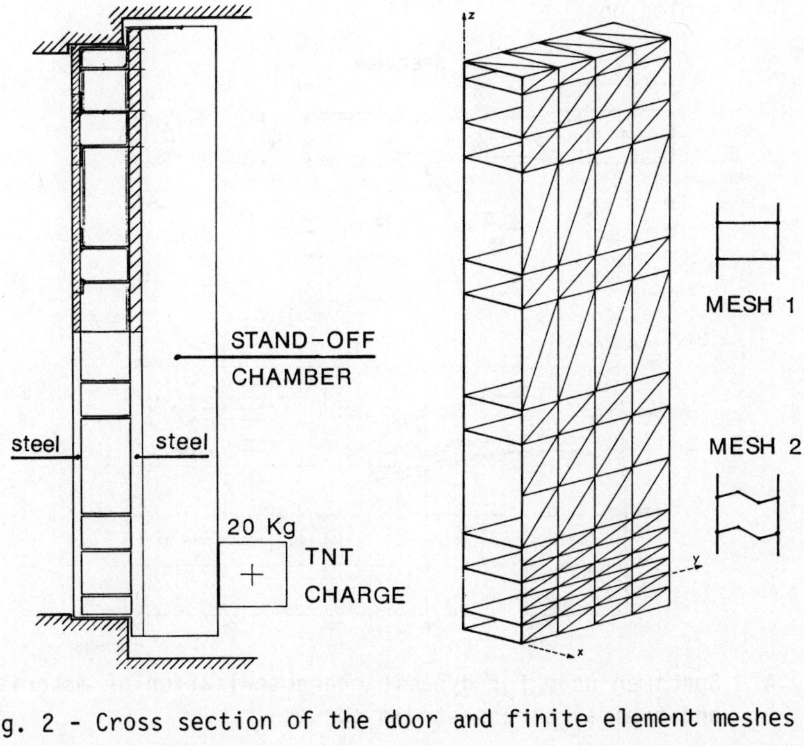

Fig. 2 - Cross section of the door and finite element meshes

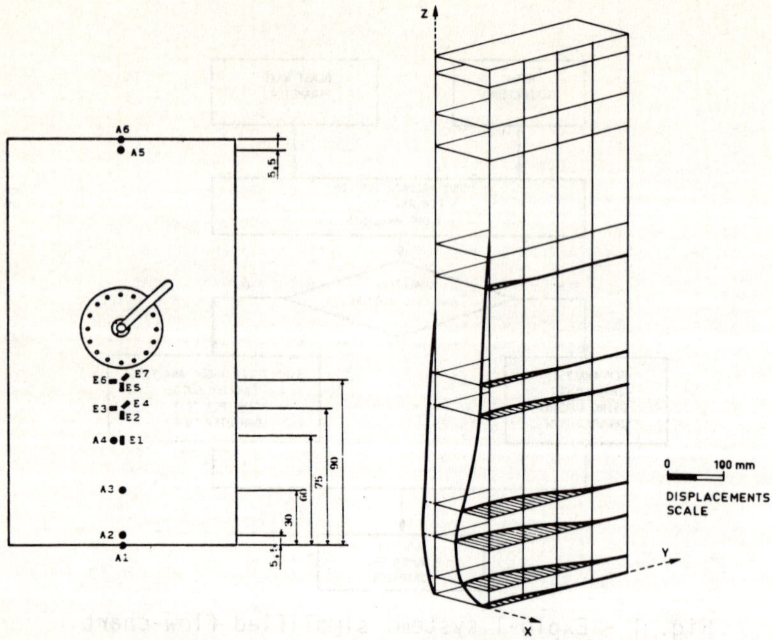

Fig. 3 - Set of instruments installed on the back of the door and residual plastic displacements measured after explosions

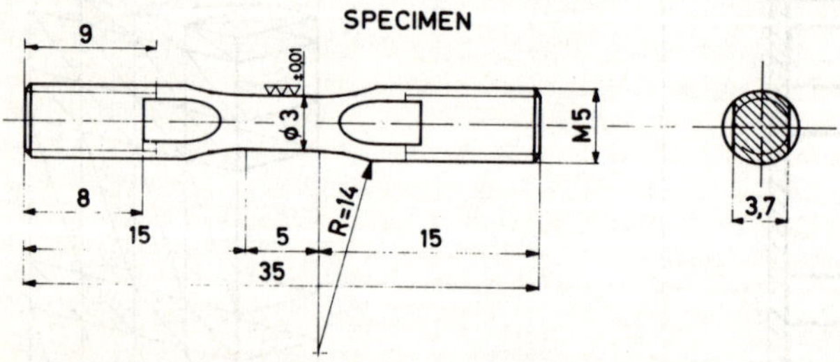

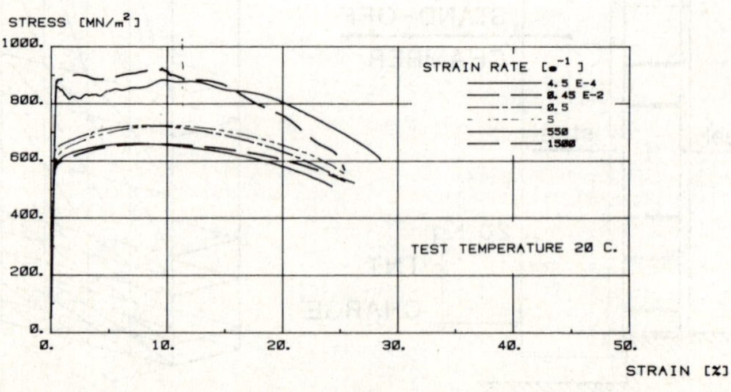

Fig. 4 - Specimen used for dynamic characterization of materials and typical stress-strain curves

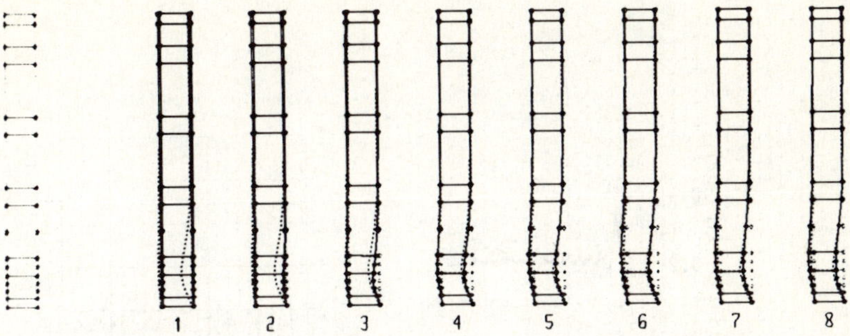

TITLE : LIPS VAGO DOOR - TNT 20Kg - B.W.P. (HENRYCH) - 4M C/5
DRAWING SCALE 1.0 m

TIMES : 1. 0.0 2. 3.6E-04 UNDEFORMED CROSS SECTION
3. 6.8E-04 4. 1.0E-03 —— DEFORMED CROSS SECTION *
5. 1.3E-03 6. 1.5E-03 .... EXPERIMENTAL DATA *
7. 3.1E-03 8. 3.7E-03 * (amplification factor = 1.0)

Fig. 5 - Superposition of residual plastic displacements with values computed by FEM analysis carried out using the mesh 1

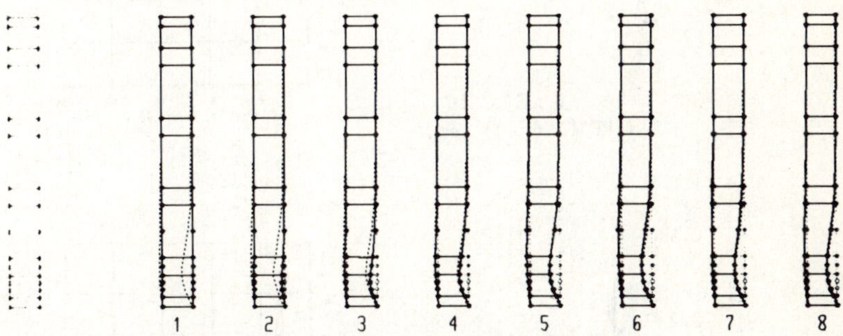

TITLE : LIPS VAGO DOOR - TNT 20Kg - B.W.P. (HENRYCH) - 4M/4ss
DRAWING SCALE 1.0 m

TIMES : 1. 0.0 2. 3.6E-04 UNDEFORMED CROSS SECTION
3. 6.8E-04 4. 1.0E-03 —— DEFORMED CROSS SECTION *
5. 1.3E-03 6. 1.6E-03 .... EXPERIMENTAL DATA *
7. 3.1E-03 8. 3.6E-03 * (amplification factor = 1.0)

Fig. 6 - Superposition of residual plastic displacements with values computed by FEM analysis carried out using the mesh 2

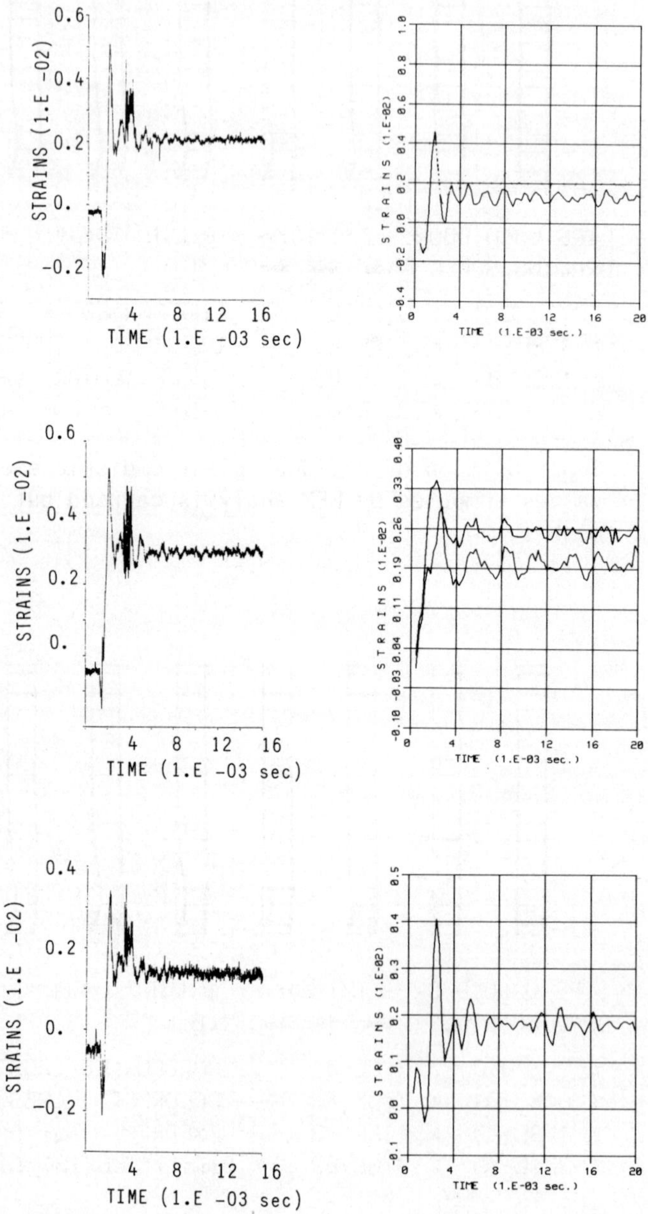

Fig. 7 - Comparison between experimental and numerical values of strains for strain-gauges El, E3, E4

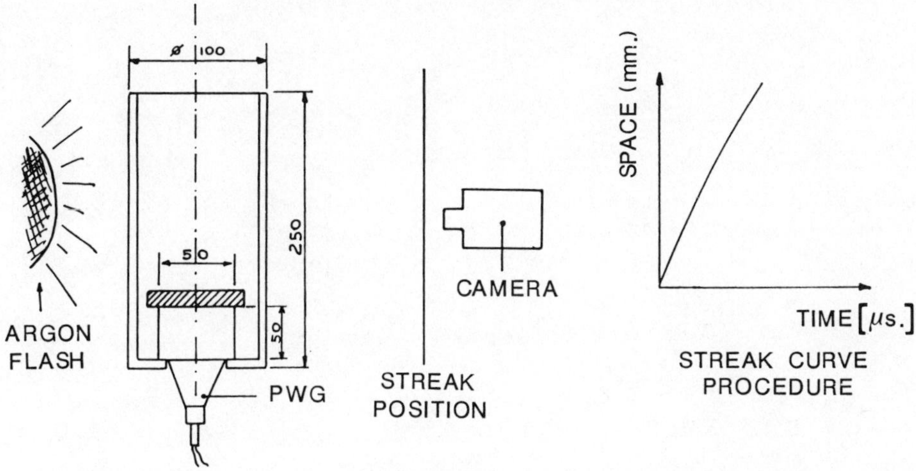

Fig. 8 - Aquarium technique: experimental test arrangement

Fig. 9 - Example of a streak record

How Reliable is Your Design Against the Impact of Impulsive Loads?

V.M. Trbojevic* and J. Marti**

* *Technica Dynamics, Lynton House, 7/12 Tavistock Square, London, WC1H 9LT, UK*

** *ETS de Ingenieros de Minas, Universidad Politecnica de Madrid, Rios Rosas 21, 28003 Madrid, Spain*

ABSTRACT

Assessing the vulnerability/integrity of structures and facilities in hazardous industries under extreme or accidental loading conditions requires advanced analysis techniques. Traditional design methods do not always account for dynamic phenomena in structural behaviour. Two examples are given in this paper.

The first example deals with the consequences of an assumed axisymmetric mode of failure of a liquid storage tank. In the case of an LNG tank with a bound wall as a secondary containment, it is shown that the hydrodynamic pressure distribution is not only different from expected, but indicates a 2.5 times increase over the conventionally employed hydrostatic head design calculation.

In the second example assessment of the integrity of a subsea production system under accidental impact load conditions is presented. Detailed impact dynamic analysis of this load case shows that, contrary to general belief, the structural response is not greatly affected by the overall characteristics of the system (well stiffness, soil properties, etc.). Instead, other factors such as impact velocity and ability to deform locally in the impact region are more crucial.

INTRODUCTION

The aim of this paper is to present an advanced methodology for the integrity assessment of structures and facilities in hazardous industries under accidental load conditions. There are two main areas of interest; in the selection of accidental loads which the system should be able to withstand, and in the actual design of hardware to withstand these loads.

In a hazardous plant many of the accidental loads will be of a dynamic nature, i.e. a mechanical impact from a dropped object, or from a blast pressure wave. The modelling of such phenomena may be outside the designers field of experience, and also lacking in engineering codes which apply to the particular design. The result may be that dynamic accidental loads are not treated adequately in the design process.

This paper outlines two examples which clearly demonstrate the significance of proper modelling of dynamic phenomena. In such cases, a conventional design approach may yield results which cause an unintended risk increase, without the designer being aware that he has created such a situation in the system. The first example is taken from landbased industry and concerns the design of a LNG tank. The second example is from a subsea production system and concerns the impact resistance of a wellhead against a dropped object.

ACCIDENTAL RUPTURE OF AN LNG TANK

Introduction

Considerable interest has been aroused in recent years in the loading that would occur on the secondary protection system of a liquid fuel storage tank, if the tank were to fail. There has been much discussion on possible load time histories and the effects of such a failure and there are many possible alternative solutions to ensuring the overall integrity of the storage system should this failure occur. The purpose of this study is to find a design load that can be used with confidence to cope with the consequences and determine what stresses are developed by the liquid on the secondary protection system following failure.

Economy of scale dictates that typically large tanks are constructed with a volume between 5000 m^3 and 50,000 m^3. In the event of an accident, the consequences of such large volumes of fuel being released and the devastating results of fire, explosion and pollution are such that the probabilities of failure or release are required to be very low. The provision of the current design of guard and bund walls or dikes alone may not be able to provide the necessary protection, especially as plants tend to be located close to populated areas. There is clearly a need to reassess the overall integrity of such systems.

There is also a diversity of views about the types of credible accident that should be considered in the design which could lead to release of liquid from the primary containment. A scenario that is now accepted as a real possibility is the rapid propagation or cracks in the primary containment, leading to rapid dynamic liquid loads being applied to the secondary containment. The problem facing the designer is often how best to estimate the loading on the secondary containment in such an event and the consequence of it. In this section, a numerical approach available for solving this problem is outlined and the preliminary results shown. It is demonstrated that the use of complex numerical methods can be crucial in obtaining a true understanding of the system behaviour and the design objectives that are to be met.

Formulation of the Failure Modes

Considerable disagreement exists as to the speeds at which cracks will actually propagate in the primary containment (steel tank) and the approach adopted is to conduct a sensitivity analysis to determine a pattern of crack propagation that will result in a "worst case" design condition. In general, vertical crack propagation is considered in conjunction with crack propagation around the base of the steel tank. Clearly the tank will move with the liquid in some way. From some work already carried out (Cuperus [1]), it is apparent that this is not a simple problem to solve and in many ways an experimental solution is intuitively the most satisfactory approach.

Numerical modelling, however, offers perhaps more flexibility and demands a mathematical representation of the physical problem and, provided that agreement can be shown between the numerical approach and experimental tests, it can be used with confidence.

The failure mode under consideration in the event of a tank rupture is a highly complex interaction between fracture propagation and the flow of a fluid with a free surface. In order to understand the physics of the complex interaction problem, two simplifying assumptions were made.

The first assumption is based on the fact that the cracks will propagate at much higher velocity than that of the fluid motion, and that they propagate at the same time in the vertical and the circumferential (round the base) directions. The effects are that the steel tank loses its integrity instantaneously and from the structural containment point of view, can be considered removed. What is left hypothetically is the cylindrical column of fluid collapsing under gravity within the secondary containment. This greatly simplifies the problem and can be considered to represent an axisymmetric mode of failure.

The second assumption is based on some evidence that the crack will propagate much faster in the vertical than in the circumferential direction. This would result in a section of the tank being removed and the liquid flowing through the gap. This would represent an asymmetric failure mode. After some preliminary calculation of the consequences of such a failure mode reported in (Baldwin [2]), it was concluded that this asymmetric mode may be interesting from a sensitivity analysis point of view, and that in reality a mixture of two failure modes is more probable. This would in effect induce some asymmetry in loading on the secondary containment the effects of which may be marginally more severe than from the purely axisymmetric failure mode.

Methodology Used in the Analysis

The finite difference program used for the fluid dynamic problem is based on an Eulerian formulation. The volume-of-fluid technique (Hirt et al [3]) is incorporated for the treatment of the free boundaries. In an Eulerian representation, the computational grid remains fixed and the flow of fluid through the mesh is computed in small time intervals. Depending on the incompressibility condition (or on acoustic pressure representation) the integration of the Navier Stokes equations in time can be performed with implicit or explicit schemes. The volume-of-fluid techniques deals with discontinuities and the fluid boundary (free surface) can undergo large deformations. The method is therefore well suited for this type of transient problem.

LNG Tank with a Bund Wall

The axisymmetric failure scenario considered here consists of rapid loss of the primary containment and consequent fluid impact onto the secondary protection system. A typical liquified natural gas (LNG) tank failure is analysed. A tank has a radius of 24 m and a volume of 50,000 m^3. The LNG density is 583 kg/m^3. The secondary protection system consists of a reinforced concrete cylindrical bund wall, the radius of which is 30 m and the height 20 m.

The sequence of "freeze frames", in Figure 1 depicts the fluid free surface positions at different time intervals. The corresponding hydrostatic (dotted line) and hydrodynamic pressure distribution are shown in Figure 2. This clearly indicates a 2.5 times increase with respect to the conventionally employed hydrostatic head calculation. In the region between 8 m and 12 m from the ground level, the maximum dynamic pressure is more that 4 times greater than the local hydrostatic pressure.

Conclusions

The analysis carried out demonstrates that the current design of the containment bund walls may not be conservative and may not contain spillage caused by a sudden tank rupture.

The example shows that the pressures on the secondary containment wall will be higher than the usual design pressures employed. The pressure distribution is also quite different from the hydrostatic pressure profile used in the mechanical design of the walls.

Finally, the application indicates how useful this type of numerical modelling is in representing unusual or catastrophic modes of failure. Actual design configurations can be checked out on a scale that is just not feasible on a full scale experimental basis. The design changes to these protection systems can easily be incorporated and checked with the same numerical model to help the design to converge to an optimal solution from the safety point of view.

DROPPED OBJECT IMPACT

Introduction

The possibility of dropping an object on the subsea production system is only significant when a vessel is positioned over the template and intervention or inspection works are carried out. It is first of all important to determine an accidental design load, i.e. the worst accident for which a safe design is to be provided.

Due to the size of the system and its geometry two analyses were performed: a global two dimensional analysis in which the model consisted of axisymmetric and plane stress parts, and the detailed three dimensional analysis of the X-mas tree valve block under the boundary conditions obtained from the two dimensional analysis. As the result of the analyses clear design recommendations were laid out which were not previously obvious.

Description of the Subsea Production System

The central part of the structure to be considered is made primarily of steel and displays an approximate radial symmetry. Its general disposition is presented in Figure 3. As can be seen, it starts at the top with the workover riser and continues with the universal running tool and the lower riser package, thus completing the falling part of the structure. Below the lower riser package are located the X-mas tree with the re-entry hub and the wellhead. At the time of impact, it is assumed that, as designed, connections between the universal running tool and lower riser package and between the X-mas tree and the wellhead are active.

The function of the lower riser package is to (dis)connect the riser system interface from/with the X-mas tree re-entry hub while providing shear and sealing capabilities.

The main element of the X-mas tree is the valve block which holds the valves and actuators required to control the flow from the well.

The wellhead provides the interface between the well, the X-mas tree assembly, and the template. The well consists of several conductors of different lengths. The template is a frame structure which is piled to the soil at four corners and which houses a maximum of six production wells.

<u>Impact Load Cases</u>

Whilst running a typical intervention tool, three phases can be identified (vice versa for retrieval operations). The first phase involves the assembly being lowered when the vessel is "off location", i.e. not positioned above the template. As soon as the assembly is 15-30m above the seabed, the lowering operation is stopped and the vessel together with the assembly is moved in above the intervention well (second step). During the last phase guidelines attached to the top of the guide posts are tensioned, and the assembly is lowered onto the dedicated slot on the template.

A risk analysis concerning the path of dropped objects in water showed that the risk of dropping an object onto the intervention well is most significant during the last phase. In order to select a design load case, the expected intervention and installation works were considered more carefully. Each operation was characterised by a frequency of occurrence (e.g. once per year, or once during the template lifetime) and a certain mass. It was thus possible to establish a distribution of number of lifts per mass category over the lifetime of the template. Based on a historical drop probability the design load case could then be selected, i.e. dropping an assembly composed of the work-over riser, the universal running tool and the lower riser package. If such an accidental drop were to happen, the well would have at least one barrier: the primary barrier is the downhole safety valve and sometimes the x-mas tree master valve. It will be adequate to require that one barrier remains intact after impact.

The total mass of the dropped object is 45 tonnes, which is increased by 20% for safety considerations, to 54 tonnes. The stationary part of the structure is composed of the X-mas tree, wellhead, template and the foundation soil together with the casings and cement which penetrate it. Due to the guide posts which will engage the corresponding funnels of the lower riser package, the impact is well centered. The drop height is estimated at 10 m. This is sufficient for the dropped object to approach the terminal velocity of 9.5 m/s by the time of impact. This constitutes the accidental event load condition, i.e. the worst case accident for which a safe design shall be provided.

Numerical Treatment

General strategy It seems clear from the description of the system that the impact to be analysed maintains an approximate axial symmetry around the axis of the structure. There are a number of features which have four-fold rather than axial symmetry but their symmetry is sufficiently close to axial to justify the approximation without danger of significant errors.

There are, however, some exceptions to the above rule which warrant further considerations. The first and more obvious is the X-mas tree valve block, as it is a part of the object of the study to assess whether its permanent deformation will allow operation of the valves after the event. The approach taken to solve this problem has consequently been a combination of two dimensional and three dimensional models which, while treating the valve block in all its complexity, render the problem soluble by exploiting its approximate axial symmetry. More specifically the analysis then consists of two major parts:

i) Global axisymmetric modelling in which the X-mas tree valve block has been replaced with an axisymmetric structure of identical mass and stiffness distribution along the axis. Such analysis produces the results for all parts of the structure with the exception of the X-mas tree valve block. It also produces time histories of velocities and forces at the top and bottom of the X-mas tree valve block all along the impact duration.

ii) Three dimensional modelling of the X-mas tree valve block, which now entertains the multiple bores and other features of that structure. The boundary conditions to be imposed on that model as a function of time are obviously the time histories of forces and velocities saved in the axisymmetric analysis.

Numerical methodology Impact problems are characterised by a short duration and strong non-linear behaviour. The latter arises from the geometrical non-linearities established on contact between bodies and, very often, also from the levels of deformation induced in the materials under the impact stresses. For problems with the above characteristics explicit time-marching procedures have long been recognised to be the best. They easily treat the non-linearities in exchange for a drastic limitation in the size of the time step used for integration of the equations of motion.

Two dimensional model The two dimensional model represented the workover riser, universal running tool, lower riser package, X-mas tree, template beam, wellhead, casings, cement and soil. The model is shown in Figure 4.

Three dimensional model The three dimensional model of the X-mas tree valve block used for calibrating its simpler, two dimensional representation in the axisymmetric analysis, and for detailed calculation of stresses and possible permanent deformation around valve bore holes, is presented in Figure 5. The boundary conditions for the three dimensional model of the X-mas tree block were derived from the global two dimensional analysis in the form of the force time histories at the top and the bottom surface.

Results of Impact Analysis

An analysis with impact velocity of 9.5 m/s was performed for the accident event load conditions.

Two dimensional analysis The history of events can be best visualised in the history of different energy components, Figure 6. The major drop in kinetic energy (both target and missile energies included) corresponds to the sum of plastic energy dissipation and elastic energy storage. Most of the dissipation takes place in about 4 to 5 ms. This allows elastic oscillation to start occurring after this time. The contours of plastic strain at the connection between the lower riser package and the X-mas tree are shown in Figure 7. The maximum plastic strains in the mesh are of the order of 15%.

Three dimensional analysis Time histories of axial stresses at the top and the bottom of the X-mas tree block obtained in the global two dimensional analysis were integrated to produce force time histories which represent the boundary conditions for the detailed analysis of the X-mas tree. The maximum plastic strains reached are of the order of 2.3%, in the region where the main vertical 5" bore intersects with the horizontal 5" bore, and also in the region where the horizontal bore intersects with the wing valve. It should be noticed that this is not a region of the greatest importance.

The differential displacement time history across the production master valve bore hole is represented by a maximum change of 0.15 mm in the bore diameter. The production master valve region of the X-mas tree did not exhibit any plastic straining.

Conclusions

One of the main features of this well assembly design are connection flanges between the lower riser package and the X-mas tree, the X-mas tree and the wellhead, and the universal running tool and the lower riser package. These connections act as filters for the force transfer. The most important connection is that of the lower riser package and the X-mas tree. The force transmitted through the re-entry hub flange to the X-mas tree valve block is limited by the cross-sectional area of the flange and the material yield stress. This means that strengthening the flange or any increase in the cross-sectional area due to possible re-design would transmit an increased force to the X-mas tree valve block. This is an important point for designers, since it is contrary to general belief that strengthening the flange would decrease the level of plastic straining and thus be beneficial.

The simple analysis originally carried out, treating the X-mas tree as a rigid mass and the wellhead, conductors and the soil as a spring, indicated that the stress levels in the X-mas tree were influenced by the spring stiffness. It has been shown here that this was not the case and the the stress wave propagation in the impact region is of a primary importance.

The production master valve region of the X-mas tree did not exhibit any plastic straining. The analysis described here shows the way the impact region should be designed in order to maximise the absorbed energy and minimise energy transitions to more vital parts of the assembly. This would hardly be possible without the use of advanced analysis tools.

REFERENCES

[1] Cuperus, N.J., Developments in Cryogenic Storage Tanks, 6th International Conference on Liquified Natural Gas, Paper 13, Session 2, Kyoto, Japan, 1980.

[2] Baldwin, J.T., Trbojevic, V.M., Design of Concrete Tanks for Abnormal Liquid Loads, Second International Conference on Cryogenic Concrete, Amsterdam, October 1983.

[3] Hirt, C.W., Nichols, B.D., Volume of Fluid (VOF) Method for the Dynamics of Free Boundaries, J. Comp., Phys., 39, 1981.

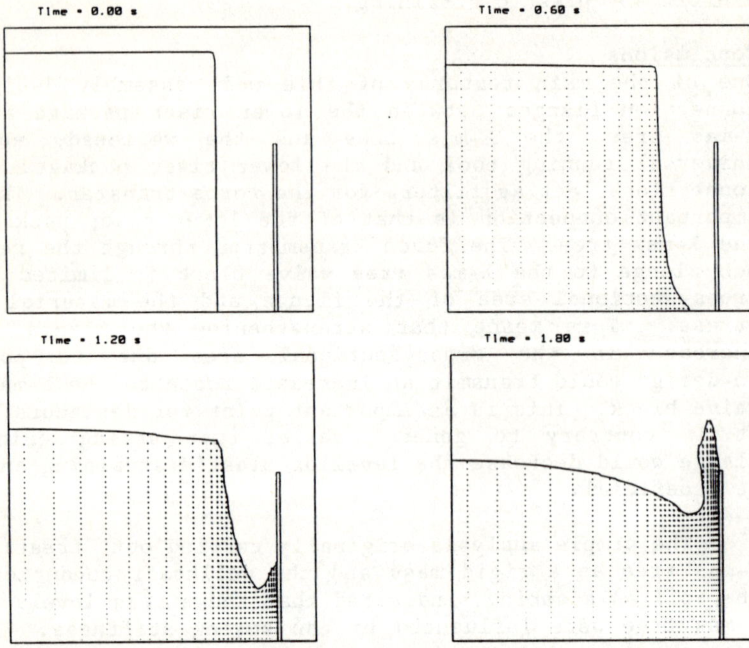

FIGURE 1. TANK WITH A SECONDARY CONTAINMENT WALL–FLUID FREE SURFACE AT DIFFERENT TIMES

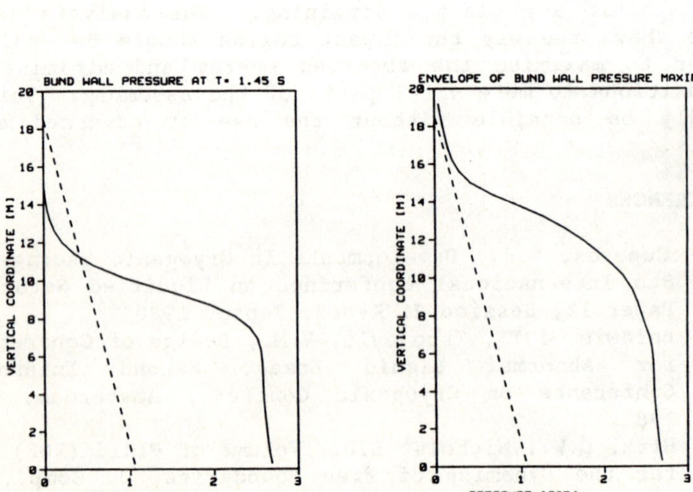

FIGURE 2. CONTAINMENT WALL PRESSURE DISTRIBUTION

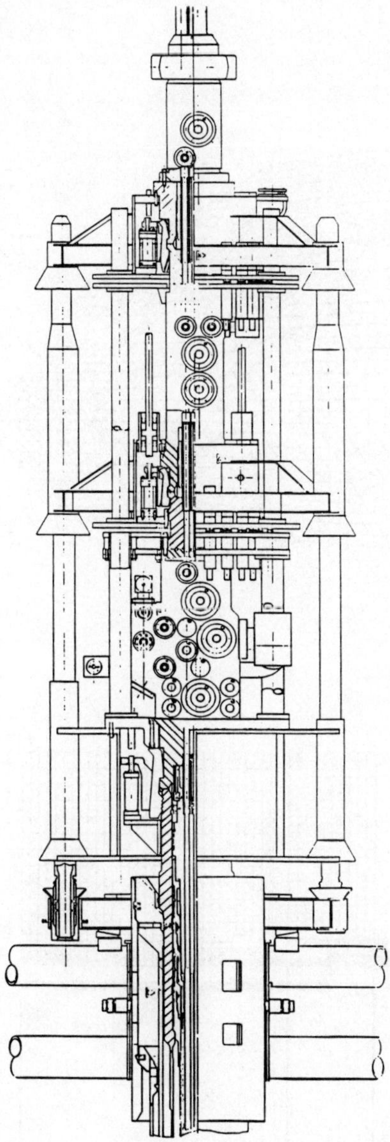

FIGURE 3. GENERAL LAYOUT OF THE WELL SYSTEM

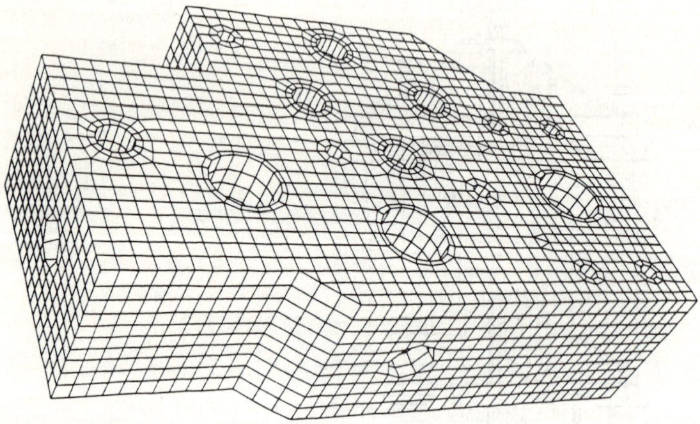

FIGURE 5. THREE-DIMENSIONAL VIEW OF XMAS TREE BLOCK

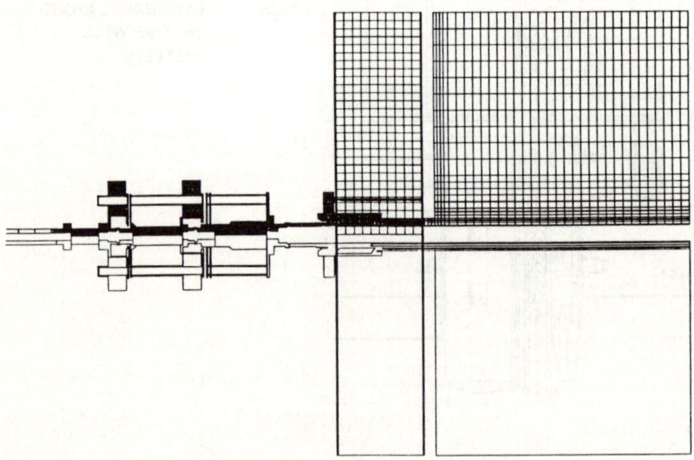

FIGURE 4. TWO-DIMENSIONAL MODEL OF THE WELL ASSEMBLY

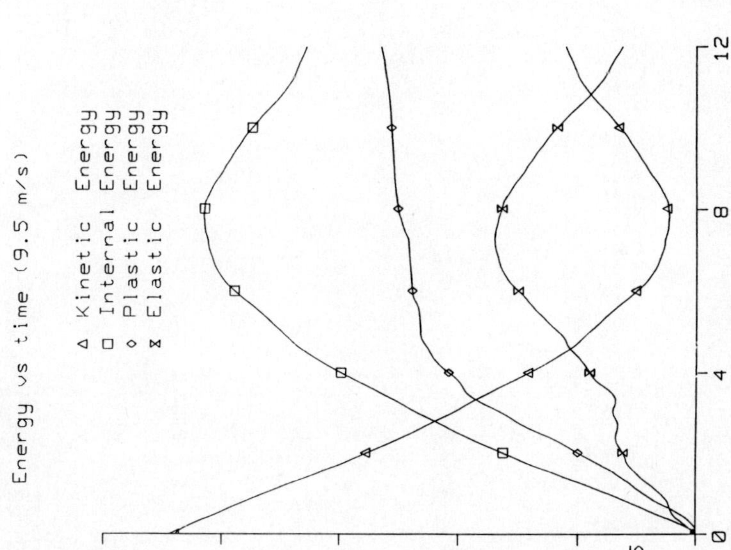

FIGURE 6. TIME HISTORY OF ENERGY COMPONENTS (9.5 m/s)

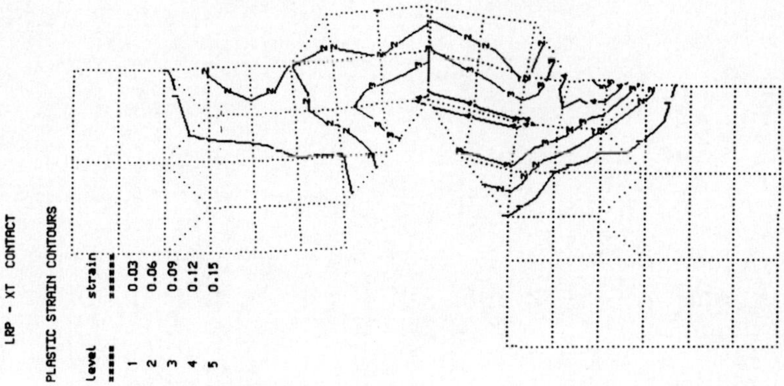

FIGURE 7. PLASTIC STRAIN CONTOURS AT LRP/XT INTERFACE (9.5 m/s)